D1357748

MATHEMATICS APPLIED
to
Deterministic Problems in the Natural Sciences

A B

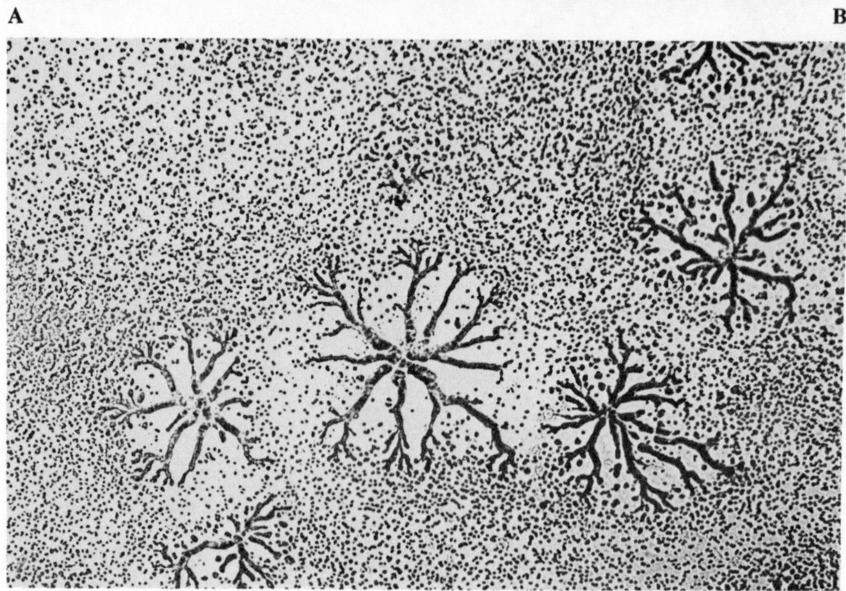

Two natural phenomena that are currently under investigation by applied mathematicians. As explained in Chapter 1, the power of mathematical modeling is well illustrated by these examples. Although stars and amebae really are discrete collections of matter, nonetheless much can be learned from models in which the matter is regarded as continuously distributed in space. [a] Whirlpool galaxy and its companion (NGC 5194/5). (Photograph courtesy of the Hale Observatories.) [b] Aggregation of cellular slime mold amebae *Polysphondylium violaceum*. (Courtesy B. Shaffer, Department of Zoology, Cambridge University.)

MATHEMATICS APPLIED to Deterministic Problems in the Natural Sciences

C. C. LIN *Massachusetts Institute of Technology*

L. A. SEGEL *Rensselaer Polytechnic Institute and Weizmann Institute of Science*

with material on elasticity by

G. H. HANDELMAN *Rensselaer Polytechnic Institute*

Macmillan Publishing Co., Inc.
New York

Collier Macmillan Publishers
London

To our parents

PRINTED IN THE UNITED STATES OF AMERICA

MACMILLAN PUBLISHING CO., INC.
866 Third Avenue, New York, New York 10022

COLLIER-MACMILLAN CANADA, LTD.

Library of Congress Cataloging in Publication Data

Lin, Chia-ch'iao, (date)
Mathematics applied to deterministic problems in the natural sciences.

Bibliography: p.
1. Mathematics—1961– I. Segel, L. A., joint author. II. Title.
QA37.2.L55 510'.2'45 73-8093
ISBN 0-02-370720-8

Printing: 1 2 3 4 5 6 7 8 Year: 4 5 6 7 8 9 0

Preface

THIS text, an introduction to applied mathematics, is concerned with the construction, analysis, and interpretation of mathematical models that shed light on significant problems in the natural sciences. It is intended to provide material of interest to students in mathematics, science, and engineering at the upper undergraduate and graduate level. Classroom testing of preliminary versions indicates that many such students do in fact find the material interesting and worthy of study.

There is little doubt that a course such as one based on this text should form part of the core curriculum for applied mathematicians. Moreover, in the last few years the professional mathematical community in the United States has emphasized the importance of some exposure to applied mathematics for all mathematics students. This exposure is recommended because of its broadening influence, and (for future university mathematicians) because of the benefits it affords in preparing for the teaching of nonspecialists. As for scientists and engineers, there is often little difference in their theoretical work and that of an applied mathematician, so they should find something of value in the approaches to problems that we espouse.

There are many books that present collections of useful mathematical techniques and illustrate the various techniques by solving classical problems of mathematical physics. Our approach is different. Typically, we select an important scientific problem whose solution will involve some useful mathematics. After briefly discussing the required scientific background, we formulate a relevant mathematical problem with some care. (The formulation step is often difficult. Not many books actually demonstrate this, but we try to give due weight to the challenges involved in constructing our mathematical models.) A new technique may then be introduced to solve the mathematical problem, or a technique known in simpler contexts may be generalized. In most instances we take care to determine what the mathematical results tell us about the scientific processes that motivated the problem in the first place.

We use a "case study" approach by and large. Such an approach is not without disadvantages. No strict logical framework girds the discussion, and the range of applicability of the methods is not precisely delineated. Heuristic and nonrigorous reasoning is often employed, so there is room for doubt concerning the results obtained. But realistic problems often require techniques that cannot at the moment be rigorously justified. There is a stimulating sense of excitement in tackling such problems. Furthermore, mathematicians and scientists frequently use heuristic reasoning and are frequently called upon to determine for themselves whether a method used to solve one problem can be adapted to solve another. Some such experience should be part of each student's education.

This work was given a lengthy title because we wished to make its limitations clear at the outset. A completely balanced introduction to applied mathematics should contain material from the social and managerial sciences, but we have restricted ourselves to the natural sciences. In rough proportion to applied mathematical research, the topics in this volume are drawn mainly from the physical sciences, but there is representation from chemistry and biology. We treat probabilistic models to a lesser extent than would be required in a balanced presentation, and our treatment emphasizes the relationship between probabilistic and deterministic points of view. Our work is also limited in its almost exclusive use of analytical methods; numerical methods are mentioned many times but are treated only briefly.

One reason for restricting the topics covered is the authors' hesitation to tread outside their areas of expertise. Another is the fact that the work is already lengthy, so a wider purview would necessarily be either overlong or superficial. In any case, much further study is required of the aspiring mathematician or scientist—we hope that our work will form a foundation and motivation for some of that study.

STYLE AND CONTENT

In our writing we have striven in most places for careful and detailed exposition, even at the risk of wordiness; for a rigorous proof can be built from its skeleton, but the reasoning of the applied mathematician often can only be mastered if it is fully described.

The nature of applied mathematics precludes an approach that is organized in a tight linear fashion. This has certain disadvantages. But among the compensations is a high degree of flexibility in a book such as this. In particular, there is a large measure of independence among the three parts of the present volume, which are the following:

PART A. An overview of the interaction of mathematics and natural science.

PART B. Some fundamental procedures—illustrated on ordinary differential equations.

PART C. Introduction to theories of continuous fields.

There is considerable further independence within each part—which we have tried to enhance, even at the expense of repetition.

Two volumes have been planned. This first volume provides ample material for a balanced and self-contained introduction to a major part of applied mathematics. The succeeding volume, described briefly in Section 1.1, penetrates further into the subject, particularly in the classical areas of fluid mechanics and elasticity.

The chapter titles, section titles, and subheadings in the table of contents give a good outline of this volume. It is not necessary to read the chapters in the given order. In Part A, for example, the only sequence which must be kept is that of the two chapters on Fourier series. It would be helpful to begin Part B by obtaining some understanding of nondimensionalization and scaling as treated in Chapter 6, but this is not strictly necessary. Chapter 8 can only be appreciated if the preceding two chapters have been covered, but this chapter can be omitted if relatively simple examples of the techniques are deemed sufficient. Chapter 9 is a prerequisite for Chapter 10, but each of the three sections of Chapter 11 is largely independent of the others and of earlier material.

Part C, too, offers various possibilities. For example, one can skip much of the material if one's goal is to obtain just enough understanding of the basic equations to permit formulation of specific problems in one-dimensional elasticity, inviscid flow, and potential theory. Or one can just study the first two sections of Chapter 12, to gain a glimpse of continuum mechanics. (Note: Section 12.1 contains many new ideas in a few pages.)

Some features of our approach are the following.

(a) We proceed from the particular to the general.

(b) For our major examples we attempt to choose physical problems that are important in their own right and also permit the illustration of major mathematical techniques. Thus the Michaelis–Menten kinetics discussed in Chapter 10 are repeatedly referred to in biochemistry, and a full treatment of the relevant mathematical problem provides an excellent illustration of singular perturbation theory. To give another example, we discuss the stability of a stratified fluid in Chapter 15—both as an illustration of inviscid flow and as a motivation for studying stability theory for a system of partial differential equations.

(c) We make a serious effort to examine the processes of deriving the equations that model certain basic scientific phenomena, rather than merely give plausibility arguments for using such equations. As an illustration of this spirit, we mention that the differential equation which governs mass conservation in a continuum is derived in four different ways in Section 14.1, and several alternative approaches are examined in the exercises of that section. One purpose of such an effort is to engender a secure understanding of the equation in question. Another is to help those readers who might some day wish to derive equations that model a phenomenon which had never before been subjected to mathematical analysis.

(d) New ideas are frequently introduced in extremely simple physical contexts. In Part B, for example, dimensional analysis, scaling, and regular perturbation theory are first met in the context of the problem of a point mass shot vertically from the surface of the earth. The qualitative features of the phenomenon are correctly guessed by most people, and the relevant differential equation is solved exactly in elementary courses. Yet, considerable effort is required to obtain a deep understanding of the problem. This effort is worthwhile because it generates a grasp of concepts that are useful in far more difficult situations.

(e) We try to make explicit various concepts and approaches that are often mastered only by inference over a period of years. Examples are provided by our discussions of the basic simplification procedure and of scaling, in Chapter 6.

(f) In historical remarks we have focused attention on humanistic aspects of science, by emphasizing that the great structures of scientific theories are gradually built by the strivings of many people. To illustrate the ongoing nature of science, we have presented certain plausible theories which are either incorrect (Newton's isothermal speed of sound—Section 15.3) or not fully in accord with observation (the galactic model of Section 1.2), or highly regarded but not yet fully accepted (as in the physiological flow problem discussed in Chapter 8).

(g) Some rather lengthy examples are worked out in detail, e.g., the perturbation calculations in Section 7.2, in response to student objections that they are often asked in exercises and in examinations to solve much harder problems than they have ever seen done in the text.

(h) We have provided a number of exercises to reinforce, test, and extend the reader's understanding. Noteworthy are multipart exercises, often based on a relatively recent journal article, which develop a major point in a step-by-step manner. (An example is Exercise 15.2.10.) Even if a student cannot do one part of the exercise, he can take its result for granted and proceed. Such exercises have been successfully used as the central part of final examinations.

PREREQUISITES

We have assumed that the potential reader has had an introductory college course in physics and is familiar with calculus and differential equations. Only a few exercises require knowledge of complex analysis. We make considerable use of such topics as directional derivatives, change of variables in multiple integrals, line and surface integrals, and the divergence theorem. Often mathematics majors will have taken an advanced calculus course that omits some of these topics, but we have found that such students are sufficiently sophisticated mathematically to be able readily to pick up by themselves what is required. [Potential readers who feel inadequacies in vector calculus and physical reasoning would profit from studying *Div, Grad, Curl, and All That* by H. M. Schey (N.Y., Norton, 1973).]

RELATIONSHIPS BETWEEN THIS TEXT AND VARIOUS COURSES

Historically, this book grew out of the union of two courses. The first was *Foundations of Applied Mathematics* introduced by G. H. Handelman at Rensselaer around 1957. (A precursor of this course was taught by A. Schild and Handelman at Carnegie Institute of Technology, now Carnegie-Mellon

University.) A second course, *Introduction to Applied Mathematics*, was introduced by C. C. Lin at Massachusetts Institute of Technology around 1960. These courses have been taught annually, many times, by the present authors in their respective institutions. In recent years preliminary drafts of the present work have been used as text material. Such drafts have also been used in applied mathematics courses taught by P. Davis at Worcester Polytechnic Institute, D. Drew at New York University, and D. Wollkind at Washington State University. Considerable improvements in the draft text have resulted; the authors welcome further suggestions from users of the printed work.

ACKNOWLEDGMENTS

The work was jointly planned. One of us (C.C.L.) wrote the initial draft of Part A, with the exception of Section 1.2, and also Chapter 16. The other of us (L.A.S.) drafted the remainder of the book, with the exception of Chapter 12, written by G. H. Handelman. There was considerable consultation on revisions. L.A.S. was responsible for the final editorial work.

In writing this book, we have drawn on a background for which we are deeply indebted to our families, colleagues, teachers, and students, and to the writers of numerous other books. We are grateful to G. H. Handelman for showing, from classroom experience, that an introduction to continuum mechanics is probably best made by starting with one-dimensional problems—and for writing out his approach as Chapter 12. Among many who have made useful suggestions concerning this volume, we must single out Roy Caplan, Paul Davis, Donald Drew, William Ling, Robert O'Malley, Jr., Edward Rothstein, Terry Scribner, Hendrick Van Ness, and especially Edith Luchins. Numbers of secretaries have performed yeoman service. The publishers have also been most helpful, particularly our editors Everett Smethurst and Elaine Wetterau.

The work of one of us (L.A.S) was partially supported in 1968–1969 by a Leave of Absence Grant from Rensselaer. Further support was received during 1971–1972 from the National Science Foundation Grant GP33679X to Rensselaer, and from a John Simon Guggenheim Foundation Fellowship. That year was spent as a visitor to the Department of Applied Mathematics, Weizmann Institute, Rehovot, Israel. The hospitality and technical assistance afforded there are gratefully acknowledged.

C. C. L.
L. A. S.

Contents

PART B
SOME FUNDAMENTAL PROCEDURES ILLUSTRATED ON ORDINARY DIFFERENTIAL EQUATIONS

PART C
INTRODUCTION TO THEORIES OF CONTINUOUS FIELDS

Conventions

EACH chapter is divided into several sections (e.g., Section 5.2 is the second section of Chapter 5). Equations are numbered consecutively within each section. Figures and tables are numbered consecutively within each chapter.

When an equation outside a given section is referred to, the section number precedes the equation number. Thus "Equation (6.3.2)" [or (6.3.2)] refers to the second numbered equation of Section 6.3. But if this equation were referred to within Section 1 of Chapter 6, then the chapter number would be assumed and the reference would be to "Equation (3.2)" [or (3.2)]. The fourth numbered equation in Appendix 3.1 is denoted by (A3.1.4).

A double dagger‡ preceding an Exercise, or a part thereof, signifies that a hint or an answer will be found in the back of this volume.

The symbol □ signifies that the proof of a theorem has concluded.

The succeeding volume is referred to as "II."

A brief bibliography of books useful to beginning applied mathematicians can be found at the end of this volume. When reference is made to one of these books, the style "Smith (1970)" is employed.

PART A

An Overview of the Interaction of Mathematics and Natural Science

CHAPTER 1
What Is Applied Mathematics?

THE PURPOSE of this book is to foster an appreciation of the nature of applied mathematics. We attempt to explain the essential processes involved and to provide experience in applied mathematics as it is used in a variety of physical problems.

The book is not intended as a compilation of methods—which cannot be complete in any case. Rather, it is believed that an appreciation of the applied mathematician's approach will provide the student with a clear framework on which to fit his knowledge, and with a creative attitude. These are much more valuable than an encyclopedia of methods.

Section 1.1 begins with some fairly detailed remarks on the nature of applied mathematics. Then follows an outline of the applied mathematical topics that are to be presented in this work. Section 1.2 consists of an introduction to the study of galactic structure, and Section 1.3 analyzes the sudden aggregation of a population of amebae. The main purpose of Sections 1.2 and 1.3 is to provide an early feeling for the type of problem now under investigation by applied mathematicians. (The authors selected the topics from among their own current interests.) On first perusal at least, the reader should not concern himself unduly with detail but should concentrate on getting a general impression.

The behavior of stars and the behavior of amebae might seem to have little in common. It is an excellent illustration of the power of applied mathematics, however, to realize that this is not the case. For example, we shall see that both stars and amebae behave *as if* their mass were continuously distributed—notwithstanding the fact that one can see (using a microscope) gaps between amebae, while the nearest stars are light-years apart. But it is of little relevance that an idealized mathematical model such as the continuum is "wrong." The real question is: Do the errors have a significant effect upon predictions concerning the phenomena under investigation?

For stars and amebae the idealization implied in the continuum model is particularly striking, since it is the particulate aspect of the phenomena that first makes itself evident. Water and air *seem* like continua—for most people the existence of molecules is hearsay—but the continuum model is, of course, also very useful for investigating phenomena involving these media.

Another similarity between the examples of Sections 1.2 and 1.3 is that both require an explanation of organized structure. The spiral galactic pattern demands an investigation for reasons of aesthetics alone, but a more basic reason is that the pattern offers an important clue to the nature of the forces involved. The **morphogenetic** (form-producing) motion of the amebae is studied because of the clues it can provide to the organized cellular movement that is a basic element of developmental biology. In both areas understanding is far from complete.

1.1 On the Nature of Applied Mathematics

Mathematics began with simple practical problems such as division of a flock of animals among family members (number theory) and the measurement of land area (geometry). Gradually, elementary ideas were organized, and they evolved into logical structures. A monumental example of early achievements is Euclid's geometry. The Greeks realized that the study of mathematical theorems can be based on certain axioms. Only much later, however, did it become clear that the axioms (such as the parallel postulate) cannot be completely and decisively verified by experience. Indeed, it was through a change in the parallel postulate that non-Euclidean geometry was created; this led to many important ramifications. There is no doubt that these pure mathematical developments are extremely important in applications.

As an increasing portion of mathematics was developed in a manner independent of theoretical sciences, the term *pure mathematics* emerged. The creative efforts of pure mathematicians are certainly impressive, but it would be unfortunate to restrict our study of mathematics in any manner. Joint study of mathematics and its applications can provide richer content and greater intellectual challenge. Moreover, such study can stimulate the development of new mathematical methods and theory. Some of these developments will in turn find applications in the sciences.

Applied mathematics is guided by the spirit of and belief in the *interdependence* of mathematics and the sciences. It would be incorrect to claim that *all* parts of mathematics involve such interdependence, but in the study of applied mathematics one must give priority to such parts. This policy is based on the assumption that the areas of mathematics that grew directly out of the study of scientific problems have a greater likelihood of again being applicable to other scientific problems. To give one example, later in this chapter we see an application to a current problem in developmental biology of stability theory for partial differential equations. This theory was developed primarily in the classical fields of fluid mechanics and elasticity.

Historically, the development of mathematics and physics had a very close connection. Classical examples may be found in the work of Newton (see Chapter 2), Gauss, Euler, Cauchy, and others. More modern examples appear throughout the study of relativity theory, Brownian motion, statistical mechanics, and the associated theories of covariance, probability, and generalized harmonic analysis. Why is there such an intimate relationship between mathematics and the physical sciences?

An answer often given* to this philosophical question is that

GOD IS A MATHEMATICIAN.

In other words, it is believed that there is a basic harmony and order in nature. Consequently, the description of natural phenomena can be organized by the logical discipline of mathematics. In social and economic contexts, by contrast, perhaps it is the logic imposed by man in his quest for optimality that allows so large a role for mathematics.

THE SCOPE, PURPOSE, AND PRACTICE OF APPLIED MATHEMATICS

The scope of applied mathematics is very broad and can be well described by borrowing the following words of Albert Einstein:

> Its realm is accordingly defined as that part of the sum total of our knowledge which is capable of being expressed in mathematical terms.†

These words were used by Einstein to define physics. Taken literally, this definition would certainly include mathematical theories of biology, economics, communication engineering, etc., and it is therefore a more adequate description of applied mathematics.

We shall now attempt to give a brief description of the objectives and the methodology of applied mathematics, and to contrast them with those of pure mathematics and of theoretical science.

The purpose of applied mathematics is to elucidate scientific concepts and describe scientific phenomena through the use of mathematics, and to stimulate the development of new mathematics through such studies. The process of using mathematics for increasing scientific understanding can be conveniently divided into the following three steps:

(i) The **formulation** of the scientific problem in mathematical terms.
(ii) The **solution** of the mathematical problems thus created.
(iii) The **interpretation** of the solution and its empirical verification in scientific terms.

There is widespread misunderstanding that the second step is the most important and that manipulative skill is the most valued asset of an applied mathematician. Generally speaking, however, all three steps are equally important. In a given class of problems, one step might stand out as more important or more difficult than another.

A knowledge of methods and a proficiency in manipulative skill are obviously necessary. A person with an encyclopedic knowledge of methods

* "God *is* number"—Pythagoras; "God ever geometrizes"—Plato; "God ever arithmetizes" —Jacobi; "The Great Architect of the Universe now begins to appear as a mathematician"—Jeans. Quotes given (without much approval) by E. T. Bell, *Men of Mathematics* (London: Penguin, 1953), p. 21.

† *Out of My Later Years* (New York: Philosophical Library, 1950), p. 98.

alone can provide valuable aid to other scientists using mathematics. But it is essential to realize that this knowledge is not sufficient for an applied mathematician working as an independent scientist. He must also be able to exercise *judgment* in the formulation of a problem, in decisions on which problems to attack, and in choices of what idealizations and approximations to adopt in order to simplify the problem without losing its essentials. An aspiring applied mathematician must try even harder to cultivate sound judgment than manipulative skill.

Finally, *it is most important to extract, from the mathematical theory, the proper scientific conclusions and implications for empirical verification.* As far as is feasible, *conclusions must be reduced to the simplest form and expressed in the most specific terms.* This step serves as a culmination of the whole effort and as a basis for future progress. A new understanding obtained, an insight gained, a new perspective attained—these are much more important and satisfying than the mere derivation of certain formulas and the compilation of certain useful numerical tables. The accumulation of specific quantitative information must be regarded as a means to an end.

It should now be apparent that an understanding of the scientific motivation of the problem and the ability to use heuristic reasoning, as well as manipulative skill, are essential to the practice of applied mathematics. Indeed, these elements stand out as being more important than the ability to carry out a rigorous proof. In many cases the rigorous formulation of a mathematical theory may take many years. In the meantime, the applied mathematician must proceed despite the incompleteness of the logical structure. However, he must strive to be correct and to be as careful as possible in his reasoning.

APPLIED MATHEMATICS CONTRASTED WITH PURE MATHEMATICS

The differences in motivation and objectives between pure and applied mathematics—and the consequent differences in emphasis and attitude—must be fully recognized. In pure mathematics, one is often dealing with such abstract concepts that logic remains the only tool permitting judgment of the correctness of a theory. In applied mathematics, empirical verification is a necessary and powerful judge.

However, there is still a close relationship between the two disciplines. In some cases (e.g., celestial mechanics), rigorous theorems can be proved that are also valuable for practical purposes. On the other hand, there are many instances in which new mathematical ideas and new mathematical theories are stimulated by applied mathematicians or theoretical scientists. The theory of distributions is a fairly recent example.

If a scientific problem cannot be adequately formulated in terms of existing mathematical concepts, new concepts have to be generated, such as the abstract "game" of von Neumann. If the mathematical problem formulated cannot be solved in terms of existing methods, or the nature of its solution cannot be adequately understood in terms of existing theory, new methods and new theory must be developed. (Many nonlinear problems are in this category.) We therefore record a fourth part of applied mathematics:*

(iv) The generation of scientifically relevant new mathematics through creation, generalization, abstraction, and axiomatic formulation.

It should be recognized that as the mathematical theory is being developed, the first few theorems proved may not produce an impact in pure mathematics but must be appreciated as an accomplishment that is useful for the purposes of applied mathematics. On the other hand, much second-rate pure mathematics is concealed beneath the trappings of applied mathematics (and vice versa). As always, knowledge and taste are needed if quality is to be assured.

APPLIED MATHEMATICS CONTRASTED WITH THEORETICAL SCIENCE

The distinction between a theoretical scientist and an applied mathematician is often blurred, because each may work in the spirit of the other. A theoretical physicist, when his problem cannot be solved by frontal attack, sometimes engages in the study of related mathematical *model problems*—even to the extent of following the practices of pure mathematicians—in order to build up confidence and judgment in understanding the mathematical aspects of the real physical problem. This type of work is often also done by applied mathematicians for their own problems. At the same time, an applied mathematician draws scientific conclusions from his theory in order to compare with empirical evidence. To do this effectively, he must command considerable scientific knowledge of the problem he is studying.

It is often the case that a theoretical scientist, from long study of his particular subject, has a deeper knowledge of a certain discipline. An applied mathematician, by contrast, may work in more than one discipline and cross-fertilize each. Indeed, *in these times of increasing specialization, cross fertilization is one of the most useful and satisfying activities of an applied mathematician.*

By and large, a theoretical physicist (for example) is more attracted to the discovery of new physical laws and principles, and therefore has more appreciation for studies related to these, even when the attempt is only partially successful. His work often tends to be more inductive and more speculative in nature. The applied mathematician is more interested in the proper mathematical descriptions of phenomena. He tends to derive the consequences of known laws and principles.

* There are those who regard this fourth part as the *only* applied *mathematics* and consider the first three parts as science, not mathematics. Others find most of the work that is classifiable under this fourth part to be of little genuine scientific relevance (so that it should be judged primarily on its mathematical merit). The authors are not doctrinaire, but they lean more toward the latter view. And with the exception of some material in Chapter 2, this book concentrates on parts (i)–(iii), for it is here that there is a gap in the literature.

To give an illustration, the origin of hurricanes would not be a subject of central interest to a modern physicist, for hurricane genesis can probably be fully described in terms of the principles of classical mechanics and thermodynamics. Yet it is a very attractive subject for the applied mathematician, who appreciates the challenge of the nonlinear problems involved as well as the inherent scientific interest of the phenomenon.

APPLIED MATHEMATICS IN ENGINEERING

Understanding of basic concepts is as useful to the engineer as it is to the scientist. But for the engineer such understanding is just a means to the end of designing structures, machines, and processes to accomplish certain tasks with efficiency and reliability. The requirement to develop detailed design criteria often makes extensive large scale numerical calculations indispensable. At the very least, however, the engineer should be able to appreciate the more qualitative model-making activities described in this work, so that he can take advantage of relevant theoretical results. And as a lofty goal, the engineer can aspire to combine applied mathematical and scientific insight with practicality, as have the masters of his profession.*

THE PLAN OF THE PRESENT VOLUME

The reader will find some others' views concerning the nature, teaching, and practice of applied mathematics in Appendix 1.1. The authors' views have been outlined above; in a sense, the remainder of this work is an elaboration of that outline.

This volume is divided into three parts. In contrast to the rest of the work, which is rather detailed, Part A provides an overview of applied mathematics. As above, it begins with an essay on the nature of applied mathematics. This chapter continues by presenting an introduction to two examples of current research in applied mathematics. Chapters 2 and 3 contrast deterministic processes, as exemplified in the equations of particle dynamics, and probabilistic processes, as exemplified in the study of random walk. Part A concludes with two chapters on Fourier series, a classical topic that is not only still useful but continues to develop. The main concept illustrated here is the principle of superposition.

In Part B an attempt is made to convey the essence of certain important applied mathematical methods and ways of thinking. Particularly in the discussion of the basic simplification procedure and of scaling in Chapter 6, this involves an attempt to make explicit certain approaches that "everyone knows" after a while, but which are rarely treated in print. The procedures

* An entertaining and informative biography of one such master is *The Wind and Beyond* by Theodore von Kármán (the subject of the book) and Lee Edson (Boston: Little, Brown, 1967). Von Kármán used fundamental applied mathematical investigations to estimate the behavior of engineering systems and thereby to play a major role in airplane design (from the earliest days through jets) and the design of turbines, pumps, tunnels, dams, and bridges.

in question have a large intuitive element, but still there is something to be learned by a detailed examination of their content.

Both regular and singular perturbation theory are discussed in Part B, as is the phase plane method. Two chapters are devoted to careful discussions of nontrivial examples that involve the various techniques and show some processes of applied mathematics at work. One example involves an osmotically driven flow that is of interest in physiology: the other considers some important aspects of biochemical kinetics. Throughout Part B only ordinary differential equations are considered so that the basic ideas are illustrated in relatively simple contexts.

Part C provides an introduction to continuum mechanics. Training in this classical theory is important to the applied mathematician for two main reasons: (a) Continuum mechanics remains an important subject in its own right, and its methods (as we shall see several times in this volume) are continually being applied in new areas; and (b) the depth that is possible in theoretical analysis must be illustrated in a well-developed field of study. Continuum mechanics is such a field and (unlike quantum mechanics for example) deals with relatively familiar physical phenomena.

Part C begins with a simple but revealing study in continuum mechanics, concerned with one-dimensional problems that arise when investigating longitudinal motions of an elastic bar. Two chapters then present careful derivations of the principal field equations that govern the behavior of general continuous media (e.g., conservation of mass, balance of linear momentum). Additional constitutive equations are added to the field equations, so "inviscid" fluids are given full mathematical description. Chapter 15 treats various examples that involve inviscid flow (stability of a stratified fluid, sound waves, flow past an obstacle). The final chapter, a study of potential theory, concludes with the use of Green's functions in an analysis of the diffraction of sound waves by a hole in a screen. Partial differential equations form the mathematical core of this part.

A PREVIEW OF THE FOLLOWING VOLUME

Many important applied mathematical ideas are illustrated only briefly in the present volume. They are deserving of much more study. Some of this is provided in a following volume (designated here as II).

The following volume begins with some prerequisites to an efficient study of three-dimensional continuum mechanics, particularly cartesian tensors. Then follows a considerable amount of material on viscous fluid flow and on elasticity. There is also a detailed discussion of dispersive waves, particularly surface waves on water. This provides the most penetrating analysis of the present work and brings the reader through much brilliant theory by classical authors to the frontiers of present-day research. The last part of II is concerned with variational methods. The more advanced theory is given a setting that involves basic ideas of functional analysis.

CONCEPTS THAT UNIFY APPLIED MATHEMATICS

Our presentation is not tightly organized like a handbook but is built (somewhat like a symphony) around certain major themes. We shall not list all the themes and their relations; it should be part of the reader's training and enjoyment to discover them for himself. To give just one illustration of what we mean, however, we note that pure mathematicians classify partial differential equations into

$$\text{elliptic}\left(\text{such as } \frac{\partial^2 u}{\partial x^2} + \frac{\partial^2 u}{\partial y^2} = 0\right),$$

$$\text{parabolic}\left(\text{such as } \frac{\partial^2 u}{\partial x^2} - \frac{\partial u}{\partial y} = 0\right),$$

$$\text{hyperbolic}\left(\text{such as } \frac{\partial^2 u}{\partial x^2} - \frac{\partial^2 u}{\partial y^2} = 0\right).$$

Applied mathematicians often prefer the complementary viewpoint of regarding natural phenomena as illustrating equilibrium, diffusion, or wave propagation. We shall see how these themes—and others, such as stability, randomness, and optimization—help to unify applied mathematics by revealing common features of superficially disparate problems.

1.2 Introduction to the Analysis of Galactic Structure

In this section we indicate how theoretical analysis can begin to explain aspects of galactic structure. As with all mathematical theories of natural phenomena, we require (i) the basic physical laws that are relevant, and (ii) the particular characteristics of the system under investigation.

Requirement (i) can be briefly disposed of, for the required laws are those of classical physics. To satisfy requirement (ii), we shall carry the reader on a brief quantitative survey of the universe. We then present in some detail the formulation and solution of a problem that provides prediction of the distribution of stars across a galactic disk. We conclude with a very brief sketch of some current ideas about the cause of the spiral structure that is a feature of many galaxies.

PHYSICAL LAWS GOVERNING GALACTIC BEHAVIOR

In studying galaxies, the basic physical laws that one uses are those of classical physics: (i) Newton's laws of motion and his law of universal gravitation, which govern all classical mechanics; (ii) the laws of electrodynamics according to the formulation of Maxwell; and (iii) the laws of thermodynamics.

The reader should notice that there are very *few fundamental empirically based laws.* However, we frequently have to introduce new *mathematical* concepts that are essential for the study of certain physical problems. In the particular example that concludes this section, for instance, only physical laws known to Newton are used. But we require mathematical concepts of much more recent vintage, such as phase space and density in phase space.

We turn now to a survey of certain facts about our universe. It should be remembered that the qualitative and quantitative information that we cite is the fruit of an enormous amount of scientific effort, over centuries of time.

BUILDING BLOCKS OF THE UNIVERSE

Our expanding universe is about 10 billion (10^{10}) years of age. At the present time, it is composed of about 10 billion galaxies as the building blocks. These galaxies, rather well-defined collections of gas and stars, are approximately uniformly distributed in the universe (which appears to have no border in space). There is a degree of nonuniformity: galaxies form clusters through mutual gravitational attraction. Our own galaxy—the Milky Way system—is a member of a cluster of 17 galaxies.

To convey a general impression of the nature of the galaxies, let us cite a few characteristics of the Milky Way system. Its shape is that of a thin disk with a roughly spherical nucleus at its center. (The sun is situated toward the outer edge of the system, about 10,000 parsecs* from the galactic center.) At a distance of about 15,000 parsecs from the center, the mass density of the galaxy becomes quite small. The galaxy is in rotation; the linear velocity is about 250 km/sec in the solar vicinity.

The nearest star is about 4 light years (1.3 pc) away. By contrast, the orbit of Pluto is about 40 A.U. (2×10^{-4} pc). Thus the solar system may be likened to an atom of a monatomic gas. This gas is relatively rarefied, at least in our vicinity.

CLASSIFICATION OF GALAXIES

By their appearance, galaxies are classified into four categories: (i) ellipticals (including sphericals), (ii) normal spirals (disk-shaped), (iii) barred spirals, and (iv) irregulars. Examples of these galaxies are shown in Figure 1.1. Another example is shown in the frontispiece.

The galaxies can be classified according to an ordered scheme, such as that shown in Figure 1.2. Galactic classification was originally based on geometrical appearance alone, and detailed study shows that the geometrical characteristics are related to other physical parameters, for example, the gas content. This indicates that there is a dynamical basis for the differences in galactic appearance, a difference that persists over considerable periods of time (a few billion years).

* A parsec (pc) $= 3 \times 10^{18}$ cm is the distance at which the radius of the orbit of the earth around the sun (called the astronomical unit, A.U.) subtends an angle of 1 second of arc. It is thus about 2×10^5 A.U. A light-year is about 0.3 pc. An object traveling at the speed of 1 km/sec covers a distance of 1 parsec in about 1 million years.

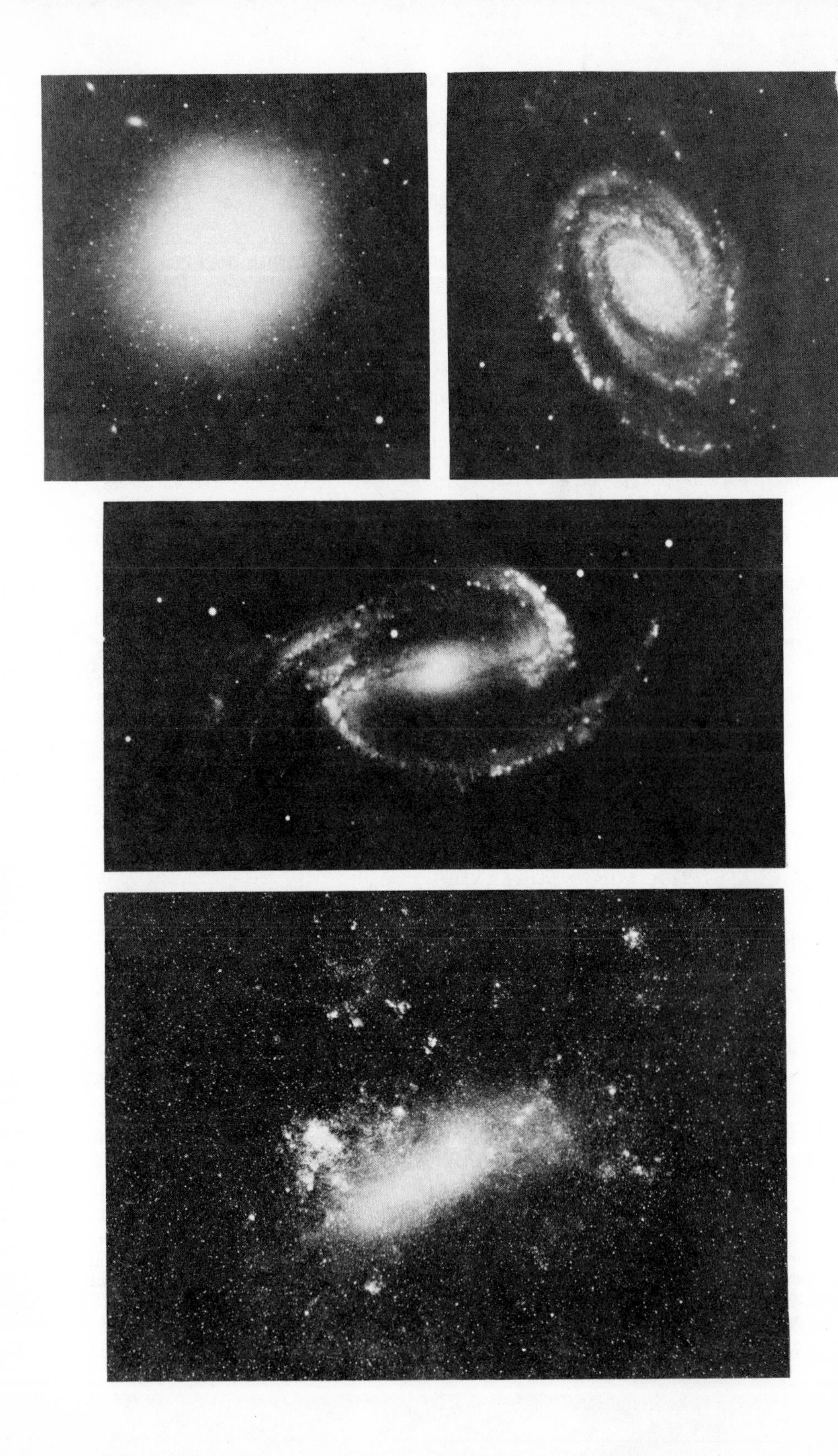

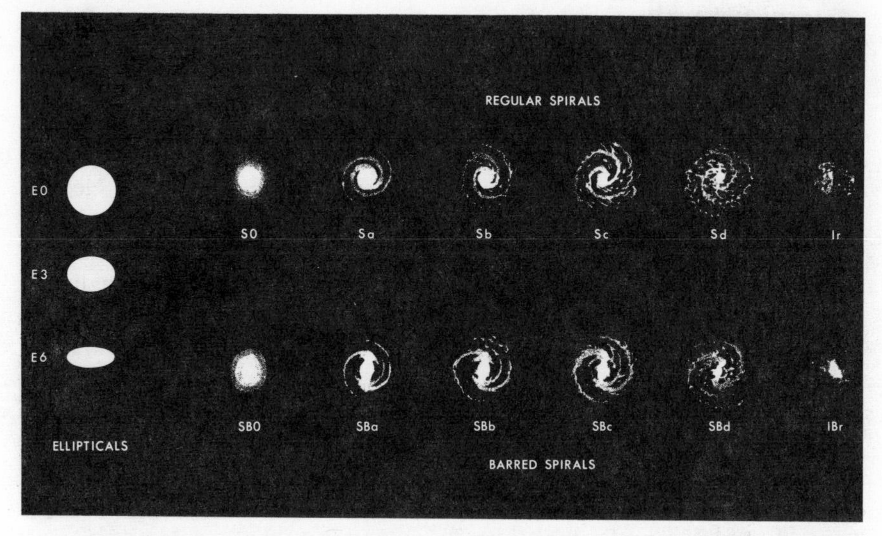

FIGURE 1.2. *Composite of Hubble and de Vaucouleurs galaxy classification systems.* (*Photograph courtesy of the Hale Observatories.*)

The disk-shaped galaxies must be in sufficiently rapid rotation to maintain their shape. Spherical galaxies presumably have no net rotation. Elliptical galaxies with different degrees of flatness have different degrees of rotation. Bar-shaped galaxies are in uniform rotation but not about an axis of symmetry. Classical examples of such rotating systems held together by gravitation were discovered by the mathematician Jacobi (1804–1851).

COMPOSITION OF GALAXIES

Each galaxy is composed of about 10 billion stars (10^9–10^{11} stars, to be more exact). Our galaxy is among the largest. The sun is an average star, about 5 billion years of age. Massive stars, e.g., those of 40 solar masses, are brilliant, and they burn out their nuclear fuel very quickly. They must therefore have been formed comparatively recently (say, within the past 10 million years). The evolution of the sun is quite slow; its nuclear energy will not be exhausted for another 10 billion years.

There is also a substantial amount of gas in the galaxies—up to more than 20 per cent in the irregular galaxies, somewhat less in the Sc spirals, down to 1 or 2 per cent in the Sa spirals, and practically none in the giant ellipticals. Our galaxy has a gaseous content of about 3–4 per cent.

FIGURE 1.1. *Examples of galaxies of various types:* (*upper left*) *elliptical galaxy EO pec. NGC* 4486; (*upper right*) *normal spiral SC NGC* 5364; (*center*) *barred galaxy SBb*(*s*), *NGC* 1300; (*lower*) *irregular galaxy Irr I, the Large Magellanic Cloud.* (*Photographs courtesy of the Hale Observatories.*)

On such a large scale, the gas behaves practically as a perfect electric conductor, even though it is only slightly ionized and the conductivity is small by terrestrial standards. The cosmic magnetic field produced by the electric current in interstellar gas is a few microgauss (10^{-6} gauss, the gauss being the cgs unit of magnetic field).

Permeating the interstellar space are cosmic ray particles and electromagnetic radiation of various wavelengths (ranging from alpha rays, to X rays, to optical radiation, to radio waves). These may exert an influence on the gas and the magnetic field. We therefore use the term "interstellar medium" to describe the whole system, which interacts only weakly with the systems of stars. Indeed, the principal dynamical interaction between the stars and the gas is gravitational.

DYNAMICS OF STELLAR SYSTEMS

To analyze the structure of the galaxy (e.g., to explain the spiral structure), one must first place it in a proper mathematical context. How does one begin to tackle such a complicated system?

As explained earlier, the first essential step is to construct an *idealized* model. This step can be taken only after we have ascertained the relevant empirical facts and the general laws governing our system.

An important set of facts about the galaxy is given in Table 1.1. We see

TABLE 1.1. Energy densities in our galaxy[a].

Source	Energy density (Units: 10^{-12} erg cm^{-3})
Total radiation (star light)	0.7
Turbulent gas motion	0.5
Total energy of galactic rotation	1300
Cosmic rays	1
Magnetic field (10^{-5} gauss)	4

[a] From P. Morrison, *Rev. Mod. Phys. 29*, 235 (1957).

that the following constituents make successively smaller contributions to the energy density: (i) the stellar system (by far the most prominent), (ii) the gas, (iii) the magnetic field, and (iv) the cosmic ray particles and other forms of radiation. Thus we should begin by formulating the dynamics of the most energetic component, the stellar system. This formulation will now be described.*

A collection of stars, separated by distances of the order of 10^6–10^7 times their radii (distances that are many times the radii of their outermost planets)

* A full system of equations governing the dynamics of a galaxy can be found in a survey by C. C. Lin, *J. Appl. Math. 14*, 876–921 (1966).

may be treated as a set of masses subject to gravitational attraction. It is possible to write down the system of dynamical equations governing their behavior according to Newton's laws of motion and universal gravitation. Such systems will be discussed in detail in Chapter 2. However, this is clearly not a suitable way to describe the galactic system, for we cannot hope to follow the motion of each of the billions of stars. We must consider them in some sort of collective manner, in the so-called statistical description.

For simplicity, we shall consider the idealized situation of stars of uniform mass. (A discussion of the validity of this idealization must be left to more specialized treatments of the topic.) Since the motion of a star is specified by its position and velocity, we consider the number of stars within a given range of position and velocity. For precise mathematical formulation, we introduce the phase space of six dimensions $(x, y, z; \ , v, w)$, combining positional coordinates and velocity coordinates. The **phase space number density** or **distribution function** $\Psi(x, y, z; u, v, w; t)$ of the stars in this phase space at time t is defined so that $\Psi(x, y, z; u, v, w; t)\, dx\, dy\, dz\, du\, dv\, dw$ gives the number of stars whose three spatial coordinates are, respectively, in the ranges $(x, x + dx)$, $(y, y + dy)$, $(z, z + dz)$, and whose three velocity components are in the ranges $(u, u + du)$, $(v, v + dv)$, and $(w, w + dw)$. The distribution function Ψ then gives a complete description of the stellar system at any instant.* To describe the dynamical process, we must consider the change of this distribution function in the course of time.

In terms of the distribution function Ψ, the mass density $\rho(x, y, z, t)$ of the stars is given by

$$\rho(x, y, z, t) = m_* \iiint_{-\infty}^{\infty} \Psi(x, y, z; u, v, w; t)\, du\, dv\, dw, \tag{1}$$

where m_* is the mass of an individual star. This equation merely states that to get the total mass density, we must sum up the mass of all the stars with various velocities.

Each of the stars is moving in the combined gravitational field of all the others. Except for stars coming into close encounter† with a given star, the motion of an individual star will be essentially influenced only by the gravitational field due to the smeared-out distribution of matter with density $\rho(x, y, z, t)$ given by (1). This gravitational field is given by the negative gradient of the gravitational potential $V(x, y, z, t)$, which is related to the density ρ by

$$V(x, y, z, t) = -G \iiint \frac{\rho(x_0, y_0, z_0, t)\, dx_0\, dy_0\, dz_0}{\sqrt{(x - x_0)^2 + (y - y_0)^2 + (z - z_0)^2}}, \tag{2}$$

* From material in Sections 13.1 and 14.5 one can understand that we work with the distribution function Ψ because interstellar distances are small on a galactic scale, but typical distances traveled between collisions are not.

† Close encounters require specialized treatment, but their net effect is negligible in the present problem. See S. Chandrasekhar, *Principles of Stellar Dynamics* (New York: Dover, 1960).

where G is the constant of universal gravitation ($G = 6.7 \times 10^{-8}$ cgs units).

The equations of motion of each star can now be written as

$$\frac{dx}{dt} = u, \quad \frac{dy}{dt} = v, \quad \frac{dz}{dt} = w, \quad \frac{du}{dt} = -\frac{\partial V}{\partial x}, \quad \frac{dv}{dt} = -\frac{\partial V}{\partial y}, \quad \frac{dw}{dt} = -\frac{\partial V}{\partial z}. \tag{3}$$

Furthermore, by invoking mass conservation and the assumed absence of collisions, one can assert that any increase of mass in a portion of phase space must be due to flow across its boundaries. As in Exercise 14.1.12 or Exercise 14.1.13, this can be shown to imply that

$$\frac{\partial \Psi}{\partial t} + u\frac{\partial \Psi}{\partial x} + v\frac{\partial \Psi}{\partial y} + w\frac{\partial \Psi}{\partial z} - \left(\frac{\partial V}{\partial x}\frac{\partial \Psi}{\partial u} + \frac{\partial V}{\partial y}\frac{\partial \Psi}{\partial v} + \frac{\partial V}{\partial z}\frac{\partial \Psi}{\partial w}\right) = 0. \tag{4}$$

Note that the gravitational field $-\nabla V$ is related to the distribution function Ψ through (1) and (2). Hence the terms within the parentheses in (4) are quadratic in Ψ; i.e., if Ψ is replaced by $\alpha\Psi$ (α being a constant), these terms are multiplied by a factor α^2. Thus (4) is a nonlinear equation.

Equations (3) which govern the motion of the individual stars, and Equation (4), which governs the collective behavior of all the stars, are closely related to each other. In the language of the theory of partial differential equations, the system of equations (3) describes the characteristic curves of (4).

From (4) one can calculate $\partial\Psi/\partial t$ at any instant provided Ψ is known, since the gravitational potential can be calculated through the use of (1) and (2). Heuristically speaking, we may then regard (4) as sufficient for the description of the stellar system at all subsequent times, once Ψ is specified at a given instant. The mathematical theory for this statement is much harder to prove, especially when one raises such questions as stability—does the change in the solution remain small if there is a slight change in the initial specification of the distribution function? Such questions often cannot be rigorously settled; yet the applied mathematician must proceed with his work. He must, of course, have reasonable assurance that his work is meaningful by appealing to other careful considerations beyond the desirable but unavailable mathematical theorems.

DISTRIBUTION OF STARS ACROSS A GALACTIC DISK

As a relatively simple but realistic example of the use of the above system of equations, let us consider the distribution of stars across a galactic disk. This is a problem that has yet not been fully solved. There appears to be "missing matter" in our galaxy in the solar neighborhood.* The solution to be presented below is usually used as the essential basis to test the observations.

* See, e.g., p. 15 of L. Woltjer's article in *Relativity and Astrophysics*, Vol. II (Providence, R.I.: American Mathematical Society, 1967).

The galactic disk is 10 kpc ($= 10^4$ pc) in radius but only 600 pc in thickness. Thus, to a good approximation, we should be able to consider the galaxy to vary locally only in the direction perpendicular to the disk. We shall take this to be the direction of the z axis with $z = 0$ at the center of the disk. Then, in the stationary state, $\Psi = \Psi(z, w)$ and (4) becomes

$$\boxed{w \frac{\partial \Psi}{\partial z} - \frac{\partial V}{\partial z} \frac{\partial \Psi}{\partial w} = 0.} \tag{5a}$$

It can be shown (Section 16.1) that the potential $V(z)$ satisfies the Poisson differential equation

$$\boxed{\frac{\partial^2 V}{\partial z^2} = 4\pi G \rho(z).} \tag{5b}$$

Our considerations now focus on a definite *mathematical problem*, that posed by (5a) and (5b). (Certain additional requirements on the solution will also be invoked later.) We shall treat these equations in two steps.

Equation (5a) can be solved by the standard method of Lagrange for integrating partial differential equations of the first order. Let us briefly recall the salient features of this method.

For a partial differential equation of the form

$$P(z, w, \Psi) \frac{\partial \Psi}{\partial z} + Q(z, w, \Psi) \frac{\partial \Psi}{\partial w} = R(z, w, \Psi), \tag{6a}$$

Lagrange's **method of characteristics** is based on the **characteristic equations**

$$\frac{dz}{P} = \frac{dw}{Q} = \frac{d\Psi}{R}. \tag{6b}$$

Let $f_1(z, w, \Psi) = c_1$ and $f_2(z, w, \Psi) = c_2$ be solutions of any two independent ordinary differential equations implied by (6b). (Here c_1 and c_2 are constants.) Then f_1 and f_2 can be shown to define functions Ψ that satisfy (6a). Thus f_1 and f_2 provide constants or **integrals** of the phenomenon described by (6a). The intersections of the surfaces $f_1 = c_1, f_2 = c_2$ give the **characteristic curves**.*

The general solution to (6a) is given by

$$F(f_1, f_2) = 0.$$

In the present case, $R = 0$, so that one integral relation is $\Psi = c_1$ [which obviously provides a solution of (5a)]. A second characteristic equation is

$$\frac{dz}{w} = \frac{dw}{-\partial V/\partial z}, \tag{7}$$

* A slight extension of these remarks shows that (3) gives the characteristics of (4), as stated above.

which leads at once to the integral

$$\tfrac{1}{2}w^2 + V(z) = E, \tag{8}$$

where E is a constant, that gives the total energy per unit mass of the star. Thus the general solution of (5a) is

$$\Psi = f(E) = f[\tfrac{1}{2}w^2 + V(z)]. \tag{9}$$

A special case of (9) is compatible with the observed Gaussian distribution of velocities. We take

$$f(E) = A_0 e^{-\beta E}, \tag{10}$$

where A_0 and β are suitable constants. The constant β is prescribed by the distribution of velocities. Furthermore, by the one-dimensional version of (1), A_0 must be related to the local density by

$$\rho = m_* \int_{-\infty}^{\infty} \Psi \, dw. \tag{11}$$

Indeed, we find the density distribution to be of the form

$$\rho(z) = \rho_0 e^{-\beta V(z)}, \tag{12}$$

where ρ_0 is the density where $V(z) = 0$, which we shall take to be at $z = 0$, the middle of the galactic plane.* The quantity ρ_0 is related to A_0 and β by

$$\rho_0 = A_0 m_* \sqrt{\frac{2\pi}{\beta}}. \tag{13}$$

The observational determination of ρ_0 is based on (12), which follows from (5a), without the use of the Poisson equation (5b). From (12) one can determine $V(z)$, and hence its second derivative, as

$$\frac{d^2V}{dz^2} = -\frac{1}{\beta}\frac{d^2}{dz^2}\ln\left(\frac{\rho}{\rho_0}\right). \tag{14}$$

One can determine β from the distribution of stellar velocities, by using (10), and one can observe the number of stars per unit volume, which is proportional to ρ. Evaluation of the expression in (14) is then possible. Furthermore, this can be done for any species of stars.† Thus we have the possibility of determining the total stellar mass by combining (14) with (5b). As mentioned before, existing knowledge reveals a considerable descrepancy. The mass determined by this dynamical procedure is about 1.7 times larger than that obtained by direct observational determination.

* A potential is defined only up to an additive constant, since just its gradient is significant.

† From observations of their spectral lines, stars are classified into a number of different species. These species basically correspond to stars of different properties such as mass, composition, and age.

Equation (5b) has yet to be satisfied. Thus, to complete the analysis, one must solve the equation

$$\frac{d^2V}{dz^2} = 4\pi G\rho_0 e^{-\beta V(z)}. \tag{15}$$

To do this, multiply both sides by dV/dz and integrate, with the specification that

$$V = \frac{dV}{dz} = 0 \qquad \text{at } z = 0, \tag{16}$$

since the gravitational force also vanishes at the middle of the galactic plane. Then (Exercise 1) there is little difficulty in showing that the distribution of density finally takes on the form

$$\frac{\rho}{\rho_0} = \operatorname{sech}^2 \frac{z}{z_0}, \tag{17}$$

where z_0 is given by

$$z_0^2 = (2\pi G\rho_0 \beta)^{-1}. \tag{18}$$

What have we gained from our theoretical analysis? Of most importance is the discrepancy between theory and observation concerning the value of ρ_0. A naïve person might think that this discrepancy proves our analysis to be of little worth, but just the opposite is true. Rough agreement between experiment and a plausible theory could be largely a coincidence, but *disagreement* means that there is *more to the phenomenon than meets the eye.* The italicized phrase is doubly relevant here, for there must be some invisible mass present to account for the excess of the predicted mass compared to that observed. The two principal conjectures are that the excess mass is due either to molecular hydrogen or to dark stars. The matter has not been resolved at the time of writing.

DENSITY WAVE THEORY OF GALACTIC SPIRALS

It is beyond the scope of this book to carry out much further analysis of stellar systems beyond that given above. We shall, however, report here briefly some recent developments in the theory of galactic spirals, based on stellar dynamics and gas dynamics, as another example of the power of mathematical analysis.

The bright spiral arms of galaxies are marked by massive young stars. These stars are known to be rotating faster in the interior parts of a galaxy than in the outer parts. If a spiral arm is "material," i.e., if it always consists of the same stars, we must then have what is known as the "winding dilemma." For trailing spiral patterns,* the spiral arms would tend to wind into a

* A spiral pattern is called "trailing" if it opens up in the direction behind that of the rotation of the stars; "leading," if in the opposite sense. A leading material spiral pattern would loosen up under the effect of differential rotation.

tighter structure. However, the Hubble classification is based not only on the tightness of winding, but on other physical parameters as well. Furthermore, these physical parameters cannot change as rapidly as the geometry of a material spiral arm. Thus we must accept the hypothesis that the spiral pattern remains quasi-permanent.

In the 1920s the Swedish astronomer Bertil Lindblad suggested that there are "density waves" in the stellar system which underlie the observed spiral structure. These are periodic temporary concentrations of stars caused by their self-gravitation. Qualitatively, they are similar to acoustic waves, but the physical mechanisms involved are different in the two cases. Lindblad did not carry out any detailed mathematical analysis of the *collective* behavior of these stars. He was therefore able to make only general conjectures about these waves, and some of them were inevitably incorrect. Thus his ideas remained unaccepted for a long time.

In the 1960s the density wave theory was revived by C. C. Lin and Frank H. Shu when they succeeded in calculating a spiral structure for the Milky Way system, in reasonable agreement with observations. The method was also applied to other galaxies, and a number of other observable consequences were deduced and checked against astronomical data. (See Figure 1.3.) For a summary report of the status of the theory, see the general discourses of B. J. Bok and C. C. Lin at the 14th General Assembly of the International Astronomical Union.* Suffice it to note here that the theoretical studies of these density waves offer very challenging mathematical problems. Indeed, the problems are similar to those encountered in the study of waves in electromagnetic plasmas. (By analogy, a stellar system may be called a "gravitational plasma.") Thus the understanding of galactic systems requires extensive mathematical analysis of a rather subtle kind. At the same time, it stimulates the study of certain mathematical methods and theories that are applicable to other important problems.

EXERCISES

1. (a) Deduce the relation (17) for the density distribution of stars across a galactic disk. Sketch.

(b) Obtain the distribution of gravitational potential

$$V(z) = 4\pi G\rho_0 z_0^2 \ln \cosh \frac{z}{z_0}. \tag{19}$$

(c) Find the gravitational field at "large" distances from the galactic plane. Discuss what is meant by "large."

* Published in *Highlights of Astronomy*, Vol. II (Dordrecht, Holland: D. Reidel Publishing Company, 1971). For general information on spiral structure, including a description of the density wave theory, see the article "Updating Galactic Spiral Structure" by B. J. Bok in *Amer. Scientist 60*, 708–22 (1972).

FIGURE 1.3. *The same photograph as in the frontispiece with added curves indicating the location of the peak of synchrotron emission as observed by D. S. Mathewson, P. C. van der Kruit, and W. N. Brouw. Coincidence of this curve with the dust lane is strong evidence in favor of the density wave theory of galactic spirals. Reprinted with permission from* Astron. and Astrophys. **17**, 468 (1972).

1.3 Aggregation of Slime Mold Amebae

In this section we shall formulate a mathematical model to describe the aggregation of slime mold amebae. We shall thereby gain some understanding of a phenomenon of considerable current interest in developmental biology. (*Developmental biology* includes *embryology*, the study of how a single cell develops into a complicated embryo, but also encompasses the development of simpler organisms that cannot be said to have an embryo.) We begin with some facts about slime mold amebae that provide a basis for our mathematical model.

SOME FACTS ABOUT SLIME MOLD AMEBAE*

One can begin a description of the life cycle of the slime mold amebae at the spore stage, where each ameba is dormant within a protective covering. When conditions are favorable, an ameba emerges from its spore. Of the order of 10 micrometers (10^{-3} cm) across, the amebae are rather shapeless one-celled organisms that move by extending contractile portions of themselves (pseudopods).

The natural habitat of the amebae is soil or dung. An important element of the food chain on earth, they feed on bacteria by engulfing them. If food is plentiful, the amebae continually feed and multiply by mitosis (dividing in two). If the food supply becomes exhausted there is an **interphase period** of random and somewhat feeble movement, where the amebae are more or less evenly distributed over the area available to them. During this period, the disappearance of the food supply triggers certain changes in the amebae. The details of these internal changes are not known, but there is no difficulty in observing the striking external phenomenon that results. After a few hours, the amebae begin to aggregate into a number of collection points. Typically,† these are more or less regularly distributed, with a spacing of a few hundred micrometers. (See the frontispiece. Figure 1.4 gives an idea of what the beginnings of aggregation look like under conditions of lower cell density.)

After aggregation has been completed, the amebae that have collected at a given point (ranging in numbers from a few in laboratory experiments, up to 200,000) form a multicelled slug. This moves as a unit, although the formerly free living amebae retain their cell walls within the slug. Then the slug stops and erects a stalk, on top of which is a roundish container of spores. The cycle is thus completed.

* "There is no philosophy which is not founded upon knowledge of the phenomena, but to get any profit from this knowledge it is absolutely necessary to be a mathematician."—Daniel Bernoulli. Quoted by C. Truesdell on p. 318 of *Essays in the History of Mechanics* (New York: Springer, 1968).

† There are several species of cellular slime mold, with various corresponding differences in behavior. [See J. T. Bonner, *The Cellular Slime Molds* (Princeton, N. J.: Princeton U.P., 1967).] Our description is appropriate for the most studied species, *Dictyostelium discoideum*.

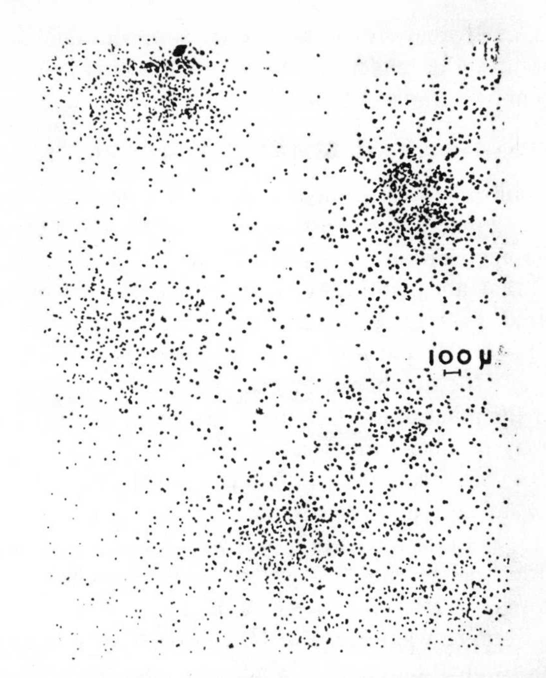

FIGURE 1.4. *Population of amebae just beginning to aggregate. The scale is indicated by a* 100-*micrometer line segment.* (*Courtesy of B. Shaffer.*)

What is responsible for the organized aggregation of the amebae? This is an important question, for "purposeful" movements occur frequently in developmental processes. Usually, these movements take place rather inaccessibly within a developing organism, but the cells of the slime mold amebae will obligingly perform in a laboratory dish so that it is easy to examine them and to experiment with them.

It has been discovered that the amebae move preferentially toward relatively high concentrations of a chemical which they themselves secrete. In some species the attractant has recently been identified as cyclic 3′,5′-adenosine monophosphate (AMP) a chemical with many important biological functions. (E. W. Sutherland was awarded the Nobel Prize for medicine in 1971 for his work in elucidating some of the roles of cyclic AMP.) It is also known that a given quantity of attractant loses its potency in a matter of minutes. This has been traced to the activity of an enzyme that alters the nature of the AMP.

Presumably, aggregation is caused by the fact that the amebae move up a gradient of attractant, but what determines the time of onset? What determines the spacing of aggregation centers? Can one quantify the process? A mathematical model is needed to answer such questions. We shall proceed to devise the simplest model that could reasonably be supposed to bear on the circumstances.* If analysis of this model is encouraging, one can add detail later. For this, the reader is referred to the paper just cited and to subsequent papers by various authors in the same journal.

FORMULATION OF A MATHEMATICAL MODEL

Since the distance between amebae is small compared to the typical distance between aggregation centers, we shall employ a continuum model. Suppose that the aggregation takes place in the (x, y) plane. For simplicity we shall assume that no quantities change with y, so that only x variation need be considered. Exercise 3 shows that this assumption of unidimensionality is not at all essential, but it makes exposition easier.

Let $a(x, t)$ denote the number of amebae per unit area at position x and time t. Consider the amebae located in the region $x_0 \le x \le x_0 + \Delta x$, where Δx is an arbitrary number (not necessarily small). We shall now write a **general balance law** for this region; this states that the rate of change of the net amount of a in the region equals the rate at which a flows across the boundary, plus the net rate of creation of a within the region. In the present case, a stands for amebae, and the net creation of amebae is equal to the excess of births over deaths. But the balance law is "general" because it applies to any substance whatever. For how else can a substance appear in the region except by creation or by flow across its boundaries?

To proceed, we must define the **flux density** $J(x_1, t)$. This gives the net rate per unit length at which a crosses the line $x = x_1$. Also, J is defined to be positive if there are more amebae crossing in the direction of x increasing than in the opposite direction. The term $Q(x, t)$ is the net rate per unit area at which a is being created. The desired balance law can now be written. Considering the rectangle of Figure 1.5, we obtain

$$\frac{\partial}{\partial t}\int_{x_0}^{x_0+\Delta x} a(x, t)\, dx = J(x_0, t) - J(x_0 + \Delta x, t) + \int_{x_0}^{x_0+\Delta x} Q(x, t)\, dx. \quad (1)$$

It is convenient to use the integral mean value theorem (Appendix 13.2) to write

$$\frac{\partial}{\partial t}[a(x_1, t)\,\Delta x] = J(x_0, t) - J(x_0 + \Delta x, t) + Q(x_2, t)\,\Delta x;$$

$$x_0 \le x_1 \le x_0 + \Delta x, \quad x_0 \le x_2 \le x_0 + \Delta x. \quad (2)$$

We divide by Δx and then take the limit as $\Delta x \to 0$, to obtain the **general balance law in differential equation form,**

$$\frac{\partial a}{\partial t}(x_0, t) = -\frac{\partial J}{\partial x}(x_0, t) + Q(x_0, t), \qquad x_0 \text{ arbitrary}. \quad (3)$$

* We adapt material in a paper by E. F. Keller and L. A. Segel, *J. Theoret. Biol. 26*, 399–415 (1970).

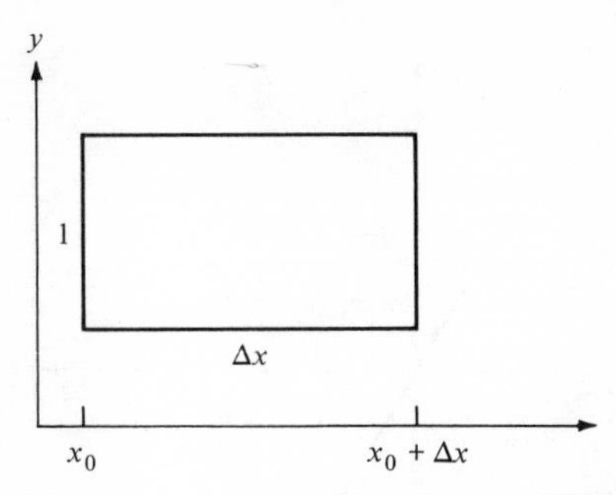

FIGURE 1.5. *The balance law* (3) *considers the rate of change of mass in a rectangle of length* Δx *and unit width, in the case where* y *variation is absent.*

In the case of the amebae, reproduction can be ignored because there is little or none of it in the absence of food. Deaths can also be ignored, since there are few during the time interval of interest. Thus we take

$$Q \equiv 0. \tag{4}$$

To obtain an expression for the flux J, let us first consider a situation when attractant is absent. Then the amebae appear to move about randomly. Owing to such "diffusionlike" movement, a concentration of amebae tends to disperse. Thus there is a random flux J_r from regions of high ameba concentration to regions of low concentration. The magnitude of the flux at x seems to depend on the concentration difference between x and nearby points. We characterize this difference by $\partial a/\partial x$ (the simplest choice) and make the hypothesis that

$$J_r(x, t) = F\left[\frac{\partial a(x, t)}{\partial x}\right] \tag{5}$$

for some function F. Now when $a \equiv$ constant, $J = 0$—for in random motion there will be as many amebae moving to the left as to the right. In this case $\partial a/\partial x \equiv 0$. Therefore, given (5), it is only sensible to assume that

$$F(0) = 0. \tag{6}$$

Thus the graph of F must have an appearance such as that depicted in Figure 1.6. For sufficiently small values of s, we can approximate the graph by a straight line. Calling the slope of the line $-\mu$, we obtain $F(s) = -\mu s$ as the simplest reasonable assumption concerning F; i.e.,

$$J_r(x, t) = -\mu \frac{\partial a(x, t)}{\partial x}. \tag{7}$$

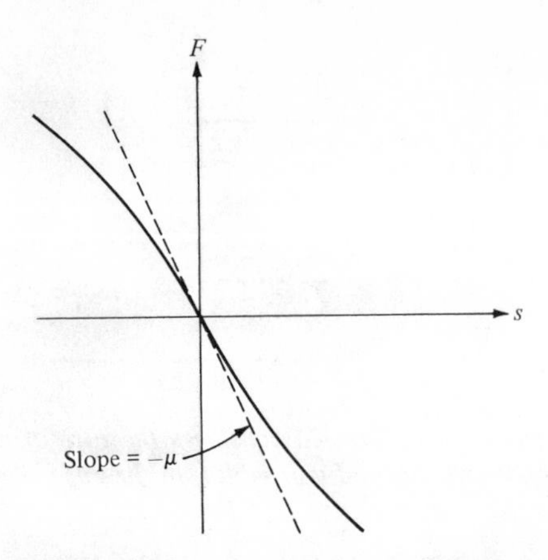

FIGURE 1.6. *Possible plot of the function F defined in* (5). *The graph provides negative values for* $s > 0$ *because there should be a leftward (negative) flux if amebae density is higher at larger values of* x. *Similar reasoning explains the positivity for* $s < 0$.

Combining (3), (4), and (7), we obtain

$$\frac{\partial a}{\partial t} = \frac{\partial}{\partial x}\left(\mu \frac{\partial a}{\partial x}\right). \tag{8}$$

Equation (8) is the **diffusion equation**, which will be studied at length beginning in Chapter 3. This equation is used to model the concentration variation of any kind of randomly moving set of particles, for example smoke particles in air (see Chapter 3). The constant μ, which governs the vigor of the random movement, is generally called the **diffusivity**. Here we call μ the **motility**, giving precise meaning to a biological term that is often used only in a qualitative manner.

Equation (8) was obtained under the assumption that no attractant was present. To account for **chemotaxis**, which is directional motion induced by variations in chemical concentration, we add to J_r of (7) an additional contribution J_c. Let $\rho(x, t)$ be the density of the attractant. Arguing as before, we pass from an initial assumption that J_c is a function of the attractant gradient $\partial\rho/\partial x$ to the assumption that J_c is proportional to the attractant gradient, at least for small values of the gradient. For a given gradient, if the amebae density is twice as great, the net flux should be twice as great. The proportionality factor should thus be a multiple of a. We are led to the hypothesis that

$$J_c = \chi a \frac{\partial \rho}{\partial x}. \tag{9}$$

The factor χ measures the strength of chemotaxis. Note that in contrast to (7), there is no negative sign in (9). This is because amebae tend to move *toward* attractant concentrations (and away from ameba concentrations).

Assuming that the total flux J in (3) is the sum of the random contribution J_r and the chemotactic contribution J_c, we are led to our final equation for the change in ameba density:

$$\boxed{\frac{\partial a}{\partial t} = \frac{\partial}{\partial x}\left(\mu \frac{\partial a}{\partial x} - \chi a \frac{\partial \rho}{\partial x}\right).} \tag{10}$$

We shall take μ and χ to be positive constants. It is not difficult to take into account a variation with ρ which is probably present, but this would only change some details of the analysis to come [Exercise 2(a)]. Note that even with constant μ and χ, (10) contains a (quadratic) nonlinear term $\chi a(\partial \rho/\partial x)$, because this term is proportional to the product of two unknown functions, a and $\partial \rho/\partial x$. [Such a quadratic nonlinearity was also found in our study of galactic structure—compare the remarks following Equation (2.4.)] Nonlinear terms generally make mathematical analysis more difficult, and hence more challenging and more rewarding.

We also need an equation for the attractant density ρ. This will be of the general form (3):

$$\frac{\partial \rho}{\partial t} = -\frac{\partial J_a}{\partial x} + Q_a.$$

(The subscript "a" refers to attractant.) The random motion of the attractant molecules will be modeled by a proportionality of flux to gradient, as in (7):

$$J_a = -D\frac{\partial \rho}{\partial x}.$$

The net creation term Q_a has a positive contribution fa as a result of the secretion of attractant by the amebae. Here f is the rate of secretion per unit amebae density. What of the decay in attractant activity? We take the rate of decay (as in radioactivity or some other spontaneous process) to be proportional to the amount of attractant present,* via the constant k. Thus $Q_a = fa - k\rho$, and the desired equation for $\partial \rho/\partial t$ is

$$\boxed{\frac{\partial \rho}{\partial t} = fa - k\rho + D\frac{\partial^2 \rho}{\partial x^2}.} \tag{11}$$

We shall take f, k, and D to be positive constants.

* As we stated above, decay is actually due to the action of an enzyme. If Michaelis–Menten kinetics (Chapter 10) are assumed, this can be modeled by making k a certain function of ρ. But the essentials of the analysis are unchanged [Exercise 2(b)].

Partial differential equations (10) and (11), for the two unknown functions $a(x, t)$ and $\rho(x, t)$, provide the basic formulation of our problem.

AN EXACT SOLUTION: THE UNIFORM STATE

It is very easy to find an exact solution of (10) and (11). This is the **uniform solution,**

$$a = a_0, \qquad \rho = \rho_0, \tag{12}$$

where a_0 and ρ_0 are constants. [When (12) holds, the system is said to be in the **uniform state**.] Upon substitution into (10) and (11), we find that (12) will indeed provide a solution if

$$fa_0 = k\rho_0 . \tag{13}$$

Equation (13) is physically reasonable. It says that in the uniform state the secretion rate of attractant must be exactly balanced by the decay rate.

ANALYSIS OF AGGREGATION ONSET AS AN INSTABILITY

We identify the uniform state with the interphase period prior to aggregation. We also model the onset of aggregation as *the breakdown of the uniform state due to the growth of inevitable small disturbances* to ameba and attractant density. That is, we identify the onset of aggregation with the sort of **instability of the uniform state** whose investigation forms a classical part of applied mathematics (see Sections 11.1 and 15.2). The instability is presumed to occur because of changes, during interphase, of the parameters μ, χ, and f that characterize ameba behavior.

The idea behind instability theory is this. Suppose that at some initial time, the state of the system is slightly disturbed from the uniform state. (Suppose that there is a local clumping of amebae, for example, and an accompanying local concentration of attractant.) Will the small disturbance tend to disappear with the passage of time or will it become more intense? In the former case we say that the uniform state is **stable to small disturbances**, in the latter **unstable**. Unstable states will not be observed, for disturbances are inevitable. In the case of instability, they will grow, so the uniform state will be replaced by some other state.

To perform a stability analysis, we introduce the variables a' and ρ' by the definitions

$$a(x, t) = a_0 + a'(x, t), \qquad \rho(x, t) = \rho_0 + \rho'(x, t). \tag{14}$$

Here $a' = a - a_0$, for example, measures departure from uniformity; therefore, it can be identified with the disturbance in ameba density.

To obtain equations for a' and ρ', we substitute (14) into (10) and (11). From the former we obtain

$$\frac{\partial a'}{\partial t} = \mu \frac{\partial^2 a'}{\partial x^2} - \chi\left[(a_0 + a')\frac{\partial^2 \rho'}{\partial x^2} + \frac{\partial a'}{\partial x'}\frac{\partial \rho'}{\partial x'}\right]. \tag{15}$$

This equation is nonlinear, owing to the presence of the quadratic terms $a'(\partial^2\rho'/\partial x^2)$ and $(\partial a'/\partial x')(\partial\rho'/\partial x')$. We shall assume that the disturbances (and their derivatives) are small, in which case we call a' and ρ' **perturbations**. The perturbations are involved quadratically in some terms and linearly in others. Products of two small terms should be negligible in comparison with the other terms of (15), which contain but a single perturbation function. We thus **linearize** the equation by *deleting all nonlinear terms*. We obtain

$$\frac{\partial a'}{\partial t} = \mu \frac{\partial^2 a'}{\partial x^2} - \chi a_0 \frac{\partial^2 \rho'}{\partial x^2} \tag{16}$$

as the linearized version of (10). As for (11), upon substituting (14) and employing (13), we obtain

$$\frac{\partial \rho'}{\partial t} = fa' - k\rho' + D\frac{\partial^2 \rho'}{\partial x^2}, \tag{17}$$

which is already linear.

We are faced with a pair of linear partial differential equations with constant coefficients. We guess that there are solutions of the form

$$a' = C_1 \sin qx\, e^{\sigma t}, \qquad \rho' = C_2 \sin qx\, e^{\sigma t}, \tag{18}$$

where C_1 and C_2 are constants.* It is easily seen [Exercise 1(a)] that there are indeed solutions of the form (18), provided that

$$(\sigma + \mu q^2)C_1 - \chi a_0 q^2 C_2 = 0, \tag{19a}$$

$$-fC_1 + (\sigma + k + Dq^2)C_2 = 0. \tag{19b}$$

This system of algebraic equations has the trivial solution $C_1 = C_2 = 0$. [From (18) we see that we have merely verified that it is possible to have an identically zero perturbation of an exact solution.] For a nontrivial solution, the determinant of the coefficients must vanish. We thus obtain the following quadratic equation for σ:

$$\sigma^2 + b\sigma + c = 0; \qquad b \equiv k + (\mu + D)q^2, \qquad c \equiv \mu q^2(k + Dq^2) - \chi a_0 f q^2. \tag{20}$$

The quadratic equation can be shown to have real roots [Exercise 1(b)]. To ensure stability, both roots must be negative, so that the exponential factor exp (σt) brings about decay of the perturbations. It is not difficult to show (Exercise 1) that a necessary and sufficient condition for stability is $c > 0$, which requires that

$$\chi a_0 f < \mu(k + Dq^2), \qquad q \neq 0. \tag{21}$$

* The reasons that lie behind such an assumption as (18) are more fully discussed in Section 15.2. We mention here that a cosine dependence in (18) will yield exactly the same results. (Compare Exercise 3.) Also, more general disturbances can be obtained by the superposition of sinusoidal solutions, using Fourier analysis (as discussed in Chapters 4 and 5).

From (18), $2\pi/q$ is the wavelength of the perturbation under investigation. Since μ and k are positive, the right side of (21) decreases monotonically as q decreases, toward a greatest lower bound of μk. Therefore, the longer the wavelength $2\pi/q$, the more "dangerous" the perturbation [for (21) is more easily violated]. We conclude that *instability* is possible whenever

$$\frac{\chi a_0 f}{\mu k} > 1, \tag{22}$$

for then (21) is violated for a range of sufficiently small values of q^2.

INTERPRETATION OF THE ANALYSIS

The view of aggregation that emerges is this. During the beginning of the interphase period, inequality (22) does not hold and the uniform state is stable. Triggered by the stimulus of starvation, the various parameters gradually change. Eventually, (22) is satisfied, and aggregation commences. We identify the size of the "aggregation territory" with the instability wavelength; our analysis predicts that the territory size is "very large" in some sense.

The instability criterion (22) has the following interpretation. Suppose that there is a concentration of amebae and attractant at some point. Random "diffusion" of amebae with motility μ tends to disperse this concentration, as does the attractant decay, whose strength is measured by k. Larger μ and k means larger stabilizing effects. It is therefore appropriate that increases in μ and k mean that the instability criterion (22) is more difficult to satisfy.

By contrast, a local concentration of attractant tends to draw amebae toward it by chemotaxis (strength χ). Also, a concentration of amebae will provide a corresponding increase of attractant concentration because of the higher local concentration of secretion sources. The strength of this effect is measured by $a_0 f$. This explains the appearance of the destabilizing factors $a_0 f$ and χ in the numerator of (22). Instability ensues if destabilizing effects outweigh stabilizing.

The explanation of aggregation provided in the previous paragraph could perhaps have been provided without doing a mathematical analysis. But, as is typical, only after carrying through some calculations does one seem able to discern the physical essence of a situation. Also, the criterion (22) is a quantitative one, capable of experimental test. No full quantitative tests have been performed as yet, but it has been discovered that both the chemotactic sensitivity χ and the attractant secretion rate f go up by about 100-fold at the time aggregation commences. As illustrated in our discussion of galactic structure, future quantitative tests might reveal defects in understanding that would never emerge from qualitative considerations.

The above analysis is based on a paper (op. cit.) that appeared in 1970 in the *Journal of Theoretical Biology*. Since that time, several other papers on slime mold aggregation, by various authors, have appeared in that journal and elsewhere. They deal with such matters as (i) a more accurate modeling of the chemical kinetics and hence (among other things) a possible refinement of the "very large territory size" prediction mentioned above; (ii) the streaming type of aggregation, as in the frontispiece; and (iii) the influence of pulsatile attractant secretion and refractory periods. Future work will doubtless examine the effects of nonlinearity. Two survey articles on slime mold aggregation and other collective chemotactic motions, written for a mathematical audience, can be found in *Some Mathematical Questions in Biology*, Vol. III (Providence, R.I.: American Mathematical Society, 1972).

EXERCISES

1. (a) Verify (19).
(b) Show that the roots of the quadratic equation (20) are real.
(c) Show that both roots are negative if and only if both b and c are positive.
(d) Show that since f, D, and χ are positive, if $c > 0$, then $b > 0$. Thereby deduce (21).

2. Investigate the changes in the analysis that would occur under the following conditions.
(a) If μ and χ were functions of ρ.
(b) If k and f were functions of ρ.

3. If two-dimensional variation is considered, the governing equations (10) and (11) are replaced by

$$\frac{\partial a}{\partial t} = \nabla \cdot (\mu \nabla a - \chi a \nabla \rho), \qquad \frac{\partial \rho}{\partial t} = fa - k\rho + D\nabla^2 \rho.$$

Assume disturbances with spatial dependence $\sin (q_1 x + q_2 y + \theta)$, where q_1, q_2, and θ are constants. Show that if $q^2 = q_1^2 + q_2^2$, then $2\pi/q$ remains the disturbance period and the instability condition remains (22).

4. Define Δ by $\chi a_0 f/\mu k = 1 + \Delta$ and suppose that Δ is small and positive. [Compare (22).] Ignoring higher order terms in Δ, find an approximation for the larger root of the quadratic. Deduce that the wavelength of the fastest growing disturbance is approximately $2\pi(2D/k\Delta)^{1/2}$.

Appendix 1.1 Some Views on Applied Mathematics

ON THE NATURE OF APPLIED MATHEMATICS

I . . . describe applied mathematics as the bridge connecting pure mathematics with science and technology. I have deliberately described this bridge as *connecting* two areas of activity rather than *leading* from one to the other, because the bridge carries two-way traffic. Its importance to science and technology is obvious, but it is not less important to pure mathematics, which would be poorer without the stimuli coming from the applications.

W. Prager, "Introductory Remarks" in the special issue, "Symposium on the Future of Applied Mathematics," *Quart. Appl. Math. 30*, 1–9 (1972).

One should not judge applied mathematicians by their contribution to mathematics any more than one judges historians by their contributions to history. Much creative effort goes into interpretation. To try to use mathematical ideas on problems which at first sight seem too complex or obscurely stated to fit any pattern is, for many people, an exciting challenge and it is this which I feel to be the essence of applied mathematics.

R. W. Brockett, "The Synthesis of Dynamical Systems," *ibid.*, pp. 41–50.

Here then are three topics on computation—algorithmic complexity, operating system modeling, and data structuring—which could use much heavier mathematical participation than they now receive. Their requirements are far from the kind of approximative analysis that has been so successful in the physical sciences. But they will be large areas of endeavor with many people engaged in their engineering development and in their use—larger than most of the uses of mathematics we now have. I think we shall have to broaden our idea of what applied mathematics is to include these subjects.

H. Cohen, "Mathematical Applications, Computation, and Complexity," *ibid.*, pp. 109–121.

I feel that 'applied mathematics' should be defined 'dynamically' and operationally as the process—and the study of the process—of application of mathematics to the other disciplines. 'Applied Mathematics' is what 'applied mathematicians' do, while 'applied mathematicians' are people who go out and talk to those in other disciplines who are attempting to use mathematics, and who know enough about the discipline itself to understand the problems on a deeper level.

P. Herman, "Comments on the COSRIMS Report," *Amer. Math. Monthly*, 517–21 (May 1970).

ON THE RELATIONSHIPS AMONG PURE MATHEMATICS, APPLIED MATHEMATICS, AND THEORETICAL SCIENCE

I think that it is a relatively good approximation to truth—which is much too complicated to allow anything but approximations—that mathematical ideas originate in empirics . . . . But, once they are so conceived, the subject begins to live a peculiar life of its own and is better compared to a creative one, governed by almost entirely aesthetical motivations . . . . As a mathematical discipline travels far from its empirical source [however] . . . . there is a grave danger that the subject will develop along the line of least resistance, that the stream, so far from its source, will separate into a multitude of insignificant branches, and that the discipline will become a disorganized mass of details and complexities.

John von Neumann, "The Mathematician," reprinted in the *World of Mathematics* (New York: Simon & Schuster, 1956), p. 2061.

'Applied Mathematics' is an insult, directed by those who consider themselves 'pure' mathematicians, at those whom they take for impure. Mathematics is, always has been, and always will be pure mathematics. The adjective 'pure' is redundant. The very essence of mathematics is abstraction, created by fancy and tempered by rigor, but 'pure mathematics' as a parricide tantrum denying growth from human sensation, as a shibboleth to cast out the impure, is a disease invented in the last century . . . .

C. Truesdell, "The Modern Spirit in Applied Mathematics," *ICSU Rev. World Sci. 6*, 195–205 (1964).

The miracle of the appropriateness of the language of mathematics for the formulation of the laws of physics is a wonderful gift which we neither understand nor deserve. We should be grateful for it, and hope that it will remain valid in future research and that it will extend, for better or for worse, to our pleasure even though perhaps also to our bafflement, to wide branches of learning.

E. P. Wigner, "The Unreasonable Effectiveness of Mathematics in the Natural Sciences," *Commun. Pure Appl. Math. 13*, 1–14 (1960).

ON THE TEACHING AND PRACTICE OF APPLIED MATHEMATICS

It has always been a temptation for mathematicians to present the crystallized product of their thoughts as a deductive general theory and to relegate the individual mathematical phenomenon into the role of an example. The reader who submits to the dogmatic form will be easily indoctrinated. Enlightenment, however, must come from an understanding of motives: live mathematical development springs from specific natural problems which can be easily understood, but whose solutions are difficult and demand new methods of more general significance.

R. Courant, Remarks from the preface of *Dirichlet's Principle* (New York: Wiley–Interscience, 1950).

There was complete unanimity on the point that presently mathematics majors do not learn enough about applications . . . . There was also agreement that every mathematician should become acquainted with at least one area of application.

Joint Report of the Committee on Graduate and Postdoctoral Education and of the Committee on Undergraduate Education, 1970–71 Annual Report, Division of Mathematical Sciences, National Research Council, p. 78.

The world is becoming more and more complicated. We are rapidly approaching an unbearable and even dangerous situation . . . . Mathematics is the only subject broad enough to prevent this. Not the mathematics actually taught but a new mathematics inspired by all fields of applications. It must be taught as *fundamental science* which provides indispensable modes of thinking and tools to cope with the real world: *the physical world*, and the *man-made world.*

A. Engel, "The Relevance of Modern Fields of Applied Mathematics for Mathematical Education," *Educ. Studies Math. 2*, 257–69 (1969).

In mathematics courses the talk is usually about problems. Not much is said about where the problems come from or about what is done with the answers. It seems to us that this missing part of the story is worth telling. It is particularly important to those who intend to be mathematicians, since it explains why the electronic computer is not going to put them on the unemployment rolls. It is also important to those who are going into the mathematically oriented sciences, since it explains why the electronic computer is not going to remove their privilege (or duty, as it may be regarded) of learning a great deal of mathematics.

R. Hooke and D. Shaffer, *Math and Aftermath* (New York: Walker and Co., 1965), p. 4.

I think we need some discipline as mathematicians trying to get into applied fields. The discipline you need is to be able to turn off the mathematics as soon as you've discovered what the really relevant physical phenomena are that go on, and then go on for the new physical phenomena.

Remarks by H. B. Keller, "Symposium on Mathematical Professional Societies and Their Present Role," reported on pp. 50–51 of the 1970–71 Annual Report, Division of Mathematical Sciences, National Research Council.

Rigor that just polishes the mathematical surface, but does not penetrate essentials of the subject matter in question, is not indispensable in applications.

U. Grenander, "Computational Probability and Statistics," *SIAM Rev. 15*, 134–91 (1973).

Too much of the attitude of many young mathematicians is that of always finding out what is wrong with what is being said [by scientists] rather than what is right; and I think one of the important things in the game of applied mathematics is to try to understand a bit of what is right.

E. W. Montroll, "Education in Applied Mathematics" (Proceedings of a conference sponsored by SIAM), *SIAM Rev. 9*, 326 (1967).

To a large extent, the community of core mathematicians has decided that it is not its responsibility to provide instruction related to the application of mathematics; to the same large extent, much of the instruction in methodology has become the responsibility of the applied mathematics community. The important question is: Will this community include in the instruction it offers illustrations of the heuristic, inventive, reasoning so necessary to progress in science, in technology, and ultimately in our whole society or will it retreat into a cloistered preoccupation with abstractions both in its research and its instruction? If it does retreat, then the present generation and possibly a few future generations of students will receive little instruction of this sort and those that enter the worlds of technology, of science, of environmental repair and of medical advancement, will do so severely handicapped by a grotesquely distorted education.

Our symposium is labeled 'The future of applied mathematics.' It is my view that if, in answer to the foregoing questions, we make the latter choice, applied mathematics has no viable future whatever—it could only become a small, rather sterile corner of mathematics, a discipline which despite its moments of greatness, already is populous enough to contain more sterility than it wants. But if we make the former choice, i.e., if among all of its other activities, the applied mathematics community provides innovative contributions to the mathematization of disciplines which are just becoming quantitative, if it concerns itself with finding approximate models for complicated phenomena (thereby sacrificing accuracy of detail for ease of interpretability), if it seeks to develop techniques whereby the implications of given mathematical models can more easily be inferred, then applied mathematicians will have inherited much of today's continuation of the challenges, the intellectual achievements, the contributions to society, and the fun which, before they abdicated, was largely the property of the mathematicians and the physicists.

G. F. Carrier, "Heuristic Reasoning in Applied Mathematics," in the special issue, "Symposium on the Future of Applied Mathematics," *Quart. Appl. Math. 30*, 11–15 (1972).

CHAPTER 2

Deterministic Systems and Ordinary Differential Equations

IN THIS chapter we shall illustrate something of the range of applied mathematics using particle dynamics as our central example. The principal purpose of our discussion will be to analyze the process of applied mathematics. We are not primarily interested at this point in enlarging the reader's facility, so a thorough exposition of certain methods will be deferred.

We first examine the relation between Kepler's laws, which describe the motion of the planets, and Newton's formulation of the theoretical law of universal gravitation. The orbits of planets and comets are deduced by the rather ingenious classical solution of the governing equations. When relativistic effects are considered, however, attempting to find an exact solution no longer is an optimal procedure. A perturbation method should be used. The same method is called for when one wishes to calculate the relatively small effect of other planets upon a given planet's orbit about the sun.

As an illustration of practical calculational procedures that have been devised to fill a definite need, we discuss perturbation theory in a certain amount of detail. Roughly speaking, this theory permits one to calculate the effects of small disturbances on otherwise known situations. Remarkably valuable information can be obtained by sophisticated variants of perturbation theory—one cannot expect its content to be summarized in a few pages. Thus our discussion is a foretaste of what is to come in the remainder of this volume.

Particle dynamics is an example of a deterministic process wherein the future is completely determined, given sufficient knowledge of the present. Whether the outcome of such processes is to be calculated in detail or just contemplated philosophically, certain purely mathematical questions arise. We conclude this chapter by stating and proving some relevant theorems. These involve existence, uniqueness, and continuous dependence on a parameter of systems of ordinary differential equations.

2.1 Planetary Orbits

People of all eras have been moved by the grandeur and the beauty of Nature when watching stars in the darkness of night. As they made careful observations, they discovered that some of the "stars" are wanderers—the planets (πλανήτης = wanderer in Greek). Indeed, records of planetary paths formed one of the earliest important collections of scientific data.

In the first comprehensive history of China "Shihchi" (*Records of the Grand Historian of China*), written about the second century B.C., there is one volume devoted to celestial observations. Five planets were recognized, and it was known that the period of Jupiter is almost exactly 12 years. This is a special case of *Bode's law*.*

In the West, accurate observations of planetary motions were made by Copernicus (1473–1543), Tycho Brahe (1546–1601), and Galileo (1564–1642). The great amount of observational data collected by Tycho Brahe formed the basis of the famous laws of Kepler (1571–1630). These laws then led Newton (1642–1727) to the discovery of the law of universal gravitation, with the help of the then newly invented differential calculus. The study of these developments provides an excellent classical case study of applied mathematics.

KEPLER'S LAWS

In their original form, Kepler's laws may be stated as follows:

I. The planets describe ellipses about the sun as a focus.

II. The areas swept over by the radius vector drawn from the sun to a planet in equal times are equal.

III. The squares of the times of describing the orbits (periods) are proportional to the cubes of the major axes.†

To accomplish a mathematical formulation, we adopt a polar coordinate system (r, θ) with the sun as the origin. We treat all the planets and even the sun itself as mass points. (Note that this is an approximation or idealization in the formulation. It is appropriate because the radius of the sun is small compared to the distance between the sun and a given planet.) The second law of Kepler then states that, following the orbit $(r(t), \theta(t))$ of a planet,

$$r^2\dot{\theta} = h, \qquad h \text{ a constant}, \tag{1}$$

* The radii of the planets of the Sun are fairly accurately given by the formula $0.4 + 0.3 \times 2^n$ in astronomical units. The values of n are $-\infty, 0, 1, 2, \ldots 7, 8$ for Mercury, Venus, Earth, Mars, minor planets, etc. Neptune and Pluto do not follow this law accurately. This fact probably implies that Pluto was originally a satellite of Neptune.

† "Newton later wrote: 'Kepler knew ye Orb to be not circular but oval, and guessed it to be Elliptical.' This is correct, but Kepler's guess was no idle guess; it came out of a hunch actively pursued, in confrontation with all previous theories and with Tycho's new data."

This quotation is from an interesting account by Curtis Wilson, in the March 1972 *Scientific American* (Vol. 227, p. 92), of how Kepler discovered the first two of his laws. Wilson's summary of his findings includes these further remarks.

"At the root of all his [Kepler's] theorizing, however, was that initial sense of the significance of the inverse relation between velocity and distance—an anticipatory glimmer of what would one day be the law of the conservation of angular momentum. It was in the light of that hunch that he was guided through 900 pages of calculation to a planetary theory better than any that had been proposed before . . . . What was . . . important through all the accident and luck was Kepler's belief in the possibility of understanding, his devotion to his task that carried him through four years of reasoning and calculation, and finally, the rightness of his initial hunch and his ability to disentangle the confused state of things before him in the light of it."

since [Exercise 1(a)] h is twice the area swept over by the radius vector in unit time. (It is appropriate to use the notation $\cdot$ for d/dt, as introduced by Newton.) To obtain a useful simple interpretation of this result, we introduce the cartesian coordinates (x, y). Then (1) can be written as

$$x\dot{y} - y\dot{x} = h, \tag{2}$$

which yields, upon differentiation with respect to t,

$$x\ddot{y} - y\ddot{x} = 0. \tag{3}$$

If we define vectors $\mathbf{r}$ and $\mathbf{s}$ by

$$\mathbf{r} = x\mathbf{i} + y\mathbf{j}, \qquad \mathbf{s} = -y\mathbf{i} + x\mathbf{j}, \tag{4}$$

then $\mathbf{r} \cdot \mathbf{s} = 0$, so $\mathbf{s}$ is perpendicular to the position vector $\mathbf{r}$. But according to (3), $\ddot{\mathbf{r}} \cdot \mathbf{s} = 0$ so that $\ddot{\mathbf{r}}$ is in the direction of $\mathbf{r}$. *That is, if the "equal areas" result holds, then the acceleration of the planet is always directed toward the sun* (or away from it). Reversal of the steps in our reasoning shows that the converse of this result is also true.

The lesson to be learned here is that the interpretation of mathematical results may depend very much on the form in which they are stated. The equivalent equations (1) and (3) enable us to see quite different aspects of the same phenomenon, with completely different interpretation and insight. Note the importance of manipulative skill.

The first law of Kepler states that the orbit can be described by the simple formula

$$r = \frac{p}{1 + e\cos\theta}, \tag{5}$$

where $p > 0$ and $0 < e < 1$. This is a standard form of an equation for the ellipse whose semimajor axis is given by $p/(1 - e^2)$. With a little manipulation [Exercise 2(a)], one can show that the acceleration in the radial direction is

$$a_r = \ddot{r} - r\dot{\theta}^2 = -\frac{h^2}{pr^2}. \tag{6}$$

Thus *the acceleration is inversely proportional to the square of the radial distance*. Note that nothing has been said about the forces between the celestial objects. [A statement, such as (6), which describes the geometric character of motion, or which follows from this character, is said to be **kinematical**.]

The quantities h and p are characteristics of the individual planets. It is by the use of the third law of Kepler that one can show that *the ratio h^2/p is the same for all the planets* [Exercise 2(b)]. Thus, for planetary motion,

$$a_r = -\frac{K}{r^2}, \tag{7}$$

where K is a universal constant.

LAW OF UNIVERSAL GRAVITATION

The laws of Kepler have led us to the inverse square law for the acceleration of all the planets, with the same constant of proportionality. Newton, by combining the above results with his second law of motion and assuming the principle of superposition (i.e., that the effect of two suns is double that of a single one), was led to formulate the present form of the law of universal gravitation,

$$F_r = \frac{GMm}{r^2}. \tag{8}$$

Here F_r is the magnitude of the force acting between two point masses of amounts M and m separated by a distance r; G is a universal constant. The force is attractive.

The simple elegance of the mathematical manipulations is striking, especially when one compares the theory with the tedious observations. But can this powerful law be dreamed up without these observations? The answer is obviously no. Neither can the required calculations be made without the use of the then newly invented discipline of calculus.

The law of universal gravitation is indeed one of the most remarkable of the laws of nature. Why is it universal? Why is the power exactly equal to 2? Even more remarkable is that, according to the general theory of relativity, such an elegant principle can be replaced by another, which has even more penetration and breadth and is at least equally elegant, if not more so. It hardly seems frivolous to say that God is a mathematician!

On the working level, we observe that in order to arrive at our conclusions *it is necessary to keep the physical interpretation of the results in mind all the time and to be quite skillful with the manipulations.*

There is another lesson to be learned from the history of the law of universal gravitation, giving further evidence of the need for adequate notation and of proficiency with analytical manipulations. Newton tried to test his law by applying it to the motion of the moon about the earth and comparing the moon's acceleration with that of a body at the surface of the earth, as directly observed. This comparison cannot be made without knowing the gravitational potential of a continuous distribution of matter over a sphere. It is now well known that the gravitational attraction due to such a distribution of mass is equivalent to that due to a concentrated mass at the center. Newton was at first not able to prove this and consequently delayed his publication of the comparison for about 20 years.

THE INVERSE PROBLEM: ORBITS OF PLANETS AND COMETS

We start now with Newton's law (8) or the kinematical law (6), and ask how the planets would move. We find that (i) we have to be able to solve differential equations to provide a satisfactory answer, and (ii) not all the orbits are ellipses. Some orbits are hyperbolas and others are parabolas—and nature has provided these examples in the cometary orbits. The solution of the relevant differential equations, as we shall now see, requires considerable skill in manipulation.

The formulation of the mathematical problem consists of asking for *periodic orbits* for the following system of two differential equations of the second order:

$$\frac{d^2r}{dt^2} - r\left(\frac{d\theta}{dt}\right)^2 = -\frac{GM}{r^2}, \tag{9a}$$

$$\frac{1}{r}\frac{d}{dt}\left(r^2\frac{d\theta}{dt}\right) = 0, \tag{9b}$$

where the expressions on the left-hand side are the components of acceleration (a_r, a_θ) in the polar coordinate system. To follow the motion of the planets, one has to integrate (9) for the pair of functions $r(t)$, $\theta(t)$. There seems to be no direct way to accomplish this analytically. (Numerical integration methods can be applied, but understanding and insight would tend to be lost.) The ingenious way out of the difficulty is to look for the orbit in the form $r = r(\theta)$. Even here, as we shall see below, a clever transformation is needed to solve the resultant differential equation.

Equation (9b) can be integrated once to yield $r^2\dot{\theta} = h$, h a constant. (The quantity $r^2\dot{\theta}$ is called a constant or **integral** of the motion.) Thus the law of uniform area (1) is recovered. We can now use (1) to replace the time derivatives in (9a) with differentiation with respect to the angular variable θ. The resultant equation is still very complicated, but it can be simplified by introducing the variable

$$u = r^{-1}.$$

With this we obtain

$$\frac{d^2u}{d\theta^2} + u = \frac{GM}{h^2}. \tag{10}$$

The general solution of (10) is [Exercise 3(b)]

$$u = h^{-2}GM[1 + e\cos(\theta - \theta_0)], \tag{11}$$

where e and θ_0 are constants of integration.

The solution (11) is periodic in θ, but not all the orbits are periodic. When $e < 1$, (11) yields the observed elliptical orbits. But when $e > 1$, u may vanish for some value of θ, and the planet would then be infinitely far away; the orbits are hyperbolas. Parabolic orbits are obtained in the marginal case $e = 1$. The integration constant θ_0 is unimportant as it merely assigns the value θ_0 to an axis of symmetry.

PLANETARY ORBITS ACCORDING TO THE GENERAL THEORY OF RELATIVITY

When modification of Newtonian mechanics is considered in the light of the general theory of relativity, it can be shown that equation (10) for the planetary orbit is changed to the form

$$\frac{d^2u}{d\phi^2} + u = \frac{GM}{h^2}(1 + \varepsilon u^2), \tag{12}$$

where ε is a small parameter (and where the influence of the other planets still has not been taken into account). The integration of this equation is more difficult, since it is nonlinear in u. However, it can still be done in terms of elliptic functions. To obtain the solution in a more convenient form, the problem should be treated by "perturbation theory," which is based on the fact that the solution of a differential equation depends continuously on a parameter that enters continuously in the differential equation. (We shall soon discuss these matters further.) In the present case, when $\varepsilon = 0$, the solution is known. Because of the continuous dependence of u upon ε, the solution for small ε can be obtained by an appropriate small modification of the solution for $\varepsilon = 0$.

The result obtained by the use of the perturbation theory shows that the orbit is still very close to an ellipse, but there is an advance of the perihelion (nearest point to sun) by the amount $2\pi\varepsilon a^2$ over one period. (A method for obtaining this result is given in Section 2.2.) The observation of this advance yields one of the principal experimental tests of the general theory of relativity.*

COMMENTS ON CHOICE OF METHODS

The advance of the perihelion can also be established by detailed numerical integration, or by examination of the solution of (12) in terms of elliptic functions. However, neither method gives the results in as transparent a form as perturbation theory, which is also extremely accurate in this case. In other instances, analysis of a well-formulated problem with the aid of modern computing machines can often yield more accurate numerical results than those obtained by the use of approximate analytical methods. It is, however, often difficult to extract useful general information from a large amount of numerical data unless one has already gained *insight* into the problem through general analytical studies. On the other hand, exploratory

* Equation (12) has been derived by K. Schwarzschild. The present situation on the relativistic perihelion advance with regard to the comparison of experiment with theory is as follows. For Mercury, where the effect is largest (43 seconds of arc in a century) observations are in accord with Einstein's theory. But some controversy remains, for there is a possible flattening of the sun that could perhaps leave room for alternatives to Einstein's theory. See K. Nordtvedt, "Gravitation Theory: Empirical Status from Solar System Experiments," *Science 178*, 1157–1164 (1972).

numerical studies sometimes are an invaluable preliminary to successful analytical work. The judicious balance in the use of all types of methods is indeed most important.

N PARTICLES: A DETERMINISTIC SYSTEM

Consider now a system of N particles in gravitational interaction; e.g., the solar system comprising the sun and the nine major planets. Let us denote the position vector and the velocity vector of the particle α by $\mathbf{r}_\alpha$ and $\mathbf{v}_\alpha$, respectively ($\alpha = 1, 2, \ldots, N$). Then the equation governing the motion of the αth particle is

$$m_\alpha \ddot{\mathbf{r}}_\alpha = \mathbf{F}_\alpha, \qquad \alpha = 1, 2, \ldots, N, \tag{13}$$

where

$$\mathbf{F}_\alpha = \sum_{\beta \neq \alpha} \mathbf{F}_{\alpha\beta}; \tag{14}$$

i.e., the resultant force acting on the αth particle is due to the contribution $\mathbf{F}_{\alpha\beta}$ of all the *other* particles. We have, according to Newton's law,

$$\mathbf{F}_{\alpha\beta} = -\frac{Gm_\alpha m_\beta}{|\mathbf{r}_\alpha - \mathbf{r}_\beta|^3}(\mathbf{r}_\alpha - \mathbf{r}_\beta) = -\mathbf{F}_{\beta\alpha}, \tag{15}$$

which depends only on the positional coordinates of the particles. In other physical situations, there may be an external force field and the forces may also depend on the velocities of the particles (e.g., when the particles are electrically charged and there is a prevailing magnetic field). But, in general, there are no true forces dependent on the acceleration of the particle.

Thus we have

$$\mathbf{F}_\alpha = \mathbf{F}_\alpha(t, \mathbf{r}_\beta, \mathbf{v}_\gamma); \qquad \alpha, \beta, \gamma = 1, 2, \ldots, N. \tag{16}$$

In this most general case, we may still write the equations of motion (13) as *a system of differential equations of the first order*:

$$m_\alpha \frac{d\mathbf{v}_\alpha}{dt} = F_\alpha(t, \mathbf{r}_\beta, \mathbf{v}_\gamma), \quad \frac{d\mathbf{r}_\alpha}{dt} = \mathbf{v}_\alpha; \qquad \alpha, \beta, \gamma = 1, 2, \ldots, N. \tag{17a, b}$$

We have $6N$ variables $(\mathbf{r}_\alpha, \mathbf{v}_\alpha)$; $\alpha = 1, \ldots, N$. We expect that the subsequent behavior of the system would be determined if we were given the instantaneous positions and velocities of all the particles at any instant. Mathematically, this means that (17) should have a well-defined solution if it is solved under the conditions

$$\mathbf{r}_\alpha = \mathbf{r}_\alpha^{(0)}, \quad \mathbf{v}_\alpha = \mathbf{v}_\alpha^{(0)} \quad \text{at } t = t_0; \qquad \alpha = 1, 2, \ldots, N. \tag{18}$$

These conditions are called **initial conditions**, and the problem we have just posed is called an **initial value problem** for a system of ordinary differential equations.

That the initial value problem has a unique solution can in fact be proved, as we shall see in Section 2.3. This proof has wide ranging consequences, for it has led some philosophers to a deterministic view of the universe. For what can be the place of "free will," it can be argued, if the evolution of the cosmos can be uniquely determined by examining the configuration at any arbitrary instant?*

But nowadays existence and uniqueness theory is also important from a practical (rather than philosophical) point of view. The advent of computing machines has made it desirable to ensure that a problem is precisely formulated before expensive calculations are attempted. At least at the present stage of development of computers, one cannot expect the machine to detect difficulties in the original formulation and then to suggest suitable modifications. In practice, trial-and-error approaches are often used, for it has certainly not been possible to prove all the required theorems. But experienced opinion is superior to guesswork, and the ability to exercise sound judgments in the absence of rigorous guidelines is perhaps one of the most valuable assets of an applied mathematician from the practical standpoint.

LINEARITY

To make sure that an important concept is understood, we close this section with a brief review of linearity. The concept will be illustrated in connection with systems of differential equations.

If $f(x) = ax + b$, a and b constants, the graph of $y = f(x)$ is a straight line. For this reason, f is called a **linear function**. When $b = 0$, the function is called **homogeneous**; otherwise, it is **inhomogeneous**.

The linear homogeneous function (often just called *linear*) has strikingly simple properties. If $L(x) = ax$, then

$$L(x_1 + x_2) = L(x_1) + L(x_2), \qquad L(cx_1) = cL(x_1); \tag{19a, b}$$

or, equivalently,

$$L(c_1 x_1 + c_2 x_2) = c_1 L(x_1) + c_2 L(x_2). \tag{20}$$

Here c_1 and c_2 are any constants; x_1 and x_2 are arbitrary values of x.

Let $\mathbf{x}(t)$ be an n-element column vector (n by 1 matrix) that depends on the scalar t. Define a vector-valued function $\mathbf{L}(\mathbf{x})$ by

$$\mathbf{L}(\mathbf{x}) = \frac{d\mathbf{x}}{dt} - \mathbf{A} \cdot \mathbf{x}, \tag{21}$$

where the derivative is (as usual) taken component by component, and where $\mathbf{A} = \mathbf{A}(t)$ is an n by n matrix that multiplies $\mathbf{x}$. Note that $\mathbf{A}$ does not depend on $\mathbf{x}$.

* Needless to say, we are only scratching the surface of the philosophical position, and we shall not pursue the matter further.

The function **L** also possesses the linearity properties. For if **x** and **y** are any (differentiable) vector functions and α is a scalar (usually a real number), then

$$\boxed{\mathbf{L}(\mathbf{x}+\mathbf{y}) = \mathbf{L}(\mathbf{x}) + \mathbf{L}(\mathbf{y}), \qquad \mathbf{L}(\alpha\mathbf{x}) = \alpha L(\mathbf{x}).} \tag{22a, b}$$

We also have the analogue of (20), which treats of what is called the **linear combination** $\alpha\mathbf{x} + \beta\mathbf{y}$ of **x** and **y** (where β is another scalar):

$$\boxed{\mathbf{L}(\alpha\mathbf{x} + \beta\mathbf{y}) = \alpha\mathbf{L}(\mathbf{x}) + \beta\mathbf{L}(\mathbf{y}).} \tag{23}$$

The reader has probably met some of the many consequences of the very important linearity properties (22) or (23). It is relevant here to give an example involving systems of ordinary differential equations. Using the definition (21),

$$\mathbf{L}(\mathbf{x}) = \mathbf{0}$$

is a **linear** (homogeneous) **system** of first order differential equations. Property (22a) implies that the *sum of two solutions is a solution* for

$$\mathbf{L}(\mathbf{x}) = \mathbf{0}, \ \mathbf{L}(\mathbf{y}) = \mathbf{0},$$

implies

$$\mathbf{L}(\mathbf{x}+\mathbf{y}) = \mathbf{L}(\mathbf{x}) + \mathbf{L}(\mathbf{y}) = \mathbf{0}. \tag{24}$$

Property (22b) implies that the *scalar multiple of a solution is a solution*, for

$$\mathbf{L}(\mathbf{x}) = \mathbf{0} \qquad \text{implies that } \mathbf{L}(\alpha\mathbf{x}) = \alpha\mathbf{L}(\mathbf{x}) = \mathbf{0}. \tag{25}$$

Similarly, *a linear combination of two solutions (or of many solutions) is a solution.*

EXERCISES

1. (a) Equation (1) is often derived in elementary treatments of polar coordinates. If the derivation is unfamiliar to you, work it out for yourself.
(b) Verify (2).
(c) Show that (1) follows from (3).

2. (a) Verify (6).
(b) Prove (7).

3. (a) Verify (10).
(b) Derive (11).

4. In the text's theoretical discussion of Kepler's laws, the sun and the planets were treated as mass points. Write a brief essay that provides justification for such a treatment.

5. (a) Verify (22) and (23) when **L** is defined by (21).
(b) Prove in general that (22) holds if and only if (23) holds.
(c) Prove that a linear combination of solutions to a linear system of equations such as (21) is also a solution.

6. Eliminate θ from the equations of (9) to obtain a differential equation for radial motion. Integrate to derive Kepler's representation of radial motion:

$$r = a(1 - e \cos E), \qquad n(t - T) = E - e \sin E.$$

Here a is the semimajor axis of the ellipse, e is the eccentricity, n is the frequency of the orbit, T is the time of perihelion passage, and E (called the *eccentric anomaly*) is a parameter that runs over the range $(0, 2\pi)$ for a complete orbit. (The position angle θ is known as the *true anomaly*; the quantity $M = E - e \sin E$, which varies linearly in time, is known as the *mean anomaly*.)

To do this, first obtain the energy equation in the form

$$\left(\frac{dr}{dt}\right)^2 = h^2 \frac{a^2e^2 - (r - a)^2}{r^2a^2(1 - e^2)}$$

by noting that the radial velocity is zero at the perihelion and aphelion (the points which are nearest and farthest from the sun, respectively).

2.2 Elements of Perturbation Theory, Including Poincaré's Method for Periodic Orbits

At the beginning of this chapter we considered a single planet moving around a fixed sun. The subsequent discussion of N-particle interactions reminds us that our earlier considerations neglected the interaction among the planets.* We can build on our previous work by "approximating" the force $\mathbf{F}_\alpha$ on the αth planet by a basic force field $\mathbf{F}_\alpha^{(0)}$ (due to the sun), and regarding the effect of the other planets as "small perturbations." To formalize this process, we can replace the right-hand side of (1.13) or (1.17a) by

$$\mathbf{F}_\alpha(\lambda) = \mathbf{F}_\alpha^{(0)} + \lambda[\mathbf{F}_\alpha - \mathbf{F}_\alpha^{(0)}]$$

and allow the parameter λ to vary in the interval $(0, 1)$.† The effect of the perturbation is then obtained by comparing the solutions for $\lambda = 0$ and $\lambda = 1$, respectively.

In practice, determination of the effect of small perturbations on the motion of planets involves special difficulties. This is because the perturbation effects accumulate over many periods of time, yet we do not expect the shape of an individual orbit to be greatly changed. The difficulty was overcome by Poincaré, whose ideas will be discussed briefly after a presentation of the theory in the ordinary form.

* Also neglected was the solar motion, whose effect can be studied only by introducing another set of dependent variables.

† Although not necessary, it is instructive to imagine that as λ varies continuously from 0 to 1, the solar system varies from one with but a single planet, through intermediate systems with other planets of small mass, to the actual system of which we are a part.

PERTURBATION THEORY: ELEMENTARY CONSIDERATIONS

To clarify the ideas just discussed, let us consider the simplest example of a single ordinary differential equation

$$\frac{dy}{dx} = f(x, y, \varepsilon) \tag{1}$$

to be solved under the initial conditions

$$y = y_0 \qquad \text{when } x = x_0 . \tag{2}$$

We wish to examine the dependence of the solution $y(x, \varepsilon)$ on the parameter ε that occurs in (1).

Suppose that the solution $y(x, \varepsilon)$ is known for $\varepsilon = 0$.* Assume that $y(x, \varepsilon)$ can be expanded into a Taylor series in ε, when the latter is sufficiently small in magnitude:

$$y(x, \varepsilon) = y^{(0)}(x, 0) + \varepsilon y^{(1)}(x, 0) + \cdots + \varepsilon^n y^{(n)}(x, 0) + \cdots . \tag{3}$$

$$y^{(n)}(x, 0) = \frac{1}{n!}\left(\frac{\partial^n y}{\partial \varepsilon^n}\right)_{\varepsilon=0} . \tag{4}$$

If we now substitute (3) into (1) and regard the equation as an identity in ε, we shall find that we can obtain a sequence of *linear* differential equations in $y^{(n)}$. For example, disregarding all terms of order ε^2 or higher, we have

$$\frac{dy^{(0)}}{dx} + \varepsilon\frac{dy^{(1)}}{dx} + \cdots = f(x, y^{(0)} + \varepsilon y^{(1)} + \cdots, \varepsilon)$$

$$= f(x, y^{(0)}, 0) + \varepsilon f_y(x, y^{(0)}, 0)y^{(1)} + \varepsilon f_\varepsilon(x, y^{(0)}, 0) + \cdots,$$

where we have used Taylor's theorem to expand f. Equating equal powers of ε, we find

$$\frac{dy^{(0)}}{dx} = f(x, y^{(0)}, 0), \tag{5a}$$

$$\frac{dy^{(1)}}{dx} = f_y(x, y^{(0)}, 0)y^{(1)} + f_\varepsilon(x, y^{(0)}, 0), \qquad \text{etc.} \tag{5b}$$

The solution for (5a) has been presumed to be known. Upon substitution of this solution into (5b), one obtains a definite equation for the correction $y^{(1)}$.

The initial conditions to be imposed on the successive correction terms are

$$y^{(1)}(x_0, 0) = 0,\ y^{(2)}(x_0, 0) = 0, \ldots, y^{(n)}(x_0, 0) = 0, \tag{6}$$

since the original condition (2) does not depend on ε.

* This entails no loss of generality. If the solution is known for $\varepsilon = \varepsilon_0$, define a new parameter $\bar{\varepsilon} = \varepsilon - \varepsilon_0$.

Note that (5b) can be explicitly solved in terms of quadratures under the appropriate initial condition of (6). Indeed, calculations will show that all the terms $y^{(n)}(x, 0)$ satisfy an equation of the form

$$\frac{dy^{(n)}}{dx} = Ay^{(n)} + B^{(n)}, \qquad n = 1, 2, \ldots, \tag{7}$$

where A is $f_y(x, y^{(0)}, 0)$ for all n, and $B^{(n)}$ depends on x and all the $y^{(k)}$'s preceding $y^{(n)}$. Thus, once $y^{(0)}$ is found, the problem of obtaining the higher approximations is relatively simple in principle, since explicit solution of (7) is possible in terms of quadratures.

A simplifying feature of this type will be found in all other cases of the perturbation method, including those involving partial differential equations.

Example 1. The solution of the differential equation

$$\frac{dy}{dx} = 1 + y^2, \tag{8}$$

subject to the initial condition

$$y(0) = 0, \tag{9}$$

is

$$y = y^{(0)} = \tan x. \tag{10}$$

Find a first approximation to the solution, subject to the same initial condition, of

$$\frac{dy}{dx} = 1 + (1 + \varepsilon)y^2, \tag{11}$$

when $|\varepsilon| \ll 1$. Compare with the exact solution.

Outline of solution. The exact solution to the perturbed problem is

$$y(x, \varepsilon) = (1 + \varepsilon)^{-1/2} \tan [(1 + \varepsilon)^{1/2}x].$$

A Taylor series expansion of this solution in powers of ε gives

$$y(x, \varepsilon) = \tan x + \varepsilon(\tfrac{1}{2}x \sec^2 x - \tfrac{1}{2} \tan x) + \cdots.$$

Forgetting the exact solution for the moment, we assume that

$$y(x, \varepsilon) = y^{(0)}(x) + \varepsilon y^{(1)}(x) + \cdots.$$

On substitution into (11), we find

$$\frac{dy^{(0)}}{dx} + \varepsilon \frac{dy^{(1)}}{dx} + \cdots = 1 + (1 + \varepsilon)[(y^{(0)})^2 + 2\varepsilon y^{(0)}y^{(1)} + \cdots],$$

so that

$$\frac{dy^{(1)}}{dx} = (y^{(0)})^2 + 2y^{(0)}y^{(1)}.$$

[It is frequently the case, as here, that direct substitution provides the desired perturbation equations more easily than substitution into a general formula such as (5b).] Rearranging the differential equation for $y^{(1)}$ into standard form and using the fact that $y^{(0)}(x) = \tan x$, we obtain

$$\frac{dy^{(1)}}{dx} - 2y^{(1)} \tan x = \tan^2 x. \tag{12}$$

This equation has an integrating factor of

$$\exp\left(-2 \int \tan x \, dx\right) = \cos^2 x.$$

Indeed, (12) can be written

$$\frac{d}{dx}[y^{(1)} \cos^2 x] = \cos^2 x \tan^2 x \equiv \sin^2 x.$$

It is now easily seen that the solution for $y^{(1)}$ is precisely as given above.

The reader should fill in the details of the above calculation (Exercise 2). Hopefully, a good idea of how straightforward perturbation calculations are accomplished has already emerged. But detailed "training" and further discussion is provided in Chapter 7.

THE SIMPLE PENDULUM

Theoreticians often find it helpful to consider the operation of an idealized device that displays interesting properties in a readily visualized form. One such device is the spring–mass–dashpot system. As most readers will have seen, in the simplest instance this system is governed by a second order ordinary differential equation with constant coefficients; it therefore provides a concrete illustration of the properties of that ubiquitous equation. Here, for the first of several times in this book, we shall consider another example of such an idealized device, the simple pendulum. In the present context, consideration of the pendulum will provide an illustration (i) of the fact that perturbation theory is often used implicitly when we formulate simplified physical models, and (ii) of the large cumulative effect of small perturbations and the method Poincaré devised to describe this effect. The same effect as that of (ii), and the same calculational method but in a less transparent form, are relevant to planetary orbits.

A standard discussion in elementary physics texts is a reduction of the pendulum problem to that of simple harmonic motion when the amplitude is small, and we vaguely feel we can improve the approximation when needed. The discussion deals with a simple pendulum as an idealized object which, by definition, consists of a *rigid, straight, weightless* rod of *fixed length L*, with *point* mass m at one end. The rod is *free to rotate* about a *horizontal* axis at the *point* of suspension.

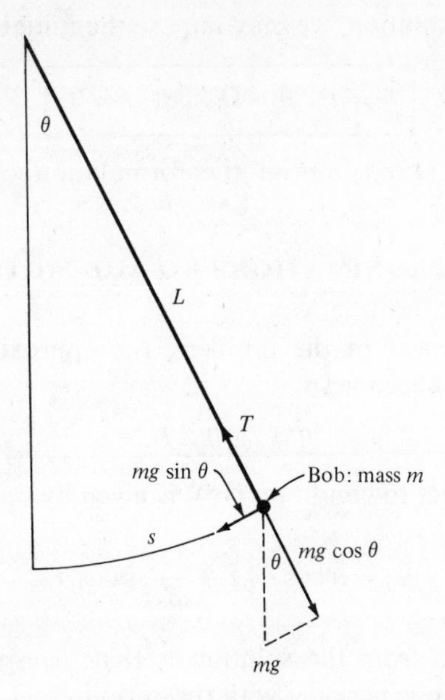

FIGURE 2.1. *Forces on a simple pendulum of mass m are represented by the dark arrows. Magnitudes of the forces are indicated. The tension T in the string is equal to mg* cos θ. *The unbalanced tangential force of magnitude mg* sin θ *causes acceleration of the bob.*

The italicized words above represent the idealizations involved in the formulation of the model. Estimation of the error introduced by these idealizations might be very difficult. (For example, if the rod is slightly elastic it becomes capable of undergoing torsional, longitudinal, and bending vibrations.) But we shall consider an approximation whose consequences can be more easily analyzed. Denote by $\theta(t)$ the inclination of the rod to the vertical at time t. If $s(t)$ denotes the distance the bob has moved from the vertical in a counterclockwise direction, then $\theta L = s$. Equating $m(d^2s/dt^2)$ to the force component which is tangential to the path of the bob, we obtain the following equation of motion:

$$mL\ddot{\theta} = -mg \sin \theta. \tag{13}$$

Here g is the acceleration due to gravity. The tension along the string does not enter, because it acts normally to the path of the bob. (See Figure 2.1.)

The equation of motion may also be written in the following slightly different form:

$$\boxed{\ddot{\theta} + \omega_0^2 \sin \theta = 0 \text{ for } t > 0, \qquad \text{where } \omega_0 = \left(\frac{g}{L}\right)^{1/2}.} \tag{14}$$

To obtain a definite solution, we may impose the initial conditions

$$\boxed{\theta = a, \quad \dot{\theta} = \Omega \quad \text{at } t = 0.} \tag{15}$$

Equations (14) and (15) comprise the formulation of our mathematical problem.

SUCCESSIVE APPROXIMATIONS TO THE MOTION OF THE PENDULUM

In the usual treatment of the problem, the approximation $\sin\theta \approx \theta$ is introduced. Then (14) becomes

$$\ddot{\theta} + \omega_0^2\theta = 0, \tag{16}$$

whose solution [subject to condition (15)] is given by

$$\theta_0 = a\cos\omega_0 t + \frac{\Omega}{\omega_0}\sin\omega_0 t. \tag{17}$$

An obvious inference from the solution is that, irrespective of the initial conditions, the motion is periodic with the period

$$P_0 = \frac{2\pi}{\omega_0} = 2\pi\left(\frac{L}{g}\right)^{1/2}. \tag{18}$$

This approximate solution is not adequate, however, if we are interested in following the motion of the mass point over many periods, for the period given by (18) is not exactly right. According to the "exact" equation (14), the motion is indeed periodic, but the correct period is

$$P = P_0 \cdot \frac{2}{\pi}\int_0^{\pi/2}(1 - k^2\sin^2\psi)^{-1/2}\,d\psi, \tag{19}$$

where $k^2 = \sin^2(\theta_m/2)$ and θ_m is the amplitude of the oscillation [i.e., the maximum value of $|\theta(t)|$]:

$$\theta_m \equiv \max_t |\theta(t)|. \tag{20}$$

With every swing of the pendulum, the approximate solution misestimates the time of maximum amplitude by a further increment of $P - P_0$. [The error in period is about 2 per cent for an amplitude $\theta_m = 10° = \pi/18$ radians (Exercise 4).] Thus the solution to the approximate equation (16) can be expected to be nearly the same as the solution to (14) only for a number of periods N such that $N(P - P_0)$ is small compared to P.

To improve on the approximation, one might write (14) in the form

$$\ddot{\theta} + \omega_0^2\theta = \omega_0^2(\theta - \sin\theta) \tag{21}$$

and then obtain a reasonably accurate value of the right-hand side of (21) by using the first approximation θ_0 of (17), writing $\theta - \sin\theta \approx \theta_0 - \sin\theta_0$. This is justified by the argument that the right-hand side of (21), namely,

$$\omega_0^2(\theta - \sin\theta) = \omega_0^2\left(\theta - \theta + \frac{\theta^3}{3!} - \frac{\theta^5}{5!} + \cdots\right) = \omega_0^2\frac{\theta^3}{3!} - \cdots, \tag{22}$$

is of the order of θ^3, while the left-hand side is of the order of θ. Thus the smaller right-hand side can be ignored altogether to obtain an initial approximation $\theta_0(t)$. A better approximation should be obtained if the substitution $\theta = \theta_0$ is made on the right side, etc. Thus the first attempt at an improved approximation, $\theta_1(t)$, might be expected to satisfy

$$\ddot{\theta}_1 + \omega_0^2\theta_1 = \omega_0^2(\theta_0 - \sin\theta_0); \qquad \theta_1(0) = a, \quad \dot{\theta}_1(0) = \Omega. \quad \text{(23a, b, c)}$$

But to make an exact evaluation of the right side of (23a) with the rough approximation θ_0 gives rise to a problem that is needlessly difficult to solve. As one might suspect, no accuracy is lost if one makes the approximation indicated in (22) and therewith replaces (23) by

$$\ddot{\theta}_1 + \omega_0^2\theta_1 = \omega_0^2\frac{\theta_0^3}{3!}; \qquad \theta_1(0) = a, \quad \dot{\theta}_1(0) = \Omega. \tag{24}$$

In principle, repetition of the process will provide successively more accurate approximations. Hence the method is known as **iteration** (Latin *iterare*, to do again), or **successive approximations**.

Problem (24) also arises in another approach to the improvement of the basic approximation θ_0, to be discussed now. We therefore defer consideration of the solution to (24).

NOTE. Chapter 7 contains a more thorough treatment of the method of iteration.

PERTURBATION SERIES APPLIED TO THE PENDULUM PROBLEM

In a perturbation series approach, it is convenient to introduce a new variable,

$$\Theta(t) = \frac{\theta(t)}{\theta_m}, \tag{25}$$

where θ_m is the maximum amplitude defined in (20). Intuitively, θ_m seems to provide a natural standard of comparison for the angular displacement. The new variable Θ satisfies $|\Theta| \le 1$, and this proves useful in estimating the size of various terms. Considerable further justification for this type of variable change is given in Section 6.3.

With (25), the governing equation (21) becomes

$$\ddot{\Theta} + \omega_0^2\Theta = \omega_0^2(\Theta - \theta_m^{-1}\sin\theta_m\Theta) = \omega_0^2\theta_m^2\frac{\Theta^3}{6} + \cdots. \tag{26}$$

We note that since $|\Theta| \leq 1$, the right-hand side of (26) is of the order* of θ_m^2, a quantity that we assume to be very small compared to unity.

For simplicity, let us consider the case where the pendulum is released from rest ($\Omega = 0$). It is physically clear that $\theta_m = a$, so that our problem becomes

$$\ddot{\Theta} + \omega_0^2 \Theta = \omega_0^2(\Theta - a^{-1} \sin a\Theta)$$

$$= \omega_0^2 \sum_{n=1}^{\infty} (-1)^{n+1}(a^2)^n \frac{\Theta^{2n+1}}{(2n+1)!}, \tag{27a}$$

$$\Theta(0) = 1, \qquad \dot{\Theta}(0) = 0. \tag{27b, c}$$

To find a solution of (27), we write

$$\Theta = \Theta(t, a) = \Theta^{(0)}(t) + a\Theta^{(1)}(t) + a^2\Theta^{(2)}(t) + \cdots \tag{28}$$

and substitute it into (27a) to obtain an infinite sequence of equations, in the spirit of our earlier discussion of perturbation theory. The initial conditions for the various approximations are

$$\Theta^{(0)}(0) = 1, \quad \dot{\Theta}^{(0)}(0) = 0; \tag{29a}$$

$$\Theta^{(n)}(0) = 0, \quad \dot{\Theta}^{(n)}(0) = 0, \qquad \text{for } n \geq 1. \tag{29b}$$

In this formulation, the dependence of the solution on the amplitude a is explicitly demonstrated. Note that if all the higher powers of the amplitude a are neglected, we have the usual simple harmonic motion.

To ensure that our method works, we might wish to satisfy ourselves that the series (28) converges. But this turns out to be an incorrect stipulation, for it is neither necessary nor sufficient. Even when such a series is divergent, it can still be useful as an *asymptotic* series (see Chapter 3). On the other hand, convergence does not assure its usefulness. An inordinately large number of terms might have to be summed to obtain reasonable accuracy.

As a matter of fact, it can be shown that the series (28) is in general convergent in any *finite* interval $0 \leq t \leq T$, provided that a is sufficiently small.† However, this does not assure us of the usefulness of the solution for many periods. One way to see this is actually to carry out the calculations,§ which show that

$$\Theta^{(0)}(t) = \cos \omega_0 t, \qquad \Theta^{(1)}(t) \equiv 0. \tag{30}$$

* "Of the order of," roughly speaking, means "about the same size as," but it is useful to give a precise meaning to "order" in this sense. See Appendix 3.1.

† This can be inferred from the *analytic* dependence of the right-hand side of (27a) on the *complex variable a*. This general theory is beyond the scope of the present discussion. See Coddington and Levinson (1955).

§ Some readers may have fully grasped the idea of straightforward perturbation theory from the single example that has been worked out. They can attempt to verify (30), (31), and (33) by themselves. Other readers may prefer, either now or later, to peruse Section 7.1, where the required calculations are carried out in detail.

Thus $\Theta^{(2)}$ satisfies

$$\ddot{\Theta}_2 + \omega_0^2 \Theta_2 = \frac{\omega_0^2}{6} \cos^3 \omega_0 t. \tag{31}$$

Using a trigonometric identity, we can rewrite (31) as

$$\ddot{\Theta}_2 + \omega_0^2 \Theta_2 = \frac{\omega_0^2}{24} (\cos 3\omega_0 t + 3 \cos \omega_0 t). \tag{32}$$

This is a nonhomogeneous linear equation with constant coefficients. The right-hand side is of the simple type that invites the use of the method of undetermined coefficients. The presence of the "resonant" $\cos(\omega_0 t)$ term leads us to anticipate a contribution to the solution with a factor of t. And indeed, the solution of (32) and the initial conditions (29b) turns out to be

$$\Theta^{(2)}(t) = \tfrac{1}{192} \cos \omega_0 t - \tfrac{1}{192} \cos 3\omega_0 t + \tfrac{1}{16}\omega_0 t \sin \omega_0 t. \tag{33}$$

Noteworthy in (33) is the term proportional to $\cos 3\omega_0 t$. This appearance of more rapidly oscillating "higher harmonics" is a typical nonlinear phenomenon.

But the term proportional to $t \sin \omega_0 t$ holds the center of our interest. Alone among the contributions to the correction term $a^2\Theta^{(2)}$, this term is not uniformly small for all time. Rather, for long times $t \approx \omega_0^{-1} a^{-2}$ we must contemplate a **secular*** **term** which, far from being a small correction, has become of the same size as $\Theta^{(0)}$. When this happens, the whole basis of the approximation is invalidated. Thus the approximation $\Theta^{(0)} + a^2\Theta^{(2)}$ is an improvement over $\Theta^{(0)}$ only for the time interval that $\Theta^{(0)}$ itself offers a reasonable approximation (Exercise 5). We have not yet penetrated to the core of the problem.

POINCARÉ'S PERTURBATION THEORY

The failure of the unmodified perturbation method occurs in all cases of periodic motions. In connection with the study of the perturbation of planetary orbits, Poincaré† developed a modified perturbation theory to overcome the difficulty. The key to his theory is the introduction of a parametric representation that recognizes the change of period indicated by (19). This permits the removal of any secular term that might arise in the calculations.

The Poincaré representation of the solution $\Theta(t, a)$ is

$$\Theta = \Theta^{(0)}(\tau) + a\Theta^{(1)}(\tau) + a^2\Theta^{(2)}(\tau) + \cdots, \tag{34a}$$

$$t = \tau + at^{(1)}(\tau) + a^2 t^{(2)}(\tau) + \cdots, \tag{34b}$$

* *Saeculum* means "generation" or "age" in Latin; hence the word "secular" applies to processes of slow change.

† Poincaré (1854–1912) was one of the world's leading mathematicians around the turn of the century.

where τ is a parameter. The idea is to let the solution $\Theta^{(0)}(\tau)$ take on the simple form (34a) in the variable τ, but to distort the time scale so that the defects in the solution are removed. We expect to eliminate all secular terms and to get the correct period (19) to a good approximation. The calculation is fairly complicated but can be simplified by noting that the series (34) should really be in a^2. It also turns out that one may take all the correction terms to be a constant multiple of τ, thus

$$t = \tau(1 + a^2 h_2 + \cdots), \qquad h_2 \text{ a constant.} \tag{35}$$

The equation for $\Theta^{(2)}$ is then found (Exercise 6) to be

$$\frac{d^2\Theta^{(2)}}{dt^2} + \omega_0^2\,\Theta^{(2)} - 2h_2\,\frac{d^2\Theta^{(0)}}{dt^2} = \frac{\omega_0^2}{24}(\cos 3\omega_0 t + 3\cos\omega_0 t), \tag{36}$$

or, using (30),

$$\frac{d^2\Theta^2}{dt^2} + \omega_0^2\,\Theta^{(2)} = \omega_0^2(\tfrac{1}{8} - 2h_2)\cos\omega_0 t + \frac{\omega_0^2}{24}\cos 3\omega_0 t.$$

As in (33), the cos $\omega_0 t$ term on the right side would lead to an undesirable secular term proportional to $t \sin \omega_0 t$. But now we can avoid this by taking

$$h_2 = \tfrac{1}{16}. \tag{37}$$

The period in τ is thus $2\pi/\omega_0$ to this approximation, but is $(2\pi/\omega_0)(1 + a^2 h_2)$ in t. This should be compared with (19).

The reader can better appreciate the need to change the time scale by obtaining the pendulum period using exact integration (Exercise 9). A more interesting example involves the advance of the perihelion of Mercury's orbit (Exercise 10). The reader may find it easier to do Exercise 10 after having carefully studied the detailed discussion in Section 7.1 of perturbation theory for the simple pendulum.

GENERALIZATION OF THE POINCARÉ METHOD

The solution of specific problems is only a prelude to understanding—a means to an end. In particular, the lessons learned from attacking a specific problem can often be formulated in terms of general principles and developed into general theory in order to deepen our understanding. But this cannot always be done in a formal manner. The Poincaré theory is a case in point.

The success of the parametric representation introduced by Poincaré has led to its application to other types of ordinary differential equations, for the removal of a variety of difficulties of a similar nature. Basically, there is a need to shift the solution surface by a mapping or distortion in the domain of the independent variables. But it is not fruitful to try to classify all the problems that can be solved by this approach.

We have not mentioned any theorems that rigorously justify the use of the Poincaré method. Such theorems are, in fact, available in a number of instances. Nevertheless, many situations in which the value of the Poincaré approach can scarcely be doubted are not subsumed under present rigorous treatments. One reason why *final formalization of valuable ideas in the form of theorems is often impracticable* is that theorems of desirable generality are very difficult to prove. A more subtle reason is that a listing of definitive theorems seems to close a subject, whereas an applied mathematician is always struggling to open new areas of application by suitable modification and generalization of known results.

EXERCISES

1. Demonstrate the validity of (7).

2. Fill in the details of the solution to Example 1.

3. Prove the validity of (19). Begin by multiplying (14) by $\dot{\Theta}$.

4. Demonstrate the correctness of the statement under (20) concerning the error in period. Assume that $\Omega = 0$.

5. Consider the time interval in which $a^2\Theta^{(2)}$, as given in (33), is a small correction to $\Theta^{(0)}$. Show that this interval is essentially the same as the number of periods N mentioned under (20).

6. Carry the Poincaré method through the derivation of (36). [The principal result (37) then follows at once.]

7. Show that the more general assumption (34b) reduces to (35) when the Poincaré method is carried through.

8. A grandfather's clock is to be designed. It is to be 5 feet long and to have a maximum swing of $\frac{1}{2}$ foot at the tip of the pendulum. Show that the linear approximation for the period $T = 2\pi(L/g)^{1/2}$ is not good enough [H. O. Pollack, *Educ. Stud. Math.* 2, 393–404 (1969)]. To do this, derive and use the inequality

$$(\alpha^2 - \Theta^2)\alpha^{-1}\sin\alpha \le 2\cos\Theta - 2\cos\alpha \le (\alpha^2 - \Theta^2)\left(1 - \tfrac{1}{3}\sin^2\frac{\alpha}{2}\right)$$

and apply it to a form of the integral giving the exact period. Obtain

$$\frac{1 - \cos\alpha}{12} \le \frac{T - T_0}{T_0} \le \left(\frac{\alpha}{\sin\alpha}\right)^{1/2} - 1,$$

where T is the exact period and α is the maximum angular displacement. *Hint:* Use the fact that $(\sin x)/x$ has a local maximum at $x = 0$ to show that

$$\int_\Theta^\alpha \sin x\,dx \ge \frac{\sin\alpha}{\alpha}\int_\Theta^\alpha x\,dx, \qquad |x| \le \alpha \le \frac{\pi}{2}.$$

Also show that

$$1 - \cos x \ge 2\left(\frac{x}{\alpha}\right)^2 \sin^2\frac{\alpha}{2}.$$

Integrate from 0 to x and then from Θ to α.

9. Show that a first integral of

$$\frac{d^2\theta}{dt^2} + \omega_0^2 \sin\theta = 0$$

is

$$\frac{1}{2}\left(\frac{d\theta}{dt}\right)^2 = \omega_0^2(\cos\theta - \cos a) = 2\omega_0^2\left(\sin^2\frac{a}{2} - \sin^2\frac{\theta}{2}\right),$$

where a is the maximum amplitude. Verify that a further transformation

$$\sin\frac{\theta}{2} = \sin\frac{a}{2}\sin\psi$$

brings the solution of the problem into the form

$$t = \frac{1}{\omega_0} x \int_0^{\psi} (1 - k^2 \sin^2\psi)^{-1/2}\, d\psi,$$

where $k^2 = \sin^2(a/2)$. Obtain the period of oscillation as a power series in k^2.

10. An experimental test of the general theory of relativity is the advance of the perihelion of Mercury. According to this theory, the equation of the orbit is (1.12):

$$\frac{d^2u}{d\phi^2} + u = a(1 + \varepsilon u^2), \qquad a \equiv GMh^{-2}.$$

This differs from the classical form by the term εu^2. In the Newtonian case ($\varepsilon = 0$), the solution is

$$u = u_0 + a[1 + e\cos(\phi - \phi_0)],$$

where e is the eccentricity and ϕ_0 determines the position of the perihelion. Show that the angle between two succeeding perihelions is $2\pi a^2 \varepsilon$ to a first approximation. One method is the following.

If the solution of the relativistic equation is given in the form

$$u = a_0 + a_1 \cos\rho\phi + a_2\cos 2\rho\phi + \cdots$$

and we assume

$$\rho = 1 + \varepsilon\rho^{(1)} + \varepsilon^2\rho^{(2)} + \cdots,$$

$$a_k = a_k^{(0)} + \varepsilon a_k^{(1)} + \varepsilon^2 a_k^{(2)} + \cdots,$$

then

$$\rho^{(1)} = -a^2.$$

2.3 A System of Ordinary Differential Equations

The purpose of this section is to demonstrate the fourth part of applied mathematics that we listed in Chapter 1: how the desire to solve scientific problems motivates the development of a mathematical theory, and the manner in which such theories aid us in the solution of scientific problems.

We shall begin the discussion with a statement of certain theorems related to the initial value problem for mechanics. The proofs will follow. (As far as it is feasible, the proofs presented here will be those actually useful for constructing the solutions.) At the end of the section we shall briefly discuss the general question of the value of such proofs to the applied mathematician.

THE INITIAL VALUE PROBLEM: STATEMENT OF THEOREMS

For a given (initial) point $P_0(\tau, \zeta_1, \zeta_2, \ldots, \zeta_n)$ in the $(n+1)$-dimensional space of the points $P(t, z_1, z_2, \ldots, z_n)$, the **initial value problem** for a system of ordinary differential equations

$$\frac{dz_k}{dt} = f_k(t; z_1, z_2, \ldots, z_n); \qquad k = 1, 2, \ldots, n \tag{1}$$

is to find functions

$$z_m = g_m(t), \qquad m = 1, 2, \ldots, n \tag{2}$$

such that (1) is satisfied, and

$$g_m(t) = \zeta_m \quad \text{at } t = \tau, \qquad m = 1, 2, \ldots, n. \tag{3}$$

We now state and subsequently prove certain theorems that are valid for the initial value problem. In the statement of these theorems and in subsequent discussions we shall frequently use z and ζ to denote collectively the set of variables $(z_1, z_2, \ldots, z_n)$ and $(\zeta_1, \zeta_2, \ldots, \zeta_n)$.

Theorem 1 (Existence). Suppose that the functions* $f_k(t, z)$ are continuous in a rectangular parallelepiped defined by

$$R: |t - \tau| \le a, \quad |z_k - \zeta_k| \le b; \qquad k = 1, 2, \ldots, n. \tag{4}$$

This implies that there exists an upper bound M such that

$$|f_k| \le M, \qquad k = 1, 2, \ldots, n, \tag{5}$$

for $P(t, z_1, z_2, \ldots, z_n)$ in R. Suppose further that each f_k satisfies the following **Lipschitz condition** in R:

$$|f(t, \bar{z}_1, \ldots, \bar{z}_n) - f(t, z_1, z_2, \ldots, z_n)| \le K[|\bar{z}_1 - z_1| + \cdots + |\bar{z}_n - z_n|]. \tag{6}$$

* Some mathematicians are always careful to distinguish the function f from $f(x)$, the value of this function at x. We do not find it profitable to emphasize this distinction.

(The Lipschitz condition is implied if each of the functions f_k has continuous partial derivatives.) Then there exists a solution of the initial value problem in an interval $|t - \tau| \leq \alpha$, where

$$\alpha = \min\left(a, \frac{b}{M}\right). \tag{7}$$

The solution functions [the functions g_m of (2)] have continuous first derivatives.

Theorem 2 (Uniqueness). The solution of the initial value problem is unique.

The significance of Theorem 2 can be appreciated only when we can produce an example in which there is more than one set of functions satisfying the same equation and initial conditions, when (of course) the conditions in Theorem 1 are not all fulfilled. This will be done below.

Next we generalize the initial value problem to

$$\frac{dz_k}{dt} = f_k(t, z, \lambda), \qquad z_k = \zeta_k \text{ at } t = \tau. \tag{8}$$

Now the right-hand side of the equation depends on the parameter λ.

Theorem 3 (Continuous dependence on parameters). Let the functions $f_k(t, z, \lambda)$ $(k = 1, 2, \ldots, n)$ satisfy the requirements prescribed in Theorem 1. Further let these functions depend continuously on the parameter λ in a certain neighborhood $|\lambda - \lambda_0| < c$, where λ_0 and c are constants. Then the solution functions are also continuous functions of λ in some neighborhood of λ_0.

Theorem 3 can be used to prove that the solutions are continuous functions of the initial values (τ, ζ). This continuous dependence can be regarded as demonstrating the **stability** of the solution with respect to changes in initial values, in the sense of the following theorem.

Theorem 3′ (Stability). Let $z_m = g_m(t; \zeta_m)$ denote the mth solution function (2) with the dependence on the initial condition explicitly noted. Then given any $\varepsilon > 0$ and fixed time T, $T \geq \tau$, there exists a $\delta = \delta(\varepsilon, T)$ such that $|g_m(t; \zeta_m) - g_m(t; \zeta'_m)| < \varepsilon$ for $\tau \leq t \leq T$ whenever $|\zeta'_m - \zeta_m| \leq \delta$.

Theorem 3 is an expression of our intuitive feeling that when the functions $f_k(t, z, \lambda)$ are changed slightly (through the parameter λ), the solutions should also be changed slightly. Our corresponding expectation for the initial conditions is expressed in Theorem 3′. Let us now look more carefully at the nature of the dependence on a parameter.

We shall take the initial conditions to be fixed and consider the dependence of the solutions $z_m = g_m(t, \lambda)$ on λ. One might expect the nature of this dependence to be the same as that of the functions $f_k(t, z, \lambda)$ in (8). Thus, if the f_k are analytic in λ in the neighborhood of $\lambda = \lambda_0$ (i.e., if they have a convergent Taylor series when $0 \leq |\lambda - \lambda_0| < r$ for some r), then the solution functions g_m should have the same property. For linear differential equations, this is generally true. That is, the solution of the equation*

$$\frac{dz}{dt} = A(t, \lambda)z + b(t, \lambda) \tag{9}$$

is analytic in λ, provided that $A(t, \lambda)$ and $b(t, \lambda)$ are analytic in λ, even though they may merely be continuous in t. (Note that the Lipschitz condition is automatically satisfied for linear equations when the coefficients are continuous.) In the nonlinear case, the situation is a little more complicated, as we can see from our experience with the development of a perturbation solution for (2.1),

$$\frac{dy}{dx} = f(x, y, \varepsilon).$$

In present terms, the formal† calculations there depended on the fact that the functions $f_k(t, z, \lambda)$ were analytic in z as well as in λ. To keep the discussion relatively simple, we shall therefore restrict our attention to the first derivative with respect to λ.

If formal differentiation is justified, we would expect the functions

$$u_m(t, \lambda) \equiv \frac{\partial}{\partial \lambda} z_m(t, \lambda) \tag{10}$$

to satisfy the differential equation obtainable formally from the differentiation with respect to λ of the original equation (8), remembering that f_k and z_m both depend on λ.§ Carrying out this differentiation, we obtain

$$\frac{du_k}{dt} = \sum_m \frac{\partial f_k}{\partial z_m} u_m + \frac{\partial f_k}{\partial \lambda}. \tag{11}$$

We note that (11) is a *linear* system in $\{u_k\}$ whose coefficient functions are *known* functions of t, since the set $\{z_m(t, \lambda)\}$ is known. If the initial values ζ_m are independent of the parameter λ, then the initial conditions on the $\{u_k\}$ are

$$u_k = 0. \tag{12}$$

* This may be regarded as a single equation or a system of linear equations. In the latter case, A is a matrix, and z and b are vectors.

† A **formal calculation** is one that is presumably valid under suitable but unspecified conditions.

§ Successive parametric differentiation of an equation forms the basis of an excellent way to perform perturbation calculations. See Section 7.2.

In view of the above discussions, we expect to have

Theorem 4. If the $n(n+1)$ functions

$$\frac{\partial f_k}{\partial z_m}, \quad \frac{\partial f_k}{\partial \lambda}; \qquad k, m = 1, 2, \ldots, n;$$

are continuous in the variables t and $\{z_m\}$ and in the parameter λ, then the solution functions $z_m(t, \lambda)$ of Theorem 3 are differentiable with respect to λ, and the partial derivatives (10) satisfy the differential equations (11) and the initial conditions (12).

Since generalization is almost immediate, the proofs of all the theorems stated above will now be presented in the case of a single dependent variable.

PROOF OF THE UNIQUENESS THEOREM

The uniqueness theorem is usually the easiest to prove. In the present case, the ideas used will also be found to be useful for the proof of the existence theorem.

We consider the equation

$$\frac{dy}{dx} = f(x, y) \tag{13}$$

and seek a solution satisfying the initial condition $y = y_0$ when $x = x_0$. Let $g(x)$ be one such solution. Then

$$g'(x) = f[x, g(x)], \tag{14}$$

and hence

$$g(x) = y_0 + \int_{x_0}^{x} f[t, g(t)]\, dt. \tag{15}$$

If there were another solution $G(x)$ satisfying the same initial conditions, then

$$G(x) = y_0 + \int_{x_0}^{x} f[t, G(t)]\, dt. \tag{16}$$

If we now subtract (15) from (16), we obtain

$$G(x) - g(x) = \int_{x_0}^{x} [f(t, G) - f(t, g)]\, dt. \tag{17}$$

The magnitude of the integral on the right-hand side can be appraised with the help of the Lipschitz condition:

$$|f(t, G) - f(t, g)| \le K|G(t) - g(t)|. \tag{18}$$

Thus

$$|G(x) - g(x)| \le K \int_{x_0}^{x} |G(t) - g(t)|\, dt. \tag{19}$$

In the interval (x_0, x), $|G(x) - g(x)|$ has a maximum value that we shall denote by $\|G(x) - g(x)\|$. It then follows from (19) that

$$|G(x) - g(x)| \leq \|G(x) - g(x)\| \cdot K|x - x_0|, \tag{20}$$

and hence

$$\|G(x) - g(x)\| \leq \|G(x) - g(x)\| \cdot K|x - x_0|, \tag{21}$$

or

$$[1 - K(x - x_0)] \cdot \|G(x) - g(x)\| \leq 0.$$

If we now take the interval (x, x_0) sufficiently small, equal to $(2K)^{-1}$ for example, we can make the first factor positive. Hence (21) requires that

$$\|G(x) - g(x)\| = 0,$$

from which the desired result follows at once. □

PROOF OF THE EXISTENCE THEOREM

We have seen that the solution of the initial value problem for the differential equation (13) satisfies the *integral equation* (15). Conversely, if we have a solution of the integral equation (15), we have a solution for the differential equation, with the initial conditions implied. This can be verified by direct calculation (Exercise 1). Thus it is sufficient to prove the existence theorem for the integral equation.

To prove that (15) has a solution, we adopt the method of successive approximations. We start with a crude approximation $y = y_0$, which at least satisfies the initial condition.* We then calculate the sequence of functions

$$\begin{aligned}
y_1(x) &= y_0 + \int_{x_0}^{x} f(t, y_0)\, dt, \\
y_2(x) &= y_0 + \int_{x_0}^{x} f[t, y_1(t)]\, dt, \\
&\cdots \\
y_n(x) &= y_0 + \int_{x_0}^{x} f[t, y_{n-1}(t)]\, dt, \\
&\cdots .
\end{aligned} \tag{22}$$

We shall prove that this sequence converges uniformly to a continuous function in a suitably restricted interval $|x - x_0| \leq \alpha$, which may be specified as in Theorem 1; i.e., $\alpha = \min(a, b/M)$. [Here M is the upper bound on $|f|$ that was introduced in (5).] The restriction to the range b/M is needed

* It is certainly not necessary to make this particular initial approximation. For example, the same final result will obviously be obtained if $y_1(x)$, defined in (22), is used as the initial approximation.

only because we wish to keep the successive approximations (22) within the bound $|y - y_0| < b$, wherein the continuity of f is assured. Otherwise, there is no further restriction on the range of x.

To prove that the sequence of functions defined by (22) uniformly approaches a limit in the interval $|x - x_0| \leq \alpha$, let us consider the differences between successive functions. For $y_1(x) - y_0$ we have

$$|y_1(x) - y_0| \leq \int_{x_0}^{x} |f(t, y_0)|\, dt \leq M|x - x_0|, \tag{23}$$

where M is the upper bound on f first mentioned in (5).

In general, we have

$$|y_n(x) - y_{n-1}(x)| \leq \int_{x_0}^{x} |f[t, y_{n-1}(t)] - f[t, y_{n-2}(t)]|\, dt. \tag{24}$$

We need to relate the difference in the right-hand side to the difference $y_{n-1}(t) - y_{n-2}(t)$. This relation is supplied by the Lipschitz condition:

$$|f[t, y_{n-1}(t)] - f[t, y_{n-2}(t)]| \leq K|y_{n-1}(t) - y_{n-2}(t)|. \tag{25}$$

We are thus able to establish the recurrence relation

$$|y_n(x) - y_{n-1}(x)| \leq K \int_{x_0}^{x} |y_{n-1}(t) - y_{n-2}(t)|\, dt. \tag{26}$$

By combining (23) and (26), we obtain [Exercise 3(c)]

$$|y_n(x) - y_{n-1}(x)| \leq \frac{MK^{n-1}|x - x_0|^n}{n!}. \tag{27}$$

But the right-hand side of (27) is M/K times the nth term in the series for $\exp(K|x - x_0|)$. Thus let us adopt the device of introducing a series expression for the limit as $n \to \infty$ of $y_n(x)$, namely,

$$y_0 + \sum_{r=1}^{\infty} [y_r(x) - y_{r-1}(x)]. \tag{28}$$

This series is absolutely and uniformly convergent, since its terms are bounded by the corresponding terms of another series with that property. We thus have $\lim_{n\to\infty} y_n(x) = y(x)$ for some function y. Further, $y(x)$ can be shown to satisfy the integral equation (15) by examining the limiting form of the sequence of equations (22). Here it is only necessary to justify the inversion of the processes of integration and taking the limit; but since the integral is over a finite range, this follows from well-known theorems. □

The key process in the proof is the conversion of the differential equation into an integral equation and *then* the application of the method of successive approximations. The integral formulation has two advantages: (i) the initial values are automatically incorporated; and (ii) the integration process makes the functions smoother, and questions about the existence of a derivative are avoided.

The proof required many technical steps. Since similar steps often occur in other problems, it is convenient to summarize them in some form so that they need not be continually repeated. This is one reason for developing the theory of *function spaces*, in which an important concept is the *norm* suitably defined (in various ways) to replace the concept of *distance* in ordinary Euclidean space. Another reason is that ordinary geometric ideas can be used to give intuitions about abstract distance properties.

In the present case, one possible norm is the quantity $\|g(x) - G(x)\|$ first used in (20). Another possible measure of "distance" is the quantity D defined by

$$D^2 = \frac{1}{b-a}\int_a^b [G(x) - g(x)]^2 \, dx. \tag{29}$$

In words, D is the root-mean-squared difference between G and g. This norm does not occur naturally here, but it will be used in Chapters 4 and 5.

Appendix 12.1 of II provides certain details concerning function spaces and related concepts. [In particular, Equation (59) of that Appendix gives an abstract characterization of distance that is exemplified by both of the norms just defined.] Material in Chapter 12 of II on variational methods furnishes examples of the unity and clarity that can be achieved with such concepts.

CONTINUOUS DEPENDENCE ON A PARAMETER OR INITIAL CONDITIONS

Before we proceed to prove Theorems 3 and 3′, let us note that if the solution of the differential equation

$$\frac{dy}{dx} = f(x, y, \lambda), \tag{30}$$

under the initial condition $y = y_0$ for $x = x_0$, depends continuously on λ, then it depends continuously on (x_0, y_0). For we can introduce the new variables $(\xi, \eta) = (x - x_0, y - y_0)$ and solve the equation

$$\frac{d\eta}{d\xi} = f(\xi + x_0, \eta + y_0, \lambda) \tag{31}$$

under the initial condition $\eta = 0$ for $\xi = 0$. The function on the right-hand side of (31) is continuous in (x_0, y_0).

Return now to (30). Let us treat the equation in the integral formulation

$$y(x, \lambda) = y_0 + \int_{x_0}^{x} f[x, y(x, \lambda), \lambda] \, dx. \tag{32}$$

Consider two solutions $y(x, \lambda)$ and $y(x, \lambda_0)$ and their difference. We have

$$y(x, \lambda) - y(x, \lambda_0) = \int_{x_0}^{x} \{f[x, y(x, \lambda), \lambda] - f[x, y(x, \lambda_0), \lambda]\}\, dx + \int_{x_0}^{x} \{f[x, y(x, \lambda_0), \lambda] - f[x, y(x, \lambda_0), \lambda_0]\}\, dx. \quad (33)$$

The first integral on the right-hand side can be appraised with the help of the Lipschitz condition and the second integral by simple continuity in λ. Thus, by using an argument similar to that used in the proof of the uniqueness theorem, we have [Exercise 3(a)]

$$\|y(x, \lambda) - y(x, \lambda_0)\| \leq K|x - x_0| \cdot \|y(x, \lambda) - y(x, \lambda_0)\| + \delta|x - x_0|, \quad (34)$$

where

$$|f(x, y(x, \lambda_0), \lambda) - f(x, y(x, \lambda_0), \lambda_0)| < \delta. \quad (35)$$

For $|x - x_0| \leq (2K)^{-1}$, one can show [Exercise 3(b)] that (34) implies that

$$|y(x, \lambda) - y(x, \lambda_0)| \leq 2\delta|x - x_0|. \quad (36)$$

The quantity δ can be made as small as desired by decreasing $|\lambda - \lambda_0|$.

The restriction on $|x - x_0|$ is not crucial, since one can cover any finite interval $|x - x_0| < \alpha$ by a *finite* number of intervals of the above type. Upon passing from one interval to another, one would have to allow for a difference in the initial condition y_0, but this difference is *of the order of* δ. Since we have only a finite number of such intervals, the cumulative difference is still of the order of δ and does not influence the essentials of our arguments. □

DIFFERENTIABILITY

We consider the pair of differential equations

$$\frac{dy}{dx} = f(x, y, \lambda) \quad (37)$$

and

$$\frac{du}{dx} = f_y(x, y, \lambda)u + f_\lambda(x, y, \lambda), \quad (38)$$

where $f(x, y, \lambda)$, $f_y(x, y, \lambda)$, and $f_\lambda(x, y, \lambda)$ are continuous functions. Note that (38) is obtained by formally differentiating (37) with respect to λ and writing u for $\partial y/\partial\lambda$.

We consider a solution $y(x, \lambda)$ of (37). We also consider, for $y = y(x, \lambda)$, a solution of (38) satisfying the initial condition $u(x_0, \lambda) = 0$. The existence of these solutions is assured by the continuity of the functions involved.* We wish to prove that

$$u(x, \lambda) = \frac{\partial y(x, \lambda)}{\partial \lambda}. \tag{39}$$

The difficult part of the proof is to establish the existence of the partial derivative $\partial y/\partial \lambda$. This will be done by direct evaluation of the limit of $\Delta y/\Delta \lambda$ as $\Delta \lambda \to 0$.

Let

$$\Delta y = y(x, \lambda) - y(x, \lambda_0), \Delta f = f[x, y(x, \lambda_0), \lambda] - f[x, y(x, \lambda_0), \lambda_0].$$

From (33)

$$\frac{\Delta y}{\Delta \lambda} = \int_{x_0}^{x} f_y(x, \bar{y}, \lambda) \frac{\Delta y}{\Delta \lambda} dx + \int_{x_0}^{x} \frac{\Delta f}{\Delta \lambda} dx, \tag{40}$$

where $\bar{y}$ is a value intermediate between $y(x, \lambda_0)$ and $y(x, \lambda)$. We have used the continuity of the partial derivative $f_y(x, y, \lambda)$ so that the mean value theorem can be applied. We also have

$$u(x, \lambda_0) = \int_{x_0}^{x} f_y[x, y(x, \lambda_0), \lambda_0]u(x, \lambda_0)\, dx + \int_{x_0}^{x} f_\lambda[x, y(x, \lambda_0), \lambda_0]\, dx. \tag{41}$$

By combining (40) and (41), we obtain for the difference

$$w \equiv u(x, \lambda_0) - \frac{y(x, \lambda) - y(x, \lambda_0)}{\lambda - \lambda_0} \tag{42}$$

the integral equation

$$w = \int_{x_0}^{x} f_y(x, \bar{y}, \lambda) w\, dx + D, \tag{43}$$

where

$$D = \int_{x_0}^{x} [f_y(x, y, \lambda) - f_y(x, \bar{y}, \lambda_0)]u\, dx + \int_{x_0}^{x} \left[f_\lambda(x, y(x, \lambda_0), \lambda_0) - \frac{\Delta f}{\Delta \lambda}\right] dx.$$

Because of the continuity of the functions f_y and f_λ, the quantity D can be made as small as we wish by reducing $\Delta \lambda$. Since $|f_y(x, \bar{y}, \lambda)|$ is bounded by K, one can prove, from (43), that

$$\lim_{\Delta \lambda \to 0} w = 0 \tag{44}$$

by the same sort of reasoning as that previously used. The reader should complete the proof (Exercise 4).

* The Lipschitz condition is satisfied for (38), because the right-hand side is linear in the dependent variable u.

EXAMPLE OF NONUNIQUENESS

One can better appreciate the usefulness of the uniqueness theorem under the Lipschitz condition when one sees some examples of the lack of uniqueness when the condition is not satisfied. Consider the differential equation

$$\frac{dy}{dx} = f(x, y), \tag{45}$$

where

$$f(x, y) = \frac{4x^3 y}{x^4 + y^2} \qquad \text{when } (x, y) \neq (0, 0), \tag{46a}$$

$$f(0, 0) = 0. \tag{46b}$$

It is easy to verify that $f(x, y)$ is continuous at $(x, y) = (0, 0)$ but does not satisfy the Lipschitz condition. Equation (45) admits the solution

$$y = c^2 - \sqrt{x^4 + c^4} \tag{47}$$

for all finite real values of c. Thus there is an infinity of solutions satisfying the initial condition $(x, y) = (0, 0)$. All the integral curves have zero slope at the origin.

The lack of uniqueness can easily be understood from a heuristic point of view, since (45) does not give a unique value for the curvature of the integral curve at $(x, y) = (0, 0)$. Indeed, the curvature of the solution curve at this point is undefined. For the derivative of the slope of the solution curve is given by the limit as $(x, y) \to (0, 0)$ of

$$\frac{\dfrac{dy}{dx}(x, y) - \dfrac{dy}{dx}(0, 0)}{x - 0} = \frac{f(x, y)}{x} = \frac{4x^2 y}{x^4 + y^2}.$$

An existence theorem can still be proved in this case by the method of finite differences, to be outlined in a moment. Since there is lack of uniqueness, an initial value for d^2y/dx^2 must be prescribed (implicitly or explicitly) for each solution.

METHOD OF FINITE DIFFERENCES

The method of finite differences is the natural approach to the solution of differential equations from the point of view of numerical integration. It yields at the same time a practical way of calculating the solutions, especially with the aid of modern computing machines, and a way for proving an existence theorem. To obtain a glimpse of this method, we may start again with the integral formulation

$$y = y_0 + \int_{x_0}^{x} f[\xi, y(\xi)]\, d\xi, \tag{48}$$

Let us divide the interval (x_0, x) into n subintervals (each of length h) and write

$$y = y_0 + \int_{x_0}^{x_1} f[\xi, y(\xi)]\, d\xi + \cdots + \int_{x_{n-1}}^{x} f[\xi, y(\xi)]\, d\xi. \tag{49}$$

This is still exact, but we shall now evaluate each of the integrals in (49) approximately by writing

$$\int_{x_k}^{x_{k+1}} f[\xi, y(\xi)]\, d\xi = hf(x_k, y_k). \tag{50}$$

Here y_k is the value of y obtained by using the approximation (50) in (49) up to the point x_k; i.e., we have

$$\begin{aligned} y_1 &= y_0 + \int_{x_0}^{x_1} f(x_0, y_0)\, d\xi, \\ y_2 &= y_1 + \int_{x_1}^{x_2} f(x_1, y_1)\, d\xi, \\ &\cdots \\ y(x) = y_n &= y_{n-1} + \int_{x_{n-1}}^{x} f(x_{n-1}, y_{n-1})\, d\xi. \end{aligned} \tag{51}$$

The solution is then the broken line C_L in Figure 2.2. This is taken as an approximation to the true integral curve C. Since there is an error in y at each approximation, the accumulation of errors would, in general, make the approximating curve C_L deviate farther and farther from the true curve C

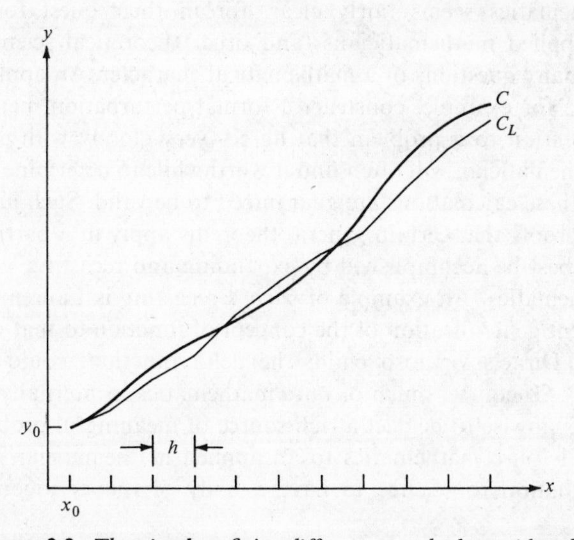

FIGURE 2.2. *The simplest finite difference method provides the broken line C_L as an approximation to the actual solution C of a differential equation.*

as the calculation process goes on according to (51). One might think at first sight that this would be intolerable. Actually, since the error in y is of the order of h^2 in each step [local formula error is $O(h^2)$], it can be shown that the cumulative error is of the order of h [cumulative formula error is $O(h)$]. Indeed, one can prove the convergence of the approximate solution to a unique true solution when the Lipschitz condition is satisfied. Even when the Lipschitz condition is not satisfied, a *partial sequence* can always be chosen to converge to a solution; but there is no longer any assurance of uniqueness.

An $O(h)$ cumulative formula error can be made as small as desired by taking h small enough. But then many calculations will be required to find the solution in a given x interval, and the cumulative roundoff error may be unacceptably large. To gain better accuracy, one can adopt higher order approximations when evaluating the integrals in (49). But then, calculations for a given interval are longer, so that they take more time and possibly introduce more roundoff error. The reader will find extensive discussions of such issues in books on numerical methods.

FURTHER REMARKS ON THE RELATION BETWEEN "PURE" AND "APPLIED" MATHEMATICS

In this section we have gone into mathematical theory that generally would be regarded as being of a "pure" nature. We conclude with some remarks on the interaction between the pure and the applied aspects of mathematics. (Also see Chapter 1.)

That most pure mathematicians would benefit from studying works in applied mathematics seems fairly clear, for in their quest for scientific knowledge, applied mathematicians (and other theoretical scientists) leave unanswered many questions of a mathematical character. An applied mathematician may, for example, construct a formal perturbation method which provides a solution to a problem that agrees very closely with experiment. The pure mathematician will often find it worthwhile to determine conditions under which these calculations are guaranteed to be valid. Such justifications may require proofs that certain general theorems apply in a particular case. Or they may best be accomplished by expanding and recasting a whole segment of mathematics. An example of such a recasting is Laurent Schwarz's relatively recent generalization of the concept of function to that of distribution, so that Dirac's virtuoso child, the delta function, could at last be legitimatized.* Of course, much of pure mathematics is internally motivated —but it seems unwise to neglect a rich source of meaningful problems.

What help is pure mathematics to an applied mathematician? Certainly, it is a contribution to science to have a body of theory incontrovertibly

* A good reference is M. J. Lighthill's *Fourier Analysis and Generalized Functions* (New York: Cambridge U.P., 1962). The dedication of this work is a capsule history: "To Paul Dirac who saw that it must be true, Laurent Schwartz who proved it, and George Temple who showed how simple it could be made."

established. The magnitude of the contribution is an increasing function of existing doubt in the theory. Indeed, if the doubt is serious, the applied mathematician himself will try to resolve the controversy by fashioning appropriate proofs.

At the educational level, the future applied mathematician should be exposed to a considerable body of mathematical theory and proof (although care must be taken not to overemphasize this aspect). He should learn typical conditions under which the operations he may have to perform are valid. Moreover, he should become aware of additional mathematical concepts that may someday form a suitable framework for his theories or calculations. For example, some applied mathematicians find use for the unifying geometric character of the function space concepts, developed in recent decades. To give a more classical example, the Gibbs phenomenon, which can arise in practical calculations, cannot be appreciated without a knowledge of the distinction between pointwise and uniform convergence. (See Section 4.3.) And, to cite another possibility, it could be that a student will first become aware of the calculational usefulness of an integral equation formulation, or of a successive approximations approach, by virtue of having studied certain constructive existence theorems.

Examples of the relation between pure and applied mathematics are provided by our discussion of theorems for differential equations. Thus it is a beautiful theoretical result that under suitable conditions solutions depend continuously on parameters. But this result misses a very important scientific issue, for it merely shows that the solution is changed by an arbitrarily small amount *over a fixed time interval* by a sufficiently small change in the parameter. Left untouched is the question of the *ultimate* effect of a given small change in a parameter, a question that could well be missed entirely by one who accepts an impressive theorem as the last word. Poincaré began the study of long term and "ultimate" effects. Much more has since been accomplished, both formally and rigorously, but the issues are by no means resolved. [See Moser (1973).]

Another example of the relation between pure and applied mathematics stems from the fact that standard existence and analyticity theorems for systems of ordinary differential equations are valid only for a sufficiently short interval of time. But if these equations describe the trajectories of interacting particles, it seems likely that the theorems should usually hold for arbitrarily long time intervals. For a discussion of classical results on such matters see Chapter 16 of E. T. Whittaker's *Analytical Dynamics* (New York: Cambridge U.P., 1927).

To achieve a degree of balance, we presented a few formal proofs in this chapter. But such proofs can be found in many fine books. Hence proofs are mainly omitted in the remainder of the present work, since we concentrate on the relatively unexplored interaction between science and mathematics that is the core of an applied mathematician's profession.

EXERCISES

1. Show that a solution of (15) satisfies (13) and the initial condition $y(x_0) = y_0$. Carefully justify all your steps.
2. Prove that $y(x)$, the sum of the series (28), satisfies the integral equation (15).
3. (a) Verify (34).
   (b) Verify (36).
   (c) Verify (27).
4. Complete the proof of (39).
5. (a) Verify the statement under (46b).
   (b) Verify (47).
   (c) Feigning ignorance of (47), see if you can derive it.

CHAPTER 3
Random Processes and Partial Differential Equations

IN CHAPTER 2 we considered the dynamics of individual particles, and we showed that interaction of the particles can be modeled by a system of ordinary differential equations. Existence of a solution to these equations under prescribed conditions (usually specified values of the initial position and velocity of each particle) is assured by a classical mathematical theory. Actual determination of an adequate approximation to the solution can be carried out, if necessary, by numerical integration. Study of the algorithms used in numerical analysis is necessary to provide assurance that a given approximation is accurate. Nevertheless, mathematical theory assures us of the unique existence of the solution. The future of the particles is completely determined by their past, so our model is **deterministic**.

We shall now study examples of models in which there is uncertainty in the predicted outcome of a process. In such situations we shall refer to **random processes**; the corresponding models will be termed **probabilistic**. For example, we shall see that when many particles interact, it is often useful to adopt a model in which each particle's motion is described by certain probabilities.

Although we shall give some consideration to random processes for their own sake, our primary goal is to investigate the connection between probabilistic and deterministic models of the same phenomenon. The key fact here is that even though a process is regarded as random, nevertheless in many circumstances we can write down deterministic problems in partial differential equations that yield a definite distribution function for the probabilities associated with the process. The partial differential equation for the distribution function is identical with the equation one would write down in a deterministic characterization of the phenomenon.

It might appear paradoxical that a random process can be characterized by a definite equation. But we know from experience that when individually unpredictable events are repeated a large number of times, there usually emerges a definite regularity. For example, when a "fair" coin is tossed once, we have no idea whether head or tail will turn up. But after a large number of tosses we can be almost certain that the proportion of heads will differ only a little from $\frac{1}{2}$.* With this example and others like it in mind, it is perhaps not so surprising that there is a determinable distribution of probabilities which characterizes a random process.

* Precise statements of this nature fall under the heading "laws of large numbers."

The characterization of situations in which probabilistic models are appropriate is rather a subtle matter, but typically there is an element of incompleteness of information. Let us once again consider the example of coin tossing. Once a coin is tossed, its motion is determined. The process of tossing, however, involves many parameters, such as the coin's initial orientation, velocity, and spin; the properties of the surface on which it bounces; etc. Usually, we have no precise knowledge of these parameters.

An even more important characteristic of situations where probabilistic models are useful is illustrated by the fact that a slight change of the parameters could shift the final configuration of a coin from "head" to "tail," or vice versa. Another familiar example of this extreme **sensitivity** is found in the kinetic theory of gases. Here each molecule can be regarded as following exact mechanical laws of motion and exact laws of interaction (if the hard sphere model is adopted, for instance). But a slight change in the initial conditions would probably result in a tremendous change in the condition after many collisions.

An averaging of the solutions of a number of problems with varying initial conditions and a consequent blurring of the sensitivity of a single problem can be regarded as underlying the successful modeling of sensitive systems by random processes.

In Section 3.1 we formulate a model in which Brownian motion is regarded as being caused by an assemblage of particles that move randomly on a line, taking steps of equal length with equal probability of moving right and left. An explicit expression is obtained for the probability $w(m, N)$ that such a particle has moved m steps to the right after taking a total of N steps. The section concludes with a sketch of how theory enables one to determine molecular properties from observations of Brownian motion.

The important concept of an asymptotic approximation is developed in Section 3.2. This is necessary in order to understand the sense in which a certain simplified expression approximates $w(m, N)$ when N is large, but the methods introduced in Section 3.2 are of far wider applicability.

In Section 3.3 we introduce a difference equation for $w(m, N)$ and then approximate this difference equation by a limiting differential equation. This equation is none other than the diffusion equation commonly used in macroscopic descriptions of the motion of many particles. We show that the "unit source" or "fundamental" solution of the diffusion equation provides the same approximation to w as that mentioned in the previous paragraph. The usefulness of the differential equation approach is illustrated by consideration of random walk in the presence of reflecting or absorbing barriers.

Further explorations of the connection between probability and differential equations are made in Section 3.4. For example, it is shown how the method of images, used in differential equations, leads to the solutions of some probability problems. The fundamental solution, first met in a probability context, is shown to provide the key to a solution of the general initial value problem for the diffusion equation. Section 3.4 concludes with a discussion that links (among other things) the collision of DNA molecules, the twisting of a beam, the probability that a randomly walking drunk will some day return to his starting point, and the irreversible character of an assemblage of particles in quasi-periodic motion.

3.1 Random Walk in One Dimension; Langevin's Equation

In Brownian motion, small particles move about irregularly in a liquid or gas. Perhaps the most common manifestation of this phenomenon is the dancing of dust particles in a beam of sunlight.

The relatively large dust particles move because of myriads of impacts by the molecules that comprise the surrounding medium. In making a mathematical model of this situation, we should without doubt regard the impacts as occurring randomly. There is no possibility of computing the trajectory of each molecule, and no interest in doing so, if one wishes to understand the broad outlines of the phenomenon.

THE ONE-DIMENSIONAL RANDOM WALK MODEL

Let us try to construct the simplest possible model that bears on the situation. Thus, although the motion is three-dimensional, let us consider a projection of the particle, which moves randomly along a line. A reasonable idealization of the motion consists of a sequence of steps, with randomly determined lengths and durations. We shall simplify the picture still further by considering steps of a fixed length Δx that occur at a fixed time interval Δt. We are thus led to the classical one-dimensional model of Brownian motion, in which a particle moves according to the following rules:

(i) During the passage of a certain fixed interval of time Δt, a particle takes one step of a certain fixed length Δx along the x axis.

(ii) It is equally probable that the step is to the right or to the left.

We shall study the characteristics of a particle that executes the **unbiased*** **random walk** described by (i) and (ii). To do this we need only the rudiments of probability theory since a very simple type of probability problem is involved. In a given situation a certain set of possible events is selected for attention. These events are often labeled "favorable." Because all events are regarded as "equally likely," the probability of a selected "favorable" event is given by the ratio

$$\frac{\text{number of favorable events}}{\text{total number of events}}.$$

We wish to determine the probability $w(m, N)$ that a particle will arrive at a point m steps to the right of its starting point after taking N steps. (Points

* The walk is biased unless steps in all permitted directions are equally probable.

to the left are associated with a negative integer m.) Every possible N-step walk is equally likely, and the "favorable" event is arriving exactly m steps to the right in such a walk.

In this section we shall derive an explicit formula for $w(m, N)$. We shall then introduce the device of a generating function to compute the average and root-mean-squared values of m.

The concept of "random walk" is colorful and it is particularly interesting when regarded as describing the fate of a drunk staggering down a street, with equal probabilities of stepping forward or backward. As one might expect, random walks can be found concealed in a wide variety of situations. To give just one further example, consider a gambling game in which a "fair" coin is tossed and player A gives a dollar to player B, or takes a dollar from him, depending on whether the coin is heads or tails, respectively. If a particle is stepped to the right (left) when A wins (loses) a dollar, then $w(m, N)$ provides the probability that A has won m dollars after N tosses. The reader will find it worthwhile to keep this additional example in mind as he reads this chapter.*

The most important physical implication of the study of Brownian motion resides in the fact that such motion provides a visible macroscopic manifestation of individual molecular forces. (By contrast, only the bulk effects of molecular motion are manifested by observations of such variables as temperature and pressure.) The most direct theoretical approach to these matters involves a differential equation with a random forcing term, the Langevin equation. We sketch the essence of the relevant theory at the end of this section.

EXPLICIT SOLUTION

We now work out an explicit formula for $w(m, N)$, the probability that a particle will arrive at a point m steps to the right of its starting point after taking N steps. Suppose that the particle takes a total of p steps to the right, $p \geq 0$, and hence a total of $N - p$ steps to the left. It would then reach a point m steps to the right $(-N \leq m \leq N)$, provided that

$$m = p - (N - p); \qquad \text{i.e., } p = \tfrac{1}{2}(N + m). \tag{1}$$

Note that m is a signed integer, but p is a nonnegative integer. Also $m = N - 2p$ is even (odd) when N is even (odd). Thus m ranges from $-N$ to N in steps of 2. For example, if a total of $N = 3$ steps are taken, the possible values of displacement are $m = -3, -1, 1, 3$.

The number of paths with p steps to the right and $N - p$ to the left is given by

$$\frac{N!}{p!\,(N-p)!} \equiv C_p^N. \tag{2}$$

* For other examples and a good introduction to the literature, see M. Barber and B. Ninham, *Random and Restricted Walks* (New York: Gordon and Breach, 1970).

Recall that C_p^N is called the **binomial coefficient**, because of its role in the binomial expansion

$$(x + y)^N = \sum_{p=0}^{N} C_p^N x^{N-p} y^p. \tag{3}$$

The total number of different N-step paths is 2^N. Dividing by the number of favorable events, we obtain the result

$$w(m, N) = \frac{C_p^N}{2^N}, \qquad \text{where} \quad p = \tfrac{1}{2}(N + m). \tag{4}$$

The number desired for (2) is the same as the number of full box-empty box patterns formed by p balls in N boxes, one ball per box. (A ball in the kth box is analogous to the kth step being to the right.) We start by considering p distinguishable balls. The first ball, p_1, can be placed in any of N boxes; the second, p_2, in any of $N - 1$ boxes; etc. Thus p distinguishable balls can be placed in N boxes in the following number of ways:

$$N(N - 1) \cdots [N - (p - 1)] = \frac{N!}{(N - p)!}. \tag{5}$$

Interchanging distinguishable balls permutes the order in which they appear but does not change the pattern of full and empty boxes. There are $p!$ permutations of p balls. Summing up:

$$\begin{array}{l} \text{number of ways } p \\ \text{distinguishable balls} \\ \text{can be placed in } N \\ \text{boxes} \end{array} = \begin{array}{l} \text{number of full} \\ \text{box-empty box} \\ \text{patterns} \end{array} \cdot \begin{array}{l} \text{number of} \\ \text{permutations of} \\ \text{distinguishable} \\ \text{balls within a} \\ \text{pattern} \end{array}$$

or

$$\frac{N!}{(N - p)!} = C_p^N \cdot p!,$$

yielding (2).

Let us check that the sum of all probabilities is unity. Using (3), we find that indeed

$$\sum_{m=-N}^{N}{}'' w(m, N) = \sum_{p=0}^{N} C_p^N (\tfrac{1}{2})^N = \sum_{p=0}^{N} C_p^N (\tfrac{1}{2})^{N-p} (\tfrac{1}{2})^p = (\tfrac{1}{2} + \tfrac{1}{2})^N = 1. \tag{6}$$

NOTE. The double prime on the sum is used to indicate that the summation index takes *every other* value.

MEAN, VARIANCE, AND THE GENERATING FUNCTION

Suppose that f is a function whose domain is the possible outcomes m of the random walk. Then the **expected value** of f is defined by

$$\langle f \rangle_N = \sum_{m=-N}^{N}{}'' f(m) w(m, N). \tag{7}$$

The term "expected value" or **average value** is used, because in a large number A of N-step walks, one would expect the particle to arrive m steps to the right of its starting point $Aw(m, N)$ times. The average value of $f(m)$ would be expected to be

$$A^{-1} \sum_{m=-N}^{N}{}'' f(m)[Aw(m, N)] = \langle f \rangle_N. \tag{8}$$

It is customary to omit the subscript N in (7). Also, customary usage employs $\langle f(m) \rangle$ instead of $\langle f \rangle$. For example, if $f(m) = m^k$, $\langle f \rangle$ is replaced by $\langle m^k \rangle$. To give another example, the average value of $p(m) = \frac{1}{2}(N + m)$, the number of steps to the right, can be written in the following equivalent ways:

$$\langle p \rangle = \sum_{m=-N}^{N}{}'' p(m)w(m, n) = \sum_{p=0}^{N} pC_p^N(\tfrac{1}{2})^N.$$

Here we have used formula (4) for $w(m, n)$. Passage from the first sum to the second has been made by changing variables from m to p.

The most important cases of $\langle f \rangle$ are the **average** or **mean displacement** $\langle m \rangle$ and the **mean square displacement** or **variance** $\langle m^2 \rangle$. The expected value of $f(m) = m^k$ is called the kth **moment**.

In order to evaluate the various moments, we introduce the **generating function**

$$G(u) = \sum_{p=0}^{N} C_p^N(\tfrac{1}{2})^N u^p. \tag{9}$$

This function is formed by attaching a power of a newly introduced variable u to the probability $w(m, N)$ and summing over all permissible ranges of the parameter p (which is preferable to m because p goes in steps of unity).* Alternatively, (9) can be viewed as a Fourier series of form

$$g(t) = \sum_{p} a_p e^{ipt}, \tag{10}$$

with $u = e^{it}$ and $a_p = 0$ for $p > N$. The generating function method is therefore related to the Fourier transform method to be considered later.

* It is not difficult to follow manipulations using the generating function, but it is amazing that anyone would have thought of this device. Amazement is lessened upon learning that the first person to make use of the generating function was the genius Euler. In Vol. 1, Chapter VI, of his splendid book *Induction and Analogy in Mathematics* (New York: Oxford U.P., 1954), G. Polya discusses the motivation for using a generating function. He states that "a generating function is a device somewhat similar to a bag. Instead of carrying many little objects detachedly, which could be embarrassing, we put them all in a bag, and then we have only one object to carry, the bag. Quite similarly, instead of handling each term of the sequence $a_0, a_1, a_2, \ldots, a_n, \ldots$ individually, we put them all in a power series $\sum_n ax^n$, and then we have only one mathematical object to handle, the power series." And to deal with power series, we can use the extraordinarily powerful theory of analytic functions. For a guide to the extensive literature on generating functions see E. B. McBride, *Obtaining Generating Functions* (New York: Springer, 1971) and the review of this monograph by J. Wimp in *SIAM Rev. 14*, 663–67 (1972).

To demonstrate the use of the generating function, we note that

$$G'(1) = \sum_{p=1}^{N} pC_p^N(\tfrac{1}{2})^N = \sum_{m=-N}'' pw(m, n) = \langle p \rangle. \tag{11}$$

But since $p = \frac{1}{2}(N + m)$,

$$\langle p \rangle = \tfrac{1}{2}N \sum_{m=-N}^{N}{}'' w(m, N) + \tfrac{1}{2} \sum_{m=-N}^{N}{}'' mw(m, N) = \tfrac{1}{2}N + \tfrac{1}{2}\langle m \rangle. \tag{12}$$

If $G(u)$ can be written in a simple form, we shall therefore have no difficulty in evaluating $G'(1) = \langle p \rangle$ and hence the desired average value $\langle m \rangle$. But another use of the binomial expansion (3) gives

$$G(u) = (1 + u)^N(\tfrac{1}{2})^N, \tag{13}$$

which is a closed form for G, in contrast to the series form (9). Equation (13) allows us to conclude at once that $G'(1) = N/2$. From (11) and (12), then, we see that

$$\langle p \rangle = \frac{N}{2}, \qquad \langle m \rangle = 0. \tag{14a, b}$$

It is to be expected that the average displacement $\langle m \rangle$ is zero, for motions to the left and to the right are equally probable.

Example. Find $\langle m^2 \rangle^{1/2}$ and interpret the result.

Solution. From (9)

$$G''(1) = \sum_{p=0}^{N} p(p-1)C_p^N(\tfrac{1}{2})^N = \langle p(p-1) \rangle = \langle p^2 \rangle - \langle p \rangle.$$

But, from (13),

$$G''(u) = N(N-1)(1+u)^{N-2}\left(\frac{1}{2}\right)^N \cdot G''(1) = \frac{N(N-1)}{4}.$$

Using the result $\langle p \rangle = N/2$ of (14a),

$$\langle p^2 \rangle - \langle p \rangle = \tfrac{1}{4}N^2 - \tfrac{1}{4}N, \qquad \langle p^2 \rangle = \tfrac{1}{4}N^2 + \tfrac{1}{4}N.$$

However,

$$m = 2p - N \qquad \text{so that} \qquad \langle m^2 \rangle = 4\langle p^2 \rangle - 4N\langle p \rangle + N^2.$$

That is,

$$\langle m^2 \rangle = N^2 + N - 2N^2 + N^2 = N.$$

The root-mean-squared displacement from the origin, i.e., the **variance**, is

$$\langle m^2 \rangle^{1/2} = N^{1/2}, \tag{15}$$

which is much less than the number of steps N. This is a consequence of the cancellation of efforts due to the random leftward and rightward motion of the particle.

USE OF A STOCHASTIC DIFFERENTIAL EQUATION TO OBTAIN BOLTZMANN'S CONSTANT FROM OBSERVATIONS OF BROWNIAN MOTION

We now turn to a very brief exposition of how observations of Brownian particles can lead to a determination of molecular properties. This possibility was provided by Einstein, and the experiments were carried out by Perrin. An outline of their approach follows.

Assume that the macroscopic resistance on the particle is proportional to the velocity—as is predicted by classical hydrodynamics (see Section 3.6 of II). Introduce the **ensemble average**, $\langle\ \rangle$, a mean calculated from many repetitions of the same experiment. Observe the particle for a period t to obtain either (i) the net displacement projected along a line (denoted by x), or (ii) the net displacement itself (denoted by r). Einstein showed that these observations of diffusion should obey the statistical law

$$\langle x^2\rangle = \tfrac{1}{3}\langle r^2\rangle = 2Dt, \tag{16}$$

where the diffusion coefficient D is given by

$$D = \frac{k\Theta}{f}. \tag{17}$$

Here Θ is the absolute temperature, and f is the coefficient of resistance such that the force acting on a particle moving at velocity $\mathbf{v}$ is $-f\mathbf{v}$. For spherical particles of radius a moving in a medium with viscosity coefficient μ, f is determined by Stokes' law [Equation (3.6.10) of II]:

$$f = 6\pi\mu a. \tag{18}$$

In (17), k stands for Boltzmann's constant. (It is a fundamental physical result that the average kinetic energy of a gas molecule is equal to $\frac{3}{2}k\Theta$.)

Perrin was able to verify (16) and thus to determine the Boltzmann constant k over wide ranges of Θ, μ, and a. Since k equals the previously known universal gas constant R, divided by Avogadro's number N_0, the determination of k led at once to a value for N_0.

Result (16) is essentially found in Equations (3.8) and (3.28) by examining random walk in an appropriate limit, but a fuller description requires another approach. The modern theory of the Brownian motion of a free particle generally starts with *Langevin's equation*,

$$m\frac{d\mathbf{v}}{dt} = -f\mathbf{v} + \mathbf{F}(t), \tag{19}$$

where $\mathbf{v}$ denotes the velocity of the particle and m its mass. The influence of the surrounding medium is split up into two parts, the macroscopic dynamical friction $-f\mathbf{v}$ and a random force $\mathbf{F}(t)$. It is assumed that $\mathbf{F}(t)$ is independent of the position and the velocity of the moving particle, and that it varies extremely rapidly compared to the variation of $\mathbf{v}$.

It would take us too far afield to reproduce the required theory of the Langevin equation in this volume. The reader must be referred to references such as pp. 22–27 of Chandrasekhar's article in Wax (1954). But in the next two paragraphs we shall provide a very sketchy treatment in an attempt to convey the essence of the relationship one finds between macroscopic and microscopic processes.

If we multiply (19) scalarly with **x**, the displacement of the particle from the mean position of a group of particles, and consider the ensemble average over the group, we obtain

$$m\left(\frac{d}{dt}\langle \mathbf{x}\cdot\mathbf{v}\rangle - \langle \mathbf{v}^2\rangle\right) = -f\langle \mathbf{x}\cdot\mathbf{v}\rangle - \langle \mathbf{x}\cdot\mathbf{F}\rangle. \tag{20}$$

We now assume that $\langle \mathbf{x}\cdot\mathbf{F}\rangle = 0$ and that $\langle (d\mathbf{x}/dt)\cdot(d\mathbf{x}/dt)\rangle$ becomes stationary. We can then solve (20) for the quantity $\langle \mathbf{x}\cdot\mathbf{v}\rangle$. This quantity is proportional to the diffusion coefficient; for, from (16),

$$\langle \mathbf{x}\cdot\mathbf{v}\rangle = \frac{1}{2}\left\langle\frac{d(\mathbf{x}\cdot\mathbf{x})}{dt}\right\rangle = \frac{3}{2}\,D. \tag{21}$$

The solution for $\langle \mathbf{x}\cdot\mathbf{v}\rangle$ shows that eventually $\langle \mathbf{x}\cdot\mathbf{v}\rangle$ also becomes stationary and takes on the value

$$\langle \mathbf{x}\cdot\mathbf{v}\rangle = \frac{m\langle \mathbf{v}^2\rangle}{f}. \tag{22}$$

Unfortunately, we have no way of evaluating the right-hand side of (22) in this simple derivation. We may argue, however, that the random motion of the macroscopic particle will attain an energy which is equal to that of the molecular motion. Thus we postulate that

$$\tfrac{1}{2}m\langle \mathbf{v}^2\rangle = \tfrac{3}{2}k\Theta. \tag{23}$$

Combining (21), (22), and (23), we obtain (17).

EXERCISES

1. (a) Show that $w(m, N)$ is an even function of m and interpret this result.
(b) Verify (13).

2. Given $w(m, N)$ as in (4):
(a) Find $\langle p^3\rangle$ and $\langle m^3\rangle$.
(b) Find $\langle p^4\rangle$ and $\langle m^4\rangle$.

3. Consider a random walk where the probability ρ of taking a step to the right is not necessarily the same as the probability $\lambda = 1 - \rho$ of taking a step to the left.
(a) Show that now

$$w(m, N) = C_p^N \rho^m \lambda^{N-m}.$$

(b) Introduce a generating function and use it to find the mean and variance of the displacement m.

(c) Repeat Exercise 2 in this case.

4. (a) The **Bessel function of the first kind**, of order p, is defined as follows:

$$J_p(x) = \sum_{k=0}^{\infty} \frac{(-1)^k (x/2)^{2k+p}}{k!(k+p)!}. \tag{24}$$

Show (formally) that this series provides a solution to **Bessel's equation**,

$$x^2 \frac{d^2y}{dx^2} + x\frac{dy}{dx} + (x^2 - p^2)y = 0. \tag{25}$$

(b) If p is an integer n, show that

$$J_{-n}(x) = (-1)^n J_n(x). \tag{26}$$

(c) Demonstrate that

$$e^{xr/2}e^{-x/2r} = \sum_{j=0}^{\infty}\sum_{k=0}^{\infty} \frac{(-1)^k (x/2)^{j+k}}{j!\,k!} r^{j-k} = \sum_{n=-\infty}^{\infty} r^n J_n(x), \tag{27}$$

which provides the generating function for the Bessel functions of integer index n.

(d) In (27), replace r by $\exp(i\theta)$ and deduce the following integral representation:

$$J_n(x) = \frac{1}{2\pi}\int_{-\pi}^{\pi} e^{i(x \sin\theta - n\theta)}\, d\theta; \qquad n = 0, \pm 1, \pm 2, \ldots. \tag{28}$$

(e) By a change of variable, obtain the alternative formula

$$J_n(x) = \frac{e^{in(\pi/2)}}{2\pi}\int_0^{2\pi} e^{-i(x\cos\theta_1 + n\theta_1)}\, d\theta_1. \tag{29}$$

(We shall need the formula for $n = 0$ in Section 16.3.)

3.2 Asymptotic Series, Laplace's Method, Gamma Function, Stirling's Formula

We have seen that the probability $w(m, N)$ that a Brownian particle will arrive m steps to the right of its starting point after taking a total of N steps is given by

$$w(m, N) = \frac{N!}{p!(N-p)!\,2^N}, \qquad \text{where} \quad p = \tfrac{1}{2}(N + m). \tag{1}$$

We wish to approximate this result when N, p, and $N - p$ are large; we are thus led to the problem of approximating the factorial of a large number. This is a general problem that occurs in many other situations. Years ago, Stirling showed* that the natural logarithm of a factorial can be represented by the series

$$\ln n! \sim \frac{1}{2}\ln(2\pi) + \left(n + \frac{1}{2}\right)\ln n - n + \frac{1}{12n} - \frac{1}{360n^3} + \cdots. \tag{2a}$$

The dominant terms give **Stirling's formula** or **Stirling's approximation,**

$$n! \sim (2\pi n)^{1/2} n^n e^{-n}. \tag{2b}$$

By applying (2) to the factorials in (1) [Exercise 1(a)], we obtain the approximation

$$w(m, N) \sim \left(\frac{2}{\pi N}\right)^{1/2} \exp\left(\frac{-m^2}{2N}\right). \tag{3}$$

Table 3.1 shows that this approximation is an excellent one, even for such seemingly moderate values as $m = 4$, $n = 10$. Nevertheless, the series of which the first terms are given in (2) is known to be *divergent* for all values of z. In what rigorous sense, then, is (2) a valid mathematical expression?

TABLE 3.1. $w(m, N)$ for $N = 10$.

m	Exact formula (1)	Asymptotic formula (3)
0	0.24609	0.252
2	0.20508	0.207
4	0.11715	0.113
6	0.04374	0.042
8	0.00977	0.010
10	0.00098	0.002

To understand the nature of the Stirling approximation (2), and to handle many other problems, we must consider the concept of an asymptotic series. We begin our discussion with a relatively simple example, in which an approximating series is generated by parts integration. As there is an

* James Stirling (1692–1770) was an English mathematician of the period after Newton. Stirling actually derived a series that has smaller coefficients than those in (2), but the coefficients are more difficult to obtain. "The ingenuity needed to obtain the series with the mathematical resources then available, not even the general definition of $z!$ being known, is astonishing. The usual form is most accurately described as de Moivre's form of Stirling's series, since all the principles used in finding it were given by Stirling" in a letter to de Moivre (Jeffreys and Jeffreys, 1962, p. 467).

explicit expression for the remainder, it is not difficult to see the asymptotic nature of the approximation. Following a formal definition of asymptotic series, we turn to a heuristic description of Laplace's method for generating asymptotic series for certain important integrals. The method is illustrated on the gamma function integral, an extension of the factorial function. A demonstration of the method's validity is outlined. As a result of this, the Stirling formula is proved.

AN EXAMPLE: ASYMPTOTIC EXPANSION VIA PARTS INTEGRATION

Suppose we want to evaluate the integral

$$f(x) = \int_x^\infty t^{-1}e^{x-t}\,dt \tag{4}$$

for large positive values of x. The improper integral converges [Exercise 1(b)]. Using integration by parts, we obtain the following expression:

$$f(x) = \frac{1}{x} - \int_x^\infty t^{-2}e^{x-t}\,dt. \tag{5}$$

The magnitude of the integral in (5) is smaller than that in (4) by at least a factor of $1/x$, since $t \geq x$ in the interval of integration. Consequently, we can deduce from (5) that

$$\frac{1}{x} = f(x) + \int_x^\infty t^{-2}e^{x-t}\,dt \approx f(x).$$

We can improve the approximation $f(x) \approx x^{-1}$ by successive partial integrations. One finds [Exercise 2(a)] that

$$f(x) = S_n(x) + R_n(x), \tag{6a}$$

where

$$S_n(x) = \frac{1}{x} - \frac{1}{x^2} + \frac{2!}{x^3} - \cdots + \frac{(-1)^{n-1}(n-1)!}{x^n}, \tag{6b}$$

$$R_n(x) = f(x) - S_n(x) = (-1)^n n! \int_x^\infty t^{-(n+1)}e^{x-t}\,dt. \tag{6c}$$

The sequence $S_n(x)$ diverges as $n \to \infty$, for all values of x. Yet we feel that $S_n(x)$ must be a good approximation in some sense when x is large. We shall therefore examine the remainder $R_n(x)$ as $x \to \infty$ *for a fixed value of* n.

We know that

$$\left|\int_x^\infty t^{-(n+1)}e^{x-t}\,dt\right| \leq \int_x^\infty |t^{-(n+1)}|\,|e^{x-t}|\,dt.$$

Since $t \geq x$, $|t^{-(n+1)}| \leq x^{-(n+1)}$, so that we can write

$$|R_n(x)| < n!\,x^{-(n+1)} \int_x^\infty e^{x-t}\,dt = n!\,x^{-(n+1)}. \tag{7a}$$

Thus, if $x \to \infty$, $R_n(x) \to 0$ for any fixed value of n. In fact, as long as $x \geq 2n$, one can show [Exercise 2(b)] that

$$|R_n(x)| < 2^{-(n+1)}n^{-2}, \tag{7b}$$

which is small even for moderate values of n. When $n = 3$, for example,

$$R_n < \tfrac{1}{144} \approx 0.007 \qquad \text{provided that } x \geq 6.$$

Equation (7a) shows that the remainder in our series is less than the first term neglected. This is not always the case, however. The generalizable property of (7a) is that for fixed n, *the ratio of the remainder to the last term retained approaches zero as* $x \to \infty$. Let us formalize this statement in a definition.

DEFINITIONS IN THE THEORY OF ASYMPTOTIC EXPANSIONS

Consider

$$A_0 + \frac{A_1}{x} + \frac{A_2}{x^2} + \cdots + \frac{A_n}{x^n} + \frac{A_{n+1}}{x^{n+1}} + \cdots \equiv S_n(x) + \frac{A_{n+1}}{x^{n+1}} + \cdots. \tag{8}$$

The expression $S_n(x)$ is said to provide an **asymptotic expansion** of $f(x)$ (as $x \to \infty$) if

$$\lim_{x\to\infty} x^n[f(x) - S_n(x)] = 0 \qquad (n \text{ fixed}). \tag{9a}$$

In other notation, requirement (9a) can be written

$$f(x) - S_n(x) = o(x^n) \quad \text{as } x \to \infty \qquad (n \text{ fixed}). \tag{9b}$$

In case (9) holds we write

$$f(x) \sim \sum_{i=0}^{n} A_i x^{-i}, \qquad x \to \infty. \tag{10}$$

The symbol " $\sim$ " is read "is asymptotic to." If (10) holds for every n, we write

$$f(x) \sim \sum_{i=0}^{\infty} A_i x^{-i}, \qquad x \to \infty, \tag{11}$$

and we speak of the asymptotic power series development of f (as $x \to \infty$).

In practice, the series in (11) is usually divergent for any fixed large x. On the other hand, if this series does converge to f for x sufficiently large, i.e., if

$$\lim_{n\to\infty}\left|f(x)-\sum_{i=0}^{n}A_i x^{-i}\right|=0 \qquad \text{for } x \text{ large and fixed,} \tag{12}$$

then (11) is asymptotic as $x \to \infty$ (Exercise 3).

As we have seen, in

$$\int_x^\infty t^{-1}e^{x-t}\,dt \sim \sum_{n=1}^{\infty}(-1)^{n-1}(n-1)!\,x^{-n}$$

we have an example of a series that diverges for all x but is asymptotic as $x \to \infty$. The distinction between a convergent series and an asymptotic series must be kept clearly in mind. *To prove that a series expansion is asymptotic as $x \to \infty$, we examine the remainder $f(x) - S_n(x)$, after a fixed number n of terms, when x becomes very large. By contrast, a convergence proof requires examination of the remainder at a fixed value of the independent variable when n becomes very large.*

The definition of asymptotic series can be extended to include $x \to -\infty$. Moreover, one often wishes to consider $|z| \to \infty$, where z is complex. (In the latter case, there is usually a restriction on the range of values of arg z.) Asymptotic expansions as z approaches an arbitrary point z_0 may also arise, but it is often convenient to reduce to the case where the variable approaches ∞ by introducing $1/(z - z_0)$ as a new variable.

To give one more definition,

$$f(x) \sim g(x)\sum_{n=0}^{N}A_n x^{-n}, \qquad x \to \infty, \tag{13}$$

means

$$\frac{f(x)}{g(x)} \sim \sum_{n=0}^{N}A_n x^{-n}, \qquad x \to \infty. \tag{14}$$

Asymptotic expansions can often be obtained by judicious use of parts integration. (See Exercise 7.) We shall now describe a different approach, which is of wider applicability and which provides a relatively straightforward derivation of Stirling's series.

LAPLACE'S METHOD

We present a method, due to Laplace, for obtaining the asymptotic expansion of certain integrals containing a large parameter. Suitably generalized, this method is one of the most frequently used in applied mathematics.

In its standard form, Laplace's method is used for integrals which can be placed in the form

$$F(\lambda) = \int_\alpha^\beta g(t)e^{-\lambda f(t)}\,dt, \qquad \lambda > 0. \tag{15}$$

An approximation is sought in the case in which the positive constant λ is large. The idea used in the whole class of methods of which Laplace's method is a paradigm is this. Transform the integral if necessary until, for large λ, the dominant contribution to the result comes from a small portion of the path of integration. Suppose that portion is in the neighborhood of a point t_0. Then simplify the integrand in that neighborhood by using a Taylor expansion about t_0.

In carrying out explicit calculations on (15), we shall assume that $f(t)$ has an absolute minimum at an interior point t_0 of the interval $[\alpha, \beta]$, with $f'(t_0) = 0$, $f''(t_0) > 0$. Approximating $f(t)$ by the first two nonvanishing terms in its Taylor series, we see that $\exp[-\lambda f(t)]$ behaves like $Q(t) \exp[-\lambda f(t_0)]$, where

$$Q(t) \equiv \exp\left[\frac{-\lambda f''(t_0)(t-t_0)^2}{2}\right].$$

Figure 3.1 depicts graphs of $Q(t)$ when

$$\frac{\lambda f''(t_0)}{2} = 1, 10, \text{ and } 100.$$

For large λ, $Q(t)$ is seen to be appreciable only within a very small distance of t_0. Within a narrow band about t_0 it should be permissible to

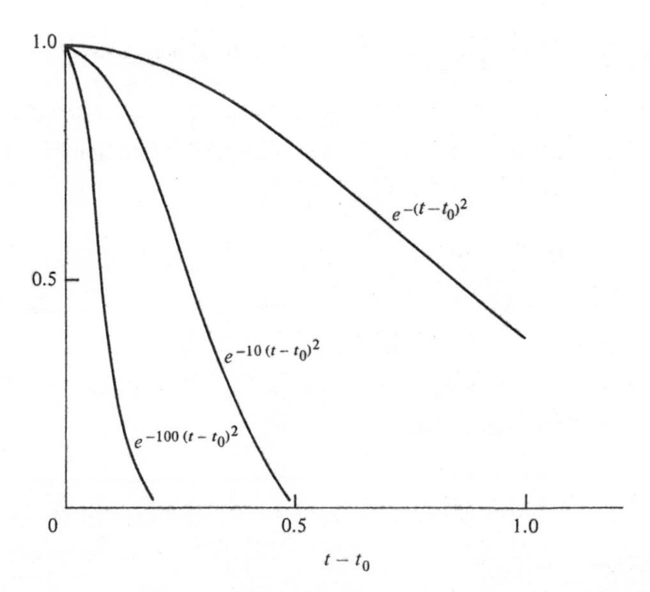

FIGURE 3.1. *Illustration of the fact that* $\exp[-\lambda(t-t_0)^2]$ *decays rapidly from its maximum value when* λ *is positive and large.*

approximate $g(t)$ of (15) by $g(t_0)$. We are thus led to the conjecture that

$$F(\lambda) \equiv \int_\alpha^\beta g(t)e^{-\lambda f(t)}\, dt$$
$$\sim g(t_0) \exp[-\lambda f(t_0)] \int_\alpha^\beta Q(t)\, dt.$$

But since only a narrow interval about t_0 makes a substantial contribution to the integral, we can obtain a further simplification at little cost in accuracy by replacing the limits of integration by $-\infty$ and ∞. With the change in variable

$$u = (t - t_0)\frac{f''(t_0)}{2}$$

our result becomes [Exercise 8(a)]

$$F(\lambda) \sim g(t_0) \exp[-\lambda f(t_0)] \left[\frac{2}{f''(t_0)}\right]^{1/2} \int_{-\infty}^{\infty} \exp(-\lambda u^2)\, du. \tag{16a}$$

Evaluating the integral by (34) below, we finally obtain*

$$F(\lambda) \sim g(t_0) \exp[-\lambda f(t_0)] \left[\frac{2\pi}{\lambda f''(t_0)}\right] \qquad \text{as } \lambda \to \infty. \tag{16b}$$

Rather than provide a proof for (16b), we pass at once to the development of Stirling's series. We shall perform manipulations that are similar in spirit to those just completed.

DEVELOPMENT OF THE ASYMPTOTIC STIRLING SERIES FOR THE GAMMA FUNCTION

Our first step, by no means an easy one, is to find an integral representation for $n!$. This is accomplished by introducing† the **gamma function** $\Gamma(s)$, where

$$\Gamma(s) = \int_0^\infty x^{s-1}e^{-x}\, dx. \tag{17}$$

It takes only a straightforward application of parts integration [Exercise 2(c)] to verify that when s is a positive integer n, then

$$\Gamma(n) = (n - 1)!. \tag{18}$$

But the integral in (17) is defined for more than positive integral values of s. An important special case is

$$\Gamma(\tfrac{1}{2}) = \sqrt{\pi}, \tag{19}$$

* Formula (16b) can be generalized so that it applies to complex integrals by choosing an integration path of "steepest descent" near "saddle points" that are complex analogues of the stationary points of f. In the present case, the path of integration is fixed, but it happens to be the path of steepest descent.

† Given our earlier remarks on the origin of the generating function, the reader will not be surprised to learn that it was Euler who obtained the integral representation (17).

a formula with a trick proof [Exercise 2(d)]. Indeed, (17) remains meaningful even if s assumes complex values, as long as $\mathrm{Re}(s) > 0$.

We are interested in the behavior of $\Gamma(s)$ for large positive s. To obtain an integrand containing an exponential with s as a factor, we write (17) as

$$\Gamma(s) \equiv \int_0^\infty x^{s-1}e^{-x}\,dx = \int_0^\infty e^{(s-1)\ln x}e^{-x}\,dx,$$

which shows that

$$\Gamma(s) = \int_0^\infty e^{-sF}\,dx, \qquad \text{where} \quad F(x,s) = -\ln x + s^{-1}\ln x + s^{-1}x.$$

Considered as a function of x, $F(x, s)$ is stationary at the point x_0, where

$$\frac{\partial F}{\partial x} = 0, \qquad \text{i.e.,} \quad x_0 = s - 1. \tag{20}$$

In contrast to the fixed location of t_0 in our treatment of (15), the point x_0 "moves" as $s \to \infty$. To fix its location, we make $x = s - 1$ correspond to the fixed point $t = 1$ by the change of variable

$$t = \frac{x}{s-1}. \tag{21}$$

With this (17) becomes

$$\Gamma(s) = (s-1)^s J, \tag{22}$$

where

$$J = \int_0^\infty \exp\left[-(s-1)f(t)\right]dt, \qquad f(t) \equiv t - \ln t. \tag{23}$$

The minimum of $f(t)$ occurs for

$$f'(t_0) = 0 \qquad \text{or} \qquad t_0 = 1. \tag{24}$$

At the minimum, f has the value $f(1) = 1$. The integrand of (23), therefore, has a maximum at $t = 1$ and drops off extremely rapidly as t goes away from this point.

To bring out the dominant features of the integrand, we write

$$\lambda \equiv s - 1, \quad f(t) = f(t_0) + [f(t) - f(t_0)], \tag{25}$$

with which our integral becomes

$$J(\lambda) = e^{-\lambda f(t_0)} \int_0^\infty e^{-\lambda[f(t)-f(t_0)]}\,dt. \tag{26}$$

(We have assumed that $s - 1 \equiv \lambda$ is positive, but since we are interested in large s, this is not an additional restriction.) To emphasize the key fact that

$[f(t) - f(t_0)]$ is a nonnegative function with a single maximum, we introduce a new variable w that satisfies

$$w^2 = f(t) - f(t_0). \tag{27}$$

We specify w completely by

$$\begin{aligned} w &= -\sqrt{f(t) - f(t_0)}, \qquad t \le t_0; \\ w &= \sqrt{f(t) - f(t_0)}, \qquad t \ge t_0. \end{aligned} \tag{28}$$

Using the fact that w increases continuously and monotonically from $-\infty$ to ∞ as t increases from 0 to ∞ [Exercise 4(a)], we rewrite (26) as

$$J(\lambda) = e^{-\lambda f(t_0)} \int_{-\infty}^{\infty} e^{-\lambda w^2} \frac{dt}{dw}\, dw. \tag{29}$$

The next task is to expand dt/dw as a power series in w. Now

$$w^2 \equiv f(t) - f(t_0) = t - 1 - \ln t, \tag{30}$$

so

$$w^2 = \tfrac{1}{2}(t-1)^2 - \tfrac{1}{3}(t-1)^3 + \cdots. \tag{31}$$

Consequently [Exercise 4(b)], we may write

$$t - 1 = \sqrt{2}\, w(1 + a_1 w + a_2 w^2 + \cdots). \tag{32}$$

The coefficients $a_1, a_2, \ldots$ may be determined by substituting (32) into (31) and comparing terms [Exercise 4(c)].

From (32) we can easily calculate a power series for dt/dw. *Assuming that term-by-term integration is permitted,* we find that the evaluation of (29) reduces to computation of integrals of the form

$$I_m(\lambda) = \int_{-\infty}^{\infty} e^{-\lambda w^2} w^m\, dw; \qquad m = 0, 1, 2, \ldots. \tag{33}$$

An elementary substitution shows that I_0 is a multiple of $\Gamma(\frac{1}{2})$. Therefore, using the expression for $\Gamma(\frac{1}{2})$ in (19), we find

$$I_0(\lambda) = \sqrt{\pi/\lambda}. \tag{34}$$

Similar substitutions reduce determination of I_m to computation of Γ integrals. Alternatively, one can formally differentiate (34) with respect to λ (Exercise 5). The result is

$$\begin{aligned} I_m(\lambda) &= 0, \qquad m \text{ odd}; \\ I_m(\lambda) &= \frac{1 \cdot 3 \cdot 5 \cdots m-1}{2^{m/2} \lambda^{m/2}} \left(\frac{\pi}{\lambda}\right)^{1/2}, \qquad m \text{ even}. \end{aligned} \tag{35}$$

Putting all our results together (Exercise 5), we obtain

$$J = \left(\frac{2\pi}{\lambda}\right)^{1/2} e^{-\lambda}[1 + O(\lambda^{-1})], \tag{36}$$

so that

$$\ln \Gamma(s) = \left(s - \frac{1}{2}\right) \ln (s-1) - (s-1) + \frac{1}{2} \ln 2\pi + O\left(\frac{1}{s-1}\right). \tag{37}$$

In particular, when $s = n + 1$, n an integer, we recover the leading terms of Stirling's series (2a).

JUSTIFICATION OF TERM-BY-TERM INTEGRATION

Series like (32), in general, have a finite radius of convergence, and thus our termwise integration is not justifiable if one insists on working with convergent series. Yet, for large λ, the only important contribution to the integral is near $w = 0$, and here the series is expected to converge. We have a strong intuition that all is well. Indeed, the result (35) can be justified (and even extended, as in Exercise 6) in the asymptotic sense. To do this, instead of using the series (32) we make a slight change of viewpoint and write $t - 1$ as a finite sum plus a remainder:

$$t - 1 \equiv \sqrt{2w}\,[1 + \cdots + a_{2n} w^{2n} + a_{2n+1} w^{2n+1} + \mathscr{R}_{2n+2}(w) w^{2n+2}]. \tag{38}$$

Coefficients a_i with odd subscripts provide no contribution to the integral in (29), since the corresponding integrand is an odd function. Thus the last nonvanishing contribution from the polynomial in (38) is proportional to $I_{2n}(\lambda)$ and this is $O(\lambda^{-n-1/2})$, by (35). The remainder makes a contribution proportional to

$$\int_{-\infty}^{\infty} \mathscr{R}_{2n+2}(w) w^{2n+2} e^{-\lambda w^2}\, dw = \lambda^{-n-3/2} \int_{-\infty}^{\infty} \mathscr{R}_{2n+2}\left(\frac{t}{\sqrt{\lambda}}\right) t^{2n+2} e^{-t^2}\, dt. \tag{39}$$

For (39) to be $O(\lambda^{-n-3/2})$, as desired, all we need is that the integral on the right side converges. This will certainly be the case if $\mathscr{R}_{2n+2}(t/\sqrt{\lambda})$ is bounded for large t, or even if it becomes infinite as long as this is not in any way comparable with the growth of $\exp(t^2)$.

The literature contains a number of theorems that give sufficient conditions for the validity of various manipulations which yield asymptotic expansions. As usual, however, we stress here the spirit of the approach, since in any case published theorems must often be extended before they apply to a case met in practice. The extension may be very difficult. If so, many applied mathematicians may not make the effort to carry it out.

This is quite typical of many manipulations in applied mathematics. The formal steps can be justified, but perhaps the usual rule is to accept the plausible, unless there is reasonable doubt to the contrary, i.e., unless a serious issue can be raised. If you wish, you may imagine that the following qualifying statement is often implied in work in applied mathematics: "The result obtained is plausible and tentative; but we have as yet no reason to doubt its validity." In many cases of extreme complexity, the mathematical justification, even of a general process, may take years to accomplish. It is obvious that one cannot wait for it before going ahead. The attitudes and practices that we adopt are consequently fundamentally different from those adopted by pure mathematicians.

EXERCISES

1. (a) Verify (3).
(b) Show that the improper integral in (4) is convergent.

2. (a) Verify (6b) and (6c).
(b) Verify (7b).
(c) Verify (18).
(d) Show that

$$I = \int_{-\infty}^{\infty} e^{-x^2}\,dx = \int_{-\infty}^{\infty} e^{-y^2}\,dy = \sqrt{\pi}$$

by considering I^2 and introducing polar coordinates in the (x, y) plane. Thereby verify (19).

3. Show that if (8) converges for x sufficiently large, then (8) is an asymptotic series as $x \to \infty$.

4. (a) Verify (29) and (31).
(b) Provide a formal proof that justifies the inversion necessary to obtain (32).
(c) Determine the coefficients a_1 and a_2 in (32).

5. (a) Verify (34) and (36).
(b) Derive the formulas in (35) by reducing the integrals to gamma functions.
(c) Derive the formulas in (35) by formal differentiation of (34).
(d) Justify the formal procedure of part (c).

6. Extend the text's derivation of (37), and obtain more terms of Stirling's series (2a).

7. Use integration by parts to obtain asymptotic expansions as $x \to \infty$ of the following integrals:

(a) The complementary error function $2\pi^{-1/2} \int_x^{\infty} e^{-t^2}\,dt$.

(b) The Fresnel integrals

$$C(x) = \int_0^x \cos\left(\frac{\pi}{2}t^2\right) dt, \qquad S(x) = \int_0^x \sin\left(\frac{\pi}{2}t^2\right) dt.$$

8. (a) Verify (16a) and (16b).
(b) Provide a formal proof that (16b) is valid if suitable restrictions are made on f and g.

3.3 A Difference Equation and Its Limit

One of the essential elements that underlies the success of theoretical analyses is the practice of approaching the same problem in a variety of ways. The strength of the various methods will then supplement each other and enable one to reach a deeper understanding. So far we have attempted to follow the motion of a particle as it started out from the origin. We have obtained exact and approximate expressions for $w(m, N)$, the probability that a particle will arrive m steps to the right after taking a total of N steps. If we now turn our attention to the particle's immediate behavior after it has reached the point i steps to the right of the origin, we shall be enlightened in an entirely different manner. We shall find that random walk can be described by a certain difference equation. This equation can be approximated by a differential equation, which turns out to be the diffusion equation. Various solutions of the diffusion equation will be shown to provide solutions for probability problems of considerable interest.

A DIFFERENCE EQUATION FOR THE PROBABILITY FUNCTION

We assert that the following **difference equation**,

$$\bar{w}(x, t + \Delta t) = \tfrac{1}{2}\bar{w}(x - \Delta x, t) + \tfrac{1}{2}\bar{w}(x + \Delta x, t), \tag{1}$$

governs the change of

$$\bar{w}(x, t) \equiv w\left(\frac{x}{\Delta x}, \frac{t}{\Delta t}\right) \tag{2}$$

in finite steps of time Δt. In (1) and (2) we have defined

$$x = m(\Delta x), \qquad t = N(\Delta t). \tag{3}$$

Thus $\bar{w}(x, t)$ is the probability that the particle which starts at the origin at time $t = 0$ is located at point x at time t. Our difference equation (1) merely states that to reach the point x at the time $t + \Delta t$, at time t the particle must either reach the point $x - \Delta x$ [probability $w(x - \Delta x, t)$] and then take a step to the right [probability $\frac{1}{2}$], or it must reach the point $x + \Delta x$ and then take a step to the left.

Equation (1) governs the evolution of $\bar{w}$. To complete the formulation of the problem in these terms, we must add the condition that the particle begins at the origin:

$$\bar{w}(0, 0) = 1; \quad \bar{w}(x, 0) = 0, \qquad x \neq 0. \tag{4}$$

We expect to be able to recover the exact solution (1.4) for w by solving the difference equation (1) subject to the initial conditions (4). Readers may not be able to provide the solution from first principles, but they can easily verify that (1.4) satisfies (1) and (4) (Exercise 1).

APPROXIMATION OF THE DIFFERENCE EQUATION BY A DIFFERENTIAL EQUATION

In our previous discussion we obtained the approximate solution (2.3) by applying Stirling's formula to the exact solution (1.4). A very useful result emerged, but surely there must be a more direct way to obtain the approximation. It should require less work to obtain a less detailed result, and indeed the present formulation provides a direct and natural approach to the approximation (2.3).

We are interested in the solution only after the particle has taken a large number of steps. Thus we should consider the limit $N \to \infty$. From (3), however, if we leave Δt fixed in this limit, then $t \to \infty$. We wish to examine the solution at an arbitrary time t, so we are led to approximate (1) in the limit

$$N \to \infty, \quad t \text{ fixed}, \qquad \text{so } \Delta t \to 0. \tag{5}$$

Recall that $m\,\Delta x$ is the coordinate of the particle after N steps, assuming that the particle starts from the origin. If the particle takes a very large total number N of steps, we expect that there is an excellent possibility that the (positive or negative) integer m may be very large in magnitude. We certainly do not wish to exclude such a possibility; therefore, we also consider the limit $m \to \infty$. Because we wish to be able to describe the situation at any fixed point x, from (3) we are led to approximate (1) when

$$m \to \infty, \quad x \text{ fixed}, \qquad \text{so } \Delta x \to 0. \tag{6}$$

The limits (5) and (6) require us to approximate (1) when Δx and Δt are small. To accomplish this, we naturally think of expanding the functions in (1) by Taylor's formula. [It gives us pause that $\bar{w}$ is so far defined only at a set of discrete points in the (x, t) plane, but we can circumvent this problem by imagining a polynomial that smoothly passes through all these values of $\bar{w}$.] We thus approximate (1) by (Exercise 2).

$$\bar{w}_t(x, t)\,\Delta t + O(\Delta t)^2 = \tfrac{1}{2}\bar{w}_{xx}(x, t)(\Delta x)^2 + O(\Delta x)^3. \tag{7}$$

We note that if we divide (7) by Δt and assume that

$$\lim_{\substack{\Delta x \to 0 \\ \Delta t \to 0}} \frac{(\Delta x)^2}{2\,\Delta t} = D, \qquad D \neq 0, \tag{8}$$

we obtain the plausible-looking result $\bar{w}_t = D\bar{w}_{xx}$. This is encouraging, but we must proceed a little more cautiously.

First, we observe that it is no longer appropriate to keep the quantity $\bar{w}$ as the dependent variable. This is because the probability of finding the particle at a particular point is expected to approach zero as the number of points becomes infinite. Since only every other point on the x axis can be occupied at any instant, we introduce the quantity

$$u = \frac{\overline{w}}{2\,\Delta x}. \tag{9}$$

In terms of u, the probability at time t of finding a particle on or between $a = i\,\Delta x$ and $b = k\,\Delta x$ can be expressed by

$$\sum_{m=i}^{k}{}'' u(m\,\Delta x, t) \cdot 2\,\Delta x. \tag{10}$$

[Remember that the sum takes in every other value of m from i to k. We take both i and k to be even (odd) if N is even (odd).]

We must also take note of the fact that assumption (8) seems at first to have little intuitive appeal. But if this assumption is not made, then our limiting process does not seem to yield a sensible result. Let us therefore accept (8) for the moment. If further examination of our limiting results confirms our impression that we are making progress, we can return to the justification of (8). On the other hand, if our formalism proves unedifying, there is no point in convincing ourselves of its legitimacy.

Using (9) and (8), then, the leading approximation to (7) for small Δx and Δt is

$$\frac{\partial u}{\partial t} = D\frac{\partial^2 u}{\partial x^2}. \tag{11}$$

Taking the limit of (10) as $\Delta x \to 0$, we find that $u(x, t)$ is related to the probability $U(a, b; t)$ of finding the particle in the interval (a, b) at time t by

$$U(a, b; t) = \int_a^b u(x, t)\,dx. \tag{12}$$

More informally, the **probability density function** u is such that $u(x, t)\,dx$ is the probability that at time t the particle is located between x and $x + dx$.*

SOLUTION OF THE DIFFERENTIAL EQUATION FOR THE PROBABILITY DISTRIBUTION FUNCTION

What if a particle starts at the origin and takes a large number of steps at random? We would expect that the distribution of probabilities would satisfy (11). Certainly,

$$\int_{-\infty}^{\infty} u(x, t)\,dx = 1, \tag{13}$$

* Note the analogy with the mass density function $\rho(x, t)$, which can be regarded as providing the mass of material located at time t between x and $x + dx$.

for the particle must be somewhere along the x axis. Moreover, since the particle started at the origin we must require that*

$$\lim_{t \downarrow 0} u(x, t) = 0, \qquad x \neq 0. \tag{14}$$

If we could solve the differential equation (11) subject to the conditions (13) and (14), we would expect to achieve our goal of proceeding directly to an approximation for large N, without first obtaining the exact solution.

Those reasonably familiar with partial differential equations can find the desired solution by looking for a function u of self-similar form. But to see whether our program is sensible, we can work backward. In present terms, the approximate solution (1.4) is [Exercise 3(a)]

$$u \approx u_0(x, t), \qquad u_0(x, t) \equiv (4\pi Dt)^{-1/2} \exp\left(-\frac{x^2}{4Dt}\right). \tag{15}$$

And it is not difficult to verify that u_0 satisfies the differential equation (11), the normalization condition (13), and the initial condition (14) [Exercise 3(b)]. It can also be shown that the solution is unique.

We have uncovered a method for *directly* obtaining an approximate solution to random walk problems: set up a difference equation, perform appropriate limiting operations, and then solve the resulting problem in partial differential equations.

Since our approach seems successful, we shall want to examine it more carefully and also to try it out on some new problems. This will be done in the remainder of this chapter. For the moment, let us look a little more closely at the solution u_0. We note that the area under the curve $u = u_0(x, t)$ is always unity, but that u_0 becomes more sharply peaked as $t \downarrow 0$. The maximum value of u_0 (at $x = 0$) increases as $t \downarrow 0$ like $t^{-1/2}$, but the "width" decreases like $t^{1/2}$.† Thus, as $t \downarrow 0$, u_0 approaches a limiting "delta function" that has unit area even though it is zero at every point except the origin. As t increases, the width of u_0 increases from its initial zero value like $t^{1/2}$, a fact already indicated by Equation (1.14a).

FURTHER EXAMINATION OF THE LIMITING PROCESS

We have seen that the approximation (15) can be obtained in two ways, by applying Stirling's formula to the exact solution and by solving the limiting differential equation (11) under appropriate conditions. But for accurate approximation via Stirling's formula, we had to require not only that the total number of steps N was large but also that p and $N - p$ were large. Recall from Section 3.1 that the latter two numbers are the total number of steps taken to the right and left, respectively. Thus if p or $N - p$ is small,

* The downward arrow in (14) means that t approaches zero through positive values.

† The width may be defined by setting $x^2 = 4Dt$, for instance, to obtain the position where the exponential factor in (15) is $1/e$ of its maximum.

most of the steps are either taken to the left or to the right, and the displacement m is comparable either to $-N$ or to N. As a consequence, we may interpret the above-mentioned requirement on the use of Stirling's formula as excluding those rare cases where the displacement is comparable to the total distance traveled. This requirement was not mentioned in our derivation of the limiting differential equation. Why not?

To see the answer, let us write out in full the limiting process we used. This was

$$\Delta x \to 0, \quad \Delta t \to 0, \quad m \to \infty, \quad N \to \infty, \quad m(\Delta x) \to x, \quad N(\Delta t) \to t,$$
$$\frac{(\Delta x)^2}{2(\Delta t)} \to D. \tag{16}$$

With this

$$\frac{\Delta x}{\Delta t} = \frac{2D}{\Delta x} \to \infty, \tag{17}$$

so that the speed with which the tiny steps are taken is infinite in the limit. Also, and more importantly,

$$\frac{m}{N} = \frac{m}{N}\frac{\Delta x}{\Delta t}\frac{\Delta t}{\Delta x} = \frac{x}{t}\frac{\Delta t}{\Delta x} \to 0. \tag{18}$$

Thus, our limiting process does, in fact, implicitly use the assumption that the displacement from the origin, $m\,\Delta x \equiv x$, is small compared to the total distance traveled, $N\,\Delta x$.

The previous discussion provides some insight into the reason why our limit must be such that $(\Delta x)^2/\Delta t$ remains finite. From (18) the requirement that m be small compared to N means that $\Delta x/\Delta t$ must approach infinity. This makes our requirement of finite $(\Delta x)^2/\Delta t$ a little less mysterious. But a better reason for this requirement is that the variance in displacement at time t is approximately $t(\Delta x)^2/\Delta t$ [Exercise 3(c)], and the variance must be fixed in a sensible limit. For if the root-mean-squared displacement is nearly zero or nearly infinite, then nearly nothing will be seen at a fixed position x.

REFLECTING AND ABSORBING BARRIERS

To give an indication of the power afforded by the limiting differential equation, let us consider a random walk problem in which a particle starts from the origin, as before, but in the presence of a perfectly **reflecting barrier** at $x = L + \frac{1}{2}\,\Delta x$, $L > 0$. The presence of this barrier means that a particle that reaches $x = L$ at time t and then moves to the right will again be found at $x = L$ at time $t + \Delta t$. A particle also can attain the position $x = L$ by moving to the right from $x = L - \Delta x$. Consequently,

$$\overline{w}(L, t + \Delta t) = \tfrac{1}{2}\overline{w}(L - \Delta x, t) + \tfrac{1}{2}\overline{w}(L, t). \tag{19}$$

Upon introducing the probability distribution function $u = \bar{w}/(2\,\Delta x)$ via (9) and letting $\Delta x \to 0$, this condition becomes

$$u_x(L, t) = 0. \tag{20}$$

All other conditions remain the same, except that (13) must be replaced by

$$\int_{-\infty}^{L} u(x, t)\, dx = 1, \tag{13a}$$

since all particles remain to the left of $x = L$.

Finding a solution $u = u_R(x, t)$ to the differential equation (11) that satisfies (13a), (14), and (20) is—as we shall see in Section 3.4—not too much more difficult than finding a solution to (11) when the last condition is omitted. The result is that the distribution function in the presence of a barrier at $x = L$ is given by

$$u = u_R(x, t); \qquad u_R(x, t) \equiv u_0(x, t) + u_0(x - 2L, t). \tag{21}$$

It is possible to find an exact solution to the random walk problem with a reflecting barrier and then to obtain (21) by using Stirling's approximation. But, given some familiarity with differential equations at least, a smaller amount of ingenuity is perhaps necessary to obtain (21) from an appropriate solution to the differential equation (11). For both the exact and the approximate problems, a reflection idea is generally used to obtain a solution.

Let us now consider the situation when there is a **perfect absorber** just to the right of $x = L$. By this we mean that the particle is removed if it moves to the right from $x = L$. Thus the only way that a particle can arrive at $x = L$ is first to arrive at $x = L - \Delta x$ and then move to the right. In this situation, then, the governing difference equation holds for $x \le L - \Delta x$, and there is the further condition that

$$\bar{w}(L, t + \Delta t) = \tfrac{1}{2}\bar{w}(L - \Delta x, t).$$

Upon introducing $u \equiv \bar{w}/(2\,\Delta x)$ and neglecting higher order terms in Δx, we obtain the boundary condition

$$u(L, t) = 0. \tag{22}$$

The solution of (11), (13), (14), and (22) is

$$u = u_A(x, t), \qquad u_A(x, t) = u_0(x, t) - u_0(x - 2L, t). \tag{23}$$

Equation (23) gives the approximate probability that a particle, having started from the origin at time zero, is located between x and $x + dx$ at time t. Also of interest is the probability $F(L, t)\, dt$ that a particle is absorbed at $x = L$ in the time interval $(t, t + \Delta t)$, given that it started at the origin when $t = 0$. Another interpretation of F is that it represents the probability that a particle makes its **first passage** to L in the time interval $(t, t + dt)$. As we shall demonstrate in Section 3.4, the required formula is remarkably simple. It is

$$F(L, t) = -D\left(\frac{\partial u_A}{\partial x}\right)_{x=L}. \tag{24}$$

Thus the first passage time is proportional to the gradient in the probability distribution function for random walk in the presence of an absorbing barrier.

COAGULATION: AN APPLICATION OF FIRST PASSAGE THEORY

We shall indicate one of the numerous applications of first passage time calculations by considering the coagulation exhibited by colloidal particles when an electrolyte is added to the solvent. [Our discussion will be very sketchy. For more details see pp. 60ff. of Chandrasekhar's article in Wax (1954).] The theory, due to Smulchowski, begins by assuming that the electrolyte causes a sphere of influence of radius R to appear around each particle, and that particles stick when they approach within a distance R. Suppose that the sphere of influence of a single particle is centered on the origin and that a sea of other particles, of concentration v, moves randomly around it. Regarding the sphere as an absorbing barrier and the random sea as associated with a diffusivity D_1, the rate at which particles arrive at $\rho = R$ (ρ the radial polar coordinate) is

$$4\pi D_1\left(\rho^2 \frac{\partial w}{\partial \rho}\right)_{\rho=R}, \tag{25}$$

where we have used the formula for the gradient in spherical coordinates. In (25), w is the solution of

$$\begin{aligned} &w_t = D_1\nabla^2 w, && \rho > R. \\ &w \equiv v \quad \text{at } t = 0, && \rho > R; \\ &w = 0 \quad \text{at } \rho = R, && t > 0. \end{aligned} \tag{26}$$

If the particle, assumed stationary in the first instance, is itself undergoing Brownian motion with diffusivity D_2, it can be shown that D_1 in (25) and (26) should be replaced by $D_1 + D_2$. Knowing the coalescence rate, one can then write down a set of kinetic equations that describe the rate of formation of k-fold particles by coalescence of i- and j-fold $(i + j = k)$, and the disappearance of k-fold particles due to their coalescence with other particles. Some simplifying assumptions allow solution of the equations, and the results are in reasonable accord with experiment.

EXERCISES

1. Verify that (1) and (4) are satisfied by the function $w(m, N)$ defined in (1.4).
2. Verify (7).
3. (a) Verify that (1.4) implies (15).
   (b) Show that the function u_0 defined in (15) satisfies (11), (13), and (14).
   (c) Show that (1.15) can be interpreted as showing that the mean squared displacement in time t is $t(\Delta x)^2/(\Delta t)$.
   (d) If, by chance, you are already familiar with similarity methods [see Section 3.4 of II], use them to derive (15).
4. Using (15), evaluate the first four moments for diffusion from a point source. Compare them with those obtained in (1.14), (1.15), and Exercise 1.2.
5. (a) Consider a particle executing an unbiased random walk on a cubic lattice, between points a distance Δx apart. Let Δt be the duration of a step. By generalizing the text's discussion of the one-dimensional problem, show that

$$\frac{\partial u}{\partial t} = D\left(\frac{\partial^2 u}{\partial x^2} + \frac{\partial^2 u}{\partial y^2} + \frac{\partial^2 u}{\partial z^2}\right), \tag{27}$$

   where

$$D = \lim_{\substack{\Delta x \to 0 \\ \Delta t \to 0}} \frac{(\Delta x)^2}{6(\Delta t)} \tag{28}$$

   and u is suitably defined to represent a probability density.
   (b) Comment on the difference between (28) and the corresponding one-dimensional formula (8).
   (c) What do you think that the two-dimensional version of (28) and (8) is? Test your conjecture.
6. Consider one-dimensional random walk in which the particle moves a step to the left, a step to the right, or stays where it is, with equal probability. Show that the limiting probability distribution again satisfies the diffusion equation. Why is the change in definition of the diffusivity D plausible?
7. Consider the random walk of a particle along the x axis. Suppose that the particle moves a distance Δ to the left or right with probability q or $1 - q$, respectively. One step takes τ time units.
   (a) Write a difference equation for $w(m, N)$, the probability that the particle is a distance $m\Delta$ from the origin at time $N\tau$ (m and N are integers).
   ‡(b) Find the limiting differential equation as $\Delta \to 0$, $\tau \to 0$, $\Delta^2/2\tau = D$, $m\Delta \to x$, $N\tau \to t$. Show that q must approach $\frac{1}{2}$ in a certain way if there is to be a limit of the type sought.

8. Consider random walk along the x axis in a field of constant force. In time τ a particle moves a distance Δ to the right or left with probability $\frac{1}{2} - \beta\Delta$ or $\frac{1}{2} + \beta\Delta$, where β is a constant. Write a difference equation for the probability that the particle is a distance $m\Delta$ from the origin at time $N\tau$, m and N integers. Find the limiting differential equation as $\Delta \to 0$, $\tau \to 0$, $\Delta^2/2\tau = D$, $m\Delta \to x$, $N\tau \to t$.

9. An *elastically bound particle* moves to the right or left, if it is at $K\Delta$, $-R \le K \le R$ (R a fixed integer), with probability $\frac{1}{2}(1 - KR^{-1})$ or $\frac{1}{2}(1 + KR^{-1})$. The duration of each step is τ. Convince yourself that "elastically bound" is a sensible description. If $\Delta \to 0$, $\tau \to 0$, $R \to \infty$, $\Delta^2/2\tau = D$, $(R\tau)^{-1} \to \gamma$, etc., derive the limiting equation

$$\frac{\partial u}{\partial t} = \gamma \frac{\partial (xu)}{\partial x} + D\left(\frac{\partial^2 u}{\partial x^2}\right), \tag{29}$$

for an appropriate probability density u.

10. It is not easy to derive honestly, but it *is* straightforward to verify, that

$$u(x, t) = (1 - e^{-2\gamma t})^{-1/2} \exp\left[\frac{-\gamma x^2}{2D[1 - \exp(-2\gamma t)]}\right] \tag{30}$$

satisfies (29) and the auxiliary conditions $u \ge 0$; $\int_{-\infty}^{\infty} u(x, t)\,dx = 1$; $\lim_{t\to 0} u(x, t) = 0$, $x \ne 0$.

(a) Verify this.

(b) Compare this result with that of the *free particle*. In particular, show that (30) approaches the free particle result as $\gamma \to 0$.

(c) Comment on the behavior of the two results as $t \to \infty$ for x fixed.

11. (a) Suppose that a particle moves in an unbiased random fashion in the plane, taking a step of length Δ every τ seconds. Justify the equation

$$\psi(x, y, t + \tau) = \frac{1}{2\pi}\int_0^{2\pi} \psi(x + \Delta\cos\theta, y + \Delta\sin\theta, t)\,d\theta. \tag{31}$$

In doing so, define ψ. Start by explaining why a particle should step in a direction between $\theta = \theta_0$ and $\theta = \theta_0 + d\theta$ with probability $d\theta/2\pi$.

(b) Introduce a probability density function. By approximating (31), show that this function satisfies the diffusion equation.

3.4 Further Considerations Pertinent to the Relationship Between Probability and Partial Differential Equations

It is quite remarkable that the process of Brownian motion is governed in the limit by a partial differential equation of a fairly simple form. It is even more remarkable that this equation is common to many other physical processes such as heat conduction or diffusion of one gas through another, for these are quite different physical processes, at least superficially. The fact that many natural processes can be described by similar or even identical equations is essential to the very existence of applied mathematics as a scientific discipline. For, if there were no such natural unifying tendency, the mathematical study of various phenomena would be hopelessly fragmented.

In the present section we continue to study the relations between random walk and partial differential equations. We begin by reviewing the derivation of the diffusion equation from a macroscopic or continuum viewpoint. We stress that different microscopic models can give rise to this same macroscopic equation. The blurring of inessential microscopic detail is one reason why the theory of a number of rather different phenomena is subsumed in a study of a few differential equations.

We shall examine various solution techniques for the diffusion equation. The interplay between probability and differential equation theory is illustrated in the process. For example, a probability problem gives rise to the source solution, which is the key to solving the general initial value problem for the diffusion equation. On the other hand, certain special solutions for the diffusion equation provide answers to probability questions.

We close with an examination of a remarkable set of relationships between various probability problems on the one hand, and Laplace's equation and the inhomogeneous Laplace equation (Poisson's equation) on the other.

MORE ON THE DIFFUSION EQUATION AND ITS CONNECTION WITH RANDOM WALK

We shall examine the limiting partial differential equation

$$\frac{\partial u}{\partial t} = D\left(\frac{\partial^2 u}{\partial x^2}\right) \tag{1}$$

more carefully. We have already encountered this equation in Chapter 1. Let us review the main points of its derivation from a macroscopic perspective.

Let $u(x, t)$ be the amount per unit length of some quantity. Let $J(x, t)$ denote the *flux* of this quantity, the net rate at which the quantity is passing from the left of x to the right of x, at time t.* If the quantity is conserved, the rate at which it accumulates in an interval must equal the rate at which it passes *into* the interval through its boundaries. Thus, for arbitrary Δx,

$$\frac{d}{dt}\int_x^{x+\Delta x} u(x, t)\,dx = J(x, t) - J(x + \Delta x, t).$$

In the limit as $\Delta x \to 0$, this yields

$$\frac{\partial u}{\partial t} = -\frac{\partial J}{\partial x}. \tag{2}$$

* It is often useful to regard u as the amount per unit *volume* of a quantity whose distribution only depends on a single coordinate x. Then J is a *flux density*, the flux per unit area placed perpendiculary to the x coordinate line.

We suppose that there is a tendency for relatively high concentrations of u to disperse, i.e., there is a flow from high concentrations to low concentrations. The simplest reasonable hypothesis one can think of to describe a process in which flow is "caused" by differences in u is to assume that the magnitude of J is proportional to $\partial u/\partial x$. Denoting the proportionality constant by $-D$, we therefore write

$$J = -D\frac{\partial u}{\partial x}. \tag{3}$$

With D positive, (3) correctly shows a flow of u to the left if u increases with x ($\partial u/\partial x > 0$) and a flow to the right if u decreases with x.

Combining (2) and (3), we obtain (1) (assuming that D is a constant). This *diffusion equation* is the one we obtained as our limiting description of random motion. As mentioned in the first paragraph of this section, (1) is known to provide an excellent description of many physical phenomena. For example, if u denotes the concentration of solute particles immersed in solvent [in which case (3) is called **Fick's Law**], then (1) describes the diffusion of solute. If u denotes temperature [in which case (3) is called the **Newton–Fourier law of cooling**], then (1) describes the diffusion of heat. And in Chapter 1 we have mentioned the application of (1) to the random motion of motile microorganisms.

When dust particles move in air they are subjected, as a result of collisions with surrounding air molecules, to a vast number of tiny nudges. Paint particles in water receive similar nudges from water molecules. When we speak of the diffusion of dust particles or paint, we have in mind a spreading over times that are very long compared with the typical time interval between nudges and a spreading over distances that are very long compared with the typical amount that a particle moves between nudges. Thus, in retrospect, we can agree with the reasonableness of our finding that the diffusion equation results from a random walk in which step size Δx and step interval Δt are regarded as small.

Heat diffusion is harder to perceive clearly. Molecular collisions should certainly cause some sort of random change in the local kinetic energy, but it is not obvious that heat "walks" in the same way that dust particles do. It is helpful in this connection to consider Exercise 3.6, where the reader is asked to show that the diffusion equation (1) results from a random walk wherein with equal probability the particle moves to the left, moves to the right, or stays where it is. This provides one illustration of a general feature: *the same* limiting or *macroscopic equation is obtained for a number of different microscopic models*. There is a *blurring of microscopic detail*. Thus a theoretical unity is obtained by ignoring differences (say between molecular details of heat and solute diffusion), which are in any case irrelevant for many purposes.

We wish to emphasize that a random walk model of Brownian motion is appropriate because of the vast disparity between the two scales involved. To fix ideas, imagine that one is examining a randomly moving dust particle through a certain magnifying glass M_1. Only sufficiently large motions can be resolved by M_1; so, too frequent observation of the particle is pointless. With this in mind, suppose that one finds that it is sensible to plot the course of the particle by determining its position every t_1 seconds. A more powerful magnifying glass M_2 might be utilized. Now one can profitably measure the position of the particle every t_2 seconds, $t_2 < t_1$. A yet more powerful glass M_3 reveals yet more detail and makes it appropriate to determine particle position every t_3 seconds, $t_3 < t_2$.

All these "macroscopic" time scales t_i are very large compared with the typical time τ between molecule-particle collisions. Thus one can pick an intermediate time Δt such that $\tau \ll \Delta t \ll t_i$; $i = 1, 2, \ldots$. Based on Δt, one can construct a random walk model of the dust particle motion in which a "step" is defined as the typical displacement after a time interval of length Δt. Two salient features of the random walk model are that successive steps are regarded as independent and analysis proceeds via the diffusion limit. Successive steps can safely be treated as independent, since there is effectively no correlation between events separated in time by Δt seconds. (By contrast, the effects of successive nudges are not necessarily independent.) The diffusion limit is appropriate, because many steps are required to produce a typical displacement.

To reiterate, diffusion limits (or, equivalently, Stirling's approximation) can be employed, because one can define a "step time" Δt that is *small* compared with the macroscopic scales t_i. And the hypothesis of independent steps can be used because Δt is *large* compared with the typical time τ between individual molecular nudges.

A final point should be made. Ever increasing detail* will become manifest as one selects successive magnifying glasses M_i. The particle path will be revealed as increasingly long and distorted, and hence the speed as increasingly great. This is the reason why one can accept the infinite particle velocity that is a formal concomitant of the Brownian model in the diffusion limit.

SUPERPOSITION OF FUNDAMENTAL SOLUTIONS: THE METHOD OF IMAGES

Equation (1), interpreted as describing the flow of heat, will be discussed at some length in Chapters 4 and 5. Here we wish to illustrate the interplay between probability and differential equations by considering the solution $u_0(x, t)$ given in (3.15) (and repeated here for reference),

* If the "magnifying glass" is powerful enough to resolve the effect of a single nudge, then no further detail is revealed by taking an even more powerful glass. But at this stage there is no longer a separation of macroscopic and molecular scales, and the entire basis for the standard random walk model has disappeared.

$$u_0(x, t) = (4\pi Dt)^{-1/2} \exp\left(\frac{-x^2}{4Dt}\right). \tag{4}$$

Remember that this solution satisfied the "delta function" initial condition (3.14) and also that the integral of u_0 over the entire x axis was unity. We can thus interpret $u_0(x, t)$ as a **unit source solution**, which gives the amount of mass that diffuses out at time t to a location between x and $x + dx$, assuming that a unit amount of mass was initially concentrated at the origin. This interpretation matches the random walk interpretation if we imagine the unit mass as composed of a large number M of tiny particles. Then the probability that a single particle is between x and $x + dx$ gives the percentage of the M particles that have this location.

The function u_0 is also called the **elementary** or **fundamental solution** to (1), since more general solutions can be built up from suitable combinations of solutions like u_0. To see this, we first observe that the diffusion equation (1) is invariant when x and t are changed into $x - \xi$ and $t - \tau$. This **translation invariance** is easily verified by direct substitution (Exercise 1). Also, it is physically obvious, since the diffusion process is the same no matter where we fix our origin of space measurement and no matter when we start our clock.

Because of the translation invariance, $u_0(x - \xi, t - \tau)$ is also a solution. This new solution represents the effect at point x and time t ($t > \tau$) of a source introduced at point ξ at time τ.

Next, we take account of the fact that the diffusion equation (1) is *linear* so that the principle of *superposition holds.* That is, if $u_1(x, t)$ and $u_2(x, t)$ are two solutions of (1), then

$$u(x, t) = c_1 u_1(x, t) + c_2 u_2(x, t) \tag{5}$$

is also a solution. So is any finite linear combination of solutions. As an example, for (1) we may consider the solution

$$u(x, t) = c_1 u_0(x - \xi_1, t) + c_2 u_0(x - \xi_2, t), \tag{6}$$

which is a superposition of the effects of two sources, one located at ξ_1 and one located at ξ_2. The strengths of the two sources, i.e., the amount of mass concentrated at ξ_1 and ξ_2, have the values c_1 and c_2, respectively.

At the end of the previous section we sought a solution to (1) with a vanishing x derivative at $x = L$. We can now solve this problem by superposition, using the **method of images.** For if one unit source is placed at $x = 0$ and another image source of equal strength is placed at $x = 2L$, then the resulting mass distribution will always be symmetric about $x = L$.* A function symmetric about $x = L$ must have a zero derivative there, and

* Note that the symmetry implies that exactly half the total mass will always remain to the left of $x = L$. Thus the total "probability mass" in $x \leq L$ is unity, as is required by (3.13a).

this is the required boundary condition. We thus obtain the required solution $u_0(x, t) + u_0(x - 2L, t)$ of Equation (3.21).

We leave to the reader the demonstration that the solution (3.23), where u is required to vanish at $x = L$, can also be readily found by the method of images. But we wish to point out how one can obtain the result (3.24) for the first passage time $F(L, t)$.

FIRST PASSAGE TIME AS A FLUX

The function $F(L, t)$ was defined so that $F(L, t)\,dt$ is the probability that a particle which starts from the origin at time zero is absorbed at $x = L$ in the time interval $(t, t + dt)$. An alternative interpretation of F can be given in terms of a situation involving a unit amount of mass that is initially concentrated at the origin, which is diffusing in the presence of a mass sink at $x = L$. In this situation F provides the rate at which mass is leaving the system at $x = L$, by being absorbed in the sink there. In other words, F is the flux of mass at $x = L$, in the direction of increasing x:

$$F(L, t) = J(L, t). \tag{7}$$

But we know that the flux is proportional to the gradient of the appropriate distribution, which in this case is the absorbing-barrier function $u_A(x, t)$ of Equation (3.23). We at once obtain the desired result of Equation (3.24),

$$F(L, t) = -D\left(\frac{\partial u_A}{\partial x}\right)_{x=L}.$$

More ingenuity would have been required to obtain this result from purely probabilistic considerations [see Chandrasekhar's article in Wax (1954).]

GENERAL INITIAL VALUE PROBLEM IN DIFFUSION

To see further use of the source solution, let us consider the general one-dimensional problem of solute diffusion. Imagine an infinite region in which the solute is initially distributed in some manner:

$$u(x, t) = g(x) \quad \text{at } t = 0, \qquad -\infty < x < \infty. \tag{8}$$

How does the solute distribute itself? What is desired is a solution to the diffusion equation (1) subject to the initial condition (8). We can obtain quite a general distribution of mass by considering a large number of sources of strength $f(\xi_i)\,\Delta\xi$ located at points $\xi_i = i\,\Delta\xi$, $i = 0, \pm 1, \ldots, \pm M$. Now $u_0(x - \xi_i, t)$ gives the mass that diffuses $x - \xi_i$ units, in time t, from a unit initial concentration. The total mass at x from all the sources is thus given by

$$\sum_{i=-M}^{M} u_0(x - \xi_i, t) f(\xi_i)\,\Delta\xi. \tag{9}$$

In the limit as $M \to \infty$, $\Delta\xi \to 0$, this would become*

$$\int_{-\infty}^{\infty} u_0(x - \xi)f(\xi)\, d\xi. \tag{10}$$

Roughly speaking, in (10), $f(\xi)\, d\xi$ is the strength of the source located at $x = \xi$. For the initial condition (8), then, we should take $g = f$. We therefore conjecture that the solution to the initial value problem is

$$u(x, t) = \int_{-\infty}^{\infty} u_0(x - \xi)g(\xi)\, d\xi. \tag{11}$$

Formula (9) can also be obtained from a probabilistic point of view. We know that $u_0(x - \xi_i, t)$ provides the probability that a particle is found $x - \xi_i$ units to the right of where it starts (at time t), given that the particle has taken many random steps. If $f(\xi_i)\,\Delta\xi$ particles started at ξ_i, then of these $u_0(x - \xi_i, t)f(\xi_i)\,\Delta\xi$ are expected at point x. Thus (9) sums the individual contributions to the particles at x from the various initial concentrations at $\xi_i \equiv i\,\Delta\xi$. We are again led to the conjecture that (11) provides the solution we seek.

To verify that the function defined in (11) does in fact solve our problem, we should first substitute it into the equation (1). There is no difficulty if $g(\xi)$ vanishes outside a finite interval, for then differentiation under the sign of integration is justified. Even for the infinite interval, the process of formal differentiation is not difficult to justify. One merely has to show that the infinite integrals, obtained *after* the formal differentiation process, are convergent.

A more urgent problem is to verify that (11) satisfies the initial condition

$$\lim_{t \downarrow 0} \int_{-\infty}^{\infty} u_0(x - \xi, t)g(\xi)\, d\xi = g(x). \tag{12}$$

To accomplish this, let us introduce the variable η by

$$\eta^2 = \frac{(x - \xi)^2}{4Dt}, \tag{13}$$

which is suggested by the exponential function in $u_0(x - \xi, t)$. [Compare (4).] Our solution (11) then becomes

$$u(x, t) = \frac{1}{\sqrt{\pi}} \int_{-\infty}^{\infty} e^{-\eta^2} g(x + \eta\sqrt{4Dt})\, d\eta. \tag{14}$$

* The strength of each individual source is made to approach zero in the limit. Otherwise, an infinite number of sources would give rise to an infinite amount of mass. As it is, we have adjusted the relation between source strength and number of sources so that the total amount of mass remains finite as the number of sources become infinite.

If we now take the *formal* limit $t \to 0$, we easily obtain the desired answer, using (2.34).

In practical applied mathematics, the discussion of the previous paragraph is often deemed sufficient verification that the initial condition is satisfied, since justification of the formal limit seems straightforward. Presumably, all one has to do is to divide the interval of integration in (14) into the three parts $(-\infty, -A)$, $(-A, A)$, $(A, +\infty)$, and take care with the double limiting process $t \to 0$, $A \to \infty$. (The reader may wish to carry out the proof.)

We have seen how the superposition principle enables us to obtain a variety of useful solutions to the diffusion equation. Superposition will be treated much more fully in the next two chapters. Here we continue our examination of the use of differential equation approximations to probability problems.

HOW TWISTING A BEAM CAN GIVE INFORMATION ABOUT DNA MOLECULES

A paper that is unusually rich in interconnections among various sciences and various branches of mathematics is "Diffusion Out of a Triangle" by W. Smith and G. S. Watson [*J. Appl. Prob. 4*, 479–88 (1967)]. We now present a brief exposition of the content of this paper.

Sometimes separate strands of long helical DNA molecules are present in solution. There are two types of such molecules, with complementary chemical structure. Thermal agitation brings these molecules together from time to time. It is presumed that only if complementary portions slide into apposition will the two separate molecules successfully "zip up" into a double helix. Otherwise, the molecules will eventually slide apart. We wish to calculate the frequency of collisions (called "successful") in which the zip-up process occurs.

The problem is idealized into a consideration of two rigid rods of unit length. Let (x, y) denote a randomly chosen intersection point, $0 \leq x \leq 1$, $0 \leq y \leq 1$. Suppose that thermal motion of the solvent models leads to a random walk of the point (x, y). Which will happen first, a coming into register $(x = y)$ or a sliding apart [(x, y) passing out of the unit square]? By symmetry, an alternative form of the question is as follows? If a point is dropped at random on an isosceles right triangle and then executes an unbiased random walk, what is the probability that it will leave the triangle along the hypotenuse (successful collision) rather than along one of the legs (unsuccessful collision)?

More generally, let us consider the following problem. A plane domain D has a boundary Γ that is composed of two parts Γ_1 and Γ_2. Let a particle execute an unbiased random walk starting from a point (x, y) in D. Determine $P(x, y)$, the probability that a particle starting at (x, y) crosses the boundary first through Γ_1.*

* In terms of concepts that were introduced above, this problem can be regarded as requiring the probability that the particle is ultimately absorbed along Γ_1 rather than along Γ_2.

In setting out to solve this problem, imagine a square grid to be superimposed on D, with a grid spacing of Δ. Suppose that the particle executes a random walk on this grid (Figure 3.2). If the particle starting from (x, y) eventually crosses the boundary along Γ_1, then it must have moved to one of the four points adjacent to (x, y) and then crossed the boundary. Thus

$$P(x, y) = \tfrac{1}{4}[P(x - \Delta, y) + P(x + \Delta, y) + P(x, y - \Delta) + P(x, y + \Delta)]. \quad (15)$$

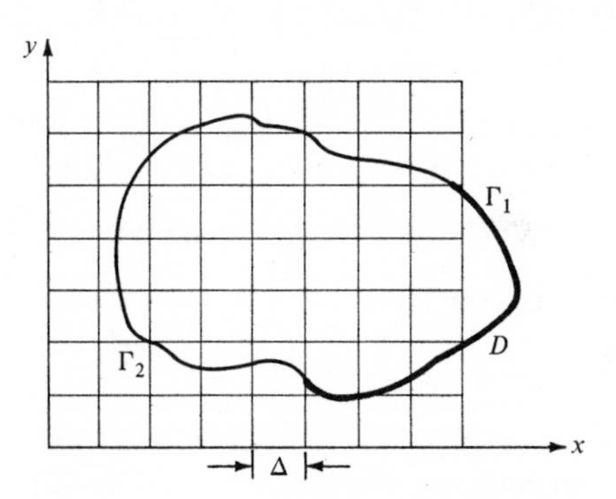

FIGURE 3.2. *Diagram for a random walk problem. It is desired to determine the probability that a particle which starts in the interior will first cross the boundary on the heavy portion* Γ_1.

As has become our custom, we assume that the step length Δ is small and retain only the lowest order terms in an approximation of the governing difference equation. After a short calculation (Exercise 7) we thereby obtain from (15) the Laplace equation

$$\frac{\partial^2 P}{\partial x^2} + \frac{\partial^2 P}{\partial y^2} = 0. \quad (16)$$

The nearer the particle starts to $\Gamma_1(\Gamma_2)$, the more (less) probable it is that it will first cross the boundary at Γ_1. This leads to the boundary conditions

$$\lim_{(x,y)\to\Gamma_1} P(x, y) = 1, \qquad \lim_{(x,y)\to\Gamma_2} P(x, y) = 0. \quad (17a, b)$$

Suppose that the particles are initially distributed in D according to the probability distribution function $f(x, y)$. That is (roughly speaking), let $f(x, y)\,dx\,dy$ be the probability that the particle is initially located between x and $x + dx$ and between y and $y + dy$. Then the probability $\bar{P}$ that a particle with leave D along Γ_1 is given by the formula

$$\bar{P} = \iint_D P(x, y) f(x, y)\,dx\,dy. \quad (18)$$

We have reduced the DNA collision question to the solution of (16) subject to the boundary conditions of (17), and then a substitution of the result $P(x, y)$ into the integral of (18). A transformation of the problem will render its solution more tractable—as it is, the discontinuous boundary conditions of (17) make the solution awkward.

The basis for the transformation is the *symmetric Green's formula*

$$\iint_D (Q\nabla^2 P - P\nabla^2 Q)\, d\sigma = \oint_\Gamma \left(Q\frac{\partial P}{\partial n} - P\frac{\partial Q}{\partial n}\right) ds, \tag{19}$$

in which $\partial/\partial n$ denotes exterior normal derivative. In (19) we take P to be the function defined above, while Q is required to satisfy

$$\nabla^2 Q = f, \qquad Q = 0 \quad \text{on} \quad \Gamma. \tag{20}$$

With this, a short calculation [Exercise 7(b)] shows that

$$\bar{P} = \int_{\Gamma_1} \left(\frac{\partial Q}{\partial n}\right) ds. \tag{21}$$

If Q can be determined, our problem is solved by (21). But the problem for Q is not only relatively simple compared to the problem for P, it is a problem that appears in many physical contexts. For example, it arises in the flow of viscous fluid down a pipe with cross section D (Exercise 3.2.14 of II) and (when f is constant) in the torsion of a prismatical bar (Section 5.3).

Using published solutions of (20) for constant f, Smith and Watson calculated $P = 0.41062$ for the case when the particle is initially dropped at random on an isosceles right triangle (the biologically interesting situation). The details need not concern us. What we wish to emphasize is how classical ideas of potential theory, and classical solutions in the theory of elasticity, illumine considerations in the seemingly distant matter of a probability calculation in biochemistry.

RECURRENCE PROPERTY IN BROWNIAN MOTION

We can now solve some elementary differential equations to answer an interesting question in probability theory. The question is, if a particle starts executing a random walk from a given point x, what is the probability that it will eventually touch another given point y? The answer depends on the dimension of the space in which the walk is executed. We shall treat the one-dimensional case, but our discussion can be readily generalized to higher dimensions.

Suppose that the particle starts at a point x in the set D described by

$$D\colon \varepsilon < |x| < R.$$

By the one-dimensional version of the argument used to obtain (16) and (17), the probability that a particle which starts at x will hit the boundary

$|x| = \varepsilon$ before the boundary $|x| = R$ is the solution $P(x)$ of the following problem:

$$\frac{d^2P}{dx^2} = 0, \qquad x \text{ in } D; \tag{22a}$$

$$P = 1 \quad \text{when } |x| = \varepsilon, \qquad P = 0 \quad \text{when } |x| = R. \tag{22b, c}$$

We seek a linear function that satisfies the boundary conditions. The result is

$$P(x) = \frac{R - |x|}{R - \varepsilon}. \tag{23}$$

As $R \to \infty$ (for x and ε fixed), $P \to 1$. This means that the particle is virtually certain to touch the interval $|x| = \varepsilon$ before it "escapes." Since the starting point is arbitrary, we can conclude that a particle is virtually certain to pass arbitrarily close to any given point on the line. And, by repeating the argument, we conclude that the particle is virtually certain to do this arbitrarily often. We say that one-dimensional random walk has the **recurrence property**. With the tools at our disposal, we can show (Exercise 9) that the recurrence property also holds in two dimensions, but not in three.

The more "directed" character of diffusion in lower dimensions is a phenomenon of importance in applications. To give an example, G. Adam and M. Delbruck* have provided quantitative results along the lines we have discussed in an effort "to propose and develop the idea that organisms handle some of the problems of timing and efficiency, in which small numbers of molecules and their diffusion are involved, by reducing the dimensionality in which diffusion takes place from three-dimensional space to two-dimensional surface diffusion."

The recurrence property seems to threaten theoreticians, however, as well as help them. It appears that random walk possesses a quasi-cyclic character, and this seems entirely out of keeping with its use as a model for such irreversible processes as diffusion. In fact, Poincaré proved that there is a stronger property of the recurrence type. A collection of interacting particles, making up a dynamical system, will return arbitrarily closely to its initial state with probability 1. (A few initial states must be excluded—the excluded states forming a set of Lebesgue measure zero.) How, asked Zermelo, can one reconcile the recurrence property of dynamical systems with the irreversibility of continuum theories?

The resolution of the "Zermelo paradox" is based on the fact that the mean time to recurrence is very long in situations that we regard as irreversible. For example, it is true that if you lower a spoonful of cream into a cup of coffee that the cream particles should eventually return to form a spoon shape,

* In A. Rich and N. Davidson (eds.), *Structural Chemistry and Molecular Biology* (San Francisco: Freeman, 1968).

but the expected time to do this is very large compared to the age of the universe. Thus there is no real contradiction with a continuum model for diffusion, which predicts that the cream will spread more or less evenly throughout the coffee in a matter of minutes. [For a more detailed exposition, see Kac's article in Wax (1954).]

The discussion of the previous paragraph returns us to our original derivation of the connection between random walk and diffusion. Both the approaches from random walk to a diffusionlike result—whether by Stirling's formula or by passage from a difference equation to a limiting differential equation—both approaches gave approximations that were inaccurate for those rare particles which almost always went to the right or almost always went to the left. Misestimating the effects of very rare events will not have a perceptible effect on over-all results, and this is why the diffusion model is so successful. But as responsible thinkers, we must not lose sight of the limitations of our models.

EXERCISES

1. Provide a formal verification of the fact that (1) remains the same if the variables (x, y) are replaced by $(x - \xi, y - \eta)$, where ξ and η are constants.

2. (a) Verify that the function u_R of (3.21) has the properties claimed for it in the text.
(b) Use the method of images to derive Equation (3.23). Verify directly that u_A has the desired properties.

3. Provide a formal proof that (11) provides a solution to (1), providing g is suitably restricted.

4. Complete the demonstration that taking the limit of (14) as $t \to 0$ indeed provides a proof that the initial condition (12) is satisfied by the function u defined in (11). Assume that $g(x)$ is a continuous function of x that is bounded for all values of x (including $|x| \to \infty$).

5. Show that if $g(x)$ has a discontinuity at a point x_0, the limiting value of the integral in (14) is

$$\tfrac{1}{2}[g(x_0 - 0) + g(x_0 + 0)],$$

the average of the values obtained when approaching x_0 from the left and right.

6. Using (11), write down the density $u(x, t)$ of a solute if initially

$$u = 1 \qquad \text{for } 0 \leq x \leq 1,$$

$$u = 0 \qquad \text{for } x < 0 \quad \text{and} \quad x > 1.$$

Express the answer in terms of the error function,

$$\operatorname{erf} \xi \equiv 2\pi^{-1/2} \int_0^{\xi} \exp(-s^2)\, ds.$$

Then verify directly that the equations and boundary conditions are satisfied.

7. (a) Expand (15) in powers of Δ and obtain (16). [Usually, in numerical analysis, the difference equation (15) is used as an approximation to (16). Here we work backward and approximate (15) by the differential equation (16).]

(b) Verify (21).

8. Derive (19) from the divergence theorem $\iint_D \nabla \cdot \mathbf{v}\, d\sigma = \oint_\Gamma \mathbf{n} \cdot \mathbf{v}\, ds$. Begin by considering the two special cases $\mathbf{v} = Q\nabla P$ and $\mathbf{v} = P\nabla Q$.

9. ‡(a) The two-dimensional analogue of (22) requires finding a function P of the polar coordinate r that satisfies

$$\frac{d^2P}{dr^2} + r^{-1}\frac{dP}{dr} = 0, \qquad \varepsilon < r < R,$$

$$P(\varepsilon) = 1, \qquad P(R) = 0.$$

Find this function and use it to show that the recurrence property holds in two dimensions.

‡(b) Extend the above calculations to the three-dimensional case by finding a suitable function P of the spherical coordinate ρ. Show that ε/ρ is the probability that a particle that starts a distance ρ from the origin will ever approach to within ε of the origin, $\varepsilon < \rho$. Deduce that the recurrence property does not hold in three dimensions.

10. (This exercise presents an alternative approach to the diffusion limit of random walk.) Suppose that a large number of particles are executing a one-dimensional random walk, with step length Δx and a time interval Δt per step. Let

$N(x, t)$ = number of particles at point x, time t;
$v(x, t)$ = net number of particles that move from x to $x + \Delta x$ in the time interval $(t, t + \Delta t)$.

Furthermore, define ρ and J by

$$\rho(x, t) = \frac{N(x, t)}{\Delta x}, \qquad J(x, t) = \frac{v(x, t)}{\Delta t}.$$

(a) What is the interpretation of ρ and v?

(b) Show that to first approximation $\partial\rho/\partial t = -\partial J/\partial x$.

(c) Justify the following assumption in unbiased random walk:

$$v(x, t) = \tfrac{1}{2}N(x, t) - \tfrac{1}{2}N(x + \Delta x, t).$$

(d) By taking an appropriate limit, derive a relation between J and ρ, and thereby show that ρ satisfies the diffusion equation.

11. Repeat Exercise 10 for the case of biased random walk, where the particle moves to the right with probability p and to the left with probability q, $q \neq p$, $p + q = 1$. (Compare Exercise 3.7.)

12. The motion of certain amebae is known to be affected by the concentration c of certain chemical attractants. Let us idealize the ameba as an organism that takes a step Δ to the right or to the left with an average frequency $f = f(c)$. Suppose that the organism has a chemical receptor at each end and that the distance between receptors is $\alpha\Delta$. Finally, let $a(x)$ be the density of amebae centered at x. (There is also a dependence on time t, but we do not indicate this in our notation.)

(a) What is implied by the assumption that the flux J is given by

$$J = \int_{x-\Delta}^{x} f[c(s + \tfrac{1}{2}\alpha\Delta)]a(s)\, ds - \int_{x}^{x+\Delta} f[c(s - \tfrac{1}{2}\alpha\Delta)]a(s)\, ds.$$

(b) From the equation above, show that to lowest order in Δ,

$$J = -\mu \frac{da}{dx} + \chi a \frac{dc}{dx},$$

where

$$\mu(c) = f(c)\Delta^2, \qquad \chi(c) = (\alpha - 1)\mu'(c).$$

(c) Try to explain why χ changes sign with $\alpha - 1$.

(d) Discuss the case $\alpha = 0$. [Chapter 1 gives some idea of the significance of the "chemotaxis" phenomenon modeled here. For further development along the lines just outlined, see E. F. Keller and L. A. Segel, *J. Theoret. Biol. 30*, 225 (1971).]

13. How might the random walk concept be used in a quantatitive discussion of polymer molecules? (These molecules are composed of a long chain of identical submolecules.)

Appendix 3.1 *O* and *o* Symbols

Definition 1. $\phi(x) = O[\psi(x)]$ as $x \to x_0$ if there exists a positive constant M such that $|\phi(x)| \leq M|\psi(x)|$ whenever x is sufficiently close to x_0.

REMARKS. "$\phi = O(\psi)$" is read "ϕ is big-oh of ψ." Most often x_0 is either zero or infinity. Reference to x_0 is frequently omitted if the x_0 is clear from the context. Sometimes the phrase "ϕ is of the order of ψ" is used instead of $\phi = O(\psi)$. If $\psi(x_0) \neq 0$, $\phi = O(\psi)$ if and only if ϕ/ψ is bounded in a sufficiently small neighborhood of x_0.

Examples. (i) $\phi(x) = O(1)$ as $x \to x_0$ means that ϕ is bounded in the neighborhood of x_0. (ii) $x^2 \sin x = O(x^3)$ as $x \to 0$. (iii) $x^2 \tanh x = O(x^2)$ as $x \to \infty$. (iv) The $O(x^5)$ term in the Maclaurin expansion for $1/(1 - 2x)$ is $(2x)^5$ ($x \to 0$ is understood).

Definition 1 is extended in various ways. Thus, for vectors, $\boldsymbol{\phi}(\mathbf{x}) = O[\boldsymbol{\psi}(\mathbf{x})]$ as $\mathbf{x} \to \mathbf{x}_0$ requires that $|\boldsymbol{\phi}(\mathbf{x})| \leq M|\boldsymbol{\psi}(\mathbf{x})|$ for $|\mathbf{x} - \mathbf{x}_0|$ small. Moreover, $\phi(x) = O[\psi(x)]$ for x in the region R means that there exists a positive constant M such that $|\phi(x)| \leq M|\psi(x)|$ for all x in R. Also, $f(x, \varepsilon) = O[g(x, \varepsilon)]$ as $x \to x_0$ means that there exists a positive quantity M, which can depend on ε but not on x, such that $|f(x, \varepsilon)| \leq M(\varepsilon)|g(x, \varepsilon)|$ whenever x is sufficiently close to x_0. If an M can be found that is independent of ε, then we say that $f = O(g)$ as $x \to x_0$, uniformly in ε.

Definition 2. $\phi(x) = o[\psi(x)]$ as $x \to x_0$ if for *every* positive constant M we have $|\phi(x)| \leq M|\psi(x)|$ whenever x is sufficiently close to x_0. In words, "ϕ is small-oh of ψ."

Particularly when $x_0 = \infty$, the above definition is often altered so that there is no requirement on ϕ and ψ when $x = x_0$. Then $\phi = O(\psi)$ if and only if $\lim_{x \to x_0} \phi/\psi = 0$.

Examples. (i) $x^2 \ln x = o(x)$ as $x \to 0$. (ii) $e^{-x} = o(1)$ as $x \to \infty$.

EXERCISES

1. Interpret and prove the following (Erdelyi, 1956, p. 8).
   (a) $O[O(\Phi)] = O(\Phi)$.
   (b) $O(\Phi)O(\Psi) = O(\Phi\Psi)$.
   (c) $O(\Phi) + O(\Phi) = O(\Phi) + o(\Phi) = O(\Phi)$.

2. Make up and prove some more lemmas like those in Exercise 1.

Chapter 4
Superposition, Heat Flow, and Fourier Analysis

The majority of scientific problems that can be treated by standard mathematical methods are linear. For linear *homogeneous* problems, the **principle of superposition** holds; i.e., if special solutions to the problem are known, then a linear combination of such solutions yields new solutions. Furthermore, the solution in the inhomogeneous case can be obtained by a linear combination of solutions to the homogeneous problem and a special solution of the inhomogeneous problem.

Perhaps the most important landmark in the development of the basic theory for solving linear problems is the publication in 1822 of Fourier's *Théorie Analytique de la Chaleur* (Mathematical Theory of Heat). Not only does this book provide an exemplary formal treatment of a general class of "boundary value problems," but it also serves to initiate the theory of a mathematical approach possessing great generality. In its original and generalized forms, Fourier's approach is indispensable to modern science and engineering. It is one of the most important landmarks in the development of applied mathematics.

The early history of Fourier analysis is typical in its demonstration of differences in emphasis between pure and applied mathematicians. E. T. Bell sums up the situation this way.*

It was while at Grenoble that Fourier composed the immortal *Théorie Analytique de la Chaleur* (Mathematical Theory of Heat), a landmark in mathematical physics. His first memoir on the conduction of heat was submitted in 1807. This was so promising that the Academy encouraged Fourier to continue by setting a contribution to the mathematical theory of heat as its problem for the Grand Prize in 1812. Fourier won the prize, but not without some criticism which he resented deeply but which was well taken.

Laplace, Lagrange, and Legendre were the referees. While admitting the novelty and importance of Fourier's work they pointed out that the mathematical treatment was faulty, leaving much to be desired in the way of rigor. Lagrange himself had discovered special cases of Fourier's main theorem but had been deterred from proceeding to the general result by the difficulties which he now pointed out. These subtle difficulties were of such a nature that their removal at the time would probably have been impossible. More than a century was to elapse before they were satisfactorily met.

* In E. T. Bell, *Men of Mathematics* (N.Y.: Simon and Schuster, 1961), pp. 197–98. Copyright © 1937 by E. T. Bell. Reprinted by permission of Simon and Schuster, Inc.

In passing it is interesting to observe that this dispute typifies a radical distinction between pure mathematicians and mathematical physicists. The only weapon at the disposal of pure mathematicians is sharp and rigid proof, and unless an alleged theorem can withstand the severest criticism of which its epoch is capable, pure mathematicians have but little use for it.

The applied mathematician and the mathematical physicist, on the other hand, are seldom so optimistic as to imagine that the infinite complexity of the physical universe can be described fully by any mathematical theory simple enough to be understood by human beings. Nor do they greatly regret that Airy's beautiful (or absurd) picture of the universe as a sort of interminable, self-solving system of differential equations has turned out to be an illusion born of mathematical bigotry and Newtonian determinism; they have something more real to appeal to at their own back door—the physical universe itself. They can experiment and check the deductions of their purposely imperfect mathematics against the verdict of experience—which, by the very nature of mathematics, is impossible for a pure mathematician. If the mathematical predictions are contradicted by experiment they do not, as a mathematician might, turn their backs on the physical evidence, but throw their mathematical tools away and look for a better kit.

This indifference of scientists to mathematics for its own sake is as enraging to one type of pure mathematician as the omission of a doubtful iota subscript is to another type of pedant. The result is that but few pure mathematicians have ever made a significant contribution to science—apart, of course, from inventing many of the tools which scientists find useful (perhaps indispensable). And the curious part of it all is that the very purists who object to the boldly imaginative attack of the scientists are the loudest in their insistence that mathematics, contrary to a widely diffused belief, is not all an affair of grubbing, meticulous accuracy, but is as creatively imaginative, and sometimes as loose, as great poetry or music can be on occasion. Sometimes the physicists beat the mathematicians at their own game in this respect: ignoring the glaring lack of rigor in Fourier's classic on the analytical theory of heat, Lord Kelvin called it "a great mathematical poem.

As Bell points out, Fourier's work was at first criticized for its lack of rigor. Yet its importance as a new approach and a powerful method was undeniable even from its very beginning, and it sparked developments in pure mathematics. The solution of specific problems, and the subsequent stimulation to more general developments, cannot be more clearly exemplified than in the study of Fourier analysis, a subject that should be familiar to every student of applied mathematics.

4.1 Conduction of Heat

Let us begin by considering a simple problem of heat conduction, involving a metal rod, insulated on the sides, with one end in a bath at temperature θ_1 and the other end in a bath at temperature θ_2. We shall formulate a mathematical representation for this problem and for its three-dimensional generalization. Next we discuss some of the mathematical issues that arise.

In particular, we prove a uniqueness theorem. We solve the one-dimensional problem by the method of separation of variables, and we show how the use of dimensionless variables facilitates interpretation of the solution.

STEADY STATE HEAT CONDUCTION

In the *steady state*, the temperature θ in the rod will be found to have a *uniform gradient*, and the rate of flow of heat will be proportional to this gradient. Thus, suppose that the end point of the rod located at $x = 0$ is at temperature θ_1 and the end located at $x = L$ is at temperature θ_2 (see Figure 4.1). Then the temperature at a distance x from one end is

$$\theta = \theta_s(x) = \theta_1 + (\theta_2 - \theta_1)\frac{x}{L}. \tag{1}$$

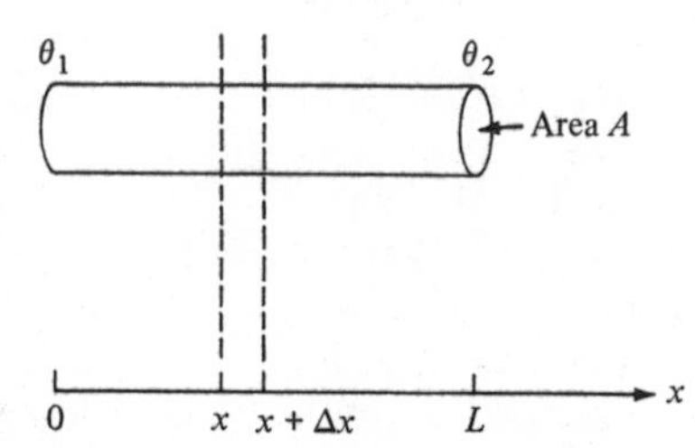

FIGURE 4.1. *One-dimensional heat conduction along a rod.*

(The subscript s is added to denote the steady state.) The temperature gradient is

$$\frac{d\theta_s}{dx} = \frac{\theta_2 - \theta_1}{L}. \tag{2}$$

The **heat flux density** J (the rate of heat flow per unit area per unit time, in the direction of increasing x) is found to be proportional to the temperature gradient. We write

$$J = -k\frac{d\theta_s}{dx}, \tag{3}$$

where k is the **coefficient of heat conductivity**. The negative sign is introduced so that k will be positive if (as expected) heat flows in the direction of decreasing temperature. Thus, with $k > 0$, J is positive in (3) when $d\theta/dx$ is negative.*

* An enormous amount of work has gone into measuring various "phenomenological coefficients" like k. For a summary of a nearly one thousand page compilation of results on pure substances alone, see "Thermal Conductivities of the Elements" by R. W. Powell and Y. S. Touloukian, *Science 181*, 999–1008 (1973).

DIFFERENTIAL EQUATION FOR ONE-DIMENSIONAL HEAT CONDUCTION

Now consider the adjustment of the temperature distribution before the steady state is reached. The temperature at any point x at any instant t is now denoted by $\theta(x, t)$. Can we find an equation governing the function $\theta(x, t)$?

We assume that (3) still holds for each instant, so that the instantaneous heat flux density is given by*

$$J = -k\frac{\partial \theta}{\partial x}. \tag{4}$$

Consider a section of a cylindrical rod between x and $x + \Delta x$ (Figure 4.1). The heat flow across the surface at x *into* this section is at the rate

$$A\left(-k\frac{\partial \theta}{\partial x}\right)_x, \tag{5}$$

where A is the (constant) cross-sectional area. The subscript x indicates that the expression within the parentheses is to be evaluated at x.

The heat flow across the surface at $x + \Delta x$ *out* of the section has the same form as (5) except that the expression is evaluated at $x + \Delta x$. Thus the *net gain* of heat is at the *rate* $\dot{q}$, where

$$\dot{q} = A\left[\left(k\frac{\partial \theta}{\partial x}\right)_{x+\Delta x} - \left(k\frac{\partial \theta}{\partial x}\right)_x\right]. \tag{6}$$

If no sources or sinks of heat are present, then the net gain of heat per unit mass is proportional to the rise of the temperature θ. Thus

$$\dot{q} = \frac{\partial}{\partial t}[c\theta\rho(A\,\Delta x)], \tag{7}$$

where ρ is the density of the material and c is its specific heat. (A unit increase of temperature increases the amount of heat in a unit mass of material by c units.) If we equate (6) and (7), divide by Δx, and take the limit as $\Delta x \to 0$, we obtain

$$\rho c\frac{\partial \theta}{\partial t} = \frac{\partial}{\partial x}\left(k\frac{\partial \theta}{\partial x}\right). \tag{8}$$

It has been implied that the rod is composed of a homogeneous material and therefore ρ, c, and k are constants; but since (8) is a local equation, it holds

* Equation (4) is an example of a **constitutive equation**. Such equations describe the behavior of a particular material, in contrast with universal physical laws, such as conservation of mass, which describe the behavior of all materials. To see that (4) is not universally valid, note that if we were dealing with a liquid, we would have to add to the heat flux density J a contribution due to convection. It is not an easy matter to establish constitutive equations. We shall discuss many such equations in the course of this work.

even for an *in*homogeneous cylindrical rod (when ρ, c, and k are functions of x).

In the case of a homogeneous material we may write (8) as

$$\frac{\partial \theta}{\partial t} = \kappa \frac{\partial^2 \theta}{\partial x^2}. \tag{9}$$

Here $\kappa = k/\rho c$ is usually called the **coefficient of thermal diffusivity** in order to recognize the similarity of (9) with the equation of diffusion of matter. Typical values of κ are 1.14 cm^2/sec for copper and 0.011 cm^2/sec for granite.

INITIAL BOUNDARY VALUE PROBLEM FOR ONE-DIMENSIONAL HEAT CONDUCTION

Having found the equation governing $\theta(x, t)$, we now turn to the complete formulation of the problem.

We are dealing with the solution of (9) in a domain D in the (x, t) plane (see Figure 4.2) defined by

$$D: 0 < x < L, \qquad 0 < t < \infty.$$

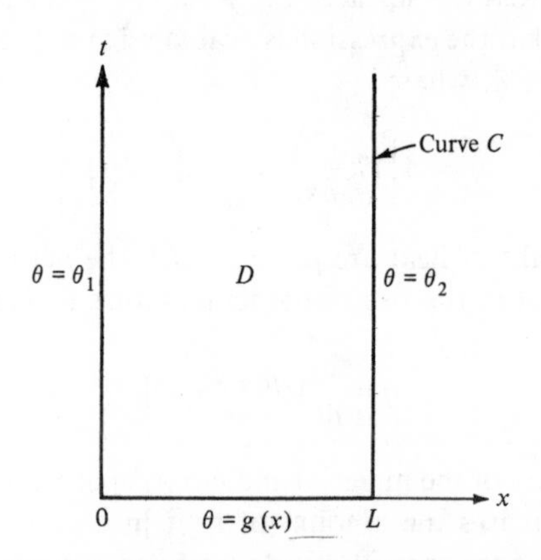

FIGURE 4.2. *Mathematical formulation of the problem of heat conduction: the domain, its boundary, and the boundary conditions.*

The boundary of this unbounded domain is a (heavily drawn) curve C that is concave (upward). From physical considerations, it is clear that the *initial* temperature of the rod must be known in order to define the problem. Let this be

$$\theta(x, 0) = g(x). \tag{10}$$

The ends of the rod are kept at fixed temperatures for all $t > 0$ so that

$$\theta(0, t) = \theta_1, \quad \theta(L, t) = \theta_2, \qquad \text{for } t > 0. \tag{11}$$

The rod may have been exposed to the baths beginning only at $t = 0$; therefore $g(0)$ and $g(L)$ need not be equal to θ_1 and θ_2, respectively.

Our mathematical problem is now formulated. To emphasize this, let us repeat the equations [(9), (10), and (11)]; find $\theta(x, t)$ for $0 < x < L, 0 < t < \infty$, where

$$\frac{\partial \theta}{\partial t} = \kappa \frac{\partial^2 \theta}{\partial x^2} \qquad \text{for } 0 < x < L, \quad 0 < t < \infty.$$

$$\theta(x, 0) = g(x) \quad (g \text{ given}) \qquad \text{for } 0 < x < L.$$

$$\theta(0, t) = \theta_1, \quad \theta(L, t) = \theta_2 \qquad \text{for } t > 0.$$

In deference to the physical interpretation of the problem, the conditions (11) are often referred to as the **boundary conditions** (being imposed at the boundary of the rod), and condition (10) is referred to as the **initial condition** (being imposed at the initial instant). Mathematically, they are both conditions prescribed on the concave boundary C of the domain D. A close look at theory (not presented here) is required to clarify this confusion in terminology.

PAST, PRESENT, AND FUTURE

Suppose that after the evolution of heat distribution in the bar has been allowed to go on for some time, the temperature at the left end is changed to θ_1 at the instant t_1, $t_1 > 0$. Physical experience indicates that the temperature distribution in the rod will be influenced in the future ($t > t_1$) but not in the past ($t < t_1$). Thus the equation must have a "sense of time," physically speaking. Indeed, it is physically clear that the solution in the domain

$$D_1 \equiv \{(x, t) | 0 < x < L, 0 < t < t_1\}$$

is entirely determined by the conditions specified on the heavily marked curve C_1 of Figure 4.3. Changes in the specification of $\theta(0, t)$ and $\theta(L, t)$ for $t > t_1$ can have no influence.

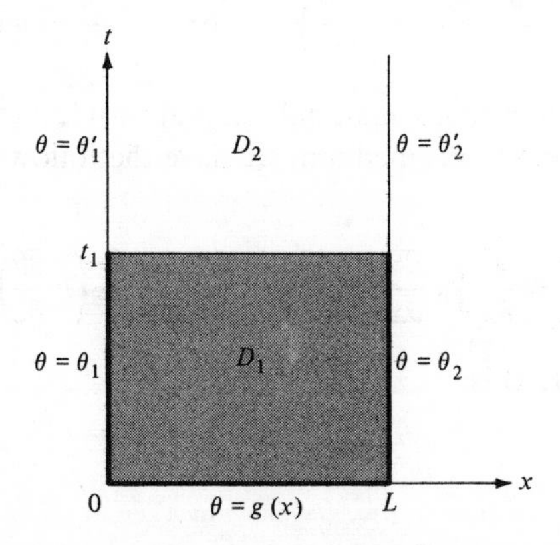

FIGURE 4.3. *Difference between past and future illustrated by a change of boundary conditions. For $t < t_1$ the solution is only affected by changes along C_1, the heavily marked portion of the boundary.*

The solution in the domain D_2 is defined by a similar specification of conditions on a concave curve. Thus we may make the *conjecture* that *the solution of the equation* (8) *in a domain D bounded by a concave curve C and a line $t =$ constant* (see Figure 4.4) *is determined by specifying the value of θ on C.* This is indeed a correct result: The unique existence of the solution can be proved for a sufficiently smooth curve C and continuous prescribed values.

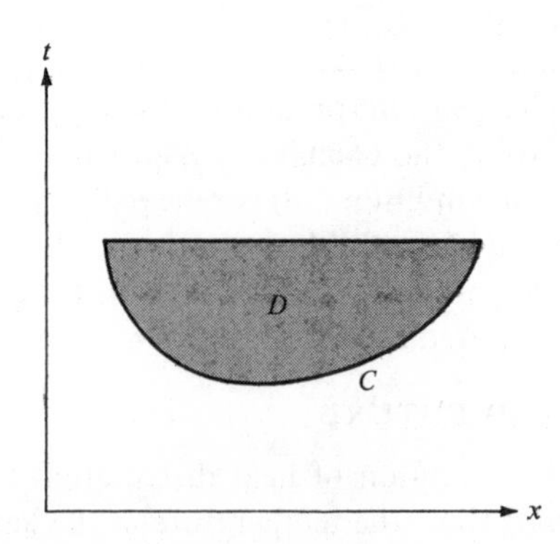

FIGURE 4.4. *General mathematical formulation of the problem of heat conduction: the domain and its boundary.*

HEAT CONDUCTION IN THREE-DIMENSIONAL SPACE

It is not difficult to generalize the formulation of the heat conduction problem to cover the case of three spatial dimensions. As we shall show in a moment, the final equation governing the change of the temperature $\theta(x, y, z, t)$ at the point (x, y, z) and time t is

$$\frac{\partial\theta}{\partial t} = \kappa\left(\frac{\partial^2\theta}{\partial x^2} + \frac{\partial^2\theta}{\partial y^2} + \frac{\partial^2\theta}{\partial z^2}\right) \qquad \text{or} \qquad \frac{\partial\theta}{\partial t} = \kappa\nabla^2\theta, \tag{12}$$

in the case of a homogeneous material. [Equation (12) generalizes (9).] In the case of an inhomogeneous medium, we have the following generalization of (8):

$$\rho c\frac{\partial\theta}{\partial t} = \frac{\partial}{\partial x}\left(k\frac{\partial\theta}{\partial x}\right) + \frac{\partial}{\partial y}\left(k\frac{\partial\theta}{\partial y}\right) + \frac{\partial}{\partial z}\left(k\frac{\partial\theta}{\partial z}\right). \tag{13a}$$

In vector notation, (13a) is

$$\rho c\frac{\partial\theta}{\partial t} = \nabla\cdot(k\nabla\theta). \tag{13b}$$

If the temperature over a surface enclosing a given volume is prescribed for all time and the temperature inside is given at the initial instant, we would expect the solution to be completely specified. Again, this is true, but the proof of the existence theorem is not easy.*

The derivation of (13) can be accomplished by considering a rectangular box with two of its corners at (x, y, z) and $(x + \Delta x, y + \Delta y, z + \Delta z)$ and using substantially the same reasoning as before.† [See Exercise 2(b).] The derivation can also be accomplished by applying the divergence theorem to the following equation for the rate of heat increase in a region R bounded by an arbitrary smooth closed surface ∂R:

$$\frac{\partial}{\partial t} \iiint_R c\rho\theta \, d\tau = \oiint_{\partial R} k \frac{\partial \theta}{\partial n} \, d\sigma = \oiint_{\partial R} k\mathbf{n} \cdot \nabla\theta \, d\sigma. \tag{14}$$

In (14) $\mathbf{n}$ is the unit outward normal. This equation uses the assumption that the heat flux through an element of surface in the direction of $\mathbf{n}$ is $-k(\partial\theta/\partial n)$ where $\partial/\partial n$ denotes the normal derivative [compare (4)]. From (14),

$$\iiint_R \left[c\rho \frac{\partial \theta}{\partial t} - \nabla \cdot (k\nabla\theta)\right] d\tau = 0. \tag{15}$$

Since R is arbitrary, the integrand must be zero if it is continuous.§ Otherwise, near a point P_0, where the integrand does not vanish, we must be able to find a small region R_0 in which the integrand has the same sign as its value at P_0. The integral (15) cannot vanish when $R \equiv R_0$, leading to an absurdity.

The complete formulation of a problem depends not only on the derivation of the equation but also on the specification of the boundary conditions. One common boundary condition is

(a) The temperature is specified for all time at each point on the boundary. (It may vary from point to point.)

Another possible boundary condition is

(b) The boundary is insulated, so that $\partial\theta/\partial n = 0$ at every point on the boundary.

A third and more general condition is

(c) The heat loss from the boundary is proportional to the amount by which the surface temperature is higher than the ambient temperature.

* Particularly noteworthy existence proofs, for the wave and Laplace equations as well as the heat equation, are supplied in a remarkable paper by R. Courant, K. Friedrichs, and H. Lewy ["On the Partial Differential Equations of Mathematical Physics," translated by P. Fox in *IBM Journal 11*, 215–34 (1967); the original paper appeared in the German journal *Math. Ann. 100*, 32–74 (1928)]. Existence emerges automatically from proofs that the solutions to appropriate finite difference schemes approach limiting values, as the mesh size decreases, which satisfy given partial differential equations and initial boundary conditions. Connections with Green's functions and random walk are also explored.

† In keeping with the "broad-brush" approach of the first three chapters, derivations are merely outlined here. For a more careful treatment of the required ideas, see Chapter 14, particularly Section 14.1.

§ This is the **Dubois Reymond lemma,** a result that we shall use many times.

There are many other possibilities such as the prescription of the time variation of the temperature over the bounding surface or a portion thereof. It is difficult to cite all the possible cases, and it is much harder to prove all the necessary existence theorems, as required by the approach of the pure mathematician. For the applied mathematician it is sufficient to adopt as a *working hypothesis* that *one* condition *analogous* to (a), (b), or (c) may be specified on the boundary. The validity of this hypothesis will be accepted unless a reasonable doubt can be demonstrated. Indeed, this hypothesis will be accepted for an inhomogeneous medium as well, and even for cases where the specific heat (say) is a function of temperature.

In general, it would be satisfying to have the unique existence of the solutions proved under prescribed conditions; but this is not always possible. Often, the applied mathematician obtains the conditions to be prescribed from consideration of the scientific problem. He sometimes checks the appropriateness of his equations by proving that they have at most one solution (uniqueness). (If more conditions are prescribed than are necessary for uniqueness, either these extra conditions are redundant or they cannot be fulfilled.) Although the existence of solutions cannot be taken for granted, proofs are surprisingly difficult. Applied mathematicians often do not consider the gain in confidence that these proofs furnish to be worth the effort.

Since uniqueness is easy to demonstrate, we now present the proof. Then we state and comment upon another general result, the maximum principle.

PROOF OF THE UNIQUENESS THEOREM

Consider the problem of heat conduction in a region R with prescribed temperature on the bounding surface S. Let $\theta_1(x, y, z, t)$ be a solution satisfying the initial condition $\theta_1 = g(x, y, z)$ at $t = 0$. To prove uniqueness, let us assume that there is another solution $\theta_2(x, y, z, t)$ satisfying the same condition. Then their difference $\theta = \theta_2 - \theta_1$ would also satisfy the heat equation (13), but with the following boundary condition and initial condition:

$$\theta = 0 \quad \text{on } S, \tag{16a}$$

$$\theta = 0 \quad \text{at } t = 0. \tag{16b}$$

Uniqueness is proved if θ can be shown to be identically equal to zero. To show this, we first multiply (13) by θ and integrate over the region R. We find [Exercise 5(a)] that

$$\frac{1}{2}\iiint_R \rho c \frac{\partial(\theta^2)}{\partial t}\, d\tau = -\varepsilon + \oint\!\!\oint_{\partial R} k\theta \frac{\partial \theta}{\partial n}\, d\sigma, \tag{17}$$

where

$$\varepsilon = \iiint_R k(\nabla\theta)^2\, d\tau.$$

The surface integral vanishes because of (16a). Since ρc is independent of t, we may integrate (17) with respect to t in the interval $(0, t_1)$ and obtain

$$\left(\frac{1}{2}\iiint_R \rho c\theta^2\, d\tau\right)_{t_1} - 0 = -\int_0^{t_1} \varepsilon\, dt \tag{18}$$

if we recall the initial condition (16b). The left-hand side of (18) is nonnegative and its right-hand side is nonpositive, if $t_1 > 0$. Thus both sides must be equal to zero. Since t_1 is *arbitrary* we must have

$$\varepsilon = 0, \qquad \text{at all times.} \tag{19}$$

Thus $\nabla\theta = \mathbf{0}$ at all times, and hence $\theta(x, y, z, t)$ can only be a function of time. The vanishing of the left-hand side of (18) then shows that $[\theta(t_1)]^2 = 0$, and hence $\theta = 0$. □

THE MAXIMUM PRINCIPLE

An important result in the theory of partial differential equations can be conjectured from the intuitive feeling that heat is never intensified as a result of diffusion. That is, we expect temperature in the interior of a domain never to exceed the largest of the temperatures that occur on the boundary or are assigned initially. With some ingenuity, one can indeed prove the following result. If $\theta(x, t)$ is defined and continuous for $0 \le x \le L$, $0 \le t \le T$, and if θ satisfies the heat equation (9) in the interior of this rectangular region, then θ assumes its maximum and minimum when $t = 0$ or when $x = 0$ or when $x = L$. (We state our results in one dimension for simplicity, but they are valid for higher dimensions, too.) For a proof, see texts on differential equations, e.g., Tychonov and Samarski (1964, pp. 165–67).

With the aid of the maximum principle, proofs of uniqueness and continuous dependence on initial and boundary conditions are easily constructed (Exercise 11). Bear in mind, however, that the maximum principle has been stated only for solutions that are continuous in the closed rectangle, so one is permitted to make deductions from this principle only for the class of solutions that are so restricted. This restriction does not apply to the uniqueness proof that was given above, which employed the energy method.

As has been illustrated to some extent, a major task in the theory of partial differential equations is to prove that the solutions to various problems exist, are unique, and depend continuously on initial and boundary data. Such problems are said to be **well posed**.

SOLUTION BY THE METHOD OF SEPARATION OF VARIABLES

There is no obvious way to approach the basic heat conduction problem posed above, namely, solution of the heat conduction equation (9), the initial condition (10), and the boundary conditions (11). The ingenious method used by Fourier is based on the idea of superposing an infinite number of appropriate simple solutions to the *linear* equation (9). The key to his method is the determination of an infinite number of such solutions by the technique of separation of variables.

For reasons that we shall give below, in problems of the type we are investigating, the Fourier method applies directly only to the part of the solution that decays to zero, leaving the steady state solution θ_s. Let us denote this decaying **transient** by $v(x, t)$; i.e.,

$$v(x, t) = \theta(x, t) - \theta_s(x). \tag{20}$$

Since θ_s is a steady solution of the heat equation (9), the transient satisfies the same equation:

$$\frac{\partial v}{\partial t} = \kappa \frac{\partial^2 v}{\partial x^2}. \tag{21}$$

The initial condition is now

$$\text{for } 0 < x < L, \qquad v(x, 0) = f(x); \qquad f(x) = g(x) - \theta_s(x). \tag{22}$$

The boundary conditions (11) become

$$v(0, t) = v(L, t) = 0 \qquad \text{for all } t > 0. \tag{23}$$

Note that, unlike (11), these boundary conditions are **homogeneous**, in the sense that they are *satisfied by the zero function.*

We follow a step-by-step procedure to solve the problem posed by (21), (22), and (23). (Exactly the same procedure is used to solve a related problem in Section 15.4. Taken together, the two illustrations should provide an adequate exposition of the method.)

STEP A. *Assume a product solution:*

$$v(x, t) = X(x)T(t). \tag{24}$$

STEP B. *Substitute into the governing differential equation.* From (24) and (21),

$$XT' = \kappa X''T \qquad \left(X' = \frac{dX}{dx}, T' = \frac{dT}{dt}\right). \tag{25}$$

STEP C. *Separate variables.* We write (25) in the form

$$\frac{X''(x)}{X(x)} = \frac{T'(t)}{\kappa T(t)}. \tag{26}$$

The variables are now separated, so that the left side is a function only of the independent variable x and the right side is a function only of the independent variable t. It appears that each side of (26) can be varied more or less at will, by altering the independent variables x and t. If this were truly the case, a contradiction would arise. Thus both sides of (26) must equal a **separation constant** k, so that we obtain the **separation equations**

$$X'' = kX, \qquad T' = k\kappa T. \tag{27a, b}$$

STEP D. *Determine permissible values of the separation constant from a problem consisting of a separated equation and suitable homogeneous boundary conditions.* The homogeneous boundary conditions are provided by (23). They give

$$X(0)T(t) = 0, \quad X(L)T(t) = 0 \qquad \text{for all } t > 0,$$

or (excluding $T \equiv 0$)

$$X(0) = X(L) = 0. \tag{28}$$

These conditions must be applied to the solution of (27a). As is usually the case, *it is helpful to give separate consideration to the possibilities* $k > 0$, $k = 0$, $k < 0$.

If $k > 0$, we write $k = \mu^2$, $\mu > 0$. The general solution of the resulting equation $X'' = X\mu^2$ is

$$X = C_1 \cosh \mu x + C_2 \sinh \mu x.$$

The boundary conditions (28) require that $C_1 = C_2 = 0$, so that $X = 0$.

If $k = 0$, we must solve $X'' = 0$ subject to (28). In this case, too, there is only the **trivial solution** $X = 0$.

If $k < 0$, we write $k = -\alpha^2$, $\alpha > 0$. The general solution of the resulting equation $X'' = -\alpha^2 X$ is

$$X = C_1 \cos \alpha x + C_2 \sin \alpha x. \tag{29}$$

The requirement $X(0) = 0$ is only satisfied if $C_1 = 0$. The requirement $X(L) = 0$ gives

$$C_2 \sin \alpha L = 0. \tag{30}$$

One possible deduction from (30) is $C_2 = 0$. But since $C_1 = 0$, this would again provide a trivial solution. Nontrivial solutions result only if the separation constant $k = -\alpha^2$ takes on certain special values. Such solutions are permitted, in the light of (30) and since $\alpha > 0$, only if

$$\alpha = \frac{m\pi}{L}, \qquad m = 1, 2, 3, \ldots. \tag{31}$$

With (31) we indeed obtain nontrivial solutions (29) of the form

$$X(x) = B_m \sin \frac{m\pi x}{L}, \qquad B_m \text{ arbitrary constants.} \tag{32}$$

STEP E. *Solve the remaining separated equation when the separation constant takes on the values determined in* (*d*). With (31), since $k = -\alpha^2$, (27b) becomes

$$T' = -\frac{m^2\pi^2}{L^2}\kappa T. \tag{33}$$

Every solution of this equation is a constant multiple of

$$\exp\left(-\frac{m^2\pi^2}{L^2}\right)\kappa t. \tag{34}$$

STEP F. *Superpose all possible product solutions and attempt to satisfy the remaining initial or boundary conditions.* "Superpose" means (roughly speaking) "add together." Thus we assume for v the infinite series

$$v(x, t) = \sum_{m=1}^{\infty} B_m \sin\frac{m\pi x}{L} \exp\left(-\frac{m^2\pi^2}{L^2}\kappa t\right). \tag{35}$$

Since the heat equation (21) is linear, a *finite* series of the form (35) would certainly provide a solution (since each term is a solution). Under "suitable conditions" infinite series have the same properties as finite sums; we proceed under the assumption that such conditions will hold here, at least in some interesting instances.

The boundary conditions (23) are homogeneous so that the series (35) should satisfy (23), since each term does separately.*

The initial condition (22) is not satisfied, but we have all the constants B_m at our disposal. If we *attempt* to satisfy this condition, we have

$$f(x) = \sum_{m=1}^{\infty} B_m \sin\frac{m\pi x}{L}, \qquad 0 < x < L, \tag{36}$$

as a *necessary* condition. From this we can proceed formally to find the value taken by a typical coefficient, say B_n, if we multiply by $\sin(n\pi x/L)$ and integrate from 0 to L. It can be shown (Exercise 8) that B_n is given by

$$B_n = \frac{2}{L}\int_0^L f(x) \sin\frac{n\pi x}{L}\, dx. \tag{37}$$

* It is now evident why we chose to apply the Fourier method to the transient solution v; for v satisfies the homogeneous boundary conditions (23), but the full solution θ satisfies the inhomogeneous boundary conditions (11). The latter conditions do not have the linearity property of being satisfied by the sum of functions that individually satisfy them. Note that the homogeneity of the X equation (27a) and the accompanying boundary conditions (28) meant that the trivial solution was always a possibility. Our efforts to obtain a nontrivial solution led to the determination of a discrete infinite set of permissible separation constants and thus to the infinite series (35)—another illustration of the key role played by homogeneity. Beginners often forget that before standard application of the Fourier separation of variables procedure (a)–(f) can be made, it must be arranged that (with two independent variables) one of the separated equations has homogeneous boundary conditions. (The general rule is that "all but one" of the separated equations must have homogeneous boundary conditions.)

This follows from

$$\left(\sin\frac{m\pi x}{L}, \sin\frac{n\pi x}{L}\right) \quad (38a)$$

$$\left(\cos\frac{m\pi x}{L}, \cos\frac{n\pi x}{L}\right) = \begin{cases} L/2 & \text{if } m = n, \\ 0 & \text{if } m \neq n, \end{cases} \quad (38b)$$

$$\left(\cos\frac{m\pi x}{L}, \sin\frac{n\pi x}{L}\right) = 0. \quad (38c)$$

where we use the notation*

$$(f, g) = \int_0^L f(x)g(x)\,dx. \quad (39)$$

With this notation, (37) takes the following form (where m is used instead of n)

$$B_m = 2L^{-1}\left(f(x), \sin\frac{m\pi x}{L}\right). \quad (40)$$

We have thus found a tentative solution to our problem as

$$\theta(x, t) = \theta_s(x) + \sum_1^{\infty} B_m\left(\sin\frac{m\pi x}{L}\right)e^{-(m\pi/L)^2\kappa t}, \quad (41)$$

where the B_m are determined from the initial condition (22) by means of Equation (40). Is (41) really a solution? To give a positive answer we must be sure that it is possible to express the function $f(x)$ in the form (36). This would certainly be true if $f(x)$ were the sum of a finite number of sinusoidal terms. Indeed, one can express any "reasonable" function in such a form, but the proof is not easy. In overcoming this difficulty to a specific problem, we shall open the door to a general mathematical theory. But before we go into this matter, let us examine some implications of the solution obtained.

INTERPRETATION; DIMENSIONLESS REPRESENTATION

Our solution (41) approaches the steady state temperature distribution $\theta_s(x)$ given by (1), as expected. Of the various terms in the infinite series, the **fundamental** ($m = 1$) decays much more slowly than the **harmonics** ($m \geq 2$). Thus the deviation from the equilibrium distribution soon has a nearly sinusoidal form, *no matter what the initial temperature distribution may be.* The amplitude of this sinusoidal term decreases exponentially in time. The

* Later we shall see how (,) can be regarded as a generalization of the scalar product of two vectors, but this need not concern us at present. Only (38a) is needed to demonstrate (37), but (38b) and (38c) are needed below.

exponent $\kappa(\pi/L)^2$ depends only on the length of the rod L and the thermal diffusivity κ of the material.*

For this problem, times should be compared to the *time scale*

$$t_0 = \frac{L^2}{\kappa}, \tag{42}$$

during which significant temporal change in the solution occurs. Significant spatial changes take place in a fraction of the *length scale* L. Measuring according to these scales is equivalent to introducing the variables

$$\xi = \frac{x}{L} \quad \text{and} \quad \tau = \frac{t}{t_0} = \frac{\kappa t}{L^2}. \tag{43}$$

These variables are **dimensionless** in that their value is independent of the unit of measurement. (For example, if $x = 10$ cm and $L = 100$ cm, then $\xi = \frac{1}{10}$, as it is if $x = 0.1$ m and $L = 1$ m.) Using these variables, we may take our tentative solution (41) to be $\Theta(\xi, \tau)$, where

$$\Theta(\xi, \tau) = \theta(L\xi, t_0\tau) = \theta_s(L\xi) + \sum_1^\infty B_m \sin m\pi\xi \, e^{-(m\pi)^2\tau}, \tag{44}$$

an equation that shows no explicit dependence on L and κ. As a matter of fact, this introduction of dimensionless parameters can be done from the beginning. Then (9) becomes [Exercise 8(b)]

$$\frac{\partial\Theta}{\partial\tau} = \frac{\partial^2\Theta}{\partial\xi^2}. \tag{45}$$

The equation is to be solved in the domain D' defined by

$$D': 0 < \xi < 1, \qquad 0 < \tau < \infty. \tag{46}$$

The initial and boundary conditions for the homogeneous problem are

$$\Theta(0, t) = \Theta(1, t) = 0,$$
$$\Theta(\xi, 0) = G(\xi), \qquad \text{where } G(\xi) = g(L\xi).$$

In the original formulation, the temperature at a point with coordinates (x, t) seemingly could depend in an arbitrary way on the two parameters κ and L. The dimensionless formulation shows that explicit dependence on κ and L disappears if ξ and τ are used as independent variables. Such a reduction of the number of parameters on which a solution explicitly depends generally follows from the use of dimensionless variables.

The dimensionless formulation can be used for making predictions via the use of experimental models. Thus, if we construct a scale model for testing

* The most slowly decaying harmonic ($m = 2$) decays like $\exp(-4\pi^2\kappa t/L^2)$, a factor that decreases to a fraction e^{-4} (i.e., less than 2 per cent) of its initial value in $L^2/\pi^2\kappa \approx 0.1\, L^2/\kappa$ time units. For a copper bar 10 cm long, this would be less than 10 sec.

a practical problem of heat conduction, using the same material, the time scale in the actual case would be N^2 times as long if the model were $1/N$ of the actual size. The reason is that the time t occurs only in the combination $\tau = \kappa t/L^2$. A given value of τ is reached in a model experiment, where L is replaced by L/N, only if t is replaced by $N^2 t$. (See Section 6.2 for further discussion concerning the use of models.)

These arguments can be applied straightforwardly to the three-dimensional equation (12) and appropriate boundary conditions and initial conditions. Consequently, for example, if we throw hot copper spheres into icy water, it would take four times as long to cool down a sphere that is twice as large in diameter (Exercise 4).

We emphasize that in dimensionless form the results are independent of the units used for our measurements. The dimensionless formulation, therefore, gives the true mathematical essence of the problem. For a fuller discussion of dimensional analysis, see Section 6.2.

ESTIMATE OF THE TIME REQUIRED TO DIFFUSE A GIVEN DISTANCE

To conclude, we deduce a **fundamental fact concerning diffusion**. This is a consequence of the result that the slowest decaying mode, with spatial dependence $\sin(\pi x/L)$, decays like $\exp(-\pi^2 \kappa t/L^2)$. Since the heat in this mode is concentrated in the middle of the bar, we can regard the decay of the initial heat distribution, roughly, as the passage of heat out of the bar through a distance $d = L/2$. In terms of d, the decay factor is $\exp(-\pi^2 \kappa t/4d^2)$. In a time $t = d^2/\kappa$ the sinusoidal temperature distribution decays to $\exp(-\pi^2/4) \approx 8$ per cent of its initial value. That is, *diffusion through a distance d takes roughly d^2/κ time units.* An alternative statement is *in time t a substance diffuses approximately $(\kappa t)^{1/2}$ distance units.*

EXERCISES

1. The general initial value problem for heat conduction is formulated for a domain D bounded by a concave curve C and a line parallel to the x axis (a characteristic curve), with the value of temperature specified on the initial curve C (Figure 4.4). Explain why this formulation is reasonable, using physical arguments.

2. (a) Check that the one-dimensional heat equation (8) holds for an inhomogeneous rod.

(b) The equation of heat conduction in three dimensions may be derived by using the same arguments employed in the text for deriving the equation for the one-dimensional problem. One considers the amount of heat flowing into a rectangular box through the six sides. Carry out this derivation.

3. By using the method used in the text for proving the uniqueness theorem, carry out the proof when the boundary condition is

$$\frac{\partial \theta}{\partial n} + h\theta = 0, \qquad h > 0,$$

on the surface S enclosing the volume occupied by the conductor. Let the initial temperature distribution be specified as in the text.

4. Formulate the problem of the cooling of a copper sphere thrown into icy water. Show that it takes four times as long to cool down a sphere that is twice as large in diameter.

5. (a) Verify (17).
   (b) The expression in (34) could be preceded by the arbitrary constant C. Show that this would introduce no new generality into the series solution (35).

6. Solve the one-dimensional heat conduction problem by Fourier series when the right end of the bar is insulated so that the second boundary condition of (11) is replaced by the condition that $\partial\theta/\partial x$ vanishes at $x = L$.

7. Suppose that both ends of the bar are insulated. Then there is no longer a unique steady state solution. Discuss this fact.

8. (a) Verify (37) and (38).
   (b) Verify (45).

9. Consider the temperature θ in a semi-infinite slab $z \geq 0$. Suppose that $\theta = 0$ for time $t < 0$, but that there is an impulsive change in temperature so that when $z = 0$, $\theta = U$ (a positive constant) for $t > 0$. Find $\theta(z, t)$. Write the answer in terms of $\operatorname{erf} \eta = 2\pi^{-1/2} \int_0^\eta \exp(-s^2)\, ds$ (so that $\operatorname{erf} \infty = 1$).

   One expects the temperature to diffuse outward a distance $(\kappa t)^{1/2}$ in time t. Furthermore, $(\kappa t)^{1/2}$ is the only other combination of variables and parameters other than z, which has the dimension of a length. For both these reasons, look for a solution to the diffusion equation of the form

$$\theta(z, t) = f[z(\kappa t)^{-1/2}].$$

10. Make a change of variable of the form $v = f(x, t)\theta$ that transforms the differential equation

$$v_t = \alpha v_{xx} + \beta v_x + \gamma v; \qquad \alpha, \beta, \gamma \text{ constants}$$

into the standard form (9).

11. Consider the class C of functions that are continuous in the closed rectangle $x \in [0, L]$, $t \in [0, T]$ and satisfy the heat equation (9) in the interior of this rectangle.
    (a) Using the maximum principle, prove uniqueness in this class for the usual initial boundary value problem.

(b) Prove that if two functions from C differ by less than ε on the three boundary lines $x = 0$, $x = L$, $t = 0$, then they differ by less than ε throughout the rectangle.

4.2 Fourier's Theorem

We have shown how the initial value problem for the heat equation can be formally solved if any reasonable function $f(x)$ can be represented by a Fourier sine series [Equation (1.36)] in the interval $0 < x < L$. Since the series gives zero as its value at the end points of the interval $(0, L)$, it would appear natural to expect a function like

$$f(x) = \frac{x}{L}\left(1 - \frac{x}{L}\right)$$

to be so represented, but not a function like $f(x) = 1$. However, we shall see that even the latter is possible in a suitable sense—indeed in more than one sense.

We shall now justify the above statement about Equation (1.36); we shall consider the limit of the sum of N terms of the series (1.36) and try to evaluate the limit as $N \to \infty$. We shall see that this limit tends to $f(x)$ in the open interval $(0, L)$, provided that $f(x)$ is **piecewise smooth** in the interval $(0, L)$. This smoothness requirement holds if the interval $(0, L)$ can be divided into a finite number of subintervals such that $f(x)$ approaches a (finite) limit as x approaches the end points of the subinterval and such that $f'(x)$ exists and is continuous within each subinterval.

Since sines are odd functions, if a sine series represents $f(x)$ on $(0, L)$ it will represent $-f(x)$ on $(-L, 0)$. Sines and cosines must be used if a general function is to be represented on the "full range" $(-L, L)$. As we shall see, the convergence results can be readily extended to the full range situation.

SUMMATION OF THE FOURIER SINE SERIES

Let us denote the sum of the first N terms of the series (1.36) by $S_N(x)$ so that

$$S_N(x) = \frac{2}{\pi}\int_0^\pi K_N(x, \xi)f(\xi)\,d\xi, \qquad 0 < x < \pi, \tag{1}$$

where

$$K_N(x, \xi) = \sum_{m=1}^{N} \sin(mx)\sin(m\xi), \qquad 0 < x, \xi < \pi. \tag{2}$$

We want to prove that $S_N(x) \to f(x)$ as $N \to \infty$. *For convenience we have set* $L = \pi$, and this will be done throughout the present section. This entails no loss of generality, for it is equivalent to choosing $\pi^{-1}L$ as a length scale in nondimensionalization.

The summation of (2) can be carried out with reasonable ease if we use the complex representation

$$\sin w = \frac{e^{iw} - e^{-iw}}{2i}.$$

The answer (Exercise 1) is

$$K_N(x, \xi) = \frac{1}{2}\left\{\frac{\sin [(N + \frac{1}{2})(x - \xi)]}{2 \sin [\frac{1}{2}(x - \xi)]} - \frac{\sin [(N + \frac{1}{2})(x + \xi)]}{2 \sin [\frac{1}{2}(x + \xi)]}\right\}. \quad (3)$$

If x is a fixed point inside the interval $(0, \pi)$, the first term within the curly brackets on the right-hand side of (3) is well defined for $0 \leq \xi \leq \pi$ except when $\xi = x$; but even here, this term has a definite limit $N + \frac{1}{2}$. The graph of this term is shown schematically in Figure 4.5. In a very narrow interval $|x - \xi| < \pi/(N + \frac{1}{2})$, the function has a very high peak. Outside this interval, it shows a highly oscillatory behavior.

The second term on the right-hand side of (3) also shows a highly oscillatory behavior for large N but no strong peak. Keeping the behavior of the two terms making up K_N in mind, we now consider $S_N(x)$, where

$$S_N(x) = S_N^{(1)}(x) + S_N^{(2)}(x), \qquad 0 < x < \pi, \quad (4)$$

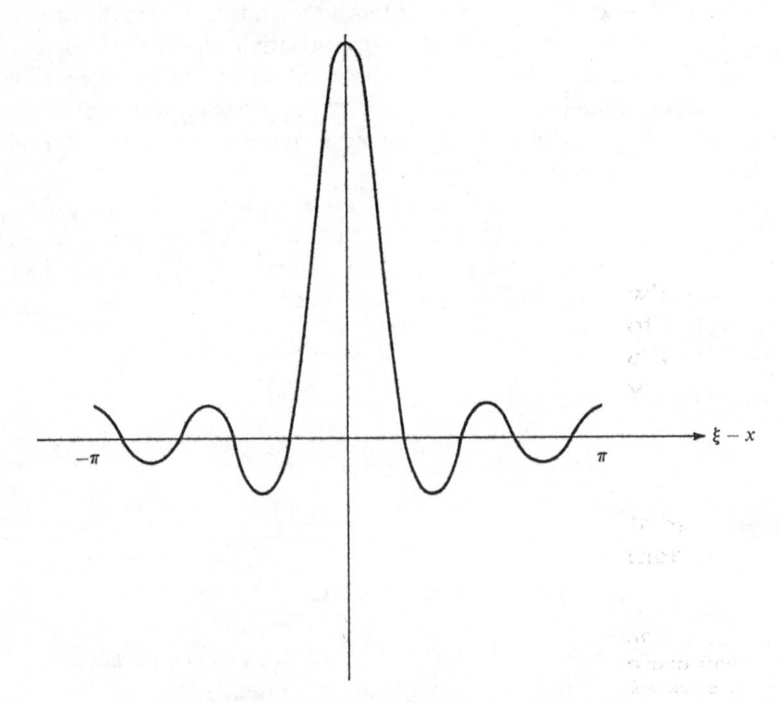

FIGURE 4.5. *Schematic graph of the function* $\sin [(N + \frac{1}{2})(x - \xi)]/2 \sin [\frac{1}{2}(x - \xi)]$.

with

$$S_N^{(1)}(x) \equiv \frac{1}{\pi}\int_0^\pi \frac{\sin[(N+\frac{1}{2})(x-\xi)]}{2\sin\frac{1}{2}(x-\xi)} f(\xi)\, d\xi, \tag{5}$$

$$S_N^{(2)}(x) \equiv -\frac{1}{\pi}\int_0^\pi \frac{\sin[(N+\frac{1}{2})(x+\xi)]}{2\sin\frac{1}{2}(x+\xi)} f(\xi)\, d\xi. \tag{6}$$

Because of the highly oscillatory nature of the integrand in (6), we can expect the integral to approach zero as $N \to \infty$. For when N is large, over each period of the sinusoidal function in the numerator the other factors are practically constant, and hence there is essentially no contribution to the integral from each period. In (5) a similar argument holds *except* when ξ is near x, where (roughly speaking) the large factor $N + \frac{1}{2}$ is "neutralized" by the small factor $x - \xi$. Thus the behavior of $f(\xi)$ is important only when ξ is near x. We state these conclusions in the form of the following lemmas, which we shall subsequently prove.

Lemma 1. If $\phi(\xi)$ is piecewise smooth in the closed interval $[a, b]$, then

$$\int_a^b \phi(\xi) \begin{matrix}\cos\\ \sin\end{matrix} \lambda\xi\, d\xi = O(\lambda^{-1}) \qquad \text{as } \lambda \to \infty. \tag{7}$$

Lemma 2. If $\phi(\xi)$ is piecewise smooth in the closed interval $[a, b]$, then at a point x_0 where ϕ is continuous,

$$\lim_{\lambda\to\infty} \int_a^b \phi(\xi) \frac{\sin \lambda(x_0 - \xi)}{x_0 - \xi}\, d\xi = \pi\phi(x_0), \qquad a < x < b. \tag{8}$$

Lemma 3. At a point x_0 where $\phi(x)$ has a jump discontinuity, (8) holds, provided that we define $\phi(x_0)$ to be the average of the left- and right-hand limits of $\phi(\xi)$ as $\xi \to x_0$:

$$\phi(x_0) = \tfrac{1}{2}[\phi(x_0 - 0) + \phi(x_0 + 0)]. \tag{9}$$

Here, by definition,

$$\phi(x_0 \pm 0) = \lim_{\varepsilon \downarrow 0} \phi(x_0 \pm \varepsilon),$$

where the downward-pointing arrow means that ε decreases to zero; i.e., ε approaches zero through positive values.

According to Lemma 1, $S_N^{(2)}(x) \to 0$. Also $S_N^{(1)}(x) \to f(x)$, according to Lemma 2, if we define

$$\phi(\xi) = f(\xi) \frac{x - \xi}{2\sin\frac{1}{2}(x - \xi)}. \tag{10}$$

Furthermore, if there is a discontinuity of $f(x)$ at $x = x_0$, the sum of the sine series at x_0 would be

$$f(x_0) = \tfrac{1}{2}[f(x_0 + 0) + f(x_0 - 0)], \tag{11}$$

in accordance with (9). This completes the process of summing the Fourier sine series—except for proving the lemmas.

PROOF OF THE LEMMAS

To prove Lemma 1, we break the integral into a finite number of similar integrals over intervals in each of which the function $\phi'(x)$ is continuous. Take a typical interval (c, d); we integrate by parts, obtaining

$$\int_c^d \phi(\xi)e^{i\lambda\xi}\,d\xi = [(i\lambda)^{-1}\phi(\xi)e^{i\lambda\xi}]_c^d - (i\lambda)^{-1}\int_c^d \phi'(\xi)e^{i\lambda\xi}\,d\xi.$$

Since all the functions involved are bounded, Lemma 1 follows immediately. □ (We have used a complex exponential function in the integrand for convenience. Because the real and imaginary parts in an equation must be separately equal to each other, both cosine and sine terms are included.)

To prove Lemma 2, let (c, d) denote an interval, containing x_0, in which ϕ is smooth. We need only consider the integral (8) over this interval, since the remaining contribution approaches zero as $\lambda \to \infty$, according to Lemma 1. Using the identity

$$\phi(x) = \phi(x_0) + [\phi(x) - \phi(x_0)],$$

and the definition

$$F(x, x_0) = \frac{\phi(x) - \phi(x_0)}{x - x_0}, \tag{12}$$

we write the integral in question as

$$I(\lambda) = \phi(x_0)\int_c^d \frac{\sin \lambda(x - x_0)}{x - x_0}\,dx + \int_c^d F(x, x_0)\sin \lambda(x - x_0)\,dx. \tag{13}$$

When $\lambda \to \infty$, the first term approaches $\pi\phi(x_0)$, since

$$\int_{-\infty}^{\infty} \frac{\sin v}{v}\,dv = \pi. \tag{14}$$

Formula (14) can be proved most easily by the method of contour integration in the complex plane. Alternatively, it is not difficult to prove the convergence of the integral, and the calculations in this section would assure us that the value of the integral is π.

The second term in (13) can be shown to approach zero as $\lambda \to \infty$ with the help of Lemma 1. Given any preassigned positive number ε, we can choose a number $\delta(\varepsilon)$ such that

$$\left| \int_{x_0-\delta}^{x_0+\delta} F(x, x_0) \sin \lambda(x - x_0)\, dx \right| < \frac{\varepsilon}{2},$$

since the integrand is bounded. In the intervals $(c, x_0 - \delta)$ and $(x_0 + \delta, d)$, the contributions to the integral can be made as small as we wish by choosing λ sufficiently large, according to Lemma 1: hence the desired result. □

To prove Lemma 3, we break the interval (a, b) into the intervals (a, x_0) and (x_0, b), and repeat the arguments for Lemma 2. We are led to the integrals

$$\int_0^{\infty} \frac{\sin v}{v}\, dv = \int_{-\infty}^{0} \frac{\sin v}{v}\, dv,$$

whose values are $\pi/2$, according to (14). The rest of the derivation is easy (Exercise 3). □

A FORMAL TRANSFORMATION

A formal method that is useful for quickly obtaining the right answer, but which is not convenient for proving its correctness, proceeds as follows.

If we introduce into (8) the new variable u by the relation

$$\lambda(\xi - x_0) = u, \tag{15}$$

we obtain an integral of the form

$$I(\lambda) = \int_{\lambda(a-x_0)}^{\lambda(b-x_0)} \phi\left(x_0 + \frac{u}{\lambda}\right) \frac{\sin u}{u}\, du. \tag{16}$$

From a formal point of view it is clear that

$$\lim_{\lambda\to\infty} I(\lambda) = \int_{-\infty}^{\infty} \phi(x_0) \frac{\sin u}{u}\, du = \pi\phi(x_0). \tag{17}$$

To justify the formal limiting process, however, we have to write

$$\phi\left(x_0 + \frac{u}{\lambda}\right) = \phi(x_0) + \left[\phi\left(x_0 + \frac{u}{\lambda}\right) - \phi(x_0)\right]$$

and use essentially the same arguments as above.

FOURIER SERIES IN THE FULL RANGE

Each term of the sine series is an odd function in x and hence the series defines an odd function $f(x)$. That is, if $f(x)$ is given by the series in $(0, \pi)$, then the series defines $f(x)$ in the interval $(-\pi, 0)$ by the identity

$$f(x) = -f(-x), \qquad -\pi < x < 0. \tag{18}$$

Beyond the interval $(-\pi, \pi)$, the function defined by the Fourier series is periodic:

$$f(x + 2\pi) = f(x). \tag{19}$$

A periodic function satisfying (19) does not have to be odd. If we want to represent a general function $f(x)$ in an interval $(-\pi, \pi)$, we can adopt the "full range" Fourier series*

$$f(x) \sim \tfrac{1}{2}a_0 + \sum_{n=1}^{\infty} (a_n \cos nx + b_n \sin nx); \qquad -\pi < x < \pi. \tag{20}$$

Formal use of Equation (1.38) yields (Exercise 4) the following expressions for the coefficients:

$$a_n = \frac{1}{\pi} \int_{-\pi}^{\pi} f(x) \cos nx \, dx, \qquad b_n = \frac{1}{\pi} \int_{-\pi}^{\pi} f(x) \sin nx \, dx. \tag{21}$$

We must now prove that if (21) holds, the series (20) converges to $f(x)$ under appropriate conditions.

The series (20) is also often used in the complex form

$$f(x) \sim \sum_{-\infty}^{\infty} c_n e^{inx}, \tag{22}$$

where

$$c_n = \frac{1}{2\pi} \int_{-\pi}^{\pi} f(\xi) e^{-in\xi} \, d\xi. \tag{23}$$

Note that

$$c_0 = \frac{a_0}{2}, \qquad c_n = \tfrac{1}{2}(a_n - ib_n). \tag{24}$$

Note also that if $f(x)$ is real, then $c_{-n} = \bar{c}_n$, where the bar denotes complex conjugate. For an application of complex Fourier series see Chapter 8 of II.

SUMMATION OF FOURIER SERIES

Consider the series in the form (22) and (23). Let

$$S_N(x) = \sum_{n=-N}^{N} c_n e^{inx}. \tag{25}$$

Then, by using (23), we see that

$$S_N(x) = \frac{1}{2\pi} \int_{-\pi}^{\pi} K_N(x, \xi) f(\xi) \, d\xi, \tag{26}$$

where now

$$K_N(x, \xi) = \sum_{-N}^{N} e^{in(x-\xi)} = \frac{\sin (N + \frac{1}{2})(x - \xi)}{\sin \frac{1}{2}(x - \xi)}. \tag{27}$$

* The symbol $\sim$ indicates that the series on the right of (20) "corresponds" to f. An equals sign is not used (when one is being careful) because sometimes the series does not converge to the function, at least not in the usual sense of "converge."

But (27) contains only the one significant term in (3), so that the lemmas proved before lead immediately to the desired result. We thus have the following theorem.

Theorem. If $f(x)$ is piecewise smooth in the interval $(-\pi, \pi)$, then its Fourier series, as defined by (20) and (21), converges to

$$f(x_0) = \tfrac{1}{2}[f(x_0 - 0) + f(x_0 + 0)] \tag{28}$$

at $x = x_0$. At the end points, the series converges to $\frac{1}{2}[f(-\pi + 0) + f(\pi - 0)]$.

HALF-RANGE SERIES

The behavior of a Fourier sine series in the interval $(0, \pi)$ can now be understood. In view of the "oddness" relationship (18), the theorem above shows that the series must converge to zero at the end points $(0, \pi)$. (This is not surprising, since each term of the series is zero at the end points.) Thus, when a Fourier sine series in the interval $(0, \pi)$ is regarded as a complete Fourier series in the interval $(-\pi, \pi)$, it always represents an odd function that vanishes at $x = 0$ or whose mean value is zero at $x = 0$.

EXERCISES

1. Verify (3).
2. (a)‡ Prove that the infinite integral of (14) is convergent.
   (b) If you are familiar with contour integration, use this method to demonstrate (14).
   (c) Discuss the difficulties inherent in obtaining by numerical methods a value for the integral in (14).
3. Complete the proof of Lemma 3.
4. Verify (21).

‡5. If a function of period 2π has a continuous third derivative in $[-\pi, \pi]$ (including the end points), show that its Fourier coefficients are $O(n^{-3})$ as $n \to \infty$.

4.3 On the Nature of Fourier Series

The principal topics to be treated in this section are (i) integration and differentiation of Fourier series; (ii) the Gibbs phenomenon, a concommitant of the nonuniform convergence of the series; (iii) the relation between Fourier series and least squares approximation; and (iv) Parseval's theorem on the sum of the squares of the Fourier coefficients. But before we discuss the general nature of Fourier series, let us examine a few simple examples. We consider, in particular, the Fourier series of some simple powers and polynomials. In each case we shall consider the three possible developments:

(i) The full range series in the complete interval $(-\pi, \pi)$.
(ii) The sine series in the interval $(0, \pi)$.
(iii) The cosine series in the interval $(0, \pi)$.

The reader should verify all the necessary calculations that yield (1), (2), (4), (5), (7), (8), and (9) below (Exercise 1).

FOURIER SERIES FOR THE CONSTANT FUNCTION

For the complete interval, the series for the constant function $f(x) = 1$ has only one term, the constant term. All the higher terms vanish:

$$a_n = b_n = 0; \qquad n \geq 1. \tag{1}$$

The cosine series takes on the same form, since $f(x)$ is even.

The sine series is much more interesting. It is

$$1 = \frac{4}{\pi}\left\{\sin x + \frac{1}{3}\sin 3x + \frac{1}{5}\sin 5x + \cdots\right\}, \qquad 0 < x < \pi. \tag{2}$$

Because the sines are odd, the above series represents -1 for the interval $(-\pi, 0)$. We shall designate by $S(x)$ the **square wave function** represented by the series (2). Thus

$$S(x) = \begin{cases} -1, & -\pi < x < 0, \\ +1, & 0 < x < \pi; \end{cases} \tag{3}$$

and S is periodic with period 2π. The function takes a step 2 upward at $x = 0$ and a step downward at $x = \pi$ [Figure 4.6(a)].

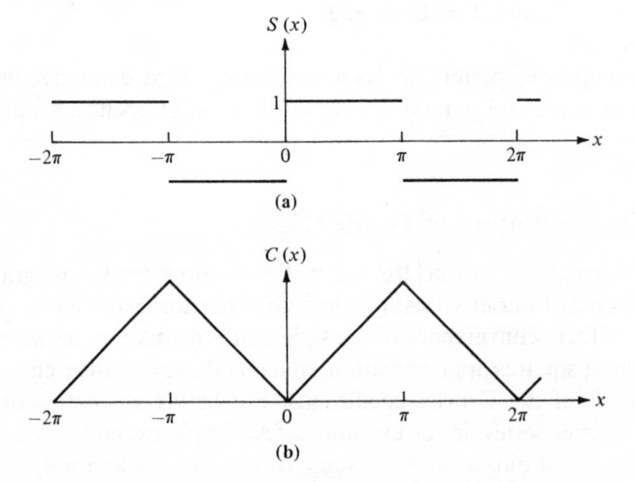

FIGURE 4.6. *The functions S(x) and C(x).*

FOURIER SERIES FOR THE LINEAR FUNCTION

For the complete interval, the Fourier series for $f(x) = x$ is

$$x = 2\{\sin x + \tfrac{1}{2}\sin 2x + \tfrac{1}{3}\sin 3x + \cdots\}, \qquad -\pi < x < \pi. \tag{4}$$

For the interval $(0, \pi)$, the sine series is obviously still given by (4); but the cosine series, which represents the function $|x|$ in $[-\pi, \pi]$, is

$$|x| = \pi - \frac{4}{\pi}\left\{\frac{\cos x}{1^2} + \frac{\cos 3x}{3^2} + \cdots\right\}, \qquad -\pi \le x \le \pi. \tag{5}$$

Let us designate the function represented by the series (5) by $C(x)$. We note that $C(x)$ is continuous and has corners at $x = 0, \pm 2\pi, \ldots$ [Figure 4.6(b)]. We also note that

$$C'(x) = S(x). \tag{6}$$

This follows by comparing (2) with (5), or by a direct examination of the functions C and S. Using (6), if we integrate the series in (2) from 0 to x, we obtain

$$\frac{\pi}{4} C(x) = (1 - \cos x) + \frac{1}{3^2}(1 - \cos 3x) + \frac{1}{5^2}(1 - \cos 5x) + \cdots. \tag{7}$$

At $x = \pi$, this gives

$$\frac{\pi^2}{8} = 1 + \frac{1}{3^2} + \frac{1}{5^2} + \cdots \tag{8}$$

The reader should verify that one can obtain the same answer by formally calculating the average of the square of the series (2) over the interval $(0, \pi)$. This is a special case of Parseval's theorem (to be considered below).

FOURIER SERIES FOR THE QUADRATIC FUNCTION

We can discuss $f(x) = x^2$ in the same spirit, but we know from our experience that its sine series is of primary interest. We proceed indirectly by formal integration of (5), obtaining, not x^2, but the polynomial

$$\frac{\pi}{8}(\pi x - x^2) = \sin x + \frac{1}{3^3}\sin 3x + \frac{1}{5^3}\sin 5x + \cdots. \tag{9}$$

The sine series for x^2 can now be obtained by (4) and (9). By putting $x = \pi/2$ into this series, we can obtain a series for π^3 [Exercise 8(b)]. This series converges faster than series (8) for π^2 and can be used as a fairly practical way to compute π.

INTEGRATION AND DIFFERENTIATION OF FOURIER SERIES

Let $f(x)$ be a piecewise smooth function with the Fourier series

$$f(x) = \frac{a_0}{2} + \sum_{n=1}^{\infty}(a_n \cos nx + b_n \sin nx), \tag{10}$$

convergent in the interval $(-\pi, \pi)$. Since the convergence is not uniform, standard theorems give no assurance that the series can be integrated term by term. Term-by-term integration is nevertheless permissible, as we shall see. There is little reason to expect that (10) can be differentiated term by term; indeed, series (2) and (4) are obvious cases in which the differentiated series do not converge.*

Let us examine these questions a little more closely. The formal integration of (10) from 0 to x leads to

$$F(x) - \frac{a_0}{2}x \sim \sum_{n=1}^{\infty} \left\{\frac{a_n}{n} \sin nx + \frac{b_n}{n}[1 - \cos nx]\right\}, \tag{11}$$

where

$$F(x) = \int_0^x f(t)\,dt. \tag{12}$$

Formal differentiation of (10) yields

$$f'(x) \sim 0 + \sum_{n=1}^{\infty} (-na_n \sin nx + nb_n \cos nx). \tag{13}$$

We shall now verify (11) and (13) directly by (i) calculating the Fourier series of the relevant functions, and then (ii) by examining their convergence. For (11), the calculation is straightforward, and the reader can easily verify (Exercise 3) that if we write

$$F(x) - \frac{a_0}{2}x = \frac{A_0}{2} + \sum_{n=1}^{\infty} (A_n \cos nx + B_n \sin nx), \tag{14}$$

then

$$A_n = \frac{-b_n}{n}, \quad B_n = \frac{a_n}{n}, \qquad n \neq 0. \tag{15}$$

The convergence of the series (14) is assured, since F is continuous (because it is the integral of a piecewise smooth function). In particular, $F(x)$ is continuous at $x = 0$, and $F(0) = 0$, so that

$$0 = \tfrac{1}{2}A_0 + \sum_{n=1}^{\infty} A_n. \tag{16}$$

Equation (11) then follows from (14) and (16).

Direct verification of (13) constitutes much more of a problem. There is no assurance that the series converges at all, much less that it converges to $f'(x)$. To see what happens, let

$$\frac{a_0'}{2} + \sum_{n=1}^{\infty} (a_n' \cos nx + b_n' \sin nx) \tag{17}$$

* The differentiated series can be "summed" in an appropriate manner, but this matter is beyond the present discussion.

denote the Fourier series for $f'(x)$. Upon partial integration of the defining integrals of a'_n and b'_n and comparison with the corresponding integrals for a_n and b_n, we find (Exercise 3) that

$$\pi a'_n = \pi n b_n + (-1)^n[f(\pi) - f(-\pi)], \tag{18}$$

$$\pi b'_n = -\pi n a_n. \tag{19}$$

Reference to the formal series (13) shows that if

$$f(-\pi) = f(\pi), \tag{20}$$

i.e., if the function is periodic and has no discontinuity anywhere, including $-\pi$ or $+\pi$, then formal differentiation is justified if the resultant series converges. An example is afforded by the function $C(x)$ with Fourier series (5). Note that $C(-\pi) = C(+\pi) = \pi$. In this example, $C'(x)$ can be shown to have a convergent Fourier series (Exercise 4).

Considerations similar to the above may be made for sine expansions and cosine expansions in the range $(0, \pi)$. Referring to equations analogous to (18) and (19), one can see that if $f'(x)$ has a sine expansion in $(0, \pi)$, it can be obtained by differentiating the cosine expansion of $f(x)$. On the other hand, if we start out with a sine expansion of $f(x)$, formal differentiation is in general not justified.

GIBBS PHENOMENON

Just before the turn of the century, the American physicist A. Michelson constructed a machine by which the first 80 Fourier components of a graphically given function could be determined. This analysis could be checked by adding up the components obtained and verifying that the function so synthesized was close to the original function. In most cases it was, but when the square wave function $S(x)$ was analyzed, at its point of discontinuity "a peculiar protuberance appeared which was not present in the original function. Michelson was puzzled and thought that perhaps a hidden mechanical defect of the machine might cause the trouble. He wrote about his observation to Josiah Gibbs, the eminent mathematical physicist, asking for his opinion. Gibbs investigated the phenomenon and explained it (in a letter to *Nature* in 1899)."*

The unusual behavior of the partial sum near a discontinuity is called the **Gibbs phenomenon**. Some such behavior should perhaps have been anticipated from the outset, in view of the fact that the trigonometric functions are infinitely differentiable, yet the Fourier series may converge to a discontinuous function.

* The quotation is from a book that is an excellent source of further reading on Fourier series. The spirit is similar to that we have tried to convey, but much more material is presented. The book is C. Lanczos, *Discourse on Fourier Series* (Edinburgh: Oliver & Boyd, 1966).

As a simple example of the Gibbs phenomenon, consider the square wave function $S(x)$ of (3) and the behavior of its Fourier series (2) near $x = 0$. The series converges, partly as a result of the decrease in size of the successive terms and partly as a result of the change of sign associated with the sine function. However, for very small values of x, this oscillatory behavior is not evident in the first k terms if $(2k + 1)x < \pi$. All these terms are positive. Indeed, we shall now show that as long as $(2k + 1)x$ is finite, there will be a deviation from the final value. One would therefore expect the partial sum of (2) to approach the limiting value only when $(2k + 1)x \to \infty$.

Using formulas (2.26) and (2.27) for $S_N(x)$, particularized for the square wave function $S(x)$ of (3), we see that

$$S_N(x) = \frac{1}{2\pi}\left(-\int_{-\pi}^{0} + \int_{0}^{\pi}\right) \frac{\sin (N + \frac{1}{2})(x - \xi)}{\sin \frac{1}{2}(x - \xi)}\, d\xi. \tag{21}$$

In the intervals $(-\pi, 0)$ and $(0, \pi)$, we introduce the transformations $\theta = x - \xi$ and $\theta = \xi - x$, respectively. Then (21) becomes

$$S_N(x) = \frac{1}{2\pi}\left(\int_{x+\pi}^{x} + \int_{-x}^{\pi-x}\right) \frac{\sin (N + \frac{1}{2})\theta}{\sin \frac{1}{2}\theta}\, d\theta. \tag{22}$$

[The reader is asked in Exercise 6(a) to verify the above equation and also equations (23), (25), (26), and (27) below.] Equation (22) may also be written as

$$S_N(x) = \frac{1}{2\pi}\left(\int_{-x}^{x} + \int_{\pi+x}^{\pi-x}\right) \frac{\sin (N + \frac{1}{2})\theta}{\sin \frac{1}{2}\theta}\, d\theta. \tag{23}$$

One of the discontinuities of $S(x)$ is at $x = 0$. Let us focus on the behavior of the partial sum $S_N(x)$ in the neighborhood of this point, *for positive values of* x. When x is small, the effect of the second term is small, since the denominator in the integrand is near 1. For the first term, we write

$$N + \tfrac{1}{2} = m, \qquad m\theta = \eta, \tag{24}$$

with which

$$S_N(x) \approx \frac{1}{\pi}\int_0^{mx} \frac{\sin \eta}{m \sin (\eta/2m)}\, d\eta \equiv I_N(x). \tag{25}$$

By writing

$$I_N(x) = \frac{2}{\pi}\int_0^{mx} \frac{\sin \eta}{\eta}\,\frac{(\eta/2m)}{\sin (\eta/2m)}\, d\eta,$$

we see that for any fixed positive value of x, no matter how small, we have

$$\lim_{N\to\infty} I_N(x) = \frac{2}{\pi}\int_0^{\infty} \frac{\sin \eta}{\eta}\, d\eta = 1. \tag{26}$$

On the other hand, for $y = mx = (N + \frac{1}{2})x$ fixed, we have

$$\lim_{N\to\infty} I_N(x) = \frac{2}{\pi}\, Si[(N + \tfrac{1}{2})x], \tag{27a}$$

where the **sine integral function** Si is defined by*

$$Si(y) = \int_0^y \frac{\sin \eta}{\eta}\, d\eta. \tag{27b}$$

The first maximum of $Si(y)$ occurs at $y = \pi$ [Exercise 5(a)]. Therefore $I_N(x)$ and $S_N(x)$ reach approximately the peak value

$$S_{\max} = \frac{2}{\pi}\int_0^\pi \frac{\sin \eta}{\eta}\, d\eta \approx 1.179 \tag{28}$$

at $x \approx \pi/(N + \frac{1}{2})$, or $Nx \approx \pi$. This means that the maximum value of the partial sum S_N of the series (2) is reached if we include only the *first* set of *positive* terms, in agreement with our general expectations.

REMARK. This comparison between the sum S_N and the integral allows us to deduce (27) by the following heuristic reasoning. If we denote $2x$ by h, and introduce $y = (N + \frac{1}{2})x$ as before, we may write $S_N(x)$ as a sum approximating to the integral in (27). The reader is asked to carry out this reasoning in Exercise 6(b).

Formula (27a) shows that, for N large, the values of x for which there is deviation of $I_N(x)$ from its limiting value unity [for $x > 0$, $(N + \frac{1}{2})x \to \infty$] is restricted to $x = O(N^{-1})$. *The maximum error remains finite at about* 18 *per cent as* $N \to \infty$, *but the range over which there is an appreciable error is squeezed down to zero as* $N \to \infty$. This is typical of Gibbs phenomenon. It is a splendid illustration of a series that converges but does not converge uniformly (Figure 4.7).

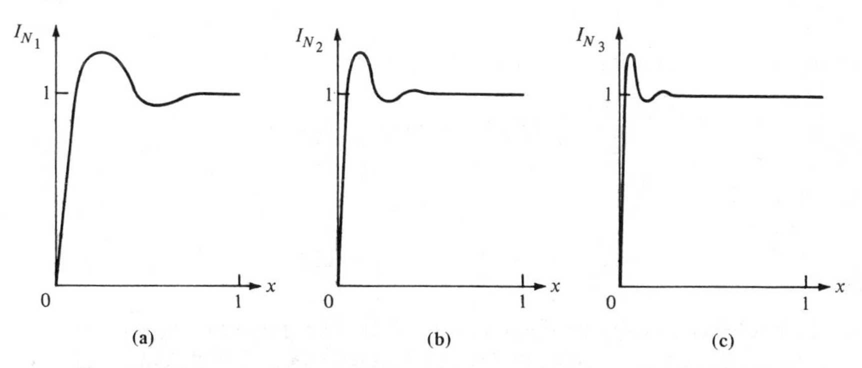

FIGURE 4.7. *Qualitative graphs of* $I_N(x)$. *The values of* N *are taken successively larger in* (a), (b) *and* (c), *respectively. As* N *increases, the peak continually narrows and passes to the left of any fixed* x. *Consequently,* $I_N(x) \to 1$ *as* $N \to \infty$, *for any fixed* x *such that* $0 < x < 1$. *But the peak is always present; so for any* N *there are always small positive values of* x *for which* $|I_N(x) - 1|$ *is not small, and thus convergence is nonuniform for* $0 < x < 1$.

* For formulas, graphs, and tables of Si, see Abramovitz and Stegun (1964, pp. 231 ff.).

APPROXIMATION WITH LEAST SQUARED ERROR

While the study of the pointwise convergence of the Fourier series is extremely interesting, one cannot help wishing for a simpler perspective of the approximation procedure. A simple approach, which has much greater generality, is the method of least squares, due originally to Gauss.

Consider a finite sum of trigonometric functions

$$\tilde{S}_N(x) = \sum_{n=-N}^{N} \gamma_n e^{inx}, \qquad \gamma_{-n} = \bar{\gamma}_n. \tag{29}$$

If we want this sum to approximate a given function $f(x)$, we would like to make the difference

$$\varepsilon_N(x) = f(x) - \tilde{S}_N(x) \tag{30}$$

as small as possible in some sense. This may be accomplished by minimizing the **mean square error:**

$$M = \frac{1}{2\pi} \int_{-\pi}^{\pi} \varepsilon_N^2 \, dx \tag{31}$$

by a proper choice of the undetermined coefficients γ_n.

The quantity M defined by (31) is a quadratic function of the coefficients γ_n, and hence a necessary condition for it to be a minimum is to require the simultaneous *linear* equations

$$\frac{\partial M}{\partial \gamma_n} = 0 \tag{32}$$

to be satisfied. Equation (32) yields the condition

$$\frac{1}{2\pi} \int_{-\pi}^{\pi} [f(x) - \tilde{S}_N(x)] e^{inx} \, dx = 0,$$

or the conditions

$$\gamma_n = \frac{1}{2\pi} \int_{-\pi}^{\pi} f(x) e^{-inx} \, dx, \qquad n = 0, 1, 2, \ldots, N, \tag{33}$$

as the reader can easily verify [Exercise 7(a)]. The quantities γ_n are thus the Fourier coefficients defined by (2.23) and denoted by c_n. In fact, a little calculation shows that

$$M = \langle f^2 \rangle - \sum_{n=-N}^{N} |c_n|^2 + \sum_{n=-N}^{N} |c_n - \gamma_n|^2, \tag{34}$$

where $\langle f^2 \rangle$ denotes the mean value of f^2 (over the interval $[-\pi, \pi]$):

$$\langle f^2 \rangle = \frac{1}{2\pi} \int_{-\pi}^{\pi} f^2(x) \, dx.$$

If we note that only the last term is at our disposal when γ_n is varied, it is clear that we must necessarily have $\gamma_n = c_n$ in order to minimize the error, and that this condition is also sufficient.

Note that the coefficients γ_n are *finally* decided by (33), in that they are not to be changed if we choose to include more terms in the approximating sum (29).

The method of least squares can be generalized in various ways. For example, suppose that we have an infinite set of functions $\phi_0(x), \ldots, \phi_n(x), \ldots,$ which satisfy the following condition involving some given nonnegative function w:

$$\int_{-\pi}^{\pi} \phi_m(x)\phi_n(x)w(x)\,dx = \delta_{mn}. \tag{35}$$

Before indicating the uses of such a set of functions, let us mention some terminology that is very often employed. Condition (35) when $m \neq n$ is by definition a property of functions that are **orthogonal on the interval** $[-\pi, \pi]$ **with respect to the weighting function** w. Condition (35) with $m = n$ is said to provide a **normalization** of the set of functions, for it specifies an otherwise arbitrary constant multiple of the ϕ_n. Mutually orthogonal and normalized functions are said to be **orthonormal**. (Subsequent discussion will justify the use of the term "orthogonal.")

Given a set of functions that satisfy the orthonormality condition (35), we may attempt to approximate an arbitrary function $f(x)$ in the interval $(-\pi, \pi)$ by a finite sum

$$\tilde{S}_N(x) = \sum_{n=0}^{N} \gamma_n \phi_n(x) \tag{36}$$

by minimizing the **weighted mean square error**

$$M_w = \frac{\int_{-\pi}^{\pi} \varepsilon_N^2(x)w(x)\,dx}{\int_{-\pi}^{\pi} w(x)\,dx}. \tag{37}$$

Here $\varepsilon_N(x)$ is defined by (30). Note that errors at values of x where w is relatively large are weighted more heavily in the computation of M_w; hence the term "weighted" error.

Let $\langle\ \rangle_w$ denote a weighted mean value so that

$$\langle f^2 \rangle_w = \frac{1}{2\pi}\int_{-\pi}^{\pi} f^2(x)w(x)\,dx.$$

We now have the following generalization of (34) [Exercise 7(a)]:

$$M_w = \langle f^2 \rangle_w - \sum_{n=-N}^{N} |c_n|^2 + \sum_{n=-N}^{N} |c_n - \gamma_n|^2, \tag{38}$$

where

$$c_n = \frac{\int_{-\pi}^{\pi} f(x)\phi_n(x)w(x)\,dx}{\int_{-\pi}^{\pi} w(x)\,dx}. \tag{39}$$

The error is minimized when the γ_n's are chosen to be the same as the c_n's defined by (39).

The method for constructing the set of functions $\phi_n(x)$ for a given $w(x)$ is essentially that of the *Gram-Schmidt process*, to be discussed in Section 5.2. We shall not go into this process here. Suffice it to say that it is possible to proceed in a step-by-step fashion, even if we accept the restriction that the $\phi_n(x)$ are polynomials of degree n. As an illustration, we shall merely mention the Legendre polynomials

$$P_n(x) \equiv \frac{1}{2^n n!} \frac{d^n}{dx^n} (x^2 - 1)^n, \qquad -1 < x < 1. \tag{40}$$

These can be regarded as polynomials defined on the interval $(-1, 1)$, which are mutually orthogonal in an unweighted sense ($w = 1$). The conventional normalization requirement imposed on Legendre polynomials is not

$$\int_{-1}^{1} P_n^2 \, dx = 1$$

but rather is $P_n(1) = 1$.

BESSEL'S INEQUALITY AND PARSEVAL'S THEOREM

If we put $\gamma_n = c_n$ in (34), we obtain the **Bessel inequality**

$$\sum_{n=-N}^{N} |c_n|^2 \le \langle f^2 \rangle, \tag{41a}$$

since $M > 0$. In terms of the real coefficients [compare (2.24)] this reads

$$\tfrac{1}{2}a_0^2 + \sum_{n=1}^{N} (a_n^2 + b_n^2) \le \langle f^2 \rangle. \tag{41b}$$

If the Fourier series converges in a pointwise manner, with the exception of a *finite* number of points of discontinuity, then

$$\lim_{N \to \infty} M = 0,$$

and hence

$$\sum_{n=-\infty}^{\infty} |c_n^2| = \langle f^2 \rangle \tag{42a}$$

or

$$\tfrac{1}{2}a_0^2 + \sum_{n=1}^{\infty} (a_n^2 + b_n^2) = \langle f^2 \rangle. \tag{42b}$$

Either of these identities is known as **Parseval's theorem.** They can be formally obtained from the Fourier series (2.20) or (2.22).

Even when the mean square error M does not approach zero, the inequalities (41) guarantee the convergence of the infinite series in (42).

RIESZ–FISCHER THEOREM

Suppose that we now pose the following problem, which can be regarded as a converse to Parseval's theorem. Given a set of real numbers a_0, a_m, b_m, $m = 1, 2, \ldots, \infty$ such that the series

$$\tfrac{1}{2}a_0^2 + \sum_{m=1}^{\infty} (a_m^2 + b_m^2)$$

is convergent, is there a function $f(x)$ such that the series

$$\tfrac{1}{2}a_0 + \sum_{n=1}^{\infty} (a_n \cos nx + b_n \sin nx) \tag{43}$$

is its Fourier series?

An affirmative answer to this question depends on the introduction of the concepts of Lebesgue measure and Lebesgue integration. With these notions introduced, we have the **Riesz–Fischer theorem**,* which states that (*i*) *the series* (43) *is indeed the Fourier series of a function f, which is square integrable, and that* (ii) *the partial sums of the series converge in the mean to f.*

The problem we posed is a very natural one from a mathematical point of view. It appears that it might have a simple solution, but it is here that new mathematical concepts and theories emerge. On the other hand, for physical applications, such a mathematical question does *not* arise naturally.

APPLICATIONS OF PARSEVAL'S THEOREM

Let us now consider, in contrast to the above theoretical discussions, some elementary applications of Parseval's theorem, which are useful for practical purposes. In particular, we shall show how one can evaluate the integral that occurs when one wants to calculate Stefan's radiation constant from Planck's law of radiation. As indicated in Exercise 9, this integral is

$$I = \int_0^{\infty} \frac{x^3\, dx}{e^x - 1}. \tag{44}$$

First, we note that if we set $x = \pi$ in (2) and then apply Parseval's theorem, we obtain another derivation of (8). Second, if we apply Parseval's theorem to (5), we obtain [Exercise 7(b)]

$$1 + \frac{1}{3^4} + \frac{1}{5^4} + \cdots = \frac{\pi^4}{96}. \tag{45}$$

To evaluate (44), as requested in Exercise 9, one expands the denominator of the integrand as a power series in e^{-x} and integrates term by term. One thereby obtains a constant multiple of

$$S_4 \equiv \frac{1}{1^4} + \frac{1}{2^4} + \frac{1}{3^4} + \cdots, \tag{46}$$

* See E. C. Titchmarsh, *Theory of Functions* (Oxford: Oxford University Press, 2nd ed., 1949), p. 423.

which differs from (45) in that *all* the integers, instead of just the odd ones, are present. But S_4 can be related to the series (45) by noting that the even terms themselves make up $2^{-4} S_4$, so that [Exercise 7(b)]

$$S_4 = \frac{\pi^4}{90}. \tag{47}$$

If presented with the problem of summing S_4, few would have the ability to obtain the answer (47) by making a seemingly roundabout detour through Fourier series. Consequently, our calculations can be regarded as an illustration of possible unexpected applicability of mathematical analysis.

EXERCISES

1. Verify the following Fourier series calculations: (1), (2), (4), (5), (7), (8) [two ways—see the remark under (8)], (9).

2. Carry out the calculations outlined under (9).

3. Verify (15), (18), and (19).

4. Show that $C'(x)$ has a convergent Fourier series.

5. (a) Verify that the first maximum of the sine integral function occurs at $y = \pi$.
(b) (Project) Without consulting the literature, derive as many properties as you can of the sine integral function.

6. (a) Verify (22), (23), (25), (26), and (27).
(b) Carry out the program of heuristic reasoning that is outlined under (28).

7. (a) Verify (33), (34), and (38).
(b) Verify (45) and (47).
(c) Why would mean cubed error not be a useful concept?

8. With the aid of the results proved in this section show that

(a) $$\frac{\pi}{4} = 1 - \frac{1}{3} + \frac{1}{5} - \frac{1}{7} + \cdots.$$

(b) $$\frac{\pi^3}{32} = 1 - \frac{1}{3^3} + \frac{1}{5^3} - \frac{1}{7^3} + \cdots.$$

(c) $$\frac{\pi^4}{96} = 1 + \frac{1}{3^4} + \frac{1}{5^4} + \frac{1}{7^4} + \cdots.$$

(d) Use series (c) to calculate an approximation to π.

9. (a) Using hints given in the text, show that

$$\int_0^\infty \frac{x^3\,dx}{e^x - 1} = \frac{\pi^4}{15}.$$

(b) The integral of (a) occurs in the theory of thermal radiation. As discussed in physics texts, according to Planck's law the density of radiation $u(\nu)$ [per unit volume, per unit range of frequency ν] at temperature T is given by

$$u(\nu) = \frac{8\pi h\nu^3/c^3}{\exp(h\nu/kT) - 1}.$$

Here h is Planck's constant and k is Boltzmann's constant.

Show that the total radiation density in a cavity at temperature T is

$$\frac{8\pi^5 k^4}{15h^3c^3} T^4,$$

giving the famous proportionality to the fourth power of the temperature of Stefan's law.

10. Calculate numerically the sum of a number of terms of series (2) to observe the Gibbs phenomenon in practice. Graph your results. (Project.)

11. Calculate numerically the sum of a number of terms of the solution of the heat equation when the initial temperature of the rod is uniform at the beginning and the ends are suddenly brought to zero temperature. Notice that there is, in general, no Gibbs phenomenon. Can you imagine a situation in which the Gibbs phenomenon would be present? (Project.)

12. This exercise concerns the inhomogeneous boundary value problem

$$\frac{d^2y}{dx^2} + ky = f, \quad 0 < x < \pi; \qquad y(0) = y(\pi) = 0. \tag{48}$$

Here k is regarded as a given fixed constant, and f is a given function of x.

(a) Using integration by parts, show that if (48) is to have a solution, then it is necessary that f be orthogonal [on the interval $(0, \pi)$ with weighting function unity] to all solutions ϕ of

$$\frac{d^2\phi}{dx^2} + k\phi = 0, \quad 0 < x < \pi; \qquad \phi(0) = \phi(\pi) = 0.$$

(This necessary condition will be used in Exercise 7.2.11, which is concerned with the application of perturbation theory to determine eigenvalues.)

(b) Assume that

$$f = \sum_{n=1}^{\infty} c_n \sin nx, \qquad y = \sum_{n=1}^{\infty} k_n \sin nx.$$

Since f is known, the c_n will be known coefficients of a Fourier sine series. The problem is to determine the k_n. Do this by substituting into (48). Show that there are zero, one, or an infinite number of solutions, depending on the nature of f. (This is a special case of the "Fredholm alternative.") Are the results compatible with part (a)?

CHAPTER 5

Further Developments in Fourier Analysis

IN CHAPTER 4 we examined simple forms of the problem of heat conduction and we presented some of the mathematical theory of Fourier series that was inspired by attempts to solve this problem. Our treatment followed historical developments closely and serves as a case study to illustrate the role of applied mathematics.

In this chapter we shall again make a study in this spirit but at a somewhat deeper level. Once more starting with specific problems, we shall outline theory and methods of wide-ranging applicability and implications.

We begin by considering some specific problems of heat conduction. A study of the problem of heat conduction in inhomogeneous media will lead to the discussion of eigenfunctions more general than the harmonic functions. The case of an infinite homogeneous medium will introduce the topic of Fourier integral. It will then be pointed out that the analysis of a time sequence, such as a weather record, may often be conveniently made in terms of Fourier series or integral but that this is not sufficient. There exist time sequences that cannot be so represented. The theory of generalized harmonic analysis is required, and this will be briefly described.

5.1 Other Aspects of Heat Conduction

This section begins by reducing the study of annual temperature variation in the earth to a relatively simple heat conduction problem for a homogeneous medium. A brief calculation brings out the main features of the solution. The discussion then turns to numerical analysis of the heat equation. Finally, advantages of a mixed numerical and analytical approach are illustrated with reference to heat conduction in a nonuniform medium.

VARIATION OF TEMPERATURE UNDERGROUND

On the surface of the earth, the daily average temperature at time t, $f(t)$, may be regarded as varying with a period of one year. We may represent this variation by

$$f(t) = \sum_{n=-\infty}^{\infty} C_n e^{2\pi i n t/T}, \qquad \text{where} \quad \bar{C}_n = C_{-n}, \quad T = 1 \text{ year}. \tag{1}$$

We want to know the consequent variation of temperature underground (say, in a cellar). We shall neglect the curvature of the surface of the earth, since we shall be concerned with depths of the order of only a few meters. These two approximations—the neglect of diurnal (daily) variation and the neglect of curvature—are introduced to formulate a *simplified model.* Such approximations could in principle be justified by analyzing a model that contains curvature and diurnal variation effects. Yet this may be quite complicated, and is indeed hardly worthwhile, at least at this stage. But plausible arguments to simplify the problem can be given in terms of the smallness of certain parameters. Adopting a procedure that we recommend for general use, we shall sketch these arguments at once, before we attempt to carry out the detailed calculations.

In the simplified model the equation governing the distribution of temperature $\theta(x, t)$ at a depth x and at time t is the usual equation of heat conduction,

$$\frac{\partial \theta}{\partial t} = \kappa \frac{\partial^2 \theta}{\partial x^2}. \tag{2}$$

Here κ is the thermometric diffusivity of soil whose value is approximately

$$\kappa = 2 \times 10^{-3} \text{ cm}^2/\text{sec}. \tag{3}$$

The relevant time scale T is a year:

$$T = 3.15 \times 10^7 \text{ sec}. \tag{4}$$

It is a fundamental fact about diffusion (see the end of Section 4.1) that in time T a substance diffusing with diffusivity κ will spread a distance of order of magnitude $\sqrt{\kappa T}$. In the present case this length scale is

$$\sqrt{\kappa T} \approx 250 \text{ cm} \approx 2.5 \text{ in}. \tag{5}$$

For diurnal variation the time scale is $T/365$ so that the length scale would be smaller than (5) by a factor of $\sqrt{365} \approx 19$. Thus neglect of diurnal variation would be expected to produce errors of the order of 5 or 10 per cent, while the curvature of the earth would produce entirely negligible effects on phenomena limited to a few meters depth.

To find the temperature underground according to our simplified model, we first take note of the boundary condition $\theta(0, t) = f(t)$, where $f(t)$ is given by (1). We *try* solutions of the form

$$\theta(x, t) = \sum_{n=-\infty}^{\infty} C_n w_n(x) e^{2\pi i n t/T}, \tag{6}$$

where $\bar{C}_n = C_{-n}$ so that θ is real. (The bar denotes a complex conjugate.) We make the following stipulations:

(i) Each of the terms will satisfy (2).

(ii) $w_n(0) = 1$ so that (1) will be satisfied.

(iii) $w_n(x)$ remains bounded and presumably approaches zero as $x \to \infty$ (for $n \neq 0$), since the temperature at great depths is not expected to be sensitive to the variations of surface temperature.

Can all these stipulations be satisfied? We must carry out the calculations to find out.

The first stipulation yields the following ordinary differential equation for $w_n(x)$:

$$\frac{d^2 w_n}{dx^2} = p_n^2 w_n, \qquad p_n^2 \equiv \frac{2\pi i n}{\kappa T}. \tag{7}$$

The possible values of p_n are

$$p_n = (1 \pm i) q_n, \qquad \text{where } q_n = \left(\frac{|n|\pi}{\kappa T}\right)^{1/2} > 0 \tag{8}$$

and the $\pm$ sign is to be taken accordingly as $n \gtrless 0$ (Exercise 1). The general solution of (7) is then

$$w_n(x) = A_n e^{(1 \pm i) q_n x} + B_n e^{-(1 \pm i) q_n x}. \tag{9}$$

Because of condition (iii), we must have $A_n = 0$. To satisfy condition (ii), we must have $B_n = 1$. Thus the final form of solution (6) becomes

$$\theta(x, t) = \sum_{n=-\infty}^{\infty} C_n e^{-(1 \pm i) q_n x} e^{2\pi i n t/T}. \tag{10}$$

To emphasize that θ is real, we write the complex coefficients in polar form, putting

$$C_n = |C_n| e^{i\gamma_n} \qquad \text{so that } C_{-n} = \bar{C}_n = |C_n| e^{-i\gamma_n}. \tag{11}$$

Equation (10) then becomes

$$\theta(x, t) = |C_0| + 2 \sum_{n=1}^{\infty} |C_n| e^{-q_n x} \cos\left(2\pi n \frac{t}{T} + \gamma_n - q_n x\right). \tag{12}$$

To interpret the solution, we first note that the cosine factor represents a wave of frequency $2\pi n/T$ and wave number q_n. (If these matters are not familiar, see Section 12.2.) Thus the nth "partial wave" propagates with a speed

$$\frac{2\pi n}{q_n T} = \left(\frac{4\pi\kappa |n|}{T}\right)^{1/2}. \tag{13}$$

In addition, there is exponential damping in the direction of propagation. From (8) the damping coefficient q_n increases like $|n|^{1/2}$.

The constants $|C_n|$ and γ_n give the amplitudes and phases of the various damped waves. These can be determined from initial conditions, but this would add little to the general understanding we seek.

Since damping increases with n, the most important contribution to the solution comes from the term with $n = 1$. With the numerical values cited above, for which $\sqrt{\kappa T} \approx 250$ cm, we find from (8) that

$$q_1 \approx 0.71 \text{ m}^{-1}. \tag{14}$$

Thus, at the point $x_1 = 4.4$ m, where $q_1 x_1 = \pi$, the temperature is opposite in phase to the surface condition, but the amplitude is down by a factor of $e^{-\pi} = 0.0435$. *It is winter at a depth of 4.4 meters when it is summer on the surface.* The temperature variation, however, is only about 4 per cent of the surface variation. This shows the usefulness of a deep cellar for wine and vegetable storage.

As noted above, the diurnal variation has a much thinner penetration; all the variations take place in a thin surface layer. This phenomenon is well known in many processes, e.g., the skin effect in electromagnetism. Exercise 3.2.1 of II deals with the entirely analogous effect produced in a viscous fluid by an oscillating plane boundary.

In all these cases the basic elements are the periodic variations in time of a quantity whose distribution is governed by a diffusive process, with basic terms exhibited by (2).

NUMERICAL INTEGRATION OF THE HEAT EQUATION

With the advent of large computers, there is a new balance between the advantages of analytical methods versus numerical methods. So far, we have concentrated on the use of analytical methods. We have nevertheless inadvertently discussed the basis for the numerical integration of the heat equation. because the natural equation to be used for finding an approximate solution to the heat equation by the method of finite differences is the difference equation (3.3.1), which was the basis for our discussion of random walk.

Indeed, consider the difference equation

$$\theta(x, t+k) - \theta(x, t) = \tfrac{1}{2}[\theta(x-h, t) + \theta(x+h, t)] - \theta(x, t), \tag{15}$$

[which is equivalent to (3.3.1)]. Approximating for small h and k, we find that

$$k\theta_t(x, k) + O(k^2) = \frac{1}{2}\left[\theta(x, t) - h\theta_x(x, t) + \frac{h^2}{2}\theta_{xx}(x, t) + O(h^3)\right.$$
$$\left. + \theta(x, t) + h\theta_x(x, t) + \frac{h^2}{2}\theta_{xx}(x, t) + O(h^3)\right] - \theta(x, t)$$

or

$$\theta_t = \frac{h^2}{2k}\theta_{xx} + O(k) + O\left(\frac{h^3}{k}\right).$$

In the formal limit as $h \to 0$, $k \to 0$, with h^2/k fixed, we obtain the diffusion equation

$$\theta_t = D\theta_{xx}, \qquad D \equiv \frac{h^2}{2k}. \tag{16a, b}$$

Now if we start with the differential equation (16) and consider (15) as an "approximation" to it, we must insist that $h = \Delta x$ and $k = \Delta t$ are related by

$$\Delta t = \frac{(\Delta x)^2}{2D}.$$

Thus the increment in t is smaller by an order of magnitude when compared with the subdivisions in x. Indeed, books on numerical analysis prove an extremely important related result, namely, that if

$$\Delta t < \frac{(\Delta x)^2}{2D}, \tag{17}$$

then the difference equation method of analysis is **stable**. This means that if the initial values are changed by a small amount, the solution is changed by an amount of the same order. On the other hand, if the interval Δt is so large that (17) is reversed, then the method is **unstable**. In that case any small error in the initial condition (or inevitable small roundoff errors) would eventually cause large errors in the subsequently calculated values.

If we refer to Figure 5.1, we see that (15) allows us to calculate the value of $u(x, t)$ at an interior point P' from the values at P and its neighboring points P_1 and P_2. The same condition obtains for Q'. For boundary points R' and S', the situation is different, but if u is given along the boundaries AA_1 and BB_1, the process of numerical integration can be carried on. [If the boundary condition is that of insulation, then we take $u(R') = u(Q')$ to conform to $\partial u/\partial x = 0$.] Thus the solution in the whole domain A_1ABB_1 can be found.*

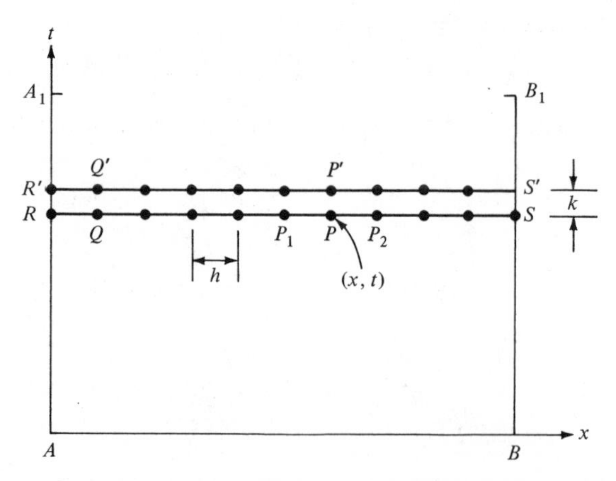

FIGURE 5.1. *Grid points involved in a numerical analysis of the heat equation. In particular, to obtain the temperature θ at $P'(x, t+k)$, we need to know θ at $P(x, t)$, $P_1(x-h, t)$, and $P_2(x+h, t)$.*

* Confidence that our problem is well posed is generated by verification that a straightforward numerical scheme, in principle at least, can provide a solution to the partial differential equation plus associated boundary and initial conditions. Another verification of such a fact, with a numerical scheme for a more complicated problem, will be found in Appendix 3.1 of II.

In the problem of heat conduction along an infinite rod ($-\infty < x < +\infty$), we would expect the temperature distribution to approach constant values with vanishing derivatives at infinity. If we take a sufficiently large interval $-A < x < A$ and impose a suitable condition at the end points, the method of finite differences can be carried through.

A comparison of the analytical versus numerical methods shows clearly that the former can give better insight into the nature of the problem. For example, it is no doubt possible to discover by numerical methods the fact that the deviation of the distribution of temperature from the steady state will eventually become sinusoidal, but the analytical solution (Equation 4.1.41) shows this immediately and clearly. Also, the source solution considered in Chapter 3 is obviously very difficult to describe numerically.

The chief advantage of the numerical solution is its greater generality, when the differential equation does not have a simple form. The analytical method works beautifully, largely because the differential equation is linear and the coefficients are constants. Indeed, the geometry has to be simple, too. Suppose that we are dealing with a problem involving inhomogeneous material. We shall see, in the next section, that analytical methods begin to run into complications. Moreover, for nonlinear problems, analytical methods are often very clumsy, if not impossible. But even for such problems, there are certain situations where analytical methods have advantages over a purely numerical approach, chiefly when singularities and other rapid variations are involved. In general, a mixed approach, using both kinds of methods, is the most sensible.

A mixed approach might begin with the analysis of a crude model to find the most important effects and to determine interesting parameter ranges. Then numerical analysis of a more complete model might provide needed detail. In a problem area where understanding is minimal, it might be wise to start with a few numerical experiments to gain a general understanding of the phenomenon. Armed with such an understanding, one can often isolate the central aspects of the phenomenon for analytical study. A return to the computer may then be useful to obtain a detailed picture.

A three-dimensional problem containing a number of parameters will probably continue to defy complete numerical analysis for a long time. And even if such an analysis were available, the resulting surfeit of data would be of little use to a person unless he had some analytically generated understanding of the problem.

HEAT CONDUCTION IN A NONUNIFORM MEDIUM

We shall now study an example that illustrates some advantages of a mixed approach and introduces an important subject called Sturm–Liouville theory.

Suppose that we are dealing with heat conduction in an alloy whose composition is not homogeneous. The density ρ, specific heat c, and heat conductivity k are then functions of position. As stated in Section 4.1, the equation for heat conduction then becomes

$$\rho c \frac{\partial \theta}{\partial t} = \frac{\partial}{\partial x}\left(k \frac{\partial \theta}{\partial x}\right) + \frac{\partial}{\partial y}\left(k \frac{\partial \theta}{\partial y}\right) + \frac{\partial}{\partial z}\left(k \frac{\partial \theta}{\partial z}\right) \tag{18}$$

or

$$\rho c \frac{\partial \theta}{\partial t} = \nabla \cdot (k \nabla \theta).$$

This equation still allows simple solutions with the variable t separated out:

$$\theta(x, y, z, t) = U(x, y, z) e^{-\lambda t}. \tag{19}$$

In the one-dimensional case, U satisfies an ordinary differential equation

$$\frac{d}{dx}\left(k \frac{dU}{dx}\right) + \lambda \rho c U = 0, \tag{20}$$

where k, ρ, and c are functions of the variable x.

In general, (20) cannot be solved explicitly in simple analytical form, but we can still gain insight into the nature of its solutions by analogy with the case of a uniform medium. For definiteness let us consider the specific problem where the end points of a rod lying along $(0, L)$ are kept at temperature zero, so that

$$U(0) = U(L) = 0. \tag{21}$$

Let us recapitulate our work on the homogeneous case. When k, ρ, and c are constants, (20) has nontrivial solutions satisfying (21) only if $\lambda = \mu_n$, where

$$\mu_n = \frac{kn^2\pi^2}{\rho c L^2}, \qquad n = 1, 2, \ldots. \tag{22}$$

The corresponding solutions are constant multiples of

$$U_n(x) = \sin \frac{n\pi x}{L}. \tag{23}$$

A solution corresponding to particular initial values is obtained by superposition.

In the present case, by analogy with the development just presented and with (6), we would expect a solution of the form

$$u = \sum_{n=1}^{\infty} U_n(x) e^{-\lambda_n t}. \tag{24}$$

Here each $U_n(x)$ is defined by the differential equation (20) (with $\lambda = \lambda_n$)

$$\frac{d}{dx}\left(k\frac{dU_n}{dx}\right) + \lambda_n \rho c U_n = 0. \tag{25a}$$

The boundary conditions are

$$U_n(0) = 0, \qquad U_n(L) = 0. \tag{25b}$$

The problem afforded by (25) has the **trivial solution** $U \equiv 0$. Nontrivial solution functions $U_n(x)$ generally exist only for certain constants λ_n, $n = 1, 2, \ldots$. The U_n are termed **eigenfunctions** belonging to the **eigenvalues** λ_n. We note that if $U_n(x)$ is a solution, so is $C_n U_n(x)$, for any constant C_n. If we now impose the initial condition $u(x, 0) = f(x)$ on the series solution (24), we see that we must have

$$f(x) = \sum_{n=1}^{\infty} C_n U_n(x) \tag{26}$$

for some choice of the constants C_n.

The solution of the heat conduction problem is now seen to depend on the possibility of expressing (in an appropriate sense) an arbitrary function as an infinite series of the form (26). The required body of theorems and techniques is known as the Sturm–Liouville theory.* We shall not pursue it in detail here, but in the next section we shall develop a few facts that can be obtained from some relatively simple arguments. We shall show how these can be useful for the explicit solution of our problem.

At this point, let us briefly compare the present approach and a purely numerical approach to the solution of (18). If the domain in question is very complicated, one might well decide to use a numerical approach right from the beginning. However, even here, it might be simpler to use superpositions of solutions of the form (19). Let us examine this in the one-dimensional case; the remarks also hold in higher dimensions.

In the case of the one-dimensional medium, we know that for the homogeneous case,

$$\lambda_n = \frac{n^2\pi^2\kappa}{L^2}, \qquad n = 1, 2, \ldots; \quad \kappa \equiv \frac{k}{\rho c}. \tag{27}$$

Since the corresponding nth spatial eigenfunction ϕ_n is multiplied by the factor $\exp(-\lambda_n t)$, only the terms for low values of n are important in the series solution (24), except for a very short initial period wherein $\kappa t\pi^2/L^2$ is not yet of magnitude unity. It is reasonable to expect that this *qualitative* feature will not be changed by the introduction of inhomogeneity, with the possible exception of cases where the inhomogeneity is "very large." Thus the practical solution of most inhomogeneous problems should reduce to

* See Courant and Hilbert (1953, Vol. 1, Chap. 5).

finding a few of the lower eigenfunctions and eigenvalues. These can be determined by numerical integration of some *ordinary* differential equations, in contrast to the *partial* differential equation (18). Thus we continue to expect that the qualitative behavior of the solution is one in which various "modes" are damped out at quite different rates. In the highly inhomogeneous case, however, the mode shapes might be far from sinusoidal.

There is a further advantage to an analytic approach. The eigenfunctions and eigenvalues thus found can be shown to be associated with other physical problems. Consider an attempt to solve the equation

$$\rho c \frac{\partial^2 w}{\partial t^2} = \frac{\partial}{\partial x}\left(k \frac{\partial w}{\partial x}\right) + \frac{\partial}{\partial y}\left(k \frac{\partial w}{\partial y}\right) + \frac{\partial}{\partial z}\left(k \frac{\partial w}{\partial z}\right) \tag{28}$$

(which occurs for vibrations of a continuous medium), by a simple *normal mode* solution of the form

$$w = W(x, y, z)e^{i\omega t}. \tag{29}$$

If $\omega^2 = \lambda$, then the equation satisfied by $W(x, y, z)$ is the same as that for $U(x, y, z)$. If the boundary conditions are also identical, we would be led to the same eigenvalue problem with the same set of eigenfunctions. Two seemingly unrelated problems, one involving diffusion and the other vibration, are thus brought together by mathematical considerations.

EXERCISES

1. Verify (7) and (8).

2. For what physical reason do we expect all the eigenvalues of (25) to be positive?

3. (Project.) Perform a numerical analysis of the problem of finding annual temperature variation in the earth. Assume that the surface temperature has only two values, a "summer" value for half the year and a "winter" value for the other half. Compare your results with the appropriate special case of (12).

4. Using Bessel functions, determine the eigenfunctions of (25) when $k = \alpha x^m$, $\rho c = \beta x^n$; α, β, m, and n are constants. Discuss.

5. Formulate the annual temperature variation problem for a spherical earth. If $u(r, t)$ is the temperature as a function of the radius r and the time t, introduce a new dependent variable $ru(r, t)$. Without actually solving any equations, discuss the correction due to sphericity.

‡6. The solution (12) shows that the annual temperature oscillates with an amplitude that depends on depth. At one observation station in the Soviet Union, the amplitude was measured at 11.5, 6.8, 4.2, and 2.6°C at 1, 2, 3, and 4 m. Use these data to determine a value for the thermometric diffusivity κ of the earth.

‡7. Measurements 2 or 3 km below the surface of the earth have revealed an increase in temperature with depth of about 3°C every 100 m. As we have seen, annual variations are insensible at these depths. A reasonable explanation for the observations, however, is that the temperature gradient is a result of the earth's cooling. Show that this is an unsatisfactory hypothesis on the basis of the following data. Rocks melt at about 1200°C, the average thermometric diffusivity of granite and basalt is 6×10^{-3} cm^2/sec, and the age of the earth is about 10^9 years.*

5.2 Sturm–Liouville Systems

We now consider a slight generalization of a mathematical problem that arose in the previous section, the problem of determining certain solutions of the **second order Sturm–Liouville differential equation**

$$\frac{d}{dx}\left[p(x)\frac{dU}{dx}\right] + [\lambda\rho(x) - q(x)]U = 0. \tag{1}$$

Here λ is a parameter; p, ρ, and q are real-valued functions of x; and the functions p and ρ are *positive*. (The function q was identically zero in the preceding example. For ordinary Fourier series, $p = \rho = 1, q = 0$.) We shall again take the boundary conditions to be

$$U(a) = 0, \qquad U(b) = 0. \tag{2}$$

It will become clear as we progress, however, that the gist of the discussion will not be greatly changed when other homogeneous† boundary conditions are introduced in place of (1). (See Exercise 7.)

We shall now show that

(i) Nontrivial solutions U_i can exist only for certain real discrete values λ_i of λ.

(ii) These solutions are orthogonal to each other with *weighting function* $\rho(x)$ in the sense that

$$\int_a^b \rho(x)U_i(x)U_j(x)\,dx = 0, \qquad \text{if } \lambda_i \neq \lambda_j. \tag{3}$$

* This problem and the preceding are adapted from an excellent discussion by Tychonov and Samarski (1964, pp. 215–55) of geophysical applications of heat diffusion theory.

† The word "homogeneous" is overworked, but its various usages have become traditional. Here are four of them: (a) A homogeneous condition on a function is one that is certainly satisfied by the zero function. Thus (1) is a homogeneous equation but $U'(0) - 5 = 0$ is not a homogeneous boundary condition. (b) A homogeneous material is one whose composition is uniform. (c) The special differential equation $dy/dx = f(y/x)$ is termed "homogeneous." (d) A function f of n variables is said to be homogeneous of degree m if

$$f(tx_1, tx_2, \ldots, tx_n) = t^m f(x_1, x_2, \ldots, x_n).$$

By analogy with our experience with Fourier series, we would also expect (iii) to have an infinite number of eigenfunctions and (iv) to be able to express a broad class of functions, in a suitable sense, as infinite series of eigenfunctions. As mentioned before, these last two propositions are not easy to prove, and we shall not attempt proofs here. However, we shall show (v) how the higher eigenfunctions and eigenvalues can be approximately calculated, even though we cannot give a rigorous proof of their existence. We shall also show (vi) how the lower eigenfunctions can be calculated by mixed numerical and analytical methods, or directly by numerical methods. [The **higher** (**lower**) **eigenvalues** are the larger (smaller) ones. The corresponding eigenfunctions are also called "higher" and "lower."]

PROPERTIES OF EIGENVALUES AND EIGENFUNCTIONS

We wish to show that the boundary conditions (2) can be satisfied only for a discrete set of values of λ. By this we mean that each eigenvalue λ can be made the center of a small circle which encloses no other eigenvalues.

Let U_α and U_β form a **fundamental set of solutions** of (1). By definition this means that any solution of (1) can be expressed in the form

$$U(x, \lambda) = c_\alpha U_\alpha(x; \lambda) + c_\beta U_\beta(x; \lambda) \tag{4}$$

for certain constants c_α and c_β. (Note that the explicit dependence of U_α and U_β on λ has been exhibited.) The boundary conditions (2) require that

$$c_\alpha U_\alpha(a; \lambda) + c_\beta U_\beta(a; \lambda) = 0,$$
$$c_\alpha U_\alpha(b; \lambda) + c_\beta U_\beta(b; \lambda) = 0.$$

The solution to this pair of homogeneous equations for c_α and c_β is nontrivial only if

$$\begin{vmatrix} U_\alpha(a; \lambda) & U_\beta(a; \lambda) \\ U_\alpha(b; \lambda) & U_\beta(b; \lambda) \end{vmatrix} = 0. \tag{5}$$

We assert that this **secular** or **characteristic equation** (5) will yield a denumerable set of eigenvalues. (This gives the **discrete spectrum**, in contrast to the continuous spectrum case where there is a continuum of eigenvalues.) A proof can be constructed by showing that $U_\alpha(x, \lambda)$ and $U_\beta(x, \lambda)$ can be chosen to be analytic functions of the parameter λ, regarded as a complex variable. Those familiar with complex variable theory will recognize that the conclusion follows from the fact that the zeros of an analytic function are isolated.

It is not hard to show (once the idea of the proof is born—see Exercise 1) that *the eigenvalues must be real numbers.* [Also, the eigenfunctions can always be taken to be real. See Exercise 1(b).] By analogy with the case of the Fourier functions, furthermore, we would anticipate that the *sequence of eigenvalues has no upper bound.* We also expect the *higher eigenfunctions* (those corresponding to larger values of λ) to *oscillate rapidly.* These expectations turn out to be correct. They form the basis of the asymptotic evaluation of the eigenvalues and of their corresponding eigenfunctions.

ORTHOGONALITY AND NORMALIZATION

We shall now show that a pair of eigenfunctions corresponding to different eigenvalues satisfy the orthogonality condition (3). Consider the two equations

$$\frac{d}{dx}\left(p\frac{dU_i}{dx}\right) + (\lambda_i \rho - q)U_i = 0, \tag{6a}$$

$$\frac{d}{dx}\left(p\frac{dU_j}{dx}\right) + (\lambda_j \rho - q)U_j = 0. \tag{6b}$$

If we multiply (6a) by U_j and (6b) by U_i, and subtract, we obtain

$$U_j \frac{d}{dx}\left(p\frac{dU_i}{dx}\right) - U_i \frac{d}{dx}\left(p\frac{dU_j}{dx}\right) + (\lambda_i - \lambda_j)\rho U_i U_j = 0.$$

If we integrate this equation between the limits (a, b), and impose the boundary conditions (2), we find using parts integration [Exercise 2(a)] that

$$\int_a^b \left[U_j \frac{d}{dx}\left(p\frac{dU_i}{dx}\right) - U_i \frac{d}{dx}\left(p\frac{dU_j}{dx}\right)\right] dx = 0, \tag{7}$$

so

$$(\lambda_i - \lambda_j)\int_a^b \rho U_i U_j \, dx = 0. \tag{8}$$

Thus, unless $\lambda_i = \lambda_j$, the orthogonality condition (3) must be satisfied.

If there are a finite number s of linearly independent eigenfunctions belonging to the same eigenvalue, we can construct (by the **Gram–Schmidt process**) a set of p mutually orthogonal eigenfunctions belonging to this eigenvalue. To accomplish this, let us assume that $U_{n1}, U_{n2}, \ldots, U_{ns}$ are *linearly independent eigenfunctions* all belonging to the eigenvalue λ_n. We shall now specify, one by one, *mutually orthogonal eigenfunctions* $V_1, \ldots, V_s$. For V_1, we shall take U_{n1}. For V_k, we shall take

$$V_k = c_{k1}V_1 + \cdots + c_{k,k-1}V_{k-1} + U_{nk}, \qquad 1 < k \leq s. \tag{9}$$

We assume that all the previous V's have been chosen, and impose the requirement that V_k shall be orthogonal to mutually orthogonal $V_1, \ldots, V_{k-1}$. We then find [Exercise 2(b)] that

$$c_{k,m}(V_n, V_m) + (U_{nk}, V_m) = 0, \tag{10}$$

where*

$$(U, V) = \int_a^b \rho(x)U(x)V(x)\,dx. \tag{11}$$

Not all the $c_{k,m}$'s are zero, unless U_{nk} is already orthogonal to all the V_m's.

We shall standardize our notation by adopting a set of eigenfunctions $\{\phi_n(x)\}$ that are orthogonal to each other and normalized so that $(\phi_n, \phi_n) = 1$. Thus

$$(\phi_m, \phi_n) = \delta_{mn}, \tag{12}$$

where

$$\delta_{mn} = 0, \quad m \neq n, \qquad \delta_{mm} = 1.$$

We should remember, however, that some of the eigenvalues may be the same. If there are two linearly independent eigenfunctions with the same eigenvalue, we speak of this eigenvalue as having a *multiplicity* of two, or being a doublet, etc. An example of multiple eigenvalues is furnished by the harmonic functions in the interval $(-\pi, \pi)$:

$$\phi_0(x) = 1,$$

$$\phi_1(x) = \frac{1}{\sqrt{\pi}} \sin x, \qquad \phi_2(x) = \frac{1}{\sqrt{\pi}} \cos x, \tag{13}$$

$$\phi_{2n-1}(x) = \frac{1}{\sqrt{\pi}} \sin nx, \qquad \phi_{2n}(x) = \frac{1}{\sqrt{\pi}} \cos nx.$$

The functions $\{\phi_{2n-1}(x), \phi_{2n}(x)\}$ form a doublet with common eigenvalue n^2.†

EXPANSION IN TERMS OF EIGENFUNCTIONS

If we try to express an arbitrary function f in a series using the set of functions $\{\phi_n\}$, as was done with Fourier series, we may evaluate the coefficients in a similar manner. Indeed, if

$$f(x) = \sum_{n=0}^{\infty} c_n \phi_n(x), \tag{14}$$

* The notation (,) emphasizes the connection with the scalar product of vectors. Indeed, the Gram–Schmidt process is often first met in the purely algebraic task of converting a linearly independent set of vectors to a mutually orthogonal set. As some readers may already know, the connection hinted at here is one of basically identical formal structure. This is exploited in Appendix 12.1 of II, where the ideas of Sturm–Liouville theory are generalized to self-adjoint operators.

† The ϕ's can be regarded as eigenfunctions of the problem

$$\phi'' + \lambda^2\phi = 0, \qquad \phi(-\pi) = \phi(\pi), \qquad \phi'(-\pi) = \phi'(\pi)$$

On comparing with the Sturm–Liouville equation (1), we confirm that the weighting function $\rho(x)$ is identically equal to unity. Orthonormality has already been asserted in (4.1.38).

we have (formally)

$$(\phi_i, f) = \sum_{n=0}^{\infty} c_n(\phi_i, \phi_n) = \sum_{n=0}^{\infty} c_n \delta_{in} = c_i,$$

i.e.,

$$c_n = (\phi_n, f) = \int_a^b \phi_n(x)\rho(x)f(x)\,dx. \tag{15}$$

It is an exceptional function that requires only a finite number of terms in (14). Thus we must have an infinite number of eigenfunctions before we can hope that (14) will be valid for a general class of functions. Certainly, this requirement alone is not sufficient. If by some accident we did not include $\phi_0(x)$ in the list, for example, we could not expect to be successful. We thus have the need to prove the **completeness** of the eigenfunctions with respect to some class of functions—in the sense that a suitable series of eigenfunctions will converge to any function of the class. This is by no means easy. We shall therefore restrict ourselves to the statement that the set of eigenfunctions of a Sturm–Liouville problem (with weak conditions on the coefficient functions) is indeed complete with respect to a broad class of functions.

To enumerate the eigenfunctions, we may depend on the number of zeros that a function has in the interval in question. For example, for the harmonic functions (13), $\phi_m(x)$ has m zeros in the open interval $(-\pi, \pi)$. Such behavior of the eigenfunctions is guaranteed by certain **oscillation theorems**. The reader is referred to books on differential equations for proofs of completeness theorems and oscillation theorems.

ASYMPTOTIC APPROXIMATIONS TO EIGENFUNCTIONS AND EIGENVALUES

One can gain an insight into the nature of the eigenfunctions ϕ_n and their associated eigenvalues λ_n by an analysis of their behavior when n and λ_n are large. For this purpose, it is convenient first to transform the original differential equation (1) into the *normal form of Liouville* by introducing a new independent variable $t = t(x)$, and a new dependent variable w, where

$$U = y(x)w. \tag{16}$$

We have two functions at our disposal, and we may aim at imposing two desirable conditions on the equations. If we choose these conditions as (i) the coefficient of dw/dt should be zero, and (ii) the coefficient of λ should be unity, we must (Exercise 3) take the functions w and t so that

$$U(x) = \frac{w}{[p(x)\rho(x)]^{1/4}}, \qquad t(x) = \int \left[\frac{\rho(x)}{p(x)}\right]^{1/2} dx. \tag{17a, b}$$

We then obtain **Liouville's normal form,**

$$\frac{d^2w}{dt^2} + [\lambda - \hat{q}(t)]w = 0, \tag{18}$$

where

$$\hat{q} = \frac{q}{p} + (p\rho)^{-1/4} \frac{d^2}{dt^2} [(p\rho)^{1/4}]. \tag{19}$$

For an equation in the form (1) it is difficult to get any idea of how the solution behaves when λ is large. For (18), however, this is an easy matter.* Assuming that λ is large compared to $\hat{q}$, we at once guess that (18) has the approximate solutions $\exp(\pm i\lambda^{1/2}t)$. Thus it is "natural" to try to solve (18), when λ is large, by the series

$$w(t) = \exp(\pm i\lambda^{1/2}t)[w_0(t) + \lambda^{-1/2}w_1(t) + \cdots]. \tag{20}$$

The reader should calculate a few terms of this series by direct substitution (Exercise 4). He will see, for example, that w_0 must be a constant.

Let us now impose the boundary conditions $w = 0$ at the end points, (t_a, t_b). We must make these points correspond to the original end points $x = a$ and $x = b$. This can be accomplished by writing (17b) in the form

$$t = \int_a^x \left[\frac{\rho(\xi)}{p(\xi)}\right]^{1/2} d\xi$$

and taking

$$t_a = 0, \qquad t_b = \int_a^b \left[\frac{\rho(\xi)}{p(\xi)}\right]^{1/2} d\xi. \tag{21}$$

We find (Exercise 5) that for large n the eigenvalues and eigenfunctions satisfy

$$\lambda_n^{1/2} = n\pi\left\{\int_a^b \left[\frac{\rho(x)}{p(x)}\right]^{1/2} dx\right\}^{-1} + O\left(\frac{1}{n}\right), \tag{22}$$

$$U_n(x) = [p(x)\rho(x)]^{-1/4} \sin\left\{n\pi \frac{\int_a^x [\rho(x)/p(x)]^{1/2}\,dx}{\int_a^b [\rho(x)/p(x)]^{1/2}\,dx}\right\}\left[1 + O\left(\frac{1}{n}\right)\right]. \tag{23}$$

These results are very useful for practical purposes. Surprisingly perhaps, but typically, they are very accurate even for fairly small values of n.

As an example, consider the *Bessel equation*

$$\frac{d}{dx}\left(x\frac{dU}{dx}\right) + \left(k^2x - \frac{m^2}{x}\right)U = 0. \tag{24}$$

* We thus learn from Liouville that if an equation contains a large parameter, we will probably accomplish something useful if we introduce new variables so that certain terms are definitely small relative to this parameter. These terms can be omitted to obtain a first approximation.

This is a special case of (1) with $p = \rho = x$, $q = m^2/x$. The Liouville transformation (17) is $U = wx^{-1/2}$ and $x = t$, and the resultant equation (18) is

$$\frac{d^2w}{dx^2} + \left(k^2 - \frac{m^2 - \frac{1}{4}}{x^2}\right)w = 0. \tag{25}$$

When k is large (and x is not near zero), the first approximation to the general solution is

$$w(x) = c_1 \cos kx + c_2 \sin kx.$$

If the boundary conditions $w(a) = w(b) = 0$ are imposed, we are led to the approximate eigenfunctions

$$w_n(x) = \sin k_n(x - a), \tag{26}$$

with

$$k_n = \frac{n\pi}{b - a}; \; n = 1, 2, 3, \ldots.$$

If $a = 1$ and $b = 2$, the lowest approximate eigenvalue is $k_1 = \pi$. When $m = 0$ (a favorable case to be sure) the actual lowest eigenvalue is between 3.1 and 3.2 (Exercise 6). Of course, when $m = \frac{1}{2}$, our "approximation" gives the exact solution.

Once the form of an approximate solution is known, it can often be obtained by a more direct method than that originally used. In the present case the calculations can be made after a simpler transformation $U = vp^{-1/2}$ followed by the formal substitution $v = e^{i\lambda^{1/2}\phi}\{v_0 + \lambda^{1/2}v_1 + \cdots\}$. With this, ϕ will automatically turn out to be the function $t(x)$ defined by (17b) (Exercise 10). *The series does not converge*, but it is asymptotic—and very useful. An asymptotic series of such a form is used to bridge classical mechanics with quantum mechanics in the Schrödinger formulation. This same form had been used long before in the theory of water waves. The procedure is also found to be useful even when $\rho(x)$ changes sign. In that case the "JWKB theory" has to be invoked to resolve the difficulties occurring near a zero of $\rho(x)$. It is, however, beyond our scope to go further into these matters. The reader is referred to such references as Cole (1968, Section 3.7).

OTHER METHODS OF CALCULATING EIGENFUNCTIONS AND EIGENVALUES

Numerical techniques can often be used advantageously to calculate eigenvalues and eigenfunctions. One way to do this is to solve initial value problems for the two linearly independent functions $U_1(x, \lambda)$ and $U_2(x, \lambda)$, where, in addition to the differential equation (1), these functions respectively satisfy the initial conditions

$$U_1(a, \lambda) = 0, \qquad U_1(b, \lambda) = 1,$$
$$U_2(a, \lambda) = 1, \qquad U_2(b, \lambda) = 0.$$

If the general solution $U(x, \lambda) = C_1 U_1(x, \lambda) + C_2 U_2(x, \lambda)$ is to satisfy the boundary conditions (2), then C_1 and C_2 must satisfy the homogeneous equations

$$C_1 U_1(a, \lambda) + C_2 U_2(a, \lambda) = 0,$$
$$C_1 U_1(b, \lambda) + C_2 U_2(b, \lambda) = 0,$$

which have a nontrivial solution if and only if

$$U_1(a, \lambda)U_2(b, \lambda) - U_1(b, \lambda)U_2(a, \lambda) = 0.$$

Roots of this transcendental equation can be determined numerically. This is not an easy task, however, particularly in more complicated (nonself-adjoint) problems with complex eigenvalues. Another difficulty arises when, say, U_1 increases much more rapidly with x than U_2. Unless precautions are taken, roundoff introduces a slight error, so instead of U_2 one is calculating $\varepsilon U_1 + U_2$. Although ε is very small, U_1 becomes so large that $\varepsilon U_1 + U_2$ becomes indistinguishable from U_1 and therefore does not supply, as required, a solution that is independent of U_1. In spite of these and other difficulties, however, numerical calculation of eigenvalues has been pursued with great success in many investigations.

Another useful approximate method makes use of the fact that eigenvalues can often be characterized as minima of certain expressions. This *Rayleigh–Ritz method* in the calculus of variations is briefly illustrated in Section 12.4 and is treated more fully in Chapter 12 of II.

EXERCISES

1. (a) Prove in the following manner that the eigenvalues of (1) are real,
 (i) Find an equation for $\overline{U}$, the complex conjugate of U, by taking the complex conjugate of (1). Remember that ρ, p, and q are real-valued functions of x, although λ may not be.
 (ii) If $\lambda \neq \bar{\lambda}$, deduce a contradiction from the orthogonality of eigenfunctions corresponding to distinct eigenvalues.
(b) Prove that if U is an eigenfunction of (1), then either Re U or Im U is a real eigenfunction.

2. (a) Verify (7) and (8).
(b) Fill in the details omitted in the text's discussion of the Gram–Schmidt process.

3. Show that to obtain Liouville's normal form (18) the change of variables should indeed be as given in (17).

4. Calculate at least two terms of the series (20) by formal substitution.

5. Verify (22) and (23).

6. Consider (25) with $m = 0$, subject to the boundary conditions $w(1) = w(2) = 0$. If you are familiar with Bessel functions, find a transcendental equation for the eigenvalues k_n. By use of a table of Bessel functions, show that k_1 lies between 3.1 and 3.2.

7. (a) Verify (3) when the boundary conditions (2) are replaced by $U'(a) = U'(b) = 0$.

(b) Are there more general boundary conditions that will still permit the derivation of (3)?

8. Verify (3) for the fourth order Sturm–Liouville problem

$$\frac{d^2}{dx^2}\left[f(x)\frac{d^2U}{dx^2}\right] + \frac{d}{dx}\left[p(x)\frac{dU}{dx}\right] + [\lambda\rho(x) - q(x)]U = 0;$$

$$U(a) = U'(a) = U(b) = U'(b) = 0.$$

9. (a) Show that the general second order equation

$$a(x)\frac{d^2U}{dx^2} + b(x)\frac{dU}{dx} + [c(x) + \lambda\, d(x)]U = 0$$

can be put into Sturm–Liouville form by an appropriate change of variables.

(b) Can the general fourth order eigenvalue problem be put into the form of Exercise 8?

10. Carry out the alternative approach to (17b) that is outlined below (26).

11. Generalize the results of Exercise 4.3.12 so that they apply to the second order Sturm-Liouville problem (1) and (2).

5.3 Brief Introduction to Fourier Transform

This brief treatment of a large topic is motivated by a desire to find the limit as $L \to \infty$ of the solution to a heat conduction problem in a rod of length L. We are led from Fourier series to Fourier transform formulas. As a check, we recover an earlier result for heat flow in the infinite rod. Further application of the Fourier transform can be found in the treatment of dispersive waves in Chapter 8 of II.

FOURIER TRANSFORM FORMULAS AND THE FOURIER IDENTITY

In Section 4.1 we considered the solution of the problem of heat conduction along a rod of length L. In the limit when $L \to \infty$, we should be able to retrieve the solution of the infinite interval heat conduction problem that we obtained in Chapter 3 with the help of the source solution. Crucial here is generalizing the representation of an arbitrary function

$$f(x) = \sum_{m=1}^{\infty} B_m \sin\frac{m\pi x}{L}$$

in the limit $L \to \infty$. (Compare Equation 4.1.36.) More generally, we shall discuss the limit as $L \to \infty$ of the complete Fourier series in complex form

$$f(x) = \sum_{-\infty}^{\infty} C_n e^{in\pi x/L}, \tag{1}$$

where

$$C_n = \frac{1}{2L} \int_{-L}^{L} f(\xi) e^{-in\pi\xi/L} \, d\xi. \tag{2}$$

There seems no obvious way to simplify (1) and (2) when L is large. It is true that for fixed n the integrand in C_n becomes highly oscillatory when $L \to \infty$. Consequently, the variation in f during a single oscillation of $\exp(-in\pi\xi/L)$ is hardly noticeable, so the positive contributions will almost cancel the negative. On the other hand, this whole idea is invalid for those values of n which are of the same magnitude as L.

From the study of the Liouville transformation we learned to attempt to clarify the role of a large parameter by changing from the given variables in the problem, in this case n and L. The previous paragraph suggests the importance of n/L; therefore, we shall introduce a multiple of this quantity as one variable:

$$k \equiv \frac{n\pi}{L}. \tag{3a}$$

(The added factor π just makes the resulting formulas a little simpler.) It turns out that the second variable (which we call Δk) should be taken as the difference between adjacent values of k (as n increases by unity):

$$\Delta k = \frac{\pi}{L}. \tag{3b}$$

With these, (2) becomes

$$C_n = C(k) = \frac{\Delta k}{2\pi} \int_{-L}^{L} f(\xi) e^{ik\xi} \, d\xi. \tag{4}$$

We now introduce (4) into (1) and obtain

$$f(x) = \sum_{k=-\infty}^{\infty} e^{ikx} \frac{\Delta k}{2\pi} \int_{-L}^{L} f(\xi) e^{-ik\xi} \, d\xi; \; k = \ldots, -2\Delta k, -\Delta k, 0, \Delta k, \ldots \tag{5}$$

The behavior of the integral in (5) as $L \to \infty$ depends crucially on the behavior of $f(\xi)$ for $|\xi|$ large. We have enough on our hands at the moment; so let us bypass this matter by considering a function $f(x)$ that vanishes identically outside an interval $(-A, A)$. If we take $L > A$, the integral in (5) may be taken between the limits $(-\infty, \infty)$ and represents a definite function

$$g(k) \equiv \int_{-\infty}^{\infty} f(\xi) e^{-ik\xi} \, d\xi. \tag{6}$$

With this (5) becomes

$$f(x) = \sum_{k=-\infty}^{\infty} e^{ikx} g(k) \frac{\Delta k}{2\pi}. \tag{7}$$

The limit of the sum as $L \to \infty$ and $\Delta k \to 0$ is

$$f(x) = \frac{1}{2\pi} \int_{-\infty}^{\infty} g(k) e^{ikx}\, dk \tag{8}$$

if the integral converges.*

In (6), the function g is known as the **Fourier transform** of the function f. Note that the **inversion formula** (8) from g back to f has substantially the same form as the Fourier transform formula (6), the main differences being the factor $\frac{1}{2}\pi$ and the change of sign of i. The numerical factor is sometimes redistributed to give more symmetry to the formulas. It is convenient to combine (6) and (8) into the **Fourier identity**

$$\boxed{f(x) = \frac{1}{2\pi} \int_{-\infty}^{\infty} dk \int_{-\infty}^{\infty} f(\xi) e^{ik(x-\xi)}\, d\xi.} \tag{9}$$

So far we have proved (9) only for the case wherein $f(x) \equiv 0$ when $|x| > A$. We shall now verify (9) for a far more general class of functions by a limiting process.

Consider the function $f(x, K)$ defined by the integral

$$f(x, K) = \frac{1}{2\pi} \int_{-K}^{K} dk \int_{-\infty}^{\infty} f(\xi) e^{ik(x-\xi)}\, d\xi. \tag{10}$$

If $\int_{-\infty}^{\infty} |f(\xi)|\, d\xi$ exists, then the order of integration may be interchanged, and we obtain

$$f(x, K) = \frac{1}{2\pi} \int_{-\infty}^{\infty} f(\xi) \frac{2i \sin K(x-\xi)}{i(x-\xi)}\, d\xi. \tag{11}$$

We now break up the interval of integration in (11) into the subintervals $(-\infty, a)$, (a, b), (b, ∞); $a < x < b$. It can be shown that if f is of bounded variation, then the contributions from the first and last of these subintervals approach zero as $K \to \infty$.† If f is piecewise smooth, the treatment of the

* Nontrivial limits involve such expressions as the quotient of two small terms, say, or (as here) a sum of very many very small terms. We have succeeded in evaluating the present limit, because by a change of variable we have cast our problem into a sum of many small terms whose limit is readily recognized as a definite integral.

† See Jeffreys and Jeffreys (1962, p. 454). Recall that $f(x)$ is of bounded variation if $\sum_{i=0}^{N} |f(x_{i+1}) - f(x_i)| < M$ for some M and for every possible subdivision $-\infty < x_0 < x_1 < \cdots < x_N < x_{N+1} < +\infty$.

middle subinterval is precisely covered by the lemmas proved in Section 4.2. In particular, according to (4.2.8),

$$\lim_{K\to\infty} f(x, K) = f(x).$$

Thus the identity (9) is established, subject to the same interpretations as before at a discontinuity.

SOLUTION OF THE HEAT EQUATION BY FOURIER TRANSFORM

Having established the Fourier transform relation, we can readily solve the problem of heat conduction over an infinite rod. Again we look for solutions to $\theta_t = \kappa\theta_{xx}$ that have the form $X(x)T(t)$. We require $X(x)$ to remain finite as $|x| \to \infty$. The most general solutions of this form are multiples of $\exp(ikx - k^2\kappa t)$, where k is any real number. From these simple solutions we construct by superposition the general solution

$$\theta(x, t) = \int_{-\infty}^{\infty} A(k) \exp(ikx - k^2\kappa t)\, dk. \tag{12}$$

Boundary conditions at $x = 0$ and L previously required us to select only a denumerable family of wave numbers k (discrete spectrum), but now k can be any real number (continuous spectrum). Previously, we took a weighted sum of permissible eigenfunctions; now, with a continuous spectrum, this sum becomes an integral weighted by the function A.

We can satisfy an arbitrary initial condition

$$\theta(x, 0) = f(x), \tag{13}$$

by (6) and (8), if $A(k) = g(k)/2\pi$. If we introduce the form (6) into (12), the solution is directly expressed in terms of the initial distribution $f(x)$ by

$$\theta(x, t) = \frac{1}{2\pi} \int_{-\infty}^{\infty} dk \int_{-\infty}^{\infty} f(\xi) \exp[ik(x - \xi) - k^2\kappa t]\, d\xi. \tag{14}$$

If we invert the order of integration and then carry out the integration with respect to k, we find [Exercise 1(a)] the solution in the form

$$\theta(x, t) = \frac{1}{\sqrt{4\pi\kappa t}} \int_{-\infty}^{\infty} f(\xi) e^{-(x-\xi)^2/4\kappa t}\, d\xi. \tag{15}$$

This was obtained before by the method of source solution in Section 3.4. We note that (15) is a valid solution of the problem *even when the infinite integral of* $|f(x)|$ *is not convergent*, as the solution (15) can be directly verified.* Here is an example that makes a strong case for heuristic reasoning and formal

* Rigorous verification that a formal solution actually satisfies all the conditions of a problem of course provides a (constructive) existence theorem. For such an approach to the heat equation see Lecture 8 of S. L. Sobolev, *Partial Differential Equations of Mathematical Physics* (New York: Pergamon, 1964).

manipulations, provided that we remember to check the validity of the final answers thus obtained. In many practical situations, however, even this check is too difficult. We must resort to other ways of assuring the correctness of the final answer, by comparison with observations, for example.

EXERCISES

1. (a) Verify (15).
(b) Consider (15) in the special case when f vanishes for negative x and has the constant value θ_0 for positive x. Derive

$$\frac{\theta}{\theta_0} = \pi^{-1/2} \int_{-\eta}^{\infty} \exp(-\xi^2)\, d\xi, \qquad \eta \equiv \frac{x}{2(\kappa t)^{1/2}}.$$

(c) Express the answer in the form of a series that is expected to be valid when η is large.

2. (a) Show that the Fourier identity can be written in the form

$$f(x) = \frac{1}{\pi} \int_0^{\infty} dk \int_{-\infty}^{\infty} f(\xi) \cos k(\xi - x)\, d\xi.$$

(b) Find a form of the Fourier identity involving only cosine (sine) functions when f is even (odd).
(c) Use (b) to show that

$$D(x) = 2\pi^{-1} \int_0^{\infty} k^{-1} \sin k \cos kx\, dk,$$

where

$$D(x) = \begin{cases} 1 \\ 0 \\ \frac{1}{2} \end{cases} \quad \text{for} \quad \begin{cases} |x| < 1, \\ |x| > 1, \\ |x| = 1. \end{cases}$$

3. If you are familiar with Bessel functions, see how far you can get in generalizing the results of this section to a situation involving cylindrical symmetry. Start with the problem of heat conduction in a cylinder of radius R, and consider the limit $R \to \infty$. Try to derive

$$f(x) = \int_0^{\infty} \xi J_0(\xi x) \int_0^{\infty} s J_0(\xi s) f(s)\, ds\, d\xi.$$

5.4 Generalized Harmonic Analysis

Although the Fourier series arose out of eigenfunctions for the solution of problems of heat conduction (and we have seen in Section 5.1 that these eigenfunctions need not be harmonic functions), it has perhaps found its greatest application for the analysis of time-dependent signals and periodic structures. To glimpse the range of these applications, just consider a few instances of periodic phenomena. The different colors perceived by man are due to electromagnetic waves of different frequencies. The ear can detect different notes when stimulated by mechanical waves of different frequencies. Periodic structures occur in inanimate crystal lattices as well as in animate millipedes. Periodic motions of planets, double stars, and variable stars are well known. The diurnal and annual variation of weather is a consequence of the periodic motion of the earth.

In spite of the prevalence of periodicity, however, once we decide to analyze time-dependent signals (such as temperature records) into harmonic components, we are introducing something artificial, since such sequences need not really be periodic. To be sure, we can try to *extract* the periodic components such as the diurnal and annual variations in weather, and the 11 year cycle in sunspot activity. Yet there may still be a great deal of the signal left over. One must therefore be prepared to deal with those nonperiodic time variations that continue indefinitely with *unabated* (though bounded) amplitude. Such problems fall outside the method of Fourier analysis discussed thus far. They require what is known as **generalized harmonic analysis**.

REMARKS ON FUNCTIONS THAT CANNOT BE ANALYZED BY STANDARD FOURIER METHODS

Although the general theory is too complicated to be presented here, one can get some feeling for the nature of the problem by considering a function that is the sum of a few periodic functions whose periods are *not commensurable*. An example is

$$f(t) = A_1 e^{i\omega_1 t} + A_2 e^{i\omega_2 t} + A_3 e^{i\omega_3 t}, \tag{1}$$

where for all integers m and n

$$m\omega_i \neq n\omega_j; \qquad i, j = 1, 2, 3; \quad i \neq j.$$

One may add to $f(t)$ a function of the type

$$\phi(t) = e^{-\omega_0{}^2 t^2}, \qquad \omega_0 > 0. \tag{2}$$

But this function decays sufficiently rapidly as $|t| \to \infty$ so that it can be resolved into harmonic components by the Fourier transform method, even though it shows no resemblance to harmonic functions. On the other hand, the function $f(t)$ of (1) does not show a singly periodic behavior and cannot be represented by a Fourier series. Neither does it die off at infinity. It can therefore not be represented by a proper Fourier integral. Of course, it is already resolved into periodic components—but how shall we analyze more general functions that do not have such simplicity?

The first step is to take a sufficiently long record, hopefully lasting over many of the periods of any significant periodic component that may be present. For example, in the case of a meteorological record we need at least 10 years' observations. To see how each truly periodic component would show up in a Fourier analysis based on such a record, let us examine the Fourier analysis of a single periodic component lasting over a finite period of time:

$$f_T(t) = \begin{cases} A_0 e^{i\omega_0 t}, & |t| \leq T, \\ 0, & |t| > T. \end{cases} \tag{3}$$

We shall refer to f_T, as a **truncated sinusoidal function.**

We shall now analyze the truncated sinusoidal function defined by (3), using both Fourier series and Fourier integral. When we consider the limit as $T \to \infty$, using the first method, we get discrete harmonic components. We may even reproduce the function exactly if the interval $2T$ happens to be an exact integral multiple of the period $2\pi/\omega_0$. In the second method, we get a continuous spectrum, but with a peak presumably at $\omega = \omega_0$. Let us examine the details.

FOURIER SERIES ANALYSIS OF A TRUNCATED SINUSOIDAL FUNCTION

Let $\Omega = 2\pi/2T = \pi/T$ be the fundamental frequency. We wish to represent (3) in the form

$$A_0 e^{i\omega_0 t} = \sum_{-\infty}^{\infty} C_n e^{in\Omega t}, \qquad \text{for } -T < t < T. \tag{4}$$

Then [Exercise 1(a)]

$$C_n = A_0 \frac{\sin(\omega_0 - n\Omega)T}{(\omega_0 - n\Omega)T}, \qquad n \neq \frac{\omega_0}{\Omega}. \tag{5}$$

The important components are associated with small values of $(\omega_0 - n\Omega)T$.

If ω_0 happens to be an integral multiple of Ω, say $\omega_0 = k\Omega$, then we have only one nonvanishing coefficient C_k in the Fourier series. If $k < \omega_0/\Omega < k + 1$, then both C_k and C_{k+1} would be important. In the worst situation, with $\omega_0/\Omega = k + \frac{1}{2}$, i.e., when the interval $2T$ happens to be a half-integral multiple of the basic period $2\pi/\omega_0$, we have [Exercise 1(b)]

$$C_{k+1} = C_k = \frac{2A_0}{\pi}. \tag{6}$$

These are the dominant coefficients, each having an amplitude of about $0.64A_0$. Their frequencies do not differ greatly from ω_0, the fractional error being only $\frac{1}{2}k$, which is just 5 per cent if the record covers a time interval of about 10 periods.

It might be puzzling that we have two fairly large components, the sum of whose amplitudes exceeds A_0. Actually, we should really consider the measure of "energy" E, where

$$E = \sum_{n=-\infty}^{\infty} |C_n|^2.$$

Then Parseval's theorem (4.3.42) guarantees that

$$E = |A|^2. \tag{7}$$

Further, Bessel's inequality (4.3.41) gives assurance that a finite number of components contains only a fraction of the energy. In the case just considered, the two dominant components contain about 80 per cent of the energy.

FOURIER INTEGRAL ANALYSIS OF A TRUNCATED SINUSOIDAL FUNCTION

The Fourier series approach just discussed has the somewhat disconcerting feature of splitting one sinusoidal component into two. We shall show that the method of Fourier integral yields only a "broadening of the spectral line"—to borrow an expression from the experimental physicist. The general Fourier transform relation

$$a(\omega) = \frac{1}{2\pi} \int_{-\infty}^{\infty} f(t) e^{-i\omega t}\, dt$$

and its inverse give for the truncated sinusoid (3)

$$a_T(\omega) = \frac{A_0 T}{\pi} \frac{\sin(\omega_0 - \omega)T}{(\omega_0 - \omega)T} \tag{8}$$

and

$$f(t) = \int_{-\infty}^{\infty} e^{i\omega t} a_T(\omega)\, d\omega \tag{9}$$

as the reader is asked to verify in Exercise 2. The square of the function $a_T(\omega)$ of (8) is plotted in Figure 5.2. The ratio of the central maximum to the first subsidiary maximum is about $(3\pi/2)^2 \approx 22.5$. Thus the energy is practically all contained in the shaded central band, which has a spread in frequency of $\Delta\omega = 2\pi/T = 2\Omega$. The fraction of energy contained is given by

$$\mathscr{F} = \frac{\int_{-\pi}^{\pi} \eta^{-2} \sin^2 \eta\, d\eta}{\int_{-\infty}^{\infty} \eta^{-2} \sin^2 \eta\, d\eta}, \tag{10}$$

which is about 90 per cent of the total energy (Exercise 3).

The advantage of the Fourier integral approach is the reproduction of the original signal as a single component, with a broadening of its line spectrum

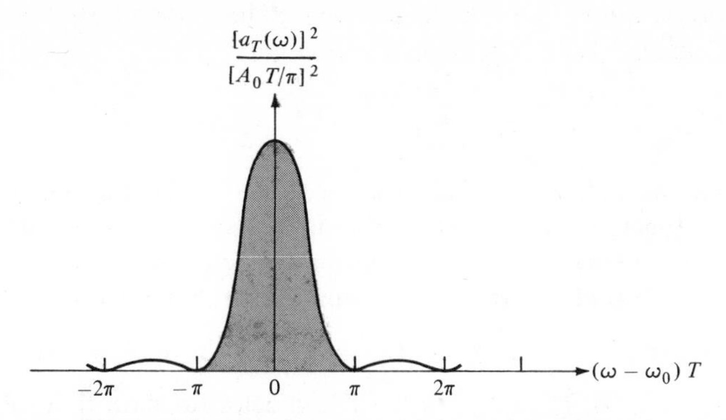

FIGURE 5.2. *Broadening of a spectral line of frequency ω_0, in the Fourier integral representation of the truncated sinusoid* (3).

but without a distortion of its frequency. This approach is also convenient for considering the case $T \to \infty$. If we consider the limiting value of (8), we obtain

$$a_\infty(\omega) \equiv \lim_{T\to\infty} a_T(\omega) = \begin{cases} \infty, & \omega = \omega_0\,; \\ \text{indeterminate}, & \omega \neq \omega_0\,. \end{cases} \tag{11}$$

To obtain definite limiting values, we introduce the definition

$$A_T(\omega) = \int_{-\infty}^{\omega} a_T(\omega)\, d\omega = \frac{A_0}{\pi}\int_{-\infty}^{\eta} \frac{\sin\eta}{\eta}\, d\eta, \tag{12}$$

where $\eta = (\omega - \omega_0)T$. When $T \to \infty$, $A_T(\omega) \to A_\infty(\omega)$, where

$$A_\infty(\omega) \equiv \begin{cases} A_0 & \text{for } \omega > \omega_0\,, \\ 0 & \text{for } \omega < \omega_0\,. \end{cases} \tag{13}$$

Following standard notation, we shall write (13) as

$$A_\infty(\omega) = A_0\, H(\omega - \omega_0), \tag{14}$$

where $H(x)$ is the **Heaviside function** defined by

$$H(x) = \begin{cases} 1 & \text{for } x > 0, \\ 0 & \text{for } x < 0. \end{cases} \tag{15}$$

We can interpret our results adequately at this point. From (9) we see that $a_T(\omega)$ is essentially the amplitude (per unit frequency) of the harmonic of frequency ω. Definition (12) shows that $A_T(\omega)$ gives the cumulative amplitude. In the limit, $A_T(\omega)$ suffers a jump as shown in (13) or (14). The entire contribution to $A_\infty(\omega)$ comes at the single frequency ω_0. Thus the Fourier integral approach comes arbitrarily close to an exact reproduction of a signal of period ω_0 if the length of time over which this signal is observed increases without bound.

More precise and usable results can be obtained upon introduction of the Stieltjes integral.* With this, (9) can be written

$$f_T(t) = \int_{\omega=-\infty}^{\omega=\infty} e^{i\omega t}\, dA_T(\omega) \tag{16}$$

and the *formal* limit $T \to \infty$ has meaning, since the Stieltjes integral can be defined for functions A_T with discontinuities. Indeed, in the limit form with $A_\infty(\omega)$ given by (14), Equation (16) yields the original function $A_0 e^{i\omega_0 t}$, for $-\infty < t < \infty$. To obtain this result from (9), we should write

$$a_T(\omega) \to a_\infty(\omega) = A_0\, \delta(\omega - \omega_0) \tag{17}$$

in the limit $T \to \infty$. Here $\delta(x)$ is the Dirac delta function that satisfies

$$\int_{-\infty}^{\infty} f(x)\delta(x-a) = f(a). \tag{18}$$

Since the relation $A_T'(\omega) = a_T(\omega)$ is implied by the definition of $A_T(\omega)$ in (12), (14) and (17) imply that

$$\delta(x) = H'(x). \tag{19}$$

This is a valid and useful formal result.

The reader will find it worthwhile to justify all the formalism by referring to the lemmas in Section 4.2.

GENERALIZATION TO STATIONARY TIME SEQUENCES

The discussion above can be formalized in the general case of a function $f(t)$ called a *stationary time sequence*.† To this end, consider a truncated function $f_T(t)$ where

$$\begin{aligned} f_T(t) &= f(t), \qquad |t| \le T; \\ &= 0, \qquad |t| > T. \end{aligned}$$

* Let F be continuous. The Riemann integral $\int_a^b F(\omega)\, d\omega$ is defined by a familiar limit of $\Sigma_i F(\omega_i^*)(\omega_{i+1} - \omega_i)$, $\omega_i \le \omega_i^* \le \omega_{i+1}$. If A is a monotonically increasing function, the Stieltjes integral $I_A = \int_a^b F(\omega)\, dA$ is defined by a similar limit of $\Sigma_i F(\omega_i^*)[A(\omega_{i+1}) - A(\omega_i)]$. If A is a step function [which we express in terms of the Heaviside function of (15)] then the integral reduces to a sum:

$$A(\omega) = \sum_{n=1}^{\infty} c_n H(\omega - \omega_n) \quad \Rightarrow \quad I_A = \sum_{n=1}^{\infty} c_n F(\omega_n).$$

(We assume that $\Sigma_{n=1}^{\infty} c_n$ is a convergent series of nonnegative constants.) If A is continuously differentiable, we obtain an ordinary Riemann integral:

$$I_A = \int_a^b F(\omega) A'(\omega)\, d\omega.$$

For proofs and further information, see texts on analysis, e.g., Chapter 5 of D. Widder's *Advanced Calculus* (Englewood Cliffs, N.J.: Prentice-Hall, 1947).

† We shall not attempt a precise definition here. The nature of a stationary time sequence is described at the beginning of the section, with the weather record as a typical example. Also see the remark below (24).

If we define in analogy to (8)

$$a_T(\omega) = \frac{1}{2\pi}\int_{-\infty}^{\infty} f_T(t)e^{-i\omega t}\,dt, \tag{20}$$

then the cumulative contribution of $a_T(\omega)$ in the interval $(\omega, \omega + \Delta\omega)$ is

$$\Delta A_T(\omega) = \int_{\omega}^{\omega+\Delta\omega} a_T(\omega)\,d\omega = \frac{1}{2\pi}\int_{-\infty}^{\infty} f_T(t)\frac{e^{-i\omega t} - e^{-i(\omega+\Delta\omega)t}}{it}\,dt, \tag{21}$$

where we have interchanged the order of integration. [Convergence difficulties prevent introduction of an integral from $-\infty$ to ω, as in the corresponding formula (12).] The appearance of the factor t in the denominator then facilitates the convergence of the integral in (21) as $T \to \infty$. If the limiting integral does exist, one may then expect to be able to introduce a function $A(\omega)$ such that $f(t)$ may be expressed as the Stieltje's integral

$$f(t) = \int_{\omega=-\infty}^{\omega=\infty} e^{i\omega t}\,dA(\omega). \tag{22}$$

Theorem 1. Equation (22) holds if $A(\omega)$ is defined by

$$A(\omega) = \frac{1}{2\pi}\int_{-\infty}^{\infty} f(t)\,\frac{1 - e^{-i\omega t}}{-it}\,dt. \tag{23}$$

Proof. We divide the infinite line $-\infty < \omega < \infty$ into intervals of length $\Delta\omega$ (with zero at one point of subdivision), and consider the sum

$$S_N(t) = \sum_{n=-N}^{N} e^{i\omega_n' t}[A(\omega_n + \Delta\omega) - A(\omega_n)],$$

where $N(\Delta\omega) = \Omega$, a fixed number. It is expected that in the limit $\Delta\omega \to 0$, $\Omega \to \infty$, $S_N(t)$ will approach $f(t)$, as indicated by (22). In the above expression, ω_n' may be taken to be any value in the interval $(\omega_n, \omega_n + \Delta\omega)$; it is convenient to take it to be $\omega_n + \Delta\omega/2$.

A little manipulation shows that

$$S_N = \frac{1}{2\pi}\int_{-\infty}^{\infty} f(t')\,\frac{2\sin[(\Delta\omega/2)t']}{t'}\sum_{n=-N}^{N} e^{i\omega_n'(t-t')}\,dt'.$$

Using (4.2.27), we see that the sum within the above integral is

$$e^{i(\Delta\omega/2)(t-t')}\,\frac{\sin[(N+\frac{1}{2})(t-t')(\Delta\omega)]}{\sin[\frac{1}{2}(t-t')(\Delta\omega)]}.$$

Thus, for fixed Ω, the limit of S_N as $N \to \infty$ is

$$S(\Omega) = \frac{1}{\pi}\int_{-\infty}^{\infty} f(t')\,\frac{\sin\Omega(t-t')}{t-t'}\,dt'.$$

The limiting process $\Omega \to \infty$ is already familiar to us (Lemma 2, Section 4.2).□

For a time sequence that combines the oscillatory behavior of (1) with an additional contribution exhibiting the rapid decay of (2) with increasing $|t|$, the function $A(\omega)$ would have three discontinuities, at $\omega = \omega_1, \omega_2, \omega_3$, respectively, with a smooth variation in between, to take care of (2),

AUTOCORRELATION FUNCTION AND THE POWER SPECTRUM

It is clear that $A(\omega)$ could be very complicated if there were an infinite number of discrete components of the type (1) such that the infinite series composed of the squares of the amplitude still converges. Moreover, a function representing noise or turbulence is generally not representable by a combination of such an infinite sum together with a function that has a Fourier transform. This is best seen by the following considerations.

We introduce the **autocorrelation function** $R(\tau)$, which is employed for analyzing noisy signals (using a bar for complex conjugate):

$$R(\tau) = \lim_{T\to\infty} \frac{1}{2T} \int_{t_0-T}^{t_0+T} \overline{f(t)} f(t+\tau)\, dt \equiv \langle \overline{f(t)} f(t+\tau) \rangle_t . \tag{24}$$

(By definition, for a stationary time sequence, the autocorrelation function is independent of the reference time t_0.) In the case of the function given in (1), the autocorrelation function is (Exercise 5)

$$R(\tau) = |A_1|^2 e^{i\omega_1\tau} + |A_2|^2 e^{i\omega_2\tau} + |A_3|^2 e^{i\omega_3\tau}, \tag{25}$$

which shows **anharmonic periodic behavior**. However, for noisy or turbulent signals,

$$R(\tau) \to 0 \qquad \text{as } \tau \to \infty,$$

since the random nature of the signal annihilates the autocorrelation for large distances.

The theory of generalized harmonic analysis has been rigorously developed by Norbert Wiener and other mathematicians. We cannot go into further details here except to mention one principal result. The theory shows that there exists a *power spectrum* defined by

$$\frac{1}{2} F(\omega) = \lim_{\Delta\omega\to 0} \frac{|\Delta A|^2}{\Delta\omega} . \tag{26}$$

Furthermore, it can be shown that this power spectrum stands in Fourier cosine transform relation to $R(\tau)$:

$$R(\tau) = \int_0^\infty F(\omega) \cos \omega\tau \, d\omega, \tag{27a}$$

$$F(\omega) = \frac{2}{\pi} \int_0^\infty R(\tau) \cos \omega\tau \, d\tau. \tag{27b}$$

For $\tau = 0$, (27a) and (24) combined yield the relation

$$\langle |f(t)|^2 \rangle_t = \int_0^\infty F(\omega)\, d\omega. \tag{28}$$

Equation (28) is the equivalent of Parseval's theorem, which justifies the term "power spectrum" [since $F(\omega)\, d\omega$ provides the contribution from the frequency interval $(\omega, \omega + d\omega)$ to the "energy" given by the mean of f^2].

VERIFICATION OF THE COSINE TRANSFORM RELATION BETWEEN THE POWER SPECTRUM AND THE AUTOCORRELATION

The derivation of (27) is quite difficult. However, in a statistical sense, where the correlation function $R(\tau)$ is defined by the **ensemble average** (an average over many repetitions of a phenomenon)

$$R(\tau) = \langle f(t) f(t + \tau) \rangle, \tag{29}$$

then the result is relatively easy to obtain. We take $f(t)$ to be real.

We start with (21) and consider its limiting form (as $T \to \infty$):

$$\Delta A(\omega) = \frac{1}{2\pi} \int_{-\infty}^{\infty} f(t) \frac{e^{-i\omega t} - e^{-i(\omega + \Delta\omega)t}}{it}\, dt. \tag{30}$$

We form $|\Delta A(\omega)|^2$ by multiplying (30) with its complex conjugate, using t' as the variable of integration in the expression for ΔA. The double integral in (t, t') may then be transformed into one using the variables (t, τ), where

$$\tau = t' - t. \tag{31}$$

We then obtain

$$(2\pi)^2 |\Delta A(\omega)|^2 = \int_{-\infty}^{\infty} e^{i\omega\tau}\, d\tau \int_{-\infty}^{\infty} dt \frac{\{1 - e^{i(\Delta\omega)(t+\tau)}\}\{1 - e^{-i(\Delta\omega)t}\}}{t(t + \tau)} f(t) f(t + \tau).$$

Let us take the statistical average. Then

$$(2\pi)^2 \langle |\Delta A(\omega)|^2 \rangle = \int_{-\infty}^{\infty} e^{i\omega\tau}\, d\tau\, R(\tau) \int_{-\infty}^{\infty} dt \frac{\{1 - e^{i(\Delta\omega)(t+\tau)}\}\{1 - e^{-i(\Delta\omega)t}\}}{t(t + \tau)}$$

and hence

$$\lim_{\Delta\omega \to 0} \frac{\langle |\Delta A(\omega)|^2 \rangle}{\Delta\omega} = \int_{-\infty}^{\infty} e^{i\omega\tau}\, d\tau\, R(\tau) \cdot \frac{1}{(2\pi)^2} \int_{-\infty}^{\infty} \frac{|1 - e^{ix}|^2}{x^2}\, dx,$$

where the variable $x = (\Delta\omega)t$ is a new integration variable. It can be shown (Exercise 7) that

$$\int_{-\infty}^{\infty} \frac{|1 - e^{ix}|^2}{x^2}\, dx = 2 \int_{-\infty}^{\infty} \frac{1 - \cos x}{x^2}\, dx = 2\pi. \tag{32}$$

Hence

$$\frac{1}{2}F(\omega) = \frac{1}{2\pi}\int_{-\infty}^{\infty} R(\tau)e^{i\omega\tau}\,d\tau.$$

Now $R(\tau)$ must be an even function of τ, since

$$R(\tau) = \langle f(t)f(t+\tau)\rangle = \langle f(t-\tau)f(t)\rangle = R(-\tau).$$

Thus the cosine transform relationship is established.

APPLICATION

Experimental verification of the Fourier transform relation (27) for turbulent signals was first accomplished by G. I. Taylor. (See Figure 5.3.)

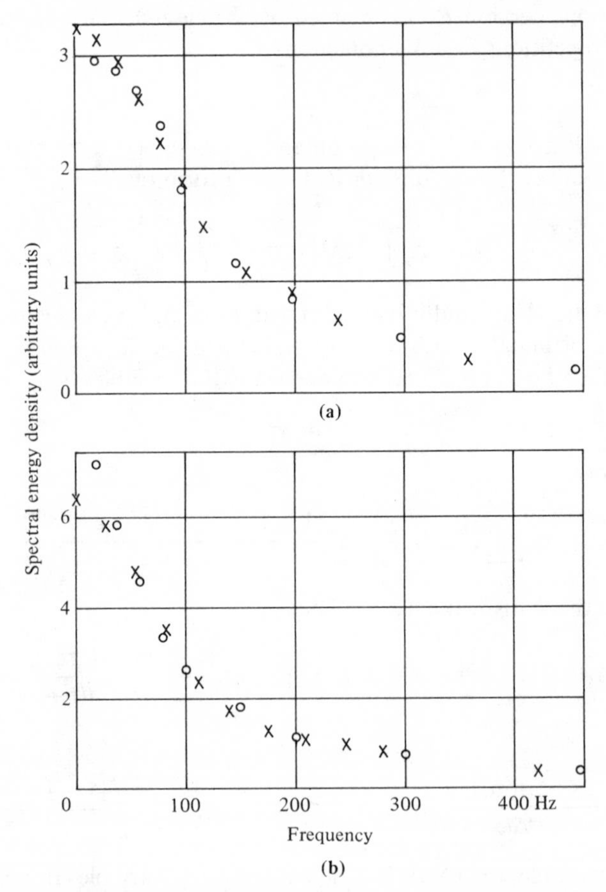

FIGURE 5.3. *One-dimensional spectra:* o, *direct measurement;* x, *transformed double correlation.* [*Reproduced with permission from R. W. Stewart and A. A. Townsend.* Phil. Trans. Roy. Soc. (London), **A234**, 359 (1951).]

In modern radio astronomical observatories, the spectral analysis of the incoming radiation is done by the use of (27). The correlation function $R(\tau)$ of the incoming signal is first computed, and the spectral function $F(\omega)$ is then obtained as its Fourier cosine transform. All this is done with automated equipment. Knowledge of F gives clues to the nature of the signaling source.

EXERCISES

1. (a) Verify (5).
(b) Verify (6).

2. Verify (8).

3. Show that the fraction $\mathscr{F}$ defined in (10) is about $\frac{9}{10}$.

4. (Project.) Justify various formal results of this section, as suggested under (19).

5. Show that the autocorrelation function for the function given in (1) is supplied in (25).

6. Fill in the details of the proof of Theorem 1.

7. If you are familiar with contour integration methods, use them to establish (32).

PART B

Some Fundamental Procedures Illustrated on Ordinary Differential Equations

CHAPTER 6
Simplification, Dimensional Analysis, and Scaling

A NUMBER of procedures that are of value to the applied mathematician will be described and illustrated in Part B of this volume. This chapter, the first of six that comprise Part B, is concerned with steps that should be taken to modify a preliminary formulation of a mathematical problem so that it assumes a more manageable character. Section 6.1 deals with the advantages and possible pitfalls of modifying a problem by deleting certain terms, terms which seem to be relatively small and whose deletion makes the problem easier to solve. In Section 2 we show that the use of dimensionless variables reduces the number of parameters which determine the solution of a problem. This is an aspect of dimensional analysis, a straightforward procedure but one that is more subtle and more useful than one first expects.

We demonstrate in Section 6.2 that dimensionless variables can be selected in a number of ways, with the same relatively few dimensionless parameters appearing in each case. In Section 6.3 we discuss a particular selection process that gives rise to what are called "scaled" dimensionless variables. When these variables are used in the dimensionless formulation of a problem, the dimensionless parameters that appear acquire added significance: they provide the relative magnitudes of various terms. Consequently, if correctly scaled variables can be chosen (and this is not an easy matter), the problem is often ripe for certain perturbation methods. There are two classes of such methods: the simpler regular perturbation methods (the subject of Chapter 8) or the more complicated singular perturbation methods (Chapter 9). Perturbation methods provide one of the most useful sets of applied mathematical tools. Thus it is well worth devoting some effort to understanding the procedure discussed in Section 6.3, which sets up a problem so that perturbation calculations become somewhat "natural."

Our pedagogical strategy in this chapter* is to present the various ideas in the context of a particular problem, that of a point mass radially projected from the earth's surface and subjected only to a gravitational force that varies with distance. (All damping effects are neglected.) Since this "projectile problem" is a very simple one, the reader should be able to focus all his

* Sections 6.1 and 6.3 follow in outline, and in many cases word for word, the article "Simplification and Scaling" by L. A. Segel [*SIAM Rev. 14*, 547–71 (1972)]. These are reprinted with permission from *SIAM Review* (1972).

energies on mastering the new concepts and techniques. These, and the material of the next chapter, are given nontrivial application in the treatment of a physiological flow problem in Chapter 8. Mathematically, what is required here is the solution of a boundary value problem for a pair of nonlinear ordinary differential equations. Many of the ideas are used later in the analysis of more complicated problems, involving partial differential equations.

6.1 The Basic Simplification Procedure

Simplification of a given set of equations may make it possible to avoid large machine calculations or massive analytic work and still to obtain useful answers. Even if the equations can be solved exactly without undue effort, simplified equations may yield a sufficiently accurate solution whose features are more readily apparent than those of the exact solution. Drastic simplification may allow rapid solution and immediate determination of whether one is on the right track.

The following **basic simplification procedure** is used time and again by applied mathematicians:

STEP A. *Somehow identify terms that are relatively small.* In practice the smallness of a given term is generally gauged in relation to other terms in the same equation.

STEP B. *Delete relatively small terms* and solve the resulting simplified problem.

STEP C. *Check for consistency.* That is, use the approximate solution just obtained to evaluate the neglected terms, to ensure that they are indeed relatively small.

We shall now present some examples of the procedure. In doing so, we shall use the symbol $\sim$ to mean "seems to be approximately equal to" and $\approx$ to mean "*is* approximately equal to."

ILLUSTRATIONS OF THE PROCEDURE

Example 1. Suppose that the problem

$$x + 10y = 21, \qquad 5x + y = 7, \tag{1a, b}$$

is nontrivial. In (1a) the coefficient of the x term is small compared to the coefficient of the y term, so it is tempting to assume that the x term may be neglected to first approximation. Omitting this term, we obtain

$$y \sim 2.1, \qquad x \sim \tfrac{1}{5}(7 - 2.1) \approx 0.98. \tag{2}$$

Approximating the unknowns by means of the values given in (2), we estimate that the ratio of the first term on the left side of (1a) to the second has magnitude $0.98/21 \approx 0.05$. This number is small compared to unity, so that our approximation appears consistent. Also, our approximate values of x and y are close to the true values $x = 1$, $y = 2$.

Example 2. (The projectile problem). Consider a body of constant mass m that is radially projected upward from the earth's surface with initial speed V. Let R denote the radius of the earth and let $x^*(t^*)$ denote radial distance from the earth's surface at time t^*. (We use starred variables here so that in later discussions of this example we can introduce unstarred dimensionless variables.) If we neglect air resistance, the governing equations and initial conditions are

$$\frac{d^2x^*}{d(t^*)^2} = -\frac{gR^2}{(x^* + R)^2}, \qquad x^*(0) = 0, \quad \frac{dx^*}{dt^*}(0) = V. \tag{3}$$

(We have merely equated acceleration with force per unit mass, using the inverse square law in a form that correctly gives acceleration $-g$ when $x^* = 0$.)

If V is small in some sense, the displacement x^* should be small compared to R. Instead of (3) we can then consider the simplified problem

$$\frac{d^2x^*}{dt^{*2}} = -g, \quad x^*(0) = 0, \qquad \frac{dx^*}{dt^*}(0) = V, \tag{4}$$

with solution

$$\frac{dx^*}{dt^*} = -gt^* + V, \qquad x^* = -\tfrac{1}{2}gt^{*2} + Vt^*. \tag{5}$$

At the time $t^* = Vg^{-1}$, when the speed vanishes, we find from (5) that x^* reaches a maximum of $\frac{1}{2}V^2/g$. Thus x^*/R is at most $\frac{1}{2}V^2/gR$, and our approximation is consistent when V^2 is small compared to gR.

There is a feeling of general well-being attached to the property of being consistent, but precisely what does consistency imply in the present context? To generate interest in the answer to this question, we now present some examples that illustrate some possibly disturbing aspects of the basic simplification procedure.

TWO CHASTENING EXAMPLES*

Example 3. Consider the equations

$$0.01x + y = 0.1, \qquad x + 101y = 11. \tag{6a, b}$$

In (6a) the coefficient of the x term is small compared to the coefficient of the y term; so let us neglect the former term to obtain

$$y \sim 0.1, \qquad x \sim 11 - 101(0.1) = 0.9. \tag{7}$$

Using (7), we estimate that the ratio of the first term on the left side of (6a) to the second has magnitude $(0.01)(0.9)/0.1 = 0.09$. Because this number is small compared to unity, our approximation appears to be consistent. But the exact answer is $y = 1$, $x = -90$; therefore, our "approximation" for y is off by a factor of 10, while our "approximation" for x is off by a factor of 100 and has the wrong sign also.

* See C. Lanczos, *Applied Analysis* (London: Pitman, 1957), pp. 198ff. and p. 276 for other examples.

To see what went wrong, let us generalize (6) and consider the following equations for $x(\varepsilon)$ and $y(\varepsilon)$:

$$\varepsilon x + y = 0.1, \qquad x + 101y = 11. \tag{8}$$

Equations (8) reduce to (6) when $\varepsilon = 0.01$. The approximation (7) corresponds to taking $\varepsilon = 0$. But is it true that $x(\varepsilon) \approx x(0)$, $y(\varepsilon) \approx y(0)$ when $\varepsilon = 0.01$? As soon as we ask this question, it becomes clear that we shall be in trouble if $x(\varepsilon)$ and $y(\varepsilon)$ change rapidly as ε varies near $\varepsilon = 0$. That this is indeed the case can be seen from the exact solution to (8):

$$x = \frac{0.9}{1 - 101\varepsilon}, \quad y = \frac{0.1 - 11\varepsilon}{1 - 101\varepsilon}; \qquad \varepsilon \neq 1/101. \tag{9}$$

Example 3 illustrates the fact that an approximation which appears consistent may not actually be so. In approximating $x(0.01)$ and $y(0.01)$ by $x(0)$ and $y(0)$, we simplified the equations by neglecting the $\varepsilon x(\varepsilon)$ term. This is a consistent procedure if $|0.01x(0.01)|$ is small compared to $|y(0.01)|$. Assuming that we only knew the approximate solutions $x(0)$ and $y(0)$, we checked for consistency as best we could by examining $|0.01x(0)/y(0)|$. This ratio has the satisfactorily small value 0.09 so that the approximation appeared consistent. We were deceived, however, since the exact result is

$$\left|\frac{0.01x(0.01)}{y(0.01)}\right| = 9.$$

We thus distinguish **apparent consistency** (the approximation to the neglected term is small) from **genuine consistency** (the term neglected is truly small). The third step in the basic simplification procedure should really be called *checking for apparent consistency.*

It is not surprising that our approximation in Example 3 was a poor one, because it was apparently consistent but not genuinely consistent. But it is disheartening that an approximation can be apparently consistent but not genuinely so. Even more disheartening is the fact that an approximation that seems genuinely consistent may be very inaccurate, as is illustrated in the next example.

Example 4. (Wilkinson, 1963, p. 41). The polynomial

$$(x - 1)(x - 2)(x - 3) \cdots (x - 20) = x^{20} - 210x^{19} + \cdots \tag{10}$$

has as zeros the first 20 positive integers. If the coefficient of x^{19} is changed by the addition of εx^{19}, where $\varepsilon = -2^{-23} \approx -1.19 \times 10^{-8}$, then the smaller zeros are almost unaltered. But the larger zeros are so radically altered that the changed equation has five pairs of complex conjugate roots, given roughly by $10 \pm 0.64i$, $12 \pm 1.7i$, $14 \pm 2.5i$, $17 \pm 2.8i$, and $20 \pm 1.9i$. Whether the added term εx^{19} is evaluated using the "approximate" zeros or the exact zeros, it is still small in magnitude compared to the retained term $-210x^{19}$. Both apparent and genuine consistency seem verified, but the approximation is again a poor one.

CONDITIONING AND SENSITIVITY

We have established the existence of **wretched consistent approximations.** It is clearly necessary to examine matters with more care. We shall find that certain general remarks can be made, but that further understanding requires a somewhat detailed analysis of various problem classes.

As was illustrated in Example 3, knowledge that a problem has been altered by a small amount is significant only if the problem is well conditioned in the sense that small changes in its statement are accompanied by small changes in its solution. The problem is termed **ill conditioned** if this is not the case. It will also be convenient here to use the terminology that the **solution to a problem is sensitive to neglect of a term** T if such neglect cannot be expected to result in a small change in the solution even when the neglected term is genuinely small. Thus Wilkinson (1963) reports that the smaller zeros of (10) are not sensitive to alterations in *any* of the coefficients (these alterations can be regarded as the neglect of certain terms in an obvious way), and even the larger zeros are not sensitive to small changes in *certain* coefficients. But the larger zeros *are* sensitive to small changes in some of the coefficients, namely, those multiplying the higher powered terms in the polynomial.

The only blanket statement that it seems possible to make concerning the value of consistency checks is that the *lack of apparent consistency is almost certainly associated with poor approximation.* To show this, let x symbolize the true solution to a problem and let $\tilde{x}$ denote the approximate solution that one obtains when the term T is neglected. Neglect of T is apparently or genuinely consistent, respectively, if $T(\tilde{x})$ or $T(x)$ is small. (The notation used here is symbolic. x could be a vector function, for example, and T could depend on the components of x and several of their derivatives.) If the solution is sensitive to the neglect of T, then, by definition, such neglect cannot be expected to lead to a good approximation. If the solution is *not* sensitive to the neglect of T, then

$$[T(x) \text{ small}] \text{ implies } [\tilde{x} \approx x] \text{ implies } [T(\tilde{x}) \text{ small}],$$

assuming that T varies continuously. The finding of a large apparent error $T(\tilde{x})$ is therefore logically inconsistent with the supposition that the term neglected is genuinely small, and neglect of a genuinely large term cannot be expected to lead to a good approximation.

The imprecise words "small" and "large" appear again and again, but these words can only be given a definite meaning in the context of a particular problem. Thus the numerical magnitude of an acceleration (say) can be changed by altering the time scale, but this change can have no effect on the question of whether the acceleration is negligible or not. To mention a slightly different example, if a certain temperature should be 1°C but was calculated as 1.1°C, the error is 10 per cent. If the Fahrenheit scale is employed, the percentage error is only about 0.5 per cent, but it is without significance that the latter percentage is much smaller than the former. Whatever temperature scale is used, the decision as to whether an error is acceptable or not rests on scientific grounds and must be independent of the units employed.

NOTE. The difference between the alterations of the time scale and the temperature scale is that the former involves only the choice of a unit, while the latter also involves the choice of a zero.

To make further progress, it is necessary to consider particular types of problems. We shall discuss the determination of the zeros of a function, and second order ordinary differential equations. See Segel (1972, op. cit.) for similar remarks on linear algebraic equations.

ZEROS OF A FUNCTION

Let a zero of $f(x, \varepsilon)$ be $x(\varepsilon)$ so that

$$f[x(\varepsilon), \varepsilon] \equiv 0. \tag{11}$$

Suppose that ε is a small parameter, and let an approximation $x^{(0)}$ to the zero $x(\varepsilon)$ be found by setting $\varepsilon = 0$, so that $f[x^{(0)}, 0] = 0$.The **equation error** in equating $f(x, 0)$ to zero rather than $f(x, \varepsilon)$ is $f(x, \varepsilon) - f(x, 0)$. The error is apparent or genuine, depending on whether this expression is evaluated when $x = x^{(0)}$ or $x = x(\varepsilon)$. Thus, for the genuine equation error g, we have

$$g \equiv f[x(\varepsilon), \varepsilon] - f[x(\varepsilon), 0] = -\varepsilon f_x^{(0)} x'(0) + O(\varepsilon^2).$$

(We continue to use a superscript zero to indicate that ε has been set equal to zero.) The apparent equation error or **residual** satisfies

$$r \equiv f[x^{(0)}, \varepsilon] - f[x^{(0)}, 0] = f[x^{(0)}, \varepsilon] = \varepsilon f_\varepsilon^{(0)} + O(\varepsilon^2).$$

We wish to relate r to $h(\varepsilon) \equiv x(\varepsilon) - x^{(0)}$, the error in the solution. We shall be retaining only the lowest powers of ε. To the lowest order,

$$h(\varepsilon) = \varepsilon x'(0).$$

But

$$0 \equiv f[x(\varepsilon), \varepsilon] - f[x_0, 0] = \varepsilon\{f_x^{(0)} x'(0) + f_\varepsilon^{(0)}\} + O(\varepsilon^2).$$

The vanishing of the expression within the curly brackets implies two things. (i) The genuine error and the apparent error are equal to first approximation. (ii) To first order, the error is given by

$$h = -\frac{r}{f_x^{(0)}}. \tag{12}$$

As is generally true, the error depends not only on the residual r but also on a measure of the condition of the problem. In this case the "condition" depends on $f_x^{(0)}$. In particular, if $f_x^{(0)}$ is small, then a small change in f is associated with an $O(1)$ change in x; in other words, a *small change in the equation is associated with a nonsmall change in the solution.* If such ill conditioning is present, one expects a large error in spite of a small residual. By contrast, for well-conditioned problems [$f_x^{(0)}$ of order of magnitude unity], (12) shows that a large residual r precludes a small error h.

If we write (12) in the form

$$\left(\frac{h}{x}\right) = -\frac{1}{f_x^{(0)}}\left(\frac{r}{x}\right),$$

we see that when relative error h/x is to be estimated, it is the relative residual r/x that is relevant, where x is the solution. Except for cases of ill conditioning, r can be compared to the approximate solution. Consequently, in Example 4 one should not examine the ratio of the neglected term to a term retained but the ratio of the neglected term to the estimated solution. The latter ratio is $\varepsilon\tilde{x}^{19}/\tilde{x}$. For $\varepsilon = -2^{-23}$, this ratio is small for $\tilde{x}$ close to unity but becomes large for $\tilde{x}$ much bigger than 2. There is a clear warning of difficulty for the larger roots. Actually, the warning is somewhat premature, for here the conditioning factor acts to keep the errors reasonable for zeros up to about 10. [See Wilkinson (1963) for a full discussion of sensitivity in this problem.] Normally, the various terms in a polynomial might be expected to be of the same size with coefficients of unit magnitude; in such a case, comparison with any term retained will give the magnitude of the desired ratio between the residual and the solution. The present polynomial is not "normal," however. Many of its coefficients are very large, but the terms nearly cancel because of sign alternations. "Abnormal" equations with such large but nearly canceling terms are not a rarity; this must be kept in mind when comparing a term neglected with an allegedly "typical" term retained.

SECOND ORDER DIFFERENTIAL EQUATIONS

Consider an initial value problem for $x(t, \varepsilon)$:

$$f(t, x, \dot{x}, \ddot{x}, \varepsilon) = 0; \qquad x(0) = A, \quad \dot{x}(0) = B, \quad |\varepsilon| \ll 1, \quad \cdot = \frac{d}{dt}. \tag{13}$$

The following reasoning closely parallels that just used in the discussion of the equation $f[x(\varepsilon), \varepsilon] = 0$.

An approximate solution $x^{(0)}(t)$ is found by setting $\varepsilon = 0$:

$$f(t, x^{(0)}, \dot{x}^{(0)}, \ddot{x}^{(0)}, 0) = 0.$$

In setting $\varepsilon = 0$, one makes an equation error of

$$f(t, x, \dot{x}, \ddot{x}, \varepsilon) - f(t, x, \dot{x}, \ddot{x}, 0).$$

To find the residual, one evaluates the equation error when $x = x^{(0)}$:

$$r \equiv f(t, x^{(0)}, \dot{x}^{(0)}, \ddot{x}^{(0)}, \varepsilon) = \varepsilon[f_5]_{\varepsilon=0} + O(\varepsilon^2).$$

The genuine equation error g satisfies

$$g = -f(t, x, \dot{x}, \ddot{x}, 0) = -\varepsilon[f_2 x_\varepsilon + f_3 \dot{x}_\varepsilon + f_4 \ddot{x}_\varepsilon]_{\varepsilon=0} + O(\varepsilon^2).$$

(Subscripts denote partial derivatives.) But

$$\begin{aligned} 0 &= f(t, x, \dot{x}, \ddot{x}, \varepsilon) - f(t, x^{(0)}, \dot{x}^{(0)}, \ddot{x}^{(0)}, 0) \\ &= \varepsilon[f_2 x_\varepsilon + f_3 \dot{x}_\varepsilon + f_4 \ddot{x}_\varepsilon + f_5]_{\varepsilon=0} + O(\varepsilon^2). \end{aligned}$$

Thus, as before, using a superscript to remind us that all terms are evaluated at $\varepsilon = 0$, we have

$$f_4^{(0)} \ddot{x}_\varepsilon^{(0)} + f_3^{(0)} \dot{x}_\varepsilon^{(0)} + f_2^{(0)} x_\varepsilon^{(0)} = -f_5^{(0)}. \tag{14}$$

We conclude from (14) that the residual and the genuine equation error are the same to lowest order. Also, we can now obtain an expression for the error in terms of the residual. To do this, we make the definitions

$$\begin{aligned} \mathbf{F}(t) &\equiv (f_2^{(0)}, f_3^{(0)}, f_4^{(0)}), \\ \mathbf{h}(t) &\equiv (x - x^{(0)}, \dot{x} - \dot{x}^{(0)}, \ddot{x} - \ddot{x}^{(0)}). \end{aligned} \tag{15}$$

We note that

$$\mathbf{h}(t) = \varepsilon(x_\varepsilon^{(0)}, \dot{x}_\varepsilon^{(0)}, \ddot{x}_\varepsilon^{(0)}) + O(\varepsilon^2).$$

Thus to lowest order, (14) implies that

$$\mathbf{F} \cdot \mathbf{h} = -r. \tag{16}$$

Now

$$|r| = |\mathbf{F}|\,|\mathbf{h}|\,|\cos(\mathbf{F}, \mathbf{h})| \le |\mathbf{F}|\,|\mathbf{h}|$$

so that

$$|\mathbf{h}| \ge \frac{|r|}{|\mathbf{F}|}.$$

We again see that a large residual precludes a small error unless the problem is ill conditioned.

Note that both the error vector **h** and the condition vector **F** involve not only the behavior of the solution, but also of its first two derivatives. Thus, even if a small change in the equation is associated with a small change in the solution itself, ill conditioning arises if, for example, the solution's second derivative is markedly affected.

It is difficult to make any general statements from (16) about whether a small residual implies a small error. Ill conditioning must certainly be excluded. Further, one must somehow satisfy oneself that the smallness of r does not arise from cancellation of nonsmall errors in $x_\varepsilon^{(0)}$, $\dot{x}_\varepsilon^{(0)}$, and $\ddot{x}_\varepsilon^{(0)}$. In particular cases one can often make stronger statements. To illustrate this, consider

$$\ddot{x} + (1 + \varepsilon x)^{-2} = 0; \qquad x(0) = 0, \qquad \dot{x}(0) = 1. \tag{17}$$

As is demonstrated in Section 6.3, (17) is the governing equation of the appropriately nondimensionalized version of the projectile problem of Example 2 when the parameter $\varepsilon \equiv V^2/Rg$ is small. In this case $\mathbf{F} = (0, 0, 1)$ and $r = \varepsilon x^{(0)}(t)$, so one can immediately deduce from (16) that

$$|\varepsilon \ddot{x}_\varepsilon^{(0)}| \leq \varepsilon \max x^{(0)}(t). \tag{18}$$

To first order in ε, the above relation shows that the error in the acceleration is small. Two integrations permit estimates of the error in x itself (Exercise 1).

An important point can be introduced via consideration of small vibrations of a simple nonlinear pendulum. As is indicated in Section 7.1 the proper formulation of this problem is

$$\ddot{\theta} + \varepsilon^{-1/2} \sin (\varepsilon^{1/2}\theta) = 0; \qquad \theta(0) = 1, \quad \dot{\theta}(0) = 0, \quad 0 < \varepsilon \ll 1. \tag{19}$$

Here $\theta(t)$ is a measure of angular displacement at time t, and ε is a measure of the initial displacement, which is assumed to be small. The zeroth approximation is $\theta^{(0)} = \cos t$. Using (16) to express the error, we find that

$$\varepsilon[\ddot{\theta}_\varepsilon^{(0)} + \theta_\varepsilon^{(0)}] = r = \varepsilon(\tfrac{1}{6})[\theta^{(0)}]^3. \tag{20}$$

We see illustrated here why it is difficult to bound the error from a knowledge of the residual, for (20) is nothing less than a differential equation for $\theta_\varepsilon^{(0)}$. We are faced with the problem of obtaining an estimate of the solution to a linear differential equation from a knowledge of the magnitude of its forcing term. There is no universal formula here, but the following simple calculation is typical of what can be done. Using the method of variation of parameters and observing that the initial conditions $\theta_\varepsilon^{(0)}(0) = \dot{\theta}_\varepsilon^{(0)}(0) = 0$ hold, we write

$$\varepsilon\theta_\varepsilon^{(0)}(t) = \int_0^t \sin (t - \xi) r(\xi)\, d\xi.$$

Since $|r| \leq \dfrac{\varepsilon}{6}$,

$$|\varepsilon\theta_\varepsilon^{(0)}(t)| \leq \frac{\varepsilon T}{6} \qquad \text{for } 0 \leq t \leq T.$$

As is well known, there is indeed an increase of error with T.

From another point of view, the current example illustrates the fact that perhaps the best way to obtain information about the validity of the first approximation is to compute the second approximation. After all, $\theta_\varepsilon^{(0)}$ is nothing more than the $O(\varepsilon)$ coefficient in the power series expansion of $\theta(t, \varepsilon)$. Equation (20), which governs $\theta_\varepsilon^{(0)}$, was obtained by what can be regarded as the parametric differentiation approach to perturbation theory.*

* See *General Sensitivity Theory* by R. Tomovic and M. Vukobratovic (N.Y.: Elsevier, 1972) for a detailed discussion of the calculation of functions like θ_ε and of the role of such functions in understanding the sensitivity of various systems to parameter variation.

RECOMMENDATIONS

In our analysis of the basic simplification procedure, we began a probe into the relation between the residual and the error. In one sense, we only scratched the surface of the subject, for an important part of numerical analysis concerns itself with how to obtain successively better approximate solutions to various problems by successively decreasing residuals.* Our main concern, however, is with problems which are so formidable that successive approximations are out of the question—one can barely obtain the first approximation. For such problems, which are frequently encountered by the applied mathematician, our discussion leads us to make the following recommendations.

(i) Use the basic simplification procedure in difficult problems. Not much extra work is required to estimate the magnitude of neglected terms, and you will at least learn where your simplification is almost certainly invalid (when the residual is large).

(ii) Although it is difficult to bound the error associated with a given residual, it may be helpful to regard the residual as an extraneous forcing and to use physical intuition to estimate its effect. Another possibility is to replace the residual by its mean or maximum value and to evaluate the effect of this constant forcing. To spot hidden ill conditioning, make small random modifications in the problem and solve again (Lanczos, 1957, op. cit., p. 170).

(iii) Beware of comparing a term neglected with a term retained. To estimate relative error, it is best to compare the neglected term with your approximate solution.

(iv) An applied mathematician studies simplified models to gain an understanding of complicated situations. In this spirit, regard a deep study of the simpler classical problems of numerical analysis and perturbation theory as not only of value in itself, but also of value in its indication of what to expect in more complicated problems.

(v) If possible, compute an "extra" term and use it to estimate the error in the first approximation. [For an illustration of this recommendation, see the discussion centered around Equation (7.1.25).]

EXERCISES

1. Use (18) to provide a bound for $|\varepsilon x_\varepsilon^{(0)}(t)|$.

2. A slab is in a steady state with temperature T_0 at $x = 0$ and T_1 at $x = 1$. The thermal conductivity is given by $k = k_0 \exp(\varepsilon x)$, where $|\varepsilon| \ll 1$,

* With the aid of Green's functions and Lipschitz conditions, the presence of a small residual can often be used to demonstrate the excellence of approximations. For example, see N. Ferguson and B. Finlayson's discussion of a mixed analytic–numerical attack on problems involving heat and mass transfer of a multicomponent system undergoing chemical reaction in catalyst pellets [*Amer. Inst. Chem. Eng. J. 18*, 1053–1059 (1972)]. In one case, 13-digit accuracy is proved.

Obtain an approximation to the temperature distribution by replacing k by its average value. Check for consistency.

3. Water flowing from a small circular hole in a container has a speed v which is approximately given by $v = 0.6\sqrt{2gh}$, where g is the gravitational acceleration and h is the height of the water above the hole. Let $A(h)$ be the area of the cross section at height h.

   (a) Derive

$$\frac{dh}{dt} = -0.6\, A(0) \frac{\sqrt{2gh}}{A(h)}.$$

   (b) Suppose that the actual shape of the container is approximated by $A(h) = h^c$, c a constant. Solve the initial value problem. Discuss the apparent consistency of the approximation. Are there circumstances where the approximation could be both apparently consistent and wretched?

6.2 Dimensional Analysis

Dimensional analysis occupies an unusual place among the techniques available to a theoretical scientist. The formal manipulations required are easy, but the ideas involved are seen on close inspection to be rather deep. (See Bridgman, 1931.) It is now known that dimensional analysis shares with the theory of similarity solutions to partial differential equations a common foundation in the theory of continuous groups. (See Birkhoff, 1960.)

It would take us too far afield to cover thoroughly the more theoretical aspects of dimensional analysis. For these the reader is referred to the literature. Rather, we shall introduce the practical aspect of the subject, largely by illustrating general ideas using the simple projectile problem formulated in Section 6.1. (At the end of Section 4.1 we discussed some aspects of dimensional analysis, in connection with heat flow.)

We shall show how to nondimensionalize a differential equation. In addition, we shall illustrate how nondimensionalization can be used even if the equations governing a phenomenon are unknown, provided that the relevant parameters can be listed. Included in our remarks on the advantages of dimensional analysis is a brief discussion of the use of scale models. We assume a rudimentary familiarity with the notion of physical dimension, e.g., in the statement "acceleration, as indicated by its typical units cm/sec^2, has dimension length/time^2."

PUTTING A DIFFERENTIAL EQUATION INTO DIMENSIONLESS FORM

We can illustrate the main ideas of dimensional analysis very simply on the projectile problem, which was already referred to in Example 2 of Section 6.1. We repeat the equations that govern the displacement $x^*(t^*)$ at time t^* of a particle which is radially projected at initial speed V:

$$\frac{d^2x^*}{d(t^*)^2} = -\frac{gR^2}{(x^*+R)^2}, \quad x^*(0)=0, \qquad \frac{dx^*}{dt^*}(0) = V. \qquad \text{(1a, b, c)}$$

Here R is the radius of the earth, g is the gravitational acceleration, and x^* is taken to be zero when the particle rests on the surface of the earth.

To put (1) into dimensionless form, we follow a two-step **procedure for nondimensionalization**, which can be used for a large class of problems.

STEP A. *List all parameters and variables, together with their dimensions* (in terms of **fundamental units** such as mass $\mathscr{M}$, length $\mathscr{L}$, time $\mathscr{T}$). In the present instance, we have the following list:

Variables	Dimension
Dependent variable x^*	$\mathscr{L}$
Independent variable t^*	$\mathscr{T}$
Parameters	
Gravitational acceleration g	$\mathscr{L}\mathscr{T}^{-2}$
Initial speed V	$\mathscr{L}\mathscr{T}^{-1}$
Earth radius R	$\mathscr{L}$

STEP B. *Let v^* be a variable. Form a combination of parameters p^* with the same dimensions as v^*. Introduce v^*/p^* as a new dimensionless variable.* Here R has the same dimensions as x^* and RV^{-1} has the same dimensions as t^*. Thus we introduce the new variables y and τ defined by

$$y = \frac{x^*}{R}, \qquad \tau = \frac{t^*}{RV^{-1}}. \qquad \text{(2a, b)}$$

These variables are **dimensionless**, because their numerical value is the same whatever standard of measurement is used. For example, if $R = 4000$ miles and $x^* = 8000$ miles, then $y = 2$. If the kilometer is used as our standard length, we have $R = 6436$ km, $x^* = 12{,}872$ km, and still $y = 2$.

Substitution of the new variables of (2) into (1) requires the chain rule. Thus (writing everything out in full)

$$\frac{d^2x^*}{d(t^*)^2} = \frac{d^2(yR)}{d(t^*)^2} = R\frac{d}{dt^*}\left(\frac{dy}{dt^*}\right) = R\frac{d}{dt^*}\left(\frac{dy}{d\tau}\frac{d\tau}{dt^*}\right) = R\frac{d}{dt^*}\left(\frac{dy}{d\tau}\frac{V}{R}\right)$$

$$= V\frac{d}{dt^*}\left(\frac{dy}{d\tau}\right) = V\frac{d}{d\tau}\left(\frac{dy}{d\tau}\right)\frac{d\tau}{dt^*} = \frac{V^2}{R}\frac{d^2y}{d\tau^2}. \qquad (3)$$

A little further calculation (Exercise 1a) gives the final result:

$$\varepsilon\frac{d^2y}{d\tau^2} = -\frac{1}{(y+1)^2}, \quad y(0)=0, \qquad \frac{dy}{d\tau}(0) = 1, \qquad (4)$$

where

$$\varepsilon = \frac{V^2}{gR}. \tag{5}$$

Note that the dimensions of the parameter ε are

$$\frac{(\mathscr{L}\mathscr{T}^{-1})^2}{\mathscr{L}\mathscr{T}^{-2}\mathscr{L}} = \mathscr{L}^{2-2}\mathscr{T}^{-2+2} = \mathscr{L}^0\mathscr{T}^0.$$

That is, ε is dimensionless.

REMARKS. (i) In (2), R is called an **intrinsic reference length**; τ is an **intrinsic reference time**. Generally speaking, *intrinsic reference quantities* are defined to be standards of measurement formed from the parameters of a given problem. (ii) It can be shown that only dimensionless combinations of parameters (like ε), sometimes called **dimensionless groups**, appear in a properly nondimensionalized problem. (iii) We have used starred quantities for dimensional variables so that we could use more convenient unstarred variables for the final dimensionless quantities. (iv) Computations like (3) are continually required in nondimensionalization. The calculations are equivalent to a formal removal of constants so that they precede the derivative, like this:

$$\frac{d^2x^*}{d(t^*)^2} = \frac{d^2(yR)}{d(RV^{-1}\tau)^2} = \frac{R}{R^2V^{-2}}\frac{d^2y}{d\tau^2} = \frac{V^2}{R}\frac{d^2y}{d\tau^2}. \tag{6}$$

(v) The temperature Θ often has to be included as a fundamental unit, in addition to mass $\mathscr{M}$, length $\mathscr{L}$, and time $\mathscr{T}$. Moreover, it can be of advantage in certain special problems to assume the existence of additional fundamental units. (See Exercise 7.)

Explicitly noting parameter dependence, we observe that (1) has a solution of the form

$$x^* = x^*(t^*; g, R, V), \tag{7}$$

while the solution to (4) can be written

$$y = y(\tau; \varepsilon). \tag{8}$$

The solution (7) depends on the three parameters g, R, and V, but in (8) only the single dimensionless parameter ε appears.

As an alternative to (2b), we could base our reference time on the acceleration g, writing

$$z = \frac{x^*}{R}, \qquad \tau_1 = \frac{t^*}{\sqrt{Rg^{-1}}}, \tag{9}$$

with which (1) becomes

$$\frac{d^2z}{d\tau_1^2} = -\frac{1}{(z+1)^2}, \qquad z(0) = 0, \quad \frac{dz}{d\tau_1}(0) = \varepsilon^{1/2}. \tag{10}$$

Now we expect a solution of the form

$$z = z(\tau_1, \varepsilon). \tag{11}$$

There are innumerable reference times, since $h(\varepsilon)RV^{-1}$ will serve, for any function h. Whichever of these reference times is employed, it will still be the case, as in (8) and (11), that the dimensionless distance from the earth depends only on a dimensionless time and the single dimensionless parameter ε.

A priori it would seem advantageous in a given problem to compare lengths, for example, with an intrinsic reference length rather than with an arbitrary length such as the distance between two scratches on a certain metal bar. As our example illustrates, this is the case—and the advantage is that *the number of parameters which appear in a problem is reduced when dimensionless variables are employed.* (To be more precise, no parameters disappear, but they occur only in certain dimensionless combinations.) Our example also illustrates the fact that in all but the simplest problems, *nondimensionalization can be carried out in a variety of ways,* each of which confers the same reduction in the number of parameters that appear.

We stress that since all parameters appear in the dimensionless version of the projectile problem only in the combination ε, any dimensionless result to be gleaned from the projectile problem depends only on the parameter ε. For example, suppose that we are interested in the dimensionless time τ_M when the projectile achieves its maximum height. Using the dimensionless formulation (4), we see that this time can be determined as the solution of the equation

$$\left.\frac{dy}{d\tau}\right|_{\tau=\tau_M} = 0.$$

But since $y = y(\tau, \varepsilon)$, the above equation shows that $\tau_M = f(\varepsilon)$ for some function f. In terms of the original variables we have that the dimensional time to maximum height, t_M^*, is given by

$$\frac{t_M^*}{RV^{-1}} = f\left(\frac{V^2}{gR}\right). \tag{12}$$

NONDIMENSIONALIZATION OF A FUNCTIONAL RELATIONSHIP

We shall now show that (12) can be deduced even if the governing differential equation of the problem is unknown, as long as all variables and parameters that should appear in such an equation are known. Again, we illustrate a general procedure on the projectile problem. (We shall give the particular example immediately below each of the six required steps.) The procedure is based on the idea that since the parameters of the problem have dimensions of the form $\mathscr{M}^a\mathscr{L}^b\mathscr{T}^c$, one should attempt to find dimensionless parameter combinations by taking products of parameters raised to a power.

STEP A. *Assume that a dimensionless quantity of interest is a function* Φ *of the dimensional parameters of the problem.**

$$\frac{t_M^*}{R^{1/2}g^{-1/2}} = \Phi(V, g, R). \tag{13}$$

STEP B. *Consider a term* π *made up of a product of powers of the quantities in the argument of* Φ.

$$\pi = V^{\alpha_1} g^{\alpha_2} R^{\alpha_3}; \qquad \alpha_1, \alpha_2, \alpha_3 \text{ constants.} \tag{14}$$

STEP C. *Insert the dimensions of the various quantities*

$$\begin{aligned}\text{Dimension of } \pi &= (\mathscr{L}\mathscr{T}^{-1})^{\alpha_1}(\mathscr{L}\mathscr{T}^{-2})^{\alpha_2}\mathscr{L}^{\alpha_3}\\ &= \mathscr{L}^{\alpha_1+\alpha_2+\alpha_3}\mathscr{T}^{-\alpha_1-2\alpha_2}.\end{aligned}$$

STEP D. *To obtain a dimensionless parameter* π, *require each exponent of* $\mathscr{M}$, $\mathscr{L}$, $\mathscr{T}$, etc., *to vanish, and thereby obtain a system of linear algebraic equations*

$$\alpha_1 + \alpha_2 + \alpha_3 = 0, \qquad -\alpha_1 - 2\alpha_2 = 0.$$

STEP E. *Find the general solution to the system of algebraic equations,* in terms of arbitrary constants $c_1, c_2, \ldots,$ etc.

$$\begin{aligned}\alpha_2 &= c_1, \qquad c_1 \text{ arbitrary;}\\ \alpha_1 &= -2c_1, \qquad \alpha_3 = -(-2c_1 + c_1) = c_1.\end{aligned} \tag{15}$$

STEP F. *Substitute back into* π *to obtain an expression of the form* $\pi_1^{c_1}\pi_2^{c_2}\cdots$. *Any dimensionless combination of the original arguments of* Φ *can be expressed as a function of the* π_i. Thus Φ *can be replaced by a function of the* π_i.

In the present case, from (14) and (15) we have

$$\pi = V^{-2c_1} g^{c_1} R^{c_1} = \left(\frac{gR}{V^2}\right)^{c_1}.$$

Since π is dimensionless for an arbitrary choice of c_1, in particular we obtain a dimensionless parameter $gR/V^2 \equiv \pi_1$ when $c_1 = 1$. Any other dimensionless combination of V, g, and R must be a function of π_1. Since the left side of (13) is dimensionless, the right side must be also. [Otherwise we could reach a contradiction by evaluating $\Phi(V, g, R)$ for two different sets of measurement standards.] Thus it must be that

$$\frac{t_M^*}{R^{1/2}g^{-1/2}} = \phi\left(\frac{gR}{V^2}\right) \tag{16}$$

for some function ϕ. The above result can be written

$$t_M^* = R^{1/2}g^{-1/2}\phi\left(\frac{gR}{V^2}\right). \tag{17}$$

* A parameter should be included if it would appear in a mathematical problem that would lead in principle to the determination of a single-valued function Φ.

It is perhaps not obvious that this is equivalent to

$$t_M^* = RV^{-1}f\left(\frac{V^2}{gR}\right), \tag{18}$$

a form of (12) that we derived from the differential equation formulation of the problem. But (16) can be written

$$t_M^* = RV^{-1}R^{-1/2}Vg^{-1/2}\phi\left(\frac{gR}{V^2}\right). \tag{19}$$

If we can choose a function ϕ such that (19) is true, we can choose a function $f(\varepsilon) = \varepsilon^{1/2}\phi(1/\varepsilon)$ to make (18) hold, and conversely.

We have validated our assertion that even without knowing the governing equation and initial conditions, one is able to show that the quotient of the time t_M^* and RV^{-1} can depend only on the single combination V^2/gR of the three parameters V, g, and R. To see the value of this ability, imagine that someone was collecting data on t_M^* from a number of experiments on various planets. An ignorant person might try to organize the data by plotting t_M^* versus V for fixed values of R and g. He would then choose several other values of R and obtain several corresponding curves on the same page (Exercise 2). He would repeat his plotting on another page, choosing another value of g. The complete presentation of his data would require a book. But if he knew about dimensional analysis, *he could plot all his data on a single curve* of t_M^*/RV^{-1} versus V^2/gR.

Taking a slightly different point of view, we remark that it is a considerable advance in understanding to know that, purely on dimensional grounds, the dimensionless time to reach maximum height can only depend on V^2/gR. To give another simple example, one can deduce by dimensional analysis alone that the period during which a simple pendulum executes a small-amplitude oscillation is independent of the mass of the bob, and in fact is a constant multiple of $\sqrt{L/g}$. (See Exercise 3.) A less trivial illustration of the use of dimensional analysis (which we shall merely mention) is provided by its role in the universal equilibrium hypothesis put forward in 1941 by the Soviet scientist Kolmogorov. This relatively simple hypothesis is a landmark in the still elusive quest to understand turbulent fluid flow. See G. K. Batchelor's *Homogeneous Turbulence* (New York: Cambridge U.P., 1960), pp. 114ff.

USE OF SCALE MODELS

We have illustrated the fact that dimensional analysis is useful to an experimenter because to determine the behavior of the system, he or she need only make measurements of the relatively few dimensionless parameters that characterize the system. There is another reason for the usefulness of dimensional analysis to scientists and engineers. With it, one can show that scale models* will duplicate the behavior of the original system provided that the governing dimensionless parameters have the same values in the two systems. Since theoreticians must have an appreciation of the nature and limitation of experiment, we shall devote a few paragraphs to a discussion of scale models.

A scale model has the same shape as the device of primary interest but is of a more convenient size. The object of the present remarks is to stress that model performance which closely mimics that of the given device can only be expected if not only size, but also other attributes, are proportioned so that the important dimensionless parameters are identical in both device and model.

Ship design provides a good example of the use of models. Suppose that one is interested in the amount of work per unit time P (for power) which must be applied to keep a certain ship of length L moving in a straight line at constant speed U. Work must be supplied primarily to replace energy that is either wasted in making waves on the water surface or that is dissipated because of the viscosity of the water. A hydrodynamicist would thus expect that

$$P = f(U, L, g, \rho, \nu), \tag{20}$$

where g is the acceleration due to gravity, ρ is the density of the water, and ν is the kinematic viscosity of water. [Material in II on water waves and on viscous fluid motion gives the background required to make the assumption (20) with confidence, but for present purposes it will do no harm to accept (20) without scepticism.] By dimensional analysis (Exercise 4) one can deduce that

$$\frac{P}{\rho L^2 U^3} = \phi(\text{Fr}, \text{Re}), \tag{21}$$

where $\text{Fr} = U/(Lg)^{1/2}$ is called the **Froude number** and $\text{Re} = UL/\nu$ is called the **Reynolds number**. These two dimensionless parameters are associated with wave resistance and viscous resistance, respectively. This should not be surprising, since only the first contains a measure of the force of gravity that is overcome by the waves and only the second contains a measure of fluid viscosity.

Suppose that one constructs a scale model of the ship in question, of exactly the same shape but one-hundredth the size. (If primes are used for model parameters, this means that $L' = 10^{-2}L$.) In order that the model experiment be associated with the same Froude number as the full scale ship, one would have to move the model one-tenth as fast as the speed U. For then

$$\text{Fr}' = \frac{U'}{(L'g)^{1/2}} = \frac{10^{-1}U}{(10^{-2}Lg)^{1/2}} = \text{Fr}.$$

* We emphasize that in this discussion the word "model" does not refer to a mathematical model but is used in the familar sense, which applies, for example, to a small replica of an airplane.

To obtain the same Reynolds number, one would have to use a fluid whose kinematic viscosity ν is one-thousandth that of water. No such fluid exists. (Air, for example, has a kinematic viscosity which is about 15 times *greater* than that of water.) It thus seems that one *cannot* make a model that reproduces the dimensionless parameters of the full scale ship.

Contrary, perhaps, to expectation, the fact that the Reynolds number for the full scale ship motion is apt to be enormous does not save the situation. [If a ship 100 m long moves 10 m/sec (about 20 miles/hr), since $\nu = 10^{-2}$ cm^2/sec for water, the Reynolds number is about 10^9.] It is reasonable to expect that if one knows how P behaves for $\text{Re} = 10^7$ and 10^8, say, that a simple extrapolation will give the behavior for $\text{Re} = 10^9$. It is a central fact of fluid mechanics, however, that "the flow characteristics do not simply flatten out when Re becomes large or small. They continue to vary and for many geometries give big and unexpected 'kicks' as R passes through values which are quite surprisingly high." (Lighthill, 1963) (The "kicks" arise from phenomena such as the sudden transition from smooth laminar to eddying turbulent flow or from other qualitative changes in the flow pattern.) In assessing ship resistance, one finds "kicks" as the Froude number varies also. An important reason for this is the sensitivity of the resistance to the nature of the interaction between bow and stern waves.

Ship design, then, serves as an example of an area in which model experiments cannot provide easy answers. Thus ship design remains more of an art than a science, although a considerable contribution to the subject has been made by workers in theoretical fluid mechanics.*

A possible source of error in model experiments is that the correctness of deductions made by dimensional analysis obviously requires the correctness of the basic assumption that the quantities of interest depend in essence only on the parameters listed. If an important parameter is ignored, model experiments cannot be expected to give accurate results. Waves from a sufficiently small ship model, for example, may be strongly influenced by surface tension T. The corresponding dimensionless parameter here can be taken to be the **Bond number** $\text{Bo} \equiv \rho g L^2/T$. If the model does not have the same Bond number as the full scale object, then errors will result if surface tension is important in either experiment.

Excellent physical understanding is clearly required to ascertain all the important parameters in complicated situations. The difficulties that arise when such understanding is lacking are illustrated by the scale-up problems which arise in chemical engineering. New chemical plants, although geometrically similar to pilot plants and appropriately similar in many other ways as well, sometimes do not work properly because the pilot plant did not

* A good review article is G. Gadd, "Understanding Ship Resistance Mathematically," *J. Inst. Math. Appl. 4*, 43–57 (1968). Also see D. H. Peregrine, "A Ship's Waves and Its Wake," *J. Fluid Mech. 49*, 353–60 (1971).

reproduce correctly some dimensionless parameter which unexpectedly has a major influence on plant operation. Chemical plants typically involve such complicated interplay between flow and chemical reaction that it is no surprise that model experiments are difficult to design.*

To cite another area of great current interest, physiological problems are similar in many ways to the problems that face chemical engineers, but are even more complicated, so that model experiments must be regarded with even greater caution. Nevertheless, such experiments can be useful. (See Section 8.2.)

SUMMARY

We have illustrated the fact that it is a routine matter to nondimensionalize the governing equations of a problem. Furthermore, we have discussed the use of dimensional analysis in situations where understanding is not sufficient to allow formulation of the governing equations. What is necessary is enough physical intuition to list the parameters that *would* appear in the governing equations if these were known. From this list, a straightforward procedure automatically yields a set of dimensionless parameters that characterize the problem in question.

We have used physical intuition and specific examples in our discussion. Our assertions and conclusions certainly bear further investigation. For example, one should be able to *prove* in some sense that it is sufficient to consider dimensionless groups which are formed by taking the products of the various parameters, each raised to a power. Similarly, one should not have to rely on "intuition" when one asserts that a dimensionless parameter can only depend on dimensionless combinations of other parameters. One approach to the required proofs is outlined in Exercises 9–12.

A detailed use of dimensional analysis on a nontrivial example will be found in Chapter 8. For the use of dimensionless variables in a complicated system of partial differential equations, initial conditions, and boundary conditions, see the treatment of water waves in Chapter 7 of II. Nondimensionalization of a less complicated system of partial differential equations is carried out in Section 15.2.

It is important to remember that there are fewer, often very significantly fewer, dimensionless parameters than dimensional (three as opposed to eight in the example of Chapter 8). Whether they are obtained experimentally,

* Although not easy, model experiments are essential in chemical engineering. This is summarized in a famous phrase of L. H. Baekeland, "Commit your blunders on a small scale and make your profits on a large scale." For an interesting treatment of the problems involved in predicting large scale plant performance from small scale experiments, see the book by R. E. Johnstone and M. W. Thring, *Pilot Plants and Scale-up Methods in Chemical Engineering* (New York: McGraw-Hill, 1957). This book begins with the quotation, "Vitruvius says that small models are of no avail for ascertaining the effects of large ones; and I here propose to prove that this conclusion is a false one"—From the Notebooks of Leonardo da Vinci.

numerically, or analytically, results can thus be found with less effort and can be displayed in a more compact and meaningful form. Nondimensionalization should generally be employed on all but the simplest problems.

EXERCISES

1. (a) Complete the derivation of (4).
(b) Exhibit explicitly the type of contradiction that is mentioned in the sentence in square brackets above (16).

2. Use your qualitative feeling for the answer and plot t_M^* versus V for several values of R. Imagine that g is fixed. Refer all your plots to a single set of axes.

3. (a) A pendulum is executing small vibrations. Show that it is impossible on dimensional grounds that the period T depends only on the length L of the pendulum and the mass m of the bob.
(b) Show by dimensional analysis that the assumption that T depends on m, L, and g, the acceleration due to gravity, leads to the relation $T = k(L/g)^{1/2}$, k a constant. [As in (2.2.18), by solving a simple differential equation, one can show that $k = 2\pi$.]
‡(c) Suppose that the pendulum is pulled out farther, so that its amplitude of oscillation is no longer small. What can you conclude by dimensional analysis?

4. Show that (20) implies (21). (The dimension of ν are $\mathscr{L}^2\mathscr{T}^{-1}$.)

5. The thrust T developed by a ship propeller in deep water (dimensions of $T = \mathscr{M}\mathscr{L}\mathscr{T}^{-2}$) depends on the radius a of the propeller, the number of revolutions per minute n, the velocity V with which the ship advances, the gravitational constant g, the density ρ, and the kinematic viscosity ν of the water (dimensions of $\nu = \mathscr{L}^2\mathscr{T}^{-1}$). Show by using the formal approach of the text that

$$\frac{T}{\rho a^2 V^2} = \phi\left(\frac{an}{V}, \frac{aV}{\nu}, \frac{ag}{V^2}\right).$$

[See E. G. Keller and R. E. Doherty, *Mathematics of Modern Engineering*, Vol. I (New York: Dover, 1961).]

6. Consider D, the diffusion constant for spherical particles of radius a in Brownian motion. Here D depends on a, on the temperature T and the viscosity μ of the surrounding gas, and on the Boltzmann constant $k = 1.38 \times 10^{-6}$ erg/degree. Show by dimensional analysis that D is a constant multiple of $kT/a\mu$. (The dimensions of D and μ are $\mathscr{L}^2\mathscr{T}^{-1}$ and $\mathscr{M}\mathscr{L}^{-1}\mathscr{T}^{-1}$, respectively.)

7. For a small sphere falling under gravity in a viscous fluid, it is observed that the speed of fall is (after a short time) a constant v. Let a, ρ_1, ρ_2, μ, and g represent the radius and density of the sphere, the density and viscosity of the liquid, and the gravitational acceleration, respectively.

(a) Using dimensional analysis with fundamental units of mass, length, and time, show that $v = (\mu/\rho_1 a) f[\rho_1^2 \mu^{-2} a^3 g, \rho_2/\rho_1]$. Use the fact that viscous stress (force/unit area) equals the product of μ and the velocity gradient (derivative of velocity with respect to length).

(b) Since motion here is unaccelerated, we need not make use of the proportionality of acceleration to force, and force can be treated as a separate fundamental unit. Hence show that $v = a^2 \rho_1 g \mu^{-1} \phi(\rho_2/\rho_1)$. [Stokes derived the formula $\phi(x) = (\frac{2}{9})(1 - x)$, a result of great utility. For example, it was used in the Millikan oil drop experiment. Stokes's result is obtained in Exercise 3.6.6 of II.]

8. (a) Consider steady nonturbulent incompressible flow of a liquid in a circular pipe. The pressure difference Δp between the two ends of the pipe should depend only on its length L, radius R, and the maximum speed U of the fluid of viscosity μ and density ρ. Show that according to dimensional analysis the situation can be described equally well by either of the following equations:

$$\frac{\Delta p}{\frac{1}{2}\rho U^2} = f\left(\frac{\rho U R}{\mu}, \frac{\rho U L}{\mu}\right), \qquad \frac{\Delta p}{\frac{1}{2}\rho U^2} = f\left(\frac{\rho U R}{\mu}, \frac{L}{R}\right).$$

[It is reasonable to suppose that when L/R is large—say $L/R > 20$ —changing L/R should have little effect on the answer. In this case the relation $\Delta p/\frac{1}{2}\rho U^2 = f(\rho U R/\mu)$ should hold to good accuracy. This illustrates the fact that physical reasoning sometimes shows how to choose the dimensionless parameters so that one of them can be neglected. For more examples of this type of reasoning, see the valuable book of Kline (1965), from which the present problem was adapted.]

Exercises 9–12 guide the reader through proofs of some of the principal results of dimensional analysis. The approach is analytic, as in Bridgman (1931). For an algebraic approach see Birkhoff (1960).

9. Given a fundamental unit of length like the meter, we know how to measure the length between two given marks. Length satisfies the **change of units rule**: *If a new fundamental unit is chosen, $1/x$ times as large as the old, then the new measurement gives a number that is x times as large as the old.* (Example: A centimeter is $\frac{1}{100}$ times as large as the meter; so, to change meter measurements to centimeter measurements, multiply by 100.) Quantities that satisfy the change of units rule, such as length, mass, and time, are called **primary quantities**.

(a) Derive the requirement of **absolute significance of relative magnitude** (ASRM): *The ratio of two measurements of a primary quantity is independent of the fundamental unit of measurement chosen.*

(Example: Mass of hydrogen atom M_H equals 1.67×10^{-24} g. Mass of electron M_e equals 9.11×10^{-28} g, $M_H/M_e = 1800$. Or $M_H = 3.69 \times 10^{-27}$ lb, $M_e = 2.01 \times 10^{-30}$ lb, and $M_H/M_e = 1800$.)

(b) Give another example of ASRM.

10. **Secondary quantities** are measured by combining the measurements of primary quantities in a prescribed way. (Example: Measurement of the secondary quantity speed equals measurement of length divided by measurement of time.)

(a) Let $(p_1, p_2, \ldots, p_n)$ and $(q_1, q_2, \ldots, q_n)$ denote two sets of measurements of primary quantities. (Example: p_1 and q_1 are distances transversed by cars p and q in times p_2 and q_2.) Let $f(p_1, p_2, \ldots, p_n)$ and $f(q_1, q_2, \ldots, q_n)$ denote corresponding measurements of a certain secondary quantity. (Example: For average car speed, $f(p_1, p_2) = p_1/p_2$.) Explain why we should postulate

$$f(x_1 p_1, \ldots, x_n p_n) = \frac{f(p_1, \ldots, p_n)}{f(q_1, \ldots, q_n)} f(x_1 q_1, \ldots, x_n q_n) \qquad (22)$$

for any x_i, p_i, q_i; $i = 1, \ldots, n$. (Note that the distinction between primary and secondary quantities is somewhat arbitrary.)

‡(b) Differentiate (22) with respect to x_1—keeping $x_1, \ldots, x_n$ fixed—and then set $x_i = 1$; $i = 1, \ldots, n$. Show that

$$f(p_1, p_2, \ldots, p_n) = g(p_2, \ldots, p_n) p_1^{a_1},$$

where a_1 is a constant and g is an arbitrary function. Deduce

$$f(p_1, p_2, \ldots, p_n) = C p_1^{a_1} p_2^{a_2} \ldots p_n^{a_n},$$

where C is a constant. Usually, we take $C = 1$. Thus the *measurement of a secondary quantity is expressible as a product of powers of measurements of primary quantities.* The secondary quantity is said to have *dimensions* $p_1^{a_1} p_2^{a_2} \ldots p_n^{a_n}$. (We use the same symbol for a primary quantity and its measurement.) [Example: Speed has dimensions $(\text{length})^1 \cdot (\text{time})^{-1}$.] A secondary quantity is **dimensionless** if all its dimensional exponents a_i are zero.

(c) Show that if the fundamental units of measurement for primary quantities are decreased by the factors $y_1, \ldots, y_n$, then S' and S, the new and old measurements of a secondary quantity, are related by

$$S' = y_1^{a_1} \cdots y_n^{a_n} S.$$

11. Let M be the net mass of water that has leaked out of a certain container, since the container was opened at time $t = 0$. Suppose that if M is carefully measured in grams and the time t in minutes, the results fit the following law:

$$M = 5 - 5 \exp(-t). \qquad (23)$$

(a) How will (23) be altered if kilograms are used as a mass scale and seconds as a time scale?

If, like (23), an equation is a true description of a situation for some given system of fundamental units, the equation is said to be **correct** (in that system of units).

(b) Show that no matter what mass and length scales are used, the measurements will be described by the formula

$$C_1 M = 5 - 5 \exp(-C_2 t), \tag{24}$$

where the constants C_1 and C_2 have the dimensions $(\text{mass})^{-1}$ and $(\text{time})^{-1}$, respectively.

Equation (24) is called **complete** because *it is true whatever fundamental units are used* (as long as the *dimensional constants* C_1 and C_2 are chosen correctly). Although not complete, (23) is a correct description of the facts. As in the transition from (23) to (24), any correct equation can be made complete by introducing a dimensional constant in front of each observed quantity.

12. This problem outlines a proof of the **Buckingham pi theorem**, which states (roughly speaking) that any complete equation implies an equation in which all variables appear in dimensionless combinations.

(a) Let $R_1, \ldots, R_m$ be measurements of m (primary or secondary) quantities. Suppose that there is one and only one functional relationship

$$\phi(R_1, \ldots, R_m) = 0 \tag{25}$$

connecting these measurements and that (25) is complete. Suppose further that there are n fundamental units and that each is decreased by a factor x_i, $i = 1, \ldots, n$. Show that there must be a relationship of the form

$$\phi(x_1^{a_{11}} x_2^{a_{12}} \cdots x_n^{a_{1n}} R_1, \ldots, x_1^{a_{m1}} \cdots x_n^{a_{mn}} R_m) = 0. \tag{26}$$

(As the proof unfolds, it may be helpful to work a particular example that illustrates the general results for which the exercise calls.)

(b) Deduce from (26) that

$$\sum_{k=1}^{m} a_{k1} R_k \phi_k(R_1, \ldots, R_m) = 0,$$

where a subscript i on ϕ denotes the partial derivative with respect to the ith argument.

(c) Assuming that $a_{k1} \neq 0$, introduce the new independent variables T_k,

$$T_k = R_k^{1/a_{k1}}, \qquad k = 1, \ldots, m, \tag{27}$$

so that

$$\Phi(T_1, \ldots, T_m) = 0,$$

where

$$\Phi(T_1, \ldots, T_m) = \phi(T_1^{a_{11}}, \ldots, T_m^{a_{m1}}).$$

Show that

$$\sum_{k=1}^{m} T_k \Phi_k = 0,$$

where the subscript k on Φ denotes a derivative with respect to the kth argument. Show also that all quantities T_k have dimensional exponent unity with respect to the first primary quantity.

(d) Introduce another new set of independent variables U_k:

$$U_k = \frac{T_k}{T_m}, \qquad k = 1, \ldots, m-1, \quad U_m = T_m,$$

so that

$$\chi(U_1, \ldots, U_m) = 0, \tag{28}$$

where

$$\chi(U_1, \ldots, U_m) = \Phi(U_1 U_m, U_2 U_m, \ldots, U_m).$$

Show that $\partial\chi/\partial U_m = 0$. Show also that the quantities U_k are dimensionless with respect to the first primary quantity. Thus (25) implies (28), a relation involving $m-1$ quantities that have a dimensional exponent zero with respect to the first primary quantity.

(e) Show that the above conclusion remains valid even if some or all of the quantities a_{k1} are zero.

(f) Deduce finally the relation

$$F(\pi_1, \pi_2, \ldots) = 0,$$

where the quantities π_k are dimensionless. Noting the change of variables involved, show also that the π_k are products of powers of $R_1, R_2, \ldots, R_m$, Show that "usually" there will be $m-n$ such dimensionless products.

REMARK. Suppose that $F = 0$ can be solved for π_1 and that the expression for π_1 actually contains R_1. If one solves for R_1, then each term in the resulting expression must be of the same dimension. This result is called the **principle of dimensional homogeneity**.

13. Some cookbooks say that a roast should be cooked x minutes per pound; others say x_1 minutes per pound for small roasts and x_2 minutes per pound for large roasts. Discuss. [See M. S. Klamkin, "On Cooking a Roast," *SIAM Review 3*, 167–69 (1961).]

6.3 Scaling

Many mathematical problems contain small or large parameters. We need only concern ourselves with small parameters, for if λ is a large parameter, then we can introduce a small parameter $\varepsilon \equiv 1/\lambda$.

Taking advantage of the smallness of a parameter is not as simple as it may seem. We saw some instances of this in Section 6.1; these were connected with ill conditioning and insensitivity.

To see another sort of difficulty that one has to contend with, consider a function of a single independent variable and a single small parameter ε. Presumably, a first approximation to the function can be obtained by letting ε tend to zero with the independent variable fixed. But the results depend on how the independent variable is chosen. As an illustration consider

$$u(x, \varepsilon) \equiv x + e^{-x/\varepsilon} \qquad \text{for } 0 < x \leq 1, \quad \varepsilon > 0. \tag{1a}$$

Here

$$\lim_{\varepsilon \to 0} u(x, \varepsilon) = x. \tag{1b}$$

If we switch to the new independent variable $\xi = x/\varepsilon$, we have

$$v(\xi, \varepsilon) \equiv u(\varepsilon\xi, \varepsilon) = \varepsilon\xi + e^{-\xi}, \qquad \lim_{\varepsilon \to 0} v(\xi, \varepsilon) = e^{-\xi}. \tag{2}$$

On the other hand, if we introduce $\eta = x/\varepsilon^2$, we obtain

$$w(\eta, \varepsilon) = u(\varepsilon^2\eta, \varepsilon) = \varepsilon^2\eta + e^{-\varepsilon\eta}, \qquad \lim_{\varepsilon \to 0} w(\eta, \varepsilon) = 1. \tag{3}$$

Which of the three limits

$$x, \quad \exp(-\xi) = \exp\left(\frac{-x}{\varepsilon}\right), \quad \text{or } 1, \tag{4a, b, c}$$

if any, gives the correct first approximation to u? This is not too difficult a question, but let us proceed slowly.

Section 6.2 revealed an important class of situations where there is a choice of independent variable. These stem from a degree of arbitrariness in selecting the intrinsic reference quantities that are used as denominators when defining dimensionless variables. This was illustrated using the projectile problem (Equation 2.1):

$$\frac{d^2x^*}{d(t^*)^2} = -\frac{gR^2}{(x^* + R)^2}, \quad x^*(0) = 0, \qquad \frac{dx^*}{dt^*}(0) = V. \tag{5}$$

Employing a reference length R and a reference time RV^{-1}, we introduced the dimensionless variables

$$y = \frac{x^*}{R}, \qquad \tau = \frac{t^*}{RV^{-1}} \tag{6}$$

and obtained (2.4):

$$\varepsilon \frac{d^2 y}{d\tau^2} = -\frac{1}{(y+1)^2}, \quad y(0) = 0, \quad \frac{dy}{d\tau}(0) = 1. \tag{7}$$

Another choice,

$$z = \frac{x^*}{R}, \quad \tau_1 = \frac{t^*}{\sqrt{Rg^{-1}}}, \tag{8}$$

led to (2.10):

$$\frac{d^2 z}{d\tau_1^2} = -\frac{1}{(z+1)^2}, \quad z(0) = 0, \quad \frac{dz}{d\tau_1}(0) = \varepsilon^{1/2}. \tag{9}$$

Here

$$\varepsilon = \frac{V^2}{gR}. \tag{10}$$

If ε is known to be very small compared to unity, then a naïve person might merely delete terms preceded by ε. But this cannot be right, for from (7) one would obtain

$$-(y+1)^{-2} = 0, \quad y(0) = 0, \quad \frac{dy}{d\tau}(0) = 0, \tag{11}$$

while from (9) one would obtain

$$\frac{d^2 z}{d\tau_1^2} = -\frac{1}{(z+1)^2}, \quad z(0) = 0, \quad \frac{dz}{d\tau_1}(0) = 0. \tag{12}$$

These "approximate" problems are different; so at least one is inappropriate. In fact, both are inappropriate. Problem (11) has no solution, while problem (12) has a negative solution (Exercise 1) and can therefore only apply beneath the surface of the earth. It is apparent that *a term cannot be neglected merely because it is preceded by a small parameter.*

Enough seeds of doubt have been sown. It is time to weed out false ideas and to reveal flourishing techniques.

In the last section we showed how to choose dimensionless variables, and we pointed out that this generally can be done in a number of ways. In this section we propose that dimensionless variables should be selected so that if a term is preceded by a small dimensionless parameter, then that term should be negligible to a first approximation.*

The process of choosing this *particular* set of dimensionless variables is called "scaling." We shall demonstrate that scaling is far from a routine matter, but it is nevertheless helpful to be clearly aware of the goal of this process and of the pitfalls that remain even if the goal can be attained.

* It does little good to know that a dimensional parameter is small, for such a parameter can be made to take an arbitrary value by an appropriate choice of units.

DEFINITION OF SCALING

Since dimensionless variables can be chosen in a variety of ways, one cannot expect that the appearance of a small dimensionless parameter will inevitably signify the presence of a relatively small term. **Scaling** amounts to nondimensionalizing so that the relative magnitude of each term *is* indicated by a dimensionless factor preceding that term. More formally, *in the process of scaling one attempts to select intrinsic reference quantities so that each term in the dimensional equations transforms into the product of a constant dimensional factor which closely estimates the term's order of magnitude and a dimensionless factor of unit order of magnitude.* (For the time being we shall use the phrase "order of magnitude" in the sense of "approximate size." We shall shortly specify the meaning of this phrase more precisely.) Intrinsic reference quantities that are selected by this process are called **scales**. Generally, scales differ for different parameter ranges. Also, we shall see that for a given range of parameters, it may be necessary to choose different scales for different ranges of independent variables.

SCALING THE PROJECTILE PROBLEM

To illustrate the scaling procedure, let us consider the projectile problem (5) in situations where the projectile's distance from the earth's surface x^* is always small compared to the earth's radius R. Such limited motion occurs only when the initial speed V is "sufficiently small." On dimensional grounds we can assert that V must be small compared to a multiple of $(Rg)^{1/2}$, the only combination of parameters other than V with dimension length/time.

When $x^* \ll R$, it is clear that the acceleration has order of magnitude g, the gravitational acceleration at the earth's surface. Now if a projectile is launched with initial speed V and is then acted on by a *uniform* deceleration of magnitude g, the projectile will momentarily come to rest (at its maximum elevation) in time V/g. Taking the average of its initial and final speeds, we estimate that in time V/g it will move a distance equal to $(\frac{1}{2}V)(V/g) = \frac{1}{2}V^2/g$. To keep the factor $\frac{1}{2}$ in this expression might imply more accuracy than we can legitimately claim—remember that we have replaced the continuously changing speed of the projectile by the average of its initial and final values, and we have ignored the change of the force of gravity with distance. We thus take V^2/g as our estimate of the order of magnitude of x^*.

REMARK. Our initial assumption that the parameters are such that x^* is always small compared to R is now seen to require that V^2/g be small compared to R. Because we have thought more deeply about the problem, it is not surprising that this is a more precise statement than the requirement "V must be small compared to a multiple of $(Rg)^{1/2}$," which is all that one can deduce on dimensional grounds.

We have completed the most difficult part of the scaling procedure, estimation of the size of various terms in the special circumstances under consideration. We now show how to take formal advantage of the above estimates, which show (i) that the displacement x^* has order of magnitude V^2/g and (ii) that the acceleration d^2x^*/dt^{*2} has order of magnitude g.

Using (i), we assert that the dimensionless displacement x should be defined by

$$x = \frac{x^*}{(V^2 g^{-1})}. \tag{13}$$

With this change of variable, x^* will be replaced according to the equation

$$x^* = \frac{V^2}{g} x. \tag{14}$$

The first factor on the right side of (14), (V^2/g), correctly and explicitly shows the order of magnitude of x^*. As x^* is of order of magnitude V^2/g, (13) shows that the dimensionless factor x must have order of magnitude unity, as required by the scaling procedure.

Because of estimate (ii) we must choose a *time scale* T such that if the dimensionless time t, defined by

$$t = \frac{t^*}{T}, \tag{15}$$

is introduced, then the term d^2x^*/dt^{*2} is transformed into the product of a dimensionless term and the constant g. But, from (13) and (15),

$$\frac{d^2x^*}{d(t^*)^2} = \frac{V^2}{gT^2} \frac{d^2x}{dt^2}.$$

The requirement $V^2/(gT^2) = g$ gives $T = V/g$ as an equation defining the time scale.

With the appropriate scaled dimensionless variables

$$x = \frac{x^*}{V^2 g^{-1}} \quad \text{and} \quad t = \frac{t^*}{V g^{-1}}, \tag{16}$$

the governing equation (5) becomes

$$g \frac{d^2x}{dt^2} = -\frac{R^2}{(xV^2 g^{-1} + R)^2}, \quad x(0) = 0, \quad \frac{dx}{dt}(0) = 1. \tag{17a, b, c}$$

To simplify, we divide both sides of (17a) by g and divide the numerator and denominator of its right side by R^2. This gives

$$\ddot{x} = -(1 + \varepsilon x)^{-2}, \qquad x(0) = 0, \qquad \dot{x}(0) = 1. \tag{18}$$

(Note that because of this division, it is *relative* orders of magnitude that are now explicitly displayed. In particular, the factor ε gives the order of magnitude of x^* divided by R.)

Since the projectile problem (5) has been nondimensionalized in the particular way called for by the scaling procedure, we are confident that when ε is small compared to unity, the term εx is small compared to 1. If the lowest approximation to $x(t)$ is denoted by $x^{(0)}(t)$, we have, from (18),

$$\ddot{x}^{(0)} = -1, \qquad x^{(0)}(0) = 0, \qquad \dot{x}^{(0)} = 1. \tag{19}$$

Thus

$$x^{(0)} = t - \tfrac{1}{2}t^2 \quad \text{and} \quad 0 \le x^{(0)} \le \tfrac{1}{2} \qquad \text{for } 0 \le t \le 2.$$

Particularly in its dimensional form,

$$\frac{d^2x^*}{d(t^*)^2} = -g, \quad x^*(0) = 0, \qquad \frac{dx^*}{dt^*}(0) = V,$$

everyone knows that (19) provides the zeroth approximation to the solution of the projectile problem. But few inexperienced individuals can without hesitation arrive at (18), a form of the problem that permits ready determination of higher approximations. (See Section 7.2.)

The inappropriateness of the previously obtained "approximate" problems (11) and (12) when $\varepsilon \ll 1$ can now be ascribed to the fact that in both (6) and (8) the choice of dimensionless variables was not in accordance with the scaling procedure. That these incorrect "approximate" problems do not have a sensible solution is comforting and typical but, unfortunately, not inevitable. (Recall Examples 3 and 4 of Section 6.1.)

We must emphasize that *the choice of scales depends on the parameter range under consideration*. As an illustration of this, observe that we have scaled the projectile problem when $\varepsilon \ll 1$, but if ε^{-1} is a small parameter, then new scaling is required. To mention one facet of the new situation, when ε^{-1} is small the initial speed is so large that the particle soon passes many earth radii from the earth's surface, and it becomes wrong to estimate that d^2x^*/dt^{*2} has magnitude g.

Another point can now be illustrated. *Full understanding of a problem requires one to have in mind a physical interpretation of each dimensionless parameter which appears in its dimensionless formulation*. In the present case, the only such parameter is $\varepsilon = V^2/gR$. As was mentioned under (18), ε can be interpreted as the ratio of two lengths. In the denominator is the earth radius R. In the numerator is V^2/g, an estimate of the maximum height achieved by a projectile shot from the surface of the earth with a nonlarge initial speed V.

ORDER OF MAGNITUDE

It is helpful to make a precise definition of the phrase "order of magnitude." A *number A* will be said to have **order of magnitude** 10^n, n an integer, if

$$3 \cdot 10^{n-1} < |A| \le 3 \cdot 10^n.$$

Since $\log_{10} 3 \approx \frac{1}{2}$, a virtually equivalent characterization of a number A with order of magnitude 10^n is

$$n - \tfrac{1}{2} \leq \log_{10} |A| \leq n + \tfrac{1}{2}.$$

By the **order of magnitude of a function** f defined over a certain region, we mean the order of magnitude of the number M, where M is the maximum (or perhaps the least upper bound) of $|f|$ over the given region. Note the distinction between the *order* of a function defined over a certain domain, denoted by the O symbol (Appendix 3.1), and the numerical *order of magnitude* of a function as defined above.

SCALING KNOWN FUNCTIONS

The scaling of *known* functions is now a straightforward matter. Consideration of a few examples will generate some intuition.

Let us at first consider a phenomenon that is governed by the first order ordinary differential equation

$$F\left(u^*, \frac{du^*}{dx^*}\right) = 0. \tag{20}$$

Suppose that the independent variable x^* is restricted to an interval I^* (which may be infinite). It will be convenient to refer to u^* as a velocity component and to imagine that x^* is a spatial variable, although our discussion applies to any dimensional dependent and independent variables u^* and x^*.

Let L be the *length scale* and U the *velocity scale*. Introduce dimensionless variables by

$$x = \frac{x^*}{L}, \qquad u = \frac{u^*}{U}. \tag{21}$$

We have

$$u^*(x^*) = Uu\left(\frac{x^*}{L}\right), \qquad \frac{du^*}{dx^*} = \frac{U}{L}\frac{du(x)}{dx}\bigg|_{x=x^*/L}. \tag{22}$$

If U and L are indeed appropriate scales, then the combinations of U and L that appear on the right side of the equations in (22) must be reasonable estimates for the maximum absolute values of the terms on the left. Now ordinary scales can usefully be regarded as estimates of *exact scales* in which U and U/L actually equal these maximum absolute values. For exact scales

$$U = \max_{x^* \text{ in } I^*} |u^*(x^*)| \tag{23}$$

and

$$\frac{U}{L} = \max_{x^* \text{ in } I^*} \left|\frac{du^*}{dx^*}\right|. \tag{24}$$

From (24), using (23),

$$L \equiv \frac{|u^*|_{\max}}{|du^*/dx^*|_{\max}}. \tag{25}$$

Equations (23) and (25) give explicit expressions for the exact velocity and length scales in problems governed by (20). Note from Figure 6.1 that L can be interpreted as the base of a right triangle whose altitude is $|u^*|_{\max}$ and whose hypotenuse has the slope $|du^*/dx^*|_{\max}$. Moreover, Figure 6.1 supports the qualitative statement that *the length scale is an estimate of the shortest distance over which the function undergoes a significant change in magnitude.*

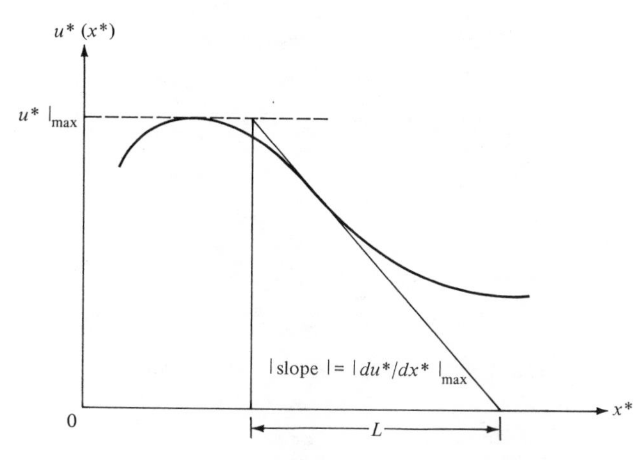

FIGURE 6.1. *The length scale L according to* (25).

Example 1. Find velocity and length scales U and L when

$$u^*(x^*) = A \sin \lambda x^*, \qquad -\infty < x^* < \infty,$$

where A and λ are positive constants.

Solution. Obviously, $U = A$. Since

$$\left|\frac{du^*}{dx^*}\right|_{\max} = |A\lambda \cos \lambda x^*|_{\max} = A\lambda,$$

(25) implies that $L = \lambda^{-1}$.

Example 2. Find U and L when

$$u^*(x^*) = A\left[x^* + \exp\left(\frac{-x^*}{\varepsilon}\right)\right], \qquad 0 \le x^* \le 1,$$

where A and ε are positive constants, $\varepsilon \ll 1$.

Solution

$$U = |u^*|_{\max} \approx A, \qquad \left|\frac{du^*}{dx^*}\right|_{\max} = \left|A + A\varepsilon^{-1} \exp\left(\frac{-x^*}{\varepsilon}\right)\right|_{\max} \approx A\varepsilon^{-1},$$

so $L = \varepsilon$. See Figure 6. 2.

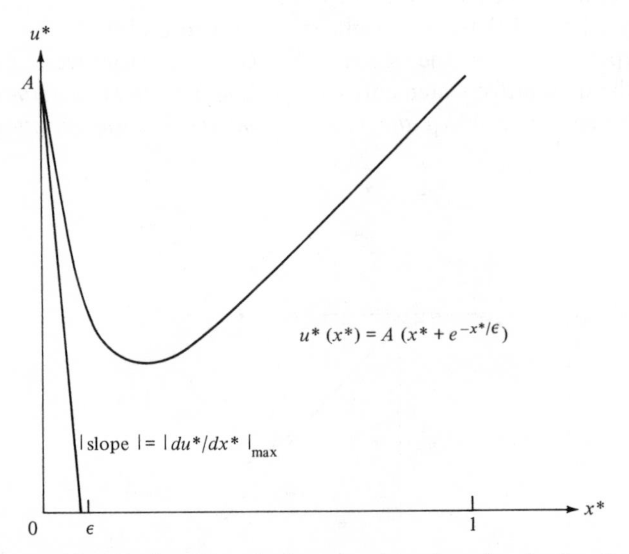

FIGURE 6.2. *The length scale is approximately* ε *for the function* u^*, *where* $u^*(x^*) = A[x^* + \exp(-x^*/\varepsilon)]$; A *and* ε *are positive constants,* $0 \le x^* \le 1$, *and* $\varepsilon \ll 1$.

Suppose that the problem is governed by

$$F\left(u^*, \frac{du^*}{dx^*}, \ldots, \frac{d^N u^*}{dx^{*N}}\right) = 0, \tag{26}$$

generalizing (20).† Equation (23) is still an appropriate definition of the velocity scale U, but in choosing the length scale L we have to consider each of the derivatives

$$\frac{d^i u^*(x^*)}{dx^{*i}} = \frac{U}{L^i}\left[\frac{d^i u(x)}{dx^i}\right]_{x = x^*/L}, \qquad i = 1, 2, \ldots, N.$$

We must choose L so that

$$\left|\frac{d^i u^*}{dx^{*i}}\right| \le \frac{U}{L^i}, \qquad i = 1, 2, \ldots, N. \tag{27}$$

† Generalization to problems in which there is more than one independent variable or more than one dependent variable is also possible. See Exercises 3 and 4.

To make our estimate as sharp as possible, we must select the largest value of L such that (27) is satisfied. It follows that L is the largest constant such that

$$L \leq \left[\frac{U}{|d^i u^*/dx^{*i}|}\right]^{1/i}$$

for $i = 1, 2, \ldots, N$; or

$$L = \text{smallest of } \frac{U}{|du^*/dx^*|_{\max}}, \left[\frac{U}{|d^2u^*/dx^{*2}|_{\max}}\right]^{1/2}, \ldots, \left[\frac{U}{|d^N u^*/dx^{*N}|_{\max}}\right]^{1/N}.$$

Example 3. If (26) is the governing equation, find the length scale L for the sinusoid of Example 1.

Solution

$$\left[\frac{U}{|d^i u^*/dx^{*i}|_{\max}}\right]^{1/i} = \left[\frac{A}{A\lambda^i}\right]^{1/i} = \lambda^{-1}$$

so that $L = \lambda^{-1}$, regardless of how many derivatives are involved in (26). (See Figure 6.3.)

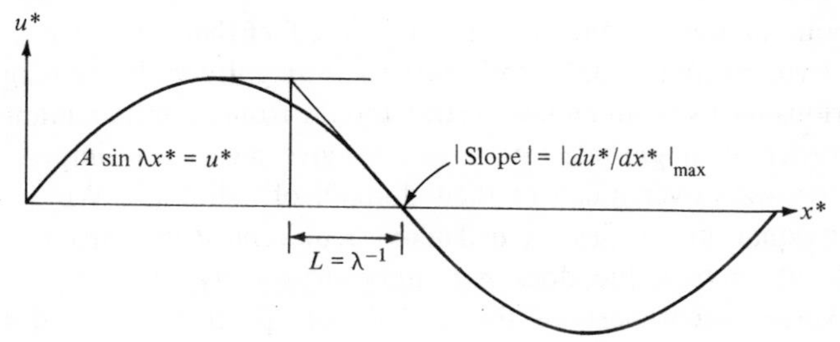

FIGURE 6.3. *The length scale* λ^{-1} *of a sinusoid is approximately one-sixth of its period* $2\pi\lambda^{-1}$.

Note that x^* does not explicitly appear in the governing equation $F(u^*, du^*/dx^*) = 0$ of (26). This is characteristic of a wide class of **spatially homogeneous** problems. For such problems "one place is just as good as another." Formally, spatial homogeneity manifests itself in the fact that the governing equation, having no explicit appearance of the independent variable x^*, retains its form when subject to the axis translation $x^* \to x^* + \text{constant}$. As would be expected, length scales for problems *not* characterized by spatial homogeneity must take into account the variations of the inhomogeneity. For example, even if the magnitudes of u^* and du^*/dx^* are correctly assessed, without further thought one cannot be sure of the magnitudes of such terms as $u^* \exp(x^*)$ and $(1/x^*)(du^*/dx^*)$.

The function considered in Example 3 has a length scale that is independent of N, the number of derivatives that must be taken into account. Here is an example of a function whose length scale does depend on N.

Example 4. If (26) is the governing equation, find U and L if

$$u^*(x^*) = M + A \sin \lambda x^*, \qquad -\infty < x^* < \infty,$$

where M, A, and λ are positive constants.

Solution

$$U = M + A, \qquad \left[\frac{U}{|d^i u^*/dx^{*i}|_{\max}}\right]^{1/i} = \frac{1}{\lambda}\left[1 + \frac{M}{A}\right]^i, \qquad \text{so } L = \frac{1}{\lambda}\left[1 + \frac{M}{A}\right]^{1/N}.$$

The dependence on N is weak providing M/A is not large compared to unity. When M is large, the length scale is large.

ORTHODOXY

Two matters of concern remain even after the process of scaling enables one to put a problem into a form that explicitly reveals the presence of terms (if any) having relatively small orders of magnitude. The first matter is that neglect of relatively small terms may have a large effect. This has been discussed in Section 6.1. In what follows, we shall assume that relatively small terms are negligible.

The second matter of concern stems from the fact that the order of magnitude of a term estimates that term's *maximum* magnitude. If the magnitudes of the various terms in an equation stray too far from their maximum values, then the order of magnitude estimates may give misleading impressions of their relative sizes over much of their domain of definition. We say that a term in an equation satisfies the **orthodoxy requirement** in a given domain if the term's absolute value does *not* differ drastically from its maximum absolute value except perhaps for a negligible portion of the domain in question. Suppose that in a certain equation the order of magnitude of a term T_1 is much greater than that of a second term T_2. If the first term fails to satisfy the orthodoxy requirement, then it may be smaller in absolute value than the second term for much of the domain in question, even though its maximum absolute value is much greater than the maximum absolute value of the second term. (See Figure 6.4.)

Both unorthodoxy and difficulties due to it are less widespread than one might think. Unorthodoxy might seem almost inevitable, e.g., when the same length scale must serve to characterize the change of several dependent variables; but this is not the case. The reason seems to be that the dependent variables are bound to combine so that certain differential equations are satisfied. It is difficult to imagine how such variables can differ wildly in behavior. Presumably, this is why it commonly occurs in practice that a length scale selected by concentrating on just one of several dependent variables frequently serves well for all of them.

"Harmless" unorthodox functions occur in the study of small amplitude water waves. (See Section 8.1 of II.) According to linear theory the velocity components and the (dynamic) pressure, and their derivatives, are unorthodox

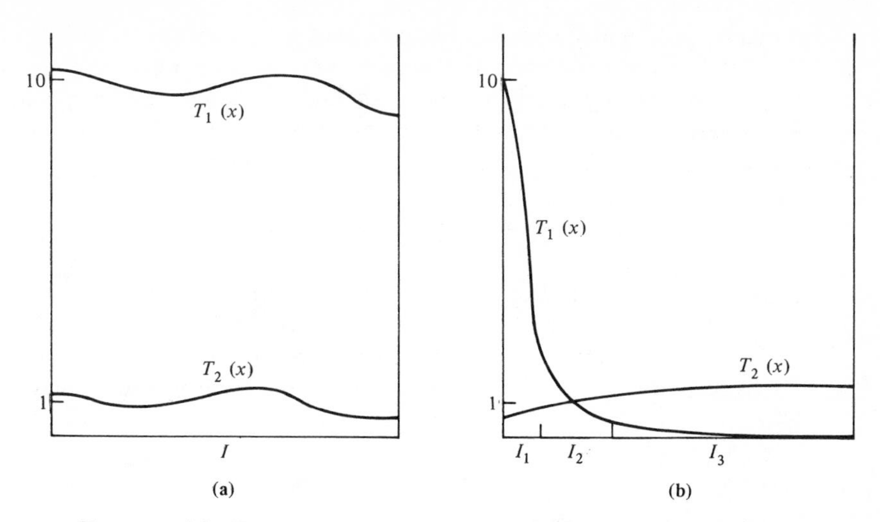

FIGURE 6.4. *Two terms T_1 and T_2 are defined on the interval I. Their orders of magnitude are* 10 *and* 1, *respectively.* (a) *T_1 satisfies the orthodoxy requirement; T_2 is small compared to T_1 throughout I.* (b) *T_1 does not satisfy the orthodoxy requirement; T_2 appears to be negligible only in the subinterval I_1, T_1 appears to be negligible in I_3, and both functions are of about the same size in the intermediate subinterval I_2.*

functions because they decay rapidly with depth. But all decay at exactly the same exponential rate so that the relative orders of magnitude of the various terms are correctly estimated by examining what goes on near the water surface. Such behavior is to be expected whenever the problem can be regarded as being governed by a system of linear equations with constant coefficients.

Another common situation in which a degree of unorthodoxy is tolerable involves oscillatory terms of relatively large maximum amplitude. Thus one would say that terms of unit order of magnitude in an equation can probably be neglected to first approximation if other terms are known to behave like sinusoids of large amplitude. This is so in spite of the fact that the sinusoidal terms are small near their zeros. Because there is almost no room for interesting behavior in a function that must pass smoothly from values specified on one side of a narrow region to *nearby* values which are specified on the other side, it would appear that these small regions of unorthodoxy can be ignored. (Compare Exercise 6.)

An important failure of orthodoxy occurs when dependent variables behave like the function

$$u^*(x^*) = A\left[x^* + \exp\left(\frac{-x^*}{\varepsilon}\right)\right], \qquad 0 < x^* \leq 1, \quad 0 < \varepsilon \ll 1, \tag{28}$$

whose graph appears in Figure 6.2. Suppose that the term du^*/dx^* must be assessed. Because ε is small, the rapid change of the exponential in (28) means that an estimate of $|du^*/dx^*|$ in an interval containing points near $x^* = 0$ is a gross overestimate for an interval not containing such points. For example, when $3\varepsilon \le x^* \le 1$,

$$\left|\frac{du^*}{dx^*}\right|_{\max} \approx A\varepsilon^{-1}e^{-3}.$$

Consequently $L \approx e^3\varepsilon$, a length scale more than 10 times as large as ε, the length scale appropriate for $0 \le x^* \le 1$. To satisfy the orthodoxy requirement, we must split [0, 1] into two parts and choose a different length scale in each part. In the *outer region*, more than a few multiples of ε from $x^* = 0$, we have

$$|u^*|_{\max} \approx A \quad \text{and} \quad \left|\frac{du^*}{dx^*}\right|_{\max} \approx A, \qquad \text{so } U = A \text{ and } L = 1.$$

Introducing these scales, we obtain

$$u = \frac{u^*}{A}, \quad x = x^*, \qquad \text{so } u(x, \varepsilon) \equiv A^{-1}u^*(x, \varepsilon) = x + e^{-x/\varepsilon}.$$

In the *inner region*, within a few multiples of ε from $x^* = 0$, we find that

$$|u^*|_{\max} \approx A \quad \text{and} \quad \left|\frac{du^*}{dx^*}\right|_{\max} \approx A\varepsilon^{-1}, \qquad \text{so } U = A \text{ and } L = \varepsilon^{-1}.$$

Using ξ and v for scaled variables in the inner region, we have

$$v = \frac{u^*}{A}, \quad \xi = \frac{x^*}{\varepsilon}, \qquad \text{so } v(\xi, \varepsilon) = A^{-1}\, u^*(\varepsilon\xi, \varepsilon) = \varepsilon\xi + e^{-\xi}.$$

To obtain a first approximation in the two regions, we let $\varepsilon \to 0$, keeping the respective independent variables fixed. We obtain in the outer region

$$u(x, \varepsilon) \approx x, \tag{29}$$

and in the inner region

$$v(\xi, \varepsilon) \approx e^{-\xi}, \qquad \text{so } u(x, \varepsilon) \approx e^{-x/\varepsilon}. \tag{30}$$

Our simple example illustrates that lack of orthodoxy in a given domain can be remedied by introducing subdomains in each of which orthodoxy is present. Different scales will be required in the different subdomains. Thus, *in the presence of unorthodoxy, one must not seek a first approximation with a single scale but rather must seek different approximations in different subdomains.* An approximation that is valid throughout the domain can actually be found, but it must vary on two or more scales at the same time.

By our last point, as illustrated in (29) and (30), we have essentially answered the question posed at the beginning of the section, as to which of the three expressions in (4) is the correct first approximation to u. It remains only to state that the approximation $u \approx 1$ of (4c) is only valid very near $x = 0$. This is to be expected, for lengths are measured with respect to ε^2, a quantity that is small, even compared to the width of the inner region. The limit $\varepsilon \to 0$, $\eta = \varepsilon^{-2}x$ fixed, gives an approximation that is valid only within a few multiples of ε^2 from $x = 0$; $u \approx 1$ is clearly the appropriate form for such an approximation. Nothing of interest is gained here by considering an "inner-inner" region of width $O(\varepsilon^2)$.

In the present instance we have advocated the introduction of different scales in different regions to overcome unorthodoxy, while in our earlier discussion of large amplitude sinusoids we suggested that the unorthodoxy be ignored. The region of unorthodoxy is narrow in both cases, but only in the present case does the function change rapidly and so accumulate considerable alteration.

An example of a situation in which more than one scale must be introduced is provided by the projectile problem when ε^{-1} is large. The scaling of (16) is still appropriate when the projectile is near the earth, but another scaling is required when the projectile is far from the earth. As above, there is an inner region (thickness $\approx V^2/g$) and a much larger outer region (projectile more than a few radii from the earth's surface).

It can happen that two scales coexist. An example is $e^{-\varepsilon x} \sin x$ which oscillates on an $O(1)$ scale with an amplitude which varies on a relatively long $O(\varepsilon^{-1})$ scale. Under some circumstances it is profitable to recognize the simultaneous existence of more than one scale, e.g., in the slow change in the frequency of a pendulum. (Compare Section 2.2.) Under other circumstances, it may be best to ignore the rapid variation or to average it out, even though this means that large errors will be made in approximating the derivatives of the answer. Averaging small scale irregularities in order to reveal major trends is the strategy of many successful phenomenological theories.*

SCALING AND PERTURBATION THEORY

Once a problem has been correctly scaled, one can in principle derive arbitrarily accurate approximations by systematic exploitation, via perturbation theory, of the presence in the equations of a small parameter. From the present point of view, "regular" perturbation theory deals with orthodox situations where one set of scales suffices. "Singular" perturbation theory has been developed to handle unorthodox situations in which more than one scale must be introduced in order to obtain an approximation that is uniformly valid throughout the domain. The elements of regular and singular perturbation theory will be discussed in Chapters 7 and 9, respectively.

* For an example see D. Drew's use of averaging methods to obtain field equations for two-phase problems in mechanics [*Stud. Appl. Math. 50*, 133–66 (1971).]

SCALING UNKNOWN FUNCTIONS

Order of magnitude estimation and scaling require a knowledge of the main features of the solution to the very problem one is trying to solve. How can one obtain the information necessary for the required order of magnitude estimates? We list six possibilities.

(i) *Utilize experimental or observational evidence* concerning the phenomenon in question. Often this indicates that a phenomenon is primarily due to a balancing of two effects, and a scaling can be obtained by balancing the terms that correspond to these effects. (The derivation of the boundary layer equations in Section 3.3 of II provides a good example of such balancing.)

(ii) *Obtain hints from experience of related problems.*

(iii) *Solve highly simplified versions of the given problem*. (This was illustrated in our discussion of the projectile problem, where we used knowledge of the answer obtained when variation of gravitational force with distance is neglected. Knowing the zeroth approximation, we employed scaling to adapt the problem for efficient determination of higher approximations.)

(iv) (Inverse procedure.) *Make certain order of magnitude assumptions merely because the concomitant neglect of terms renders the problem tractable.* (For example, assume that nonlinear terms are negligible.) Evaluate the neglected terms, perhaps using the approximate solution to do so. If possible, select parameter ranges so that these terms do indeed appear to be negligible. Hope that these parameter ranges are characteristic of interesting phenomena and that the demonstrated lack of inconsistency signals a justified neglect of small terms. (This is illustrated in the discussion of linearized water wave theory, in Section 7.3 of II.)

(v) *Use a trial and error approach.* Assume a certain scaling; solve the resulting simplified dimensionless problem; and then check to see whether the dimensionless terms are of unit order of magnitude.

(vi) *Employ the results of computer calculations* for particular but representative values of the various parameters involved in the problem.

The last item forms the basis for a joint computational analytical investigation. For problems involving ordinary differential equations, analog computers* are an invaluable tool; parameter variation is accomplished merely by twisting a dial. The use of such a computer often provides enough "feel" for the problem so that a simple and revealing analytic approximation can be determined.

* A form of scaling is required in analog computation. Units must be chosen so that various magnitudes do not exceed the capacity of the machine. Here too, one must somehow obtain information about the solution to the problem one is trying to solve. See, for example M. L. James, G M. Smith, and J. C. Wolford, *Analog Computer Simulation of Engineering Systems* (New York: International Textbook, 1966).

Models for natural phenomena often embody a system of partial differential equations containing many parameters and involving two or three spatial dimensions. To follow the temporal evolution of the solution even once by numerical methods may be extremely demanding of computer time. There may thus be no possibility of thoroughly understanding the effect of parameter variation by using a direct numerical attack. But if a few numerical solutions are available, order of magnitude estimates can often be made which provide the key to the simplifications required in an analytic investigation. If successful, such an investigation will exhibit the dependence of the solution on the various parameters and will also reveal the general features of the solution that form the basis for physical understanding.

EXERCISES

1. Show that the solution to (12) is negative for $\tau_1 > 0$. (It is not necessary to solve the problem in order to do this.)

‡2. Find the "velocity scale" U and the "length scale" L for the function $u^* = A \exp(-ax^*)$, $b \le x^* < \infty$, when the governing equation is of the form (20). Comment on the fact that L is independent of b.

3. Generalize the text's characterization of the length scale to the case where the governing equation is $F(u^*, \partial u^*/\partial x^*, \partial u^*/\partial y^*) = 0$.

4. Generalize the characterization of the length scale to the case where the problem requires determination of two dependent variables u^* and v^*. Suppose that the governing equations are

$$F\left(u^*, v^*, \frac{\partial u^*}{\partial x^*}, \frac{\partial u^*}{\partial y^*}, \frac{\partial v^*}{\partial x^*}, \frac{\partial v^*}{\partial y^*}\right) = 0,$$

$$G\left(u^*, v^*, \frac{\partial u^*}{\partial x^*}, \frac{\partial u^*}{\partial y^*}, \frac{\partial v^*}{\partial x^*}, \frac{\partial v^*}{\partial y^*}\right) = 0.$$

5. Concoct a specific example that shows an effect of spatial inhomogeneity on the length scale.

6. Solve for $y(x; \varepsilon)$:

$$\frac{d^2y}{dx^2} + \pi^2 y = \sin x + \varepsilon, \qquad y(0) = 1, \quad y'(0) = 0.$$

Compare $y(x, 0)$ with $y(x, \varepsilon)$ when $\varepsilon \ll 1$. Sketch. Thereby verify the text's remarks (concerning harmless unorthodox functions) in the paragraph preceding (28).

7. Why is the "harmless unorthodoxy" of small amplitude water waves typical of problems governed by a system of linear equations with constant coefficients?

8. This exercise concerns the projectile problem when air resistance is considered but changing gravitational attraction is not taken into account. The equation and initial conditions are

$$m\frac{d^2x^*}{dt^{*2}} + k\frac{dx^*}{dt^*} = -mg, \quad x^*(0) = 0, \quad \frac{dx^*}{dt^*}(0) = V.$$

(a) When V is sufficiently small, scaled variables are still given by (16). Why?

(b) Introduce the variables of (16) and show that the problem becomes one in which the only parameter is $\beta = kV/mg$. Give a physical interpretation of β.

9. (a) Oscillations of a pendulum released from rest are governed by Equation (7.1.1) (with $\Omega = 0$). Show that (7.1.3) defines properly scaled dimensionless variables, assuming that the initial amplitude a is not too large.

‡(b) What choice of scaled variables should be made if $a = 0$ and Ω is small?

‡(c) Discuss scaling when both a and Ω are nonzero and small.

10. Show that the limits in (1b), (2), and (3) are not uniform for, respectively, $x \in (0, 1]$ $\xi \in (0, \infty)$, $\eta \in (0, \infty)$.

CHAPTER 7
Regular Perturbation Theory

GIVEN the solution to a certain mathematical problem, how is the solution determined when the conditions of the problem are slightly altered? The systematic answer to this question forms the subject of **perturbation theory.**

Section 2.2 contains some discussion of perturbation theory. In keeping with the broad viewpoint of Part A, the treatment there is not detailed. Its principal purpose is to illustrate how the approach of Poincaré can remove bothersome secular terms that appear in a naïve approach to including the effects of other planets upon the motion of a selected planet about the sun. This is done indirectly, by discussing the effect of small nonlinearity upon the motion of a simple pendulum.

We are now in a more favorable situation regarding the study of perturbation theory. This arises through the discussion in Chapter 6 of ways to formulate a mathematical model so that a meaningful small parameter appeared in it (if this is permitted by the conditions of the problem). In the present short chapter, our object is to provide some detailed examination of regular perturbation theory. Roughly speaking, this theory is appropriate when a reasonable initial approximation is obtained by means of a straightforward attempt at simplifying a problem that involves a small parameter.

Section 7.1 provides details that were omitted in the treatment of the pendulum problem given in Chapter 2. Section 7.2 illustrates various aspects of regular perturbation calculations on the projectile problem. (The mathematical formulation of this problem was the subject of much detailed discussion in the previous chapter.) In all, three approaches to the calculations will be presented: series, parametric differentiation, and successive approximation.

Further application of regular perturbation theory, this time on a more difficult problem involving a pair of ordinary differential equations, is made in Chapter 8. Moreover, Chapter 10 in II contains a lengthy analysis of the use of this theory (with some modifications) in a water wave problem governed by a system of nonlinear partial differential equations. Taken together, all these examples provide a foundation for the use of an important technique.

7.1 The Series Method Applied to the Simple Pendulum

In this section we discuss the power series approach to perturbation calculations. The calculations are illustrated on the problem of the simple pendulum. Except toward the end of the discussion, we have avoided essential reference to Chapter 6. Thus the material to be presented here could well be read in conjunction with Section 2.2, where we presented an introduction to perturbation calculations for the pendulum.

PRELIMINARIES

In Section 2.2 we introduced the simple pendulum, a weightless rigid rod of fixed length L with a point of mass m at one end; the mass is free to rotate about a horizontal axis at the point of suspension. There, or in introductory physics texts, the governing mathematical problem is shown to consist of the following equations of motion and initial conditions for the angular displacement θ^* as a function of time t^*:

$$\frac{d^2\theta^*}{dt^{*2}} + \omega_0^2 \sin \theta^* = 0 \quad \left(\omega_0^2 \equiv \frac{g}{L}\right); \qquad t > 0.$$
$$\text{at } t = 0, \qquad \theta^* = a \quad \text{and} \quad \frac{d\theta^*}{dt^*} = \Omega. \tag{1}$$

(Here $\theta^* = 0$ when the pendulum hangs vertically downward.) With the "usual" assumption of small displacement, one replaces $\sin \theta^*$ by θ^* and obtains (Section 2.2)

$$\theta^*(t^*) = a \cos \omega_0 t^* + \Omega\omega_0^{-1} \sin \omega_0 t^*. \tag{2}$$

We wish to improve this approximation in a systematic manner. For simplicity we shall restrict our discussion to the case $\Omega = 0$ (pendulum released from rest).

The first step is to introduce dimensionless variables Θ and t:

$$t = \omega_0 t^*, \qquad \Theta = \frac{\theta^*}{a}. \tag{3}$$

The problem now takes the form

$$\frac{d^2\Theta}{dt^2} + \frac{\sin (a\Theta)}{a} = 0, \quad \Theta(0) = 1, \qquad \frac{d\Theta}{dt}(0) = 0. \tag{4}$$

The new dependent variable Θ measures angular displacement in units of the initial angular displacement a. The new independent variable t measures time in units of ω_0^{-1}, where $2\pi\omega_0^{-1}$ is the period of small oscillations. The introduction of these dimensionless variables has provided us with a problem in which there appears a single parameter a, and we wish to develop an approximate solution under the assumption that a is small compared to unity.*

* Those who have studied Section 6.3 should recognize that the new variables are scaled. The reference angle a is the largest value of $\theta^*(t^*)$. The reference time is ω_0^{-1}, in accord with Example 3 of Section 6.3.

A basic tool in our analysis is the Maclaurin expansion

$$F(a) = \sum_{i=0}^{\infty} C_i a^i; \qquad C_i \equiv \frac{d^i F(0)/da^i}{i!}. \tag{5a, b}$$

Assuming convergence, the successive partial sums of the above series provide ever more accurate approximations to F for small a. Even to write down the series, however, it is necessary that the function in question be infinitely differentiable at the origin. On the other hand, if only a few derivatives are known to exist, we can use the Maclaurin formula in the form of a finite sum plus remainder:

$$F(a) = \sum_{i=0}^{n} C_i a^i + \left[\frac{d^{n+1}F}{da^{n+1}}(\xi)\right] \frac{a^{n+1}}{(n+1)!}; \qquad \xi \text{ between 0 and } a. \tag{6}$$

No convergence questions arise here, and bounds on the $(n + 1)$st derivative permit estimates of the error.

In the material to follow, for definiteness we shall proceed under the assumption that various functions have infinite series of the Maclaurin type. If these functions have only a finite number of derivatives, however, there is a tacit understanding that the series is appropriately terminated.

To make sure that there is no confusion on the matter, we note that if a function F depends on variables other than a, partial derivatives must be used in the Maclaurin expansion, and the coefficients are no longer constant. Thus, if $F = F(t, a)$, then (5) becomes

$$F(t, a) = \sum_{i=0}^{\infty} C_i(t) a^i; \qquad C_i(t) = \frac{1}{i!}\frac{\partial^i F}{\partial a^i}(t, 0). \tag{7a, b}$$

It is usually only feasible to obtain the first few terms in the various series. To make sure this is done systematically, we must keep track of the terms we have neglected. To do this we use the O notation, where *by* $O(a^j)$ *we mean a collection of terms containing jth or higher powers of a.* For example,

$$\sin a = a - \frac{a^3}{3!} + O(a^5), \qquad \cos a = 1 + O(a^2).$$

Assuming that our formal calculations are valid, this is consistent with the precise use of the O symbol as defined in Appendix 3.1.

We now outline the series approach to perturbation theory as it is applied to a differential equation. The approach is broken down into five steps. Each step is immediately illustrated as it applies to the pendulum problem (4).

SERIES METHOD

Perturbation Method I. *Assume that the dependent variable can be expanded in a series of powers of the small parameter.*

STEP A. *Substitute the power series into the differential equation.* Here we assume

$$\Theta(t, a) = \sum_{i=0}^{\infty} \Theta_i(t) a^i. \tag{8}$$

The substitution gives

$$\sum_{i=0}^{\infty} \ddot{\Theta}_i a^i + a^{-1} \sin\left(a \sum_{i=0}^{\infty} \Theta_i a^i\right) = 0 \qquad \left(\cdot \equiv \frac{d}{dt}\right), \tag{9}$$

where the validity of term-by-term differentiation has been taken for granted.

STEP B. *Expand all quantities so that each term is written as a power series.* There is nothing further to do at present with the first term in (9); it is already written as a power series. Treatment of the second term requires that we use the Maclaurin series for the sine function to write

$$a^{-1} \sin\left(a \sum_{i=0}^{\infty} \Theta_i a^i\right)$$
$$= a^{-1}\left(a \sum_{i=0}^{\infty} \Theta_i a^i\right) - \frac{a^{-1}}{3!}\left(a \sum_{i=0}^{\infty} \Theta_i a^i\right)^3 + \frac{a^{-1}}{5!}\left(a \sum_{i=0}^{\infty} \Theta_i a^i\right)^5 - \cdots. \tag{10}$$

We aim to retain terms only through $O(a^2)$, but during intermediate calculations we retain some smaller terms, to make sure that nothing crucial has been neglected. We find

$$a^{-1} \sin\left(a \sum_{i=0}^{\infty} \Theta_i a^i\right) = \Theta_0 + a\Theta_1 + a^2\Theta_2 + O(a^3) - \frac{a^2}{6}[\Theta_0^3 + O(a)] + O(a^4)$$
$$= \Theta_0 + a\Theta_1 + a^2\left(\Theta_2 - \frac{\Theta_0^3}{6}\right) + O(a^3). \tag{11}$$

STEP C. *Collect all terms in the equation and equate to zero the successive coefficients in the series.* Using (11), we write (9) as

$$(\ddot{\Theta}_0 + \Theta_0) + a(\ddot{\Theta}_1 + \Theta_1) + a^2(\ddot{\Theta}_2 + \Theta_2 - \tfrac{1}{6}\Theta_0^3) + O(a^3) = 0. \tag{12}$$

Equation (12) has the form of (7a) with $F \equiv 0$. From (7b), this implies that $C_i \equiv 0$ for all i. Assuming the necessary convergence, we are led to require the vanishing of each coefficient in (12):

$$\begin{aligned} \ddot{\Theta}_0 + \Theta_0 &= 0, \\ \ddot{\Theta}_1 + \Theta_1 &= 0, \\ \ddot{\Theta}_2 + \Theta_2 &= \tfrac{1}{6}\Theta_0^3. \end{aligned} \tag{13}$$

Our original nonlinear differential equation has been replaced by a sequence of linear differential equations. This is a characteristic feature of the perturbation method.

STEP D. *Substitute the power series into the original initial (or boundary) conditions, expand, equate the coefficients to zero, and obtain a set of initial (or boundary) conditions to supplement the sequence of differential equations obtained in Step* (C). In the present case, $\Theta(0) = 1$ implies that

$$1 = \sum_{i=0}^{\infty} \Theta_i(0)a^i \qquad \text{or } 0 = [\Theta_0(0) - 1] + a\Theta_1(0) + a^2\Theta_2(0) + O(a^3). \quad (14)$$

Hence we require that

$$\Theta_0(0) = 1, \quad \Theta_1(0) = 0, \quad \Theta_2(0) = 0, \quad \text{etc.} \qquad (15)$$

Similarly, we require that

$$\dot{\Theta}_0(0) = 0, \quad \dot{\Theta}_1(0) = 0, \quad \dot{\Theta}_2(0) = 0, \quad \text{etc.} \qquad (16)$$

STEP E. *Successively solve the sequence of differential equations and boundary conditions obtained by Steps* (A)–(D). In the present case, we first have to solve

$$\ddot{\Theta}_0 + \Theta_0 = 0, \qquad \Theta_0(0) = 1, \qquad \dot{\Theta}_0(0) = 0. \qquad (17)$$

The differential equation has the general solution

$$\Theta_0 = A \cos t + B \sin t; \qquad A, B \text{ constants.} \qquad (18)$$

The particular solution satisfying the initial conditions is

$$\Theta_0 = \cos t. \qquad (19)$$

This is the usual approximation, in dimensionless form.

From (13), (15), and (16), $\Theta_1(t)$ satisfies

$$\ddot{\Theta}_1 + \Theta_1 = 0, \qquad \Theta_1(0) = 0, \qquad \dot{\Theta}_1(0) = 0.$$

The solution is

$$\Theta_1(t) \equiv 0. \qquad (20)$$

There is no correction at this stage.

Similarly, $\Theta_2(t)$ satisfies

$$\ddot{\Theta}_2 + \Theta_2 = \tfrac{1}{6}\Theta_0^3 = \tfrac{1}{6}\cos^3 t, \qquad \Theta_2(0) = 0, \dot{\Theta}_2(0) = 0. \qquad (21)$$

Since the solution of the homogeneous equation is known, we could use the method of variation of parameters to obtain a particular solution of the inhomogeneous equation.* The method of undetermined coefficients is frequently easier if it is applicable. It can be used here if we perform the *often helpful step of using a trigonometric identity to transform an expression involving powers of* $\cos t$ *and* $\sin t$ *into an expression involving linear combinations of these trigonometric functions*. In the present case, the appropriate identity is

$$\cos^3 t \equiv \tfrac{1}{4}(\cos 3t + 3 \cos t).$$

* This is a general feature of regular perturbation calculations for ordinary differential equations.

Using this to transform the differential equation of (21) into

$$\ddot{\Theta}_2 + \Theta_2 = \tfrac{1}{24}\cos 3t + \tfrac{1}{8}\cos t, \tag{22}$$

we can now apply the method of undetermined coefficients. In doing so, we must take note of the fact that the forcing term $\frac{1}{8}\cos t$ is a solution of the homogeneous equation. Consequently, the usual form of the corresponding particular solution must be multiplied by the independent variable t. We find (Exercise 1) for the general solution to (22),

$$\Theta_2(t) = A\cos t + B\sin t - \tfrac{1}{192}\cos 3t + \tfrac{1}{16}t\sin t. \tag{23}$$

Imposing the initial conditions, we obtain $A = \frac{1}{192}$, $B = 0$.*

From (19), (20), and (23), we see that the complete approximate solution, taking terms of $O(a^2)$ or lower into account, is

$$\Theta(t, a) \approx \cos t + a^2\left(\tfrac{1}{192}\cos t - \tfrac{1}{192}\cos 3t + \tfrac{1}{16}t\sin t\right). \tag{24}$$

Noteworthy is the higher frequency term $\cos 3t$.

DISCUSSION OF RESULTS SO FAR

Knowing a correction to the usual approximation $\cos t$, we can now determine when this correction is small. We see that the terms proportional to the bounded quantities $\cos t$ and $\cos 3t$ in the correction are small for all time provided only that a is small. It is otherwise with the term proportional to $t\sin t$: the latter is small compared to unity if and only if $\frac{1}{16}a^2 t \ll 1$. It is becoming clear that we have different situations, depending upon whether or not we examine the approximation (i) for a fixed finite time interval $[0, T]$, or (ii) for the unbounded interval $[0, \infty)$. Let us consider case (i) first.

For the time interval $[0, T]$, the correction is small and the approximation $\Theta(t, a) \approx \cos t$ appears to be consistent† provided that $\frac{1}{16}a^2 T \ll 1$. We would expect, in fact, that the approximation is **asymptotically valid** as $a \downarrow 0$, uniformly for $t \in [0, T]$. By this we mean that we anticipate that *the exact solution and the approximation can be made to differ by an arbitrarily small amount for the entire range of independent variable provided the parameter is taken sufficiently small.* Symbolically, in the present case uniform asymptotic validity of the approximation $\cos t$ requires that, given any $\varepsilon > 0$, there exists a positive constant a_0, which may depend on ε and T but must be independent of t, such that

$$|\Theta(a, t) - \cos t| < \varepsilon \quad \text{whenever } 0 \le t \le T, \qquad 0 \le |a| \le a_0.$$

* Note that a multiple of the homogeneous solution, $\cos t$, appears in the final answer (23). Beginners sometimes write down just a particular solution to equations analogous to (22), forgetting that they have no justification for discarding the homogeneous solution, which in any case is needed to satisfy the initial conditions.

† We are using the terminology of Section 6.1.

Now consider the unbounded interval $[0, \infty)$. For a given initial angular displacement a, the term $\frac{1}{16}a^2 t \sin t$ oscillates with ever-increasing amplitude. Such behavior can be rejected on physical grounds. Thus our perturbation scheme has apparently *not* provided an approximation that is asymptotically valid for $t \in [0, \infty)$. Methods for obtaining improved expansions will be discussed in Section 9.2. For now, let us be content with the fact that all appears well in $[0, T]$.

Let us look at our approximations from a slightly different point of view. At the beginning of our discussion we estimated the validity of the usual approximation $\cos t$ by examining the next term in the perturbation expansion. It is important to realize the advantage of this approach compared to the basic simplification procedure of Section 6.1. For the pendulum problem, that procedure would entail making the approximation $a^{-1} \sin a\Theta \approx \Theta$, finding the approximate solution $\Theta_0 = \cos t$, and checking for consistency by determining whether $a^{-1} \sin a\Theta_0 \approx \Theta_0$. Now

$$\lim_{a \to 0} \frac{a^{-1} \sin a\Theta_0}{\Theta_0} = 1 \qquad \text{uniformly in } t, \tag{25}$$

since $|\Theta_0| \leq 1$ for all t, so that the approximation is *apparently* consistent when a is small. But the true solution also satisfies $|\Theta(t)| \leq 1$. (This is obvious physically—a proof is requested in Exercise 4.) Thus (25) remains true when Θ is substituted for Θ_0 and the approximation is *genuinely* consistent. In the terminology of Section 6.1 the approximation $\cos t$ is nevertheless "wretched" in that after sufficient time has elapsed its misestimate of the period of oscillation will bring about grave errors in its prediction of the true angular displacement Θ. *The deficiency in the approximation was not signaled by the basic simplification procedure but it was revealed by the appearance of the next term in the perturbation expansion.* This fact serves as an antidote to the examples of wretched consistent approximations given in Section 6.1 and given again here in our discussion of the pendulum.

It is hard to think of an instance in which the failure of an approximation is not indicated by the appearance of the next term. Be that as it may, the antidote is not universal because in complicated problems a forbidding amount of additional work may be required to compute one more term. The simple substitution required by the consistency check in the basic simplification procedure may be all that it is feasible to do. In spite of its hazards, a check for consistency is well worth carrying out in such instances, for reasons that have been mentioned in Section 6.1.

HIGHER ORDER TERMS

We can obtain higher terms in our perturbation series by continuing our calculations. No new questions of principle arise, but the computations eventually appear forbiddingly difficult.

It is helpful to take advantage of the fact that the identical vanishing of $\Theta_1(t)$ is not accidental. Indeed $\Theta_i(t) = 0$ for all odd i. To see this, observe that the governing differential equation (4) is unchanged if $-a$ is substituted for a. Assuming that (4) has a unique solution, this means that $\Theta(t, a) = \Theta(t, -a)$. In other words, Θ is an even function of its second argument. (What is the physical reason behind this fact?) From this it is not difficult to show (Exercise 2) that all odd-ordered derivatives of $\Theta(t, a)$ with respect to a vanish when $a = 0$. Thus coefficients in the series (8) satisfy

$$\Theta_i(t) = \frac{1}{i!}\left[\frac{\partial^i \Theta(t, a)}{\partial a^i}\right]_{a=0} \equiv 0, \qquad i \text{ odd}.$$

We conclude that there are no odd powers of a in the series for $\Theta(t, a)$. To obtain higher corrections, it is best to take advantage of this and assume from the outset that $\Theta(t, a)$ has an expansion in powers of a^2.

REMARK. The considerations of the previous paragraph show that *before starting perturbation calculations, you should see if symmetry arguments can be used to prove that the proper expansion variable is the square of the naturally occurring small parameter*. If you miss such symmetry you will not blunder because of your error of omission; you will only have to work harder to obtain a given degree of accuracy.

By calculating more terms, we expect to improve the validity of our approximation. This is true for the bounded interval $[0, T]$. In fact, it can be shown that the series (8) is convergent when $|a| \leq a_0$, for some sufficiently small a_0. The convergence is uniform for t in $[0, T]$. That is, given any $\varepsilon > 0$ there exists a nonnegative integer N_0 (which is independent of t) such that

$$\left|\Theta(t, a) - \sum_{i=0}^{N} \Theta_i(t) a^i\right| < \varepsilon \tag{26}$$

whenever $N > N_0$, $|a| \leq a_0$, $t \in [0, T]$.

Implied by this convergence (see Section 3.2) is the fact that the series (8) is asymptotic as $a \to 0$, uniformly in t for $t \in [0, T]$. This means that for any $\varepsilon > 0$ and any fixed N, there exists a positive number a_1 (which is independent of t) such that

$$a^{-N}\left|\Theta(t, a) - \sum_{i=0}^{N} \Theta_i(t) a^i\right| < \varepsilon \tag{27}$$

whenever $|a| \leq a_1$, $t \in [0, T]$. In terms of the notation of asymptotic analysis (Appendix 3.1),

$$\Theta(t, a) = \sum_{i=0}^{N} \Theta_i(t) a^i + o(a^N) \qquad \text{uniformly for } t \in [0, T].^* \tag{28}$$

* In our formal calculations we assumed that the remainder in (28) was $O(a^{N+1})$, a stronger requirement than $o(a^N)$.

Since it is difficult to calculate very many terms of the series, the asymptotic result is apt to be more useful than the convergence proof. But remember the distinction. In a convergent approximation one considers the parameter as fixed and thinks of improving accuracy by taking more terms. In an asymptotic approximation one considers a fixed number of terms and thinks of improving accuracy by letting the parameter approach a "favorable" value.

Taking more terms does not improve the situation as far as the unbounded interval is concerned. The $O(a^n)$ terms contain contributions proportional to t^n, so *all* "correction" terms are $O(1)$ when $t = 0(a^{-1})$. In Section 2.2 we briefly considered Poincaré's method of obtaining an improved approximation. Another approach to this matter will be discussed in Chapter 11.

EXERCISES

1. Use the method of undetermined coefficients to derive (23).

2. ‡(a) Show that the derivative of an even function is an odd function and that the derivative of an odd function is an even function.

(b) Show that the second derivative of an even function is an even function.

(c) Show that all odd-ordered derivatives of an even function vanish at the origin.

3. (Lengthy.) Assume that (4) has the series solution

$$\Theta = \overline{\Theta}_0 + a_1\overline{\Theta}_1 + a_1^2\overline{\Theta}_2 + \cdots, \qquad a_1 \equiv a^2,$$

and calculate $\overline{\Theta}_2$.

‡4. Prove that (4) implies that $|\Theta(t)| \le 1$.

5. Calculate a correction to the lowest order approximation when the initial conditions of (1) are taken with $a = 0$, Ω "small."

7.2 Projectile Problem Solved by Perturbation Theory

In this section we treat by three different perturbation methods the projectile problem that was finally formulated in Equation (6.3.18). First, we employ the series method that was introduced in the previous section. By means of worked examples, we show how the somewhat lengthy calculations proceed. The student should not *read* these examples, of course, but should use them as a check on his own work.

The remainder of the section illustrates, on the same problem, the method of parametric differentiation and the method of successive approximation.

SERIES METHOD

The dimensionless distance x of a projectile from the earth at dimensionless time t is governed by the following equations and initial conditions:

$$\ddot{x} = -(1 + \varepsilon x)^{-2}, \qquad x(0) = 0, \quad \dot{x}(0) = 1. \qquad \text{(1a, b, c)}$$

This was shown in Section 6.3 (compare 6.3.18). Recall that $\varepsilon \equiv V^2/gR$, where V is the initial speed upward, g is the gravitational acceleration, and R is the earth's radius. We are interested in the effect of the variation of gravitational attraction with distance in situations where the projectile does not stray too far from earth. As discussed in Section 6.3, we should therefore examine the solution to (1) when ε is small. Following the perturbation procedure outlined above, we assume a power series expansion

$$x(t, \varepsilon) = x_0(t) + \varepsilon x_1(t) + \varepsilon^2 x_2(t) + \cdots \tag{2}$$

and substitute it into the governing equation. Next, we use (2) to write the term $(1 + \varepsilon x)^{-2}$ as a power series in ε. In principle this could be done by employing the Taylor series formula but experience teaches that *in constructing power series one should use the binomial expansion whenever possible.* It is best to perform preliminary manipulations if necessary so that the first term in the binomial is unity, in which case the expansion takes the simple form

$$(1 + J)^n = 1 + nJ + \frac{n(n-1)}{2!} J^2 + \frac{n(n-1)(n-2)}{3!} J^3 + \cdots, \qquad |J| < 1. \tag{3}$$

In the present case we have

$$\begin{aligned}(1 + \varepsilon x)^{-2} &= 1 - 2\varepsilon x + \tfrac{1}{2}(-2)(-3)\varepsilon^2 x^2 + O(\varepsilon^3) \\ &= 1 - 2\varepsilon[x_0 + \varepsilon x_1 + O(\varepsilon^2)] + 3\varepsilon^2[x_0 + O(\varepsilon)]^2 + O(\varepsilon^3) \\ &= 1 + \varepsilon(-2x_0) + \varepsilon^2(-2x_1 + 3x_0^2) + O(\varepsilon^3).\end{aligned}$$

Equating the coefficients of various powers of ε to zero, we find the following sequence of problems.

$$\ddot{x}_0 = -1, \qquad x_0(0) = 0, \qquad \dot{x}_0(0) = 1. \tag{4a}$$

$$\ddot{x}_1 = 2x_0, \qquad x_1(0) = 0, \qquad \dot{x}_1(0) = 0. \tag{4b}$$

$$\ddot{x}_2 = 2x_1 - 3x_0^2, \qquad x_2(0) = 0, \qquad \dot{x}_2(0) = 0. \tag{4c}$$

Example 1. Solve the system (4).
Solution. Equation (4a) yields upon integration

$$\dot{x}_0(t) = -t + 1, \qquad x_0(t) = t - \tfrac{1}{2}t^2. \tag{5}$$

With this (4b) becomes

$$\ddot{x}_1(t) = 2t - t^2, \qquad x_1(0) = 0, \qquad \dot{x}_1(0) = 0,$$

so

$$\dot{x}_1(t) = t^2 - \tfrac{1}{3}t^3, \qquad x_1(t) = \tfrac{1}{3}t^3 - \tfrac{1}{12}t^4. \tag{6}$$

From (4c)

$$\ddot{x}_2(t) = \tfrac{2}{3}t^3 - \tfrac{1}{6}t^4 - 3(t^2 - t^3 + \tfrac{1}{4}t^4)$$
$$= -3t^2 + \tfrac{11}{3}t^3 - \tfrac{11}{12}t^4,$$

so that

$$\dot{x}_2(t) = -t^3 + \tfrac{11}{12}t^4 - \tfrac{11}{60}t^5, \qquad x_2(t) = -\tfrac{1}{4}t^4 + \tfrac{11}{60}t^5 - \tfrac{11}{360}t^6. \tag{7}$$

Thus, the solution is

$$x = t - \tfrac{1}{2}t^2 + \varepsilon(\tfrac{1}{3}t^3 - \tfrac{1}{12}t^4) + \varepsilon^2(-\tfrac{1}{4}t^4 + \tfrac{11}{60}t^5 - \tfrac{11}{360}t^6) + \cdots. \tag{8}$$

Of interest is the time t_m at which the projectile reaches its maximum height. At this time the speed vanishes; i.e.,

$$\dot{x}(t_m) = 0. \tag{9}$$

We can find a first approximation to t_m by solving $\dot{x}_0(t) = 0$. From (5), this gives $t_m \approx 1$. To do better, we shall utilize the solution to the accuracy we have obtained. Combining (9) and (8), we thus derive the following equation for t_m:

$$1 - t_m + \varepsilon[t_m^2 - \tfrac{1}{3}t_m^3] + \varepsilon^2[-t_m^3 + \tfrac{11}{12}t_m^4 - \tfrac{11}{60}t_m^5] + O(\varepsilon^3) = 0. \tag{10}$$

This is a quintic equation, neglecting $O(\varepsilon^3)$ terms. But we are only interested in the root that is approximately unity. Furthermore, we expect that this root, like all the other quantities we have been seeking, has a series expansion in ε. Since Equation (10) for t_m is correct only through $O(\varepsilon^2)$, we assume an expansion correct to the same order:

$$t_m = 1 + a_1\varepsilon + a_2\varepsilon^2 + O(\varepsilon^3). \tag{11}$$

It is now a straightforward calculation to obtain the coefficients in this expansion.

Example 2. Calculate a_1 and a_2 in (11).
Solution. Inserting (11) into (10), we obtain

$$0 = -a_1\varepsilon - a_2\varepsilon^2 + \varepsilon[1 + 2a_1\varepsilon + O(\varepsilon^2) - \tfrac{1}{3}(1 + 3a_1\varepsilon + O(\varepsilon^2))]$$
$$+ \varepsilon^2[-1 + \tfrac{11}{12} - \tfrac{11}{60} + O(\varepsilon)]$$
$$= \varepsilon(-a_1 + 1 - \tfrac{1}{3}) + \varepsilon^2(-a_2 + 2a_1 - a_1 - \tfrac{4}{15}) + O(\varepsilon^3).$$

Upon equating each power of ε to zero, we find

$$a_1 = \tfrac{2}{3}, \qquad a_2 = a_1 - \tfrac{4}{15} = \tfrac{2}{5},$$

so

$$t_m = 1 + \tfrac{2}{3}\varepsilon + \tfrac{2}{5}\varepsilon^2 + O(\varepsilon^3). \tag{12}$$

Example 3. Determine the maximum height $x_m \equiv x(t_m)$.
Solution

$$\begin{aligned}
x_m \equiv x(t_m) &= t_m - \tfrac{1}{2}t_m^2 + \varepsilon[\tfrac{1}{3}t_m^3 - \tfrac{1}{12}t_m^4] + \varepsilon^2[-\tfrac{1}{4}t_m^4 + \tfrac{11}{60}t_m^5 - \tfrac{11}{360}t_m^6] + O(\varepsilon^3) \\
&= 1 + \tfrac{2}{3}\varepsilon + \tfrac{2}{5}\varepsilon^2 - \tfrac{1}{2}[1 + \tfrac{2}{3}\varepsilon + \tfrac{2}{5}\varepsilon^2 + O(\varepsilon^3)]^2 \\
&\quad + \varepsilon[\tfrac{1}{3}(1 + \tfrac{2}{3}\varepsilon + O(\varepsilon^2))^3 - \tfrac{1}{12}(1 + \tfrac{2}{3}\varepsilon + O(\varepsilon^2))^4] \\
&\quad + \varepsilon^2[-\tfrac{1}{4} + \tfrac{11}{60} - \tfrac{11}{360} + O(\varepsilon)] + O(\varepsilon^3) \\
&= 1 + \tfrac{2}{3}\varepsilon + \tfrac{2}{5}\varepsilon^2 - \tfrac{1}{2}[1 + \tfrac{4}{9}\varepsilon^2 + \tfrac{4}{3}\varepsilon + \tfrac{4}{5}\varepsilon^2 + O(\varepsilon^3)] \\
&\quad + \varepsilon[(\tfrac{1}{3}(1 + 2\varepsilon + O(\varepsilon^2)) - \tfrac{1}{12}(1 + \tfrac{8}{3}\varepsilon + O(\varepsilon^2))] \\
&\quad - \tfrac{7}{72}\varepsilon^2 + O(\varepsilon^3),
\end{aligned}$$

so

$$x_m(\varepsilon) = \tfrac{1}{2} + \tfrac{1}{4}\varepsilon + \tfrac{1}{8}\varepsilon^2 + O(\varepsilon^3). \tag{13}$$

The answers that we have obtained on carrying through the exercise of treating the trajectory problem by perturbation theory are, in retrospect, easy to understand qualitatively. The correction terms to t_m and x_m are positive. This is to be expected. When the weakening of gravitational attraction with distance is taken into account, the computed time for the projectile to reach its maximum distance from the earth will be longer, and the farthest excursion achieved will be larger, compared to the case when gravitational attraction is deemed invariant with distance.

In the present problem the solution is of interest only during the finite time interval $[0, 2t_m]$. Unlike the pendulum problem, there is no difficulty caused by the inevitable buildup of small effects over a sufficiently long interval. We expect our approximations to be uniformly valid for $0 \le t \le 2t_m$. Indeed, using the exact solution, one can see that, for example, the series (13) converges to the correct value of $x_m(\varepsilon)$ provided that $0 \le \varepsilon < 2$. (See Exercise 2.)

Recall that in Section 6.3 we showed that only if dimensionless variables were chosen correctly (by scaling) did the "naïve approximation" of neglecting terms multiplied by a small parameter yield an appropriate mathematical problem. From the present point of view, the "naïve approximation" is the first term in a series expansion of the solution. We draw the conclusion that *correct scaling is a prerequisite to successful use of the regular perturbation method.*

Until now, we have confined our discussion to a single method, that of series. We now treat briefly two closely related methods: parametric differentiation and successive approximations.

PARAMETRIC DIFFERENTIATION

The idea of the parametric differentiation method is that equations for the coefficients $x_i(t)$ in the expansion

$$x(t, \varepsilon) = \sum_{i=1}^{\infty} x_i(t)\varepsilon^i \tag{14}$$

can be obtained by successively differentiating the problem for $x(t, \varepsilon)$ with respect to ε. After differentiating, one sets $\varepsilon = 0$. The reason that this method works is simply that the coefficients x_i satisfy

$$x_i(t) = \frac{1}{i!}\left[\frac{\partial^i x(t, \varepsilon)}{\partial \varepsilon^i}\right]_{\varepsilon=0}. \tag{15}$$

In the series method, of course, equations for the $x_i(t)$ are obtained by substitution and collecting. The two methods are equivalent in essence, but sometimes the differentiation method yields slightly easier calculations.

We now present the method, illustrating it on the projectile problem.

Perturbation Method II

STEP A. *Differentiate equations and boundary conditions* with respect to the small parameter. Consider the projectile problem of (6.3.18):

$$\ddot{x}(t, \varepsilon) = -[1 + \varepsilon x(t, \varepsilon)]^{-2}, \qquad x(0, \varepsilon) = 0, \quad \dot{x}(0, \varepsilon) = 1. \tag{16}$$

We obtain the following upon differentiating the differential equation with respect to ε:

$$\frac{\partial \ddot{x}}{\partial \varepsilon} = 2(1 + \varepsilon x)^{-3}\left(\varepsilon \frac{\partial x}{\partial \varepsilon} + x\right). \tag{17}$$

Note that we have had to keep in mind that x depends on ε. It is convenient to introduce the notation

$$x^{(i)}(t, \varepsilon) \equiv \frac{\partial^i x(t, \varepsilon)}{\partial \varepsilon^i}; \qquad i = 0, 1, \ldots. \tag{18a}$$

In particular,

$$x^{(0)}(t, \varepsilon) \equiv x(t, \varepsilon). \tag{18b}$$

With this (17) can be written

$$\ddot{x}^{(1)} = 2[1 + \varepsilon x^{(0)}]^{-3}[\varepsilon x^{(1)} + x^{(0)}], \tag{19}$$

and a second differentiation (Exercise 3a) gives

$$\ddot{x}^{(2)} = 2[1 + \varepsilon x^{(0)}]^{-3}[\varepsilon x^{(2)} + 2x^{(1)}] - 6[1 + \varepsilon x^{(0)}]^{-4}[\varepsilon x^{(1)} + x^{(0)}]^2. \tag{20}$$

STEP B. *Set the parameter equal to zero in the various equations.* We introduce another notation:

$$y_i(t) \equiv x^{(i)}(t, 0). \tag{21}$$

[Comparison with (15) shows that the y_i differ only by a constant factor $i!$ from the x_i that are used in the series approach to regular perturbations.] Using (21), we obtain from (16), (19), and (20),

$$\ddot{y}^{(0)} = -1, \quad \ddot{y}^{(1)} = 2y^{(0)}, \quad \ddot{y}^{(2)} = 4y^{(1)} - 6[y^{(0)}]^2. \tag{22a, b, c}$$

If we repeat Steps (A) and (B) for the boundary conditions, we obtain

$$y^{(0)}(0) = 0, \qquad \dot{y}^{(0)}(0) = 1. \tag{23a}$$

$$y^{(1)}(0) = 0, \qquad \dot{y}^{(1)}(0) = 0. \tag{23b}$$

$$y^{(2)}(0) = 0, \qquad \dot{y}^{(2)}(0) = 0. \tag{23c}$$

As the reader can easily verify, the equations for the y_i differ only by the appearance of appropriate constants from the corresponding equations for the x_i, which were obtained above (Exercise 4). The same final result is attained by a different route.

SUCCESSIVE APPROXIMATIONS (METHOD OF ITERATION)

Another approach to perturbation theory can be motivated in the following way. Suppose that some problem has the form

$$F(z) = G(z). \tag{24}$$

(One can think of both F and G as polynomials for definiteness.) Suppose further that the equation $F = 0$ provides a rough approximation z_0 to the desired solution z. To obtain a second approximation, rather than ignore $G(z)$ entirely, we can approximate it by $G(z_0)$. We can then solve for a new approximation z_1, which satisfies

$$F(z_1) = G(z_0).$$

The procedure can be repeated by means of

$$F(z_n) = G(z_{n-1}), \qquad n = 2, 3, \ldots . \tag{25}$$

Example. If

$$z - 2 = 0.01z^3,$$

then an initial approximation to a solution is $z_0 = 2$. A better approximation is $z = 2 + 0.01(8) = 2.08$.

The iteration method was briefly mentioned in Section 2.2, as a way to improve upon an initial approximation in the pendulum problem. We now present a more general description of this method, using the projectile problem as an illustration.

Perturbation Method III. (For a problem involving a single ordinary differential equation for an unknown function x.)

STEP A. *Write the differential equation in the form* $F(x, \dot{x}, \ddot{x}, \ldots, t) = G(x, \dot{x}, \ddot{x}, \ldots, t)$, *where* $F = 0$ *will provide an initial approximation to the desired solution.* The initial approximation to the projectile problem (1) comes from solving $\ddot{x} = 1$. We therefore write (1a) in the form

$$\ddot{x} + 1 = -(1 + \varepsilon x)^{-2} + 1$$

or

$$\ddot{x} + 1 = \frac{\varepsilon(2x + \varepsilon x^2)}{(1 + \varepsilon x)^2}. \tag{26}$$

REMARK. In proceeding, we shall use the notation z_i for the ith approximation. This is to distinguish the present approximation from x_i and y_i, which were introduced in our discussions of Methods I and II.

STEP B. *Successively solve the equations*

$$\begin{gathered} F(z_0, \dot{z}_0, \ddot{z}_0, \ldots, t) = 0; \\ F(z_n, \dot{z}_n, \ddot{z}_n, \ldots, t) = G(z_{n-1}, \dot{z}_{n-1}, \ddot{z}_{n-1}, \ldots, t); \\ n = 1, 2, \ldots. \end{gathered} \tag{27}$$

Usually, each approximation z_i is made to satisfy the full initial conditions or boundary conditions, but this requirement may sometimes be relaxed with advantage. For the projectile problem we have the iteration scheme

$$\ddot{z}_0 + 1 = 0, \qquad z_0(0) = 0, \qquad \dot{z}_0(0) = 1. \tag{28}$$

$$\ddot{z}_n + 1 = \frac{\varepsilon(2z_{n-1} + \varepsilon z_{n-1}^2)}{(1 + \varepsilon z_{n-1})^2}, \qquad z_n(0) = 0, \qquad \dot{z}_n(0) = 1. \tag{29}$$

An example will show a useful consideration in problems that explicitly involve a small parameter. Consider the quadratic equation

$$-\varepsilon x^2 + x - 1 = 0, \qquad 0 < \varepsilon \ll 1.^* \tag{30}$$

The iteration scheme for approximations to the root near 1 is

$$x_0 - 1 = 0, \tag{31a}$$

$$x_n - 1 = \varepsilon x_{n-1}^2 \qquad \text{or} \qquad x_n = 1 + \varepsilon x_{n-1}^2, \quad n = 1, 2, \ldots. \tag{31b}$$

Thus

$$\begin{aligned} x_1 &= 1 + \varepsilon, \qquad x_2 = 1 + \varepsilon(1 + \varepsilon)^2 = 1 + \varepsilon + 2\varepsilon^2 + \varepsilon^3, \\ x_3 &= 1 + \varepsilon(1 + \varepsilon + 2\varepsilon^2 + \varepsilon^3)^2 = 1 + \varepsilon + 2\varepsilon^2 + 4\varepsilon^3 + \cdots. \end{aligned} \tag{32}$$

What is noteworthy here is that the coefficient of ε^3 in x_3 is not the same as the corresponding coefficient in x_2. The latter coefficient is wrong. This is because x_1 is correct only through terms of order ε, so that x_2 is only correct through terms of order ε^2. We can take account of this fact and save much profitless calculation of higher order terms if we replace (31) by

$$x_0 = 1, \qquad x_n = 1 + \varepsilon x_{n-1}^2 + O(\varepsilon^{n+1}) \tag{33}$$

and do not calculate the $O(\varepsilon^{n+1})$ terms.

* The symbol $\ll$ means "much less than."

Exactly the same type of considerations can be used in the analysis of (28) and (29). From (28) we find that $z_0 = t - \frac{1}{2}t^2$. Substitution into (29) yields

$$\ddot{z}_1 + 1 = \frac{2\varepsilon(t - \frac{1}{2}t^2) + \varepsilon^2(t - \frac{1}{2}t^2)^2}{[1 + \varepsilon(t - \frac{1}{2}t^2)]^2}. \tag{34}$$

With no ultimate loss of accuracy, we write this complicated equation as

$$\ddot{z}_1 + 1 = 2\varepsilon(t - \tfrac{1}{2}t^2) + O(\varepsilon^2). \tag{35}$$

Together with the initial conditions of (29) this yields

$$z_1 = t - \tfrac{1}{2}t^2 + \varepsilon(\tfrac{1}{3}t^3 - \tfrac{1}{12}t^4) + O(\varepsilon^2). \tag{36}$$

It is not difficult to show [Exercise 7(a)] that the next approximant z_2 satisfies

$$\ddot{z}_2 + 1 = 2\varepsilon(t - \tfrac{1}{2}t^2) + \varepsilon^2[\tfrac{2}{3}t^3 - \tfrac{1}{6}t^4 - 3(t - \tfrac{1}{2}t^2)^2] + O(\varepsilon^3). \tag{37}$$

Solving this equation, we find on comparison with (8) that the second approximation z_2 is correct through terms of order ε^2 as anticipated [Exercise 7(b)].

An advantage the successive approximation method is that it does not require identification of a small parameter. As the method is iterative, a minor mistake in calculation may only delay, not destroy, close approximation to the correct answer. Moreover, there are times when the form of the answer cannot readily be guessed, but yet it emerges from several iterations of the successive approximation method. Finally, this method sometimes leads to relatively easy proofs of convergence, using induction.

The chief disadvantage of the successive approximation method is that generally many hard-won terms in a new iterate only give the illusion of providing better accuracy. As we have seen, however, sometimes this disadvantage can be circumvented by a modification of the procedure.

GENERAL REMARKS ON REGULAR PERTURBATION THEORY

We have in essence been considering determination of a function f which depends on a parameter ε as well as on an independent variable t, and which satisfies a differential equation with appropriate initial or boundary conditions. When we set $\varepsilon = 0$, we obtained a solvable problem whose solution was a function f_0. We can categorize regular perturbation theory more precisely than we have, but still somewhat roughly, by saying that it consists of methods for systematically improving on the approximation f_0, when that approximation seems to be uniformly valid in the entire domain of the independent variable t. An equation lies in the province of singular perturbation theory (the subject of Chapter 9) when setting $\varepsilon = 0$ yields a problem that (i) has no solution, or (ii) has a solution that does not provide a uniformly valid approximation over the entire domain of interest, or (iii) has many solutions.

We have considered here only the case when f satisfies a single ordinary differential equation, but the method of perturbation is substantially the same when f satisfies an algebraic equation, a difference equation, a partial differential equation, an integral equation, systems of such equations, etc. Any of the three methods—series, differentiation, or successive approximations—can be used. Each yields a sequence of problems to be solved. The nature of the individual problems and the techniques required to solve them will be peculiar to the type of equations one works with, but the perturbation method provides a common framework.

The calculations we have employed are formal in that various limit operations have been interchanged without justification. For example, upon substituting an infinite series into a differential equation, we have interchanged the operations of summation and differentiation. In the differentiation approach, we exchanged the order of differentiation with respect to the parameter and differentiation with respect to the independent variable. In all cases we assumed, without rigorous justification, the existence of solutions having suitable dependence on a parameter.

The theoretically inclined mathematician might have acquired motivation for the study of theorems justifying our formal manipulations and motivation for the invention of further-reaching theorems than are currently available. Others will be content to proceed formally, knowing that the class of problems for which formal calculations can be justified is considerably narrower than the class for which it is imperative that such calculations be performed if progress is to be made.

A difficulty with perturbation methods is that the complexity of the calculations increases rapidly with the number of terms retained. Often, however, one or two terms suffices. (We see an example of this in Chapter 8.) In other cases, a computer can be used to perform the required manipulation. See, for example, "Viscous Incompressible Flow Between Concentric Rotating Spheres, Part 1, Basic Flow" by B. R. Munson and D. D. Joseph, *J. Fluid Mech. 49*, 289–303 (1971). Here terms through the seventh power of an appropriate small parameter are retained, which requires calculation of about 1000 coefficients.

EXERCISES

‡1. Find the exact solution to the projectile problem (1).

2. Use Exercise 1 to show that $x_m = \frac{1}{2}/(1 - \frac{1}{2}\varepsilon)$. Use this to check (13).

3. Verify the following equations.

(a) Equation (20).

(b) Equations (35) and (36).

4. Show that the equations of (4) are compatible with those of (22) and (23).

5. Construct an example for Perturbation Method III in which it is advantageous not to make all iterates satisfy the original initial conditions.

6. (a) Verify (32).
(b) Show that the scheme (33) also yields (32).

7. (a) Verify (37).
(b) Solve (37) and thereby demonstrate agreement with (8).

8. Treat the pendulum problem of Section 1 according to the following methods.
(a) By parametric differentiation.
(b) By successive approximations.

9. Consider the quadratic equation $m^2 + \varepsilon m - 4 = 0$, $0 < \varepsilon \ll 1$.
(a) Solve by the quadratic formula and expand the results as series in ε. Then, pretending that the problem $m^2 - 4 = 0$ is solvable but the original problem is not, develop approximate solutions.
(b) Solve by the series method.
(c) Solve by parametric differentiation.
(d) Solve by successive approximations.

10. In the projectile problem, when air resistance is taken into account but changing gravitational attraction is neglected, the equations are

$$m\frac{d^2x^*}{dt^{*2}} + k\frac{dx^*}{dt^*} = -mg, \quad x^*(0) = 0, \qquad \frac{dx^*}{dt^*}(0) = V.$$

For small damping, the change of variables

$$x = \frac{x^*}{V^2g^{-1}}, \qquad t = \frac{t^*}{Vg^{-1}},$$

is still appropriate. (Compare Exercise 6.3.8.) Assume the power series expansion

$$x(t, \beta) = x_0(t) + \beta x_1(t) + \cdots, \qquad \beta \equiv \frac{kV}{mg}.$$

Proceed as in the corresponding problem treated in the text to show that

$$t_m = 1 - \tfrac{1}{2}\beta + \tfrac{1}{3}\beta^2 + \cdots, \qquad x_m = \tfrac{1}{2} - \tfrac{1}{3}\beta + \tfrac{1}{4}\beta^2 + \cdots.$$

11. Consider the eigenvalue problem

$$y'' + \lambda(1 + \varepsilon x)y = 0, \ 0 < x < \pi; \qquad y(0) = y(\pi) = 0, \ 0 < \varepsilon \ll 1.$$

‡(a) By perturbation theory, using the result of Exercise 4.3.12(a), determine a first correction $\varepsilon\lambda_1^{(n)}$ to the eigenvalues $\lambda_0^{(n)} = n^2$ which correspond to $\varepsilon = 0$; $n = 1, 2, \ldots$.
(b) Use Exercise 4.3.12(b) to find a series expansion for the first correction to the corresponding eigenfunctions $y^{(n)}$. To obtain a unique answer, impose the normalization

$$\int_0^\pi [y^{(n)}]^2\, dx = 1.$$

12. Find approximate solutions to

$$36x^3 + (162 + 4\varepsilon)x^2 - 24\varepsilon x - 9\varepsilon = 0$$

that are valid for small $|\varepsilon|$.

13. Of interest in a certain calculation are the two roots of

$$\varepsilon + \tfrac{27}{4} = x^{-1}(x + 1)^3$$

that are near $x = \frac{1}{2}$, a root for $\varepsilon = 0$. Find the first term or two in series expansions for these roots.

CHAPTER 8
Illustration of Techniques on a Physiological Flow Problem

THIS chapter is concerned with an analysis of "standing gradient flow." The flow is osmotic in nature, being driven by solute concentration differences. The problem will be treated on a sufficiently simple level so that our discussion can be self-contained.

In our analysis we shall put to good use a number of the ideas we have studied in simpler contexts—dimensional analysis, scaling, and regular perturbation theory. We shall see how dimensional analysis reduces the number of relevant parameters from eight to three. Part of the scaling is rather challenging. Moreover, the perturbation theory is not entirely straightforward, for there are two important parameters in the problem whose relative magnitudes must be ascertained. In summary, while only involving a pair of first order ordinary differential equations, the model for standing gradient flow that we shall consider is sophisticated enough to require a panoply of our methods for its thorough analysis.

8.1 Physical Formulation and Dimensional Analysis of a Model for "Standing Gradient" Osmotically Driven Flow

In this section we outline the various physiological and biophysical considerations that are relevant to the problem we wish to investigate. We then use dimensional analysis to draw a few preliminary conclusions about the phenomenon under consideration.

SOME PHYSIOLOGICAL FACTS

The relevant physiological facts are these.

(a) Solute is secreted by various tissues into the fluid that is adjacent to them, or "bathes" them. For example, several different salts are secreted by the tissues of the kidney into the fluid that flows in kidney tubules.

(b) In the cases of interest, the solute concentration in the bathing fluid is approximately equal to, or perhaps somewhat greater than, the average concentration of solute in the secreting tissue.

(c) One possible mechanism for solute secretion is diffusion, but this cannot be the principal mechanism here. The reason is that solute is entering a region with equal or higher solute concentration than the region it is leaving.

(d) Another possible mechanism is **active transport** by the membrane that forms the boundary between the tissues and the bathing fluid. Here chemical "pumps" can expend energy to move solute across a membrane in a given direction, more or less independently of the concentrations of solute in the fluids bathing the membrane. In particular, an actively transporting membrane can pump solute from a region of low concentration to a region of high concentration.

It can be shown, however, that active transport cannot be *directly* involved in the phenomenon of interest, i.e., the solute cannot be pumped directly into the bathing fluid. The rate at which the solute is entering the bathing fluid, per unit area of membrane, has been measured. Knowing this, one can compute the dependence of solute concentration on distance normal to the membrane. (Simple diffusion theory is appropriate here.) But measurements show different results from those predicted in this calculation.*

The physiologist J. Diamond and his associates have put forth a *hypothesis* that reconciles the various facts.† As of this writing, the hypothesis can be said to be well regarded but not fully accepted. We shall now explore this hypothesis in some detail.

The physiologists proceed from the observation that long narrow channels *consistently* are a structural feature of tissues that transport fluid. Examples are the channels between adjacent cells in the wall of the gall bladder, infoldings in kidney tubules, the bile canaliculi (small canals) of the liver, and canaliculi in the cells lining the walls of the stomach. It is conjectured that solute is actively transported into the "far" end of the channels, deep in the secreting tissue (Figure 8.1). The relatively high concentration thus established in the far end would then be progressively diluted as the fluid passes toward the opening of the channel, for water will be drawn into the channel by osmosis due to this concentration. A steady or *standing gradient* of solute concentration will soon be set up. The concentration will be high at the closed "inward" end of the channel and will decrease to a relatively low value at the other "secreting" end.

We emphasize that the mechanism we are treating is postulated for a variety of secreting and absorbing tissues. For definiteness we shall confine our discussion to the secreting case. Absorbing tissues can be handled merely by running the whole mechanism backward, although a few extra complications arise (Diamond and Bossert, 1968, op. cit.).

* See J. Tormey and J. Diamond, "The Ultrastructural Route of Fluid Transport in Rabbit Gall Bladder," *J. Gen. Physiol. 50*, 2032 (1967). Also see Exercise 2.6 and the reference cited there.

† In addition to the reference just cited, see J. Diamond and W. Bossert, "Standing-gradient Osmotic Flow, a Mechanism for Coupling of Water and Solute Transport in Epithelia" *J. Gen. Physiol. 50*, 2061–2083 (1967), and, by the same authors, "Functional Consequences of Ultrastructural Geometry in 'Backwards' Fluid-transporting Epithelia," *J. Cell. Biol. 37*, 694–702 (1968).

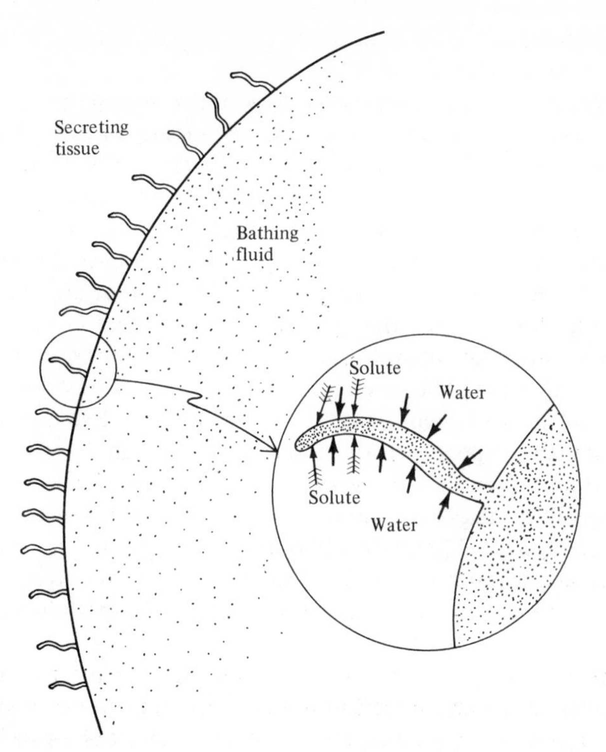

FIGURE 8.1. *Long, thin channels, as are consistently found in secreting tissues. A higher magnification view of a channel, showing the conjectured transport mechanism, appears in the insert. The solute is actively pumped into the channel near the closed end; water is drawn in by osmosis so that the solution that emerges at the open end of the channel can be considerably diluted.*

It would be out of place to discuss in depth the direct and indirect physiological evidence that supports the standing gradient hypothesis. For this we must refer the interested reader to the references cited, and to the papers quoted there. We shall confine ourselves to a study of the illumination cast on this hypothesis by the mathematical model that was devised and numerically analyzed by Diamond and Bossert in 1967, and treated analytically by L. Segel in 1970. The Diamond–Bossert paper has been cited above. We shall refer to it frequently below, often abbreviated by the initials DB. The paper by Segel* will be referred to as S; much of our discussion follows this paper very closely.

* "Standing-gradient Flows Driven by Active Solute Transport," *J. Theoret. Biol. 29*, 233–50 (1970). Material is reproduced with permission of the publisher of the journal, Academic Press Inc. (London). In particular, Figures 8.2, 8.4, and 8.5 have been redrawn from figures in the paper just cited.

OSMOSIS AND THE OSMOL

To make sure that readers feel secure about the biophysical prerequisites necessary to an understanding of the subsequent discussion, let us briefly state some important facts about osmosis.

The boundary of the tissue under consideration can be regarded as a semipermeable membrane, through which water but not salt can flow. Thus solute (e.g., salt) cannot flow through the membrane but solvent (e.g., water) can. If the number of solute particles per unit volume is different on the two sides of such a membrane, then water flows across the membrane to equalize the solute concentration. The water thus flows *from* a region of relatively high water concentration *to* a region of relatively low concentration, which is not surprising.

The rate of osmotic flow across a unit area of a semipermeable membrane turns out to depend primarily on the ratio of the *number* of solute particles per unit volume on the two sides of the membrane. In discussions of phenomena involving osmosis, it is therefore natural to use as a measure of concentration the number of particles present in a given volume of fluid. A unit of such a measure is the *osmol*,* which we now define.

Recall that a gram molecular weight or **mole** of a substance is an amount of that substance whose mass in grams is numerically equal to its molecular weight. A mole contains Avogadro's number 6.06×10^{23} molecules.

Certain molecules break apart or *dissociate* when put into solution; others do not. If a mole of a nondissociating substance like glucose is put into water, we say that the water contains 1 *osmol* of solute. A *milliosmol* is one-thousandth of an osmol.

A salt molecule in water dissociates into a sodium ion and a chloride ion. If a mole of salt is placed in an effectively infinite amount of water, the water is said to contain 2 osmols of solute. ("Effectively infinite" means "large enough to permit essentially complete dissociation.")

Because of "interference" effects, doubling the number of solute particles in a *finite* volume does not double the osmotic effect. Similarly, to give an example, adding 1 mole of salt to a solution having the concentration of human body fluid is found to increase the *effective* concentration of salt in that fluid by 1.85 osmols.† Note that the correction is only a few per cent. We need not concern ourselves with such a small factor; therefore we shall

* Introduction of this term is not strictly necessary, but it points up the lesson that an applied mathematician must learn the terminology of a new field when he starts to work in it, so that he can communicate easily with the experts. Once one has decided to make the effort, it is surprisingly easy to master a new technical vocabulary and thereby to have large areas of a new field revealed as aspects of science with which one is more or less familiar. Some further effort, also illustrated to some extent by the following discussion, shows other areas of the new field to be understandable with the aid of new but relatively simple scientific principles.

† See Chapter II of R. F. Pitts, *Physiology of the Kidney and Body Fluids* (Chicago: Yearbook Medical, 1963).

assume that the rate of solvent flow across a unit area of semi-permeable membrane is proportional to the difference of solute concentrations (mass per unit volume) across the membrane.

FACTORS THAT AFFECT STANDING GRADIENT FLOW

As a preliminary to setting up the mathematical model, let us list the physical factors that are *presumed* to govern the phenomenon. We must emphasize that the model building has already started, for a number of factors will be entirely ignored, their influence being considered secondary. Examples of such factors are the changing cross-sectional shape of the channel, radial effects, and the permeability of channel wall to *solute*. Similarly, factors that *are* considered will be taken into account in a way which is more or less approximate. For example, a *linear* relation will be assumed to connect osmotic flow of water across the channel wall and solute concentration difference across the wall, so that the permeability of the wall to water can be characterized by a single, constant, parameter. As we have mentioned, such a linear relation is probably adequate, but it should be kept in mind that a possible source of error has been introduced.

Here, then, are the *assumptions* we shall make.

(i) The channel will be idealized as a right cylinder with cross-sectional area a, circumference c, and length L.

(ii) Suppose there is a solute concentration difference across the wall of the channel. Then the rate of flow of *water* across a unit area of channel wall will be assumed to be directly proportional to this concentration difference. The constant of proportionality is the **permeability** and is denoted by the letter P.

(iii) The concentration of solute in the "bathing fluid" outside the channel is assumed to have the constant value C_0. Not only will this be assumed to be true along the sides of the channel, but contrary tendencies at the open end will be neglected so that there, too, the concentration will be required to have the value C_0.

(iv) In the channel, the solute will be carried along by a varying fluid velocity which depends only on the distance along the channel. In addition to this *convection* of solute, solute *diffusion* will take place. The diffusion constant will be denoted by D.

(v) The active solute transport will be assumed to take place only in a portion of the tube wall extending a distance δ from the closed end. The rate of solute transport per unit area of channel wall will be assumed to have the constant value N_0. The assumptions and nomenclature are summarized in Figure 8.2.

The quantity that is of primary physiological interest is the rate at which solute emerges from the open end of the channel. It is useful to think in terms of the **emergent osmolarity** Os^*, which is defined as that concentration of solute which, when multiplied by the volume of solvent that leaves the

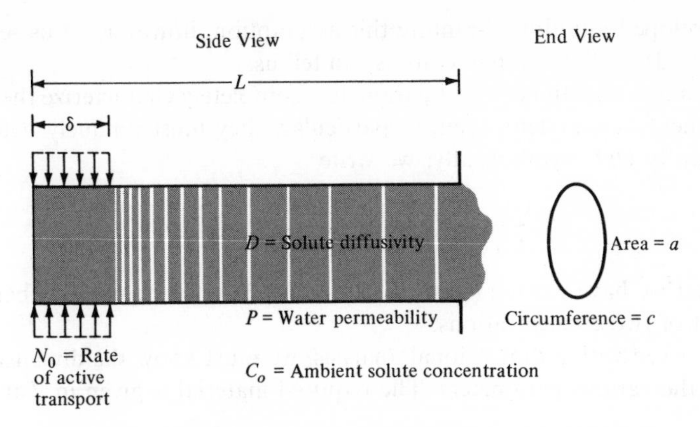

FIGURE 8.2. *The simplified standing gradient flow system.*

tube per unit time, gives the rate at which solute leaves the end of the tube per unit time.

The definition of Os^* will be clearer if we introduce some notation. Let us select an x^* axis that is parallel to the generators of the tube so that the closed end of the tube corresponds to $x^* = 0$ and the open end to $x^* = L$. Let $v^*(x^*)$ denote the speed of the fluid at position x^*, in the direction of increasing x^*. Then the volume of fluid that emerges from the open end of the tube per unit time is $av^*(L)$, since a is the cross-section area. We denote the flux density of solute at x^* by $F^*(x^*)$.† By definition, the flux density $F^*(x_0^*)$ is the rate at which solute crosses a unit area of the plane $x^* = x_0^*$ in the direction of increasing x^*. Thus solute emerges from the open end of the tube at a rate $aF^*(L)$. By definition, the emergent osmolarity Os^* satisfies $av^*(L)Os^* = aF^*(L)$, so

$$Os^* = \frac{F^*(L)}{v^*(L)}. \tag{1}$$

We now turn to a discussion of what one can learn from dimensional analysis at this point. The reader can pass at once to Section 8.2 if he wishes, for results obtained in the remainder of the present section will not be used in what follows.

DIMENSIONAL ANALYSIS OF A FUNCTIONAL RELATIONSHIP

If assumptions (i)–(v) above are correct, then a standing gradient flow is completely determined by the eight parameters δ, L, c, D, C_0, P, N_0, and a. Once these parameters are given, the idealized problem is completely specified, *assuming that we have not left out anything essential.* This is a dangerous assumption, for unerring characterization of a physical problem requires

† We are making an exception to our usual practice of employing J to denote flux density.

well-developed intuition. Granting this assumption, however, let us see what the methods of dimensional analysis can tell us.

If the above-mentioned eight parameters completely characterize the standing gradient flow system, then, in particular, they must uniquely determine the quantity Os^*. Symbolically, we write

$$\frac{Os^*}{C_0} = f(\delta, L, c, D, C_0, P, N_0, a). \tag{2}$$

Note that we have written (2) so that its left side is dimensionless, being the quotient of two concentrations.

To proceed with a dimensional analysis, we must know the dimensions of each of the various parameters. The required material is given in Table 8.1.

TABLE 8.1. Parameters and their dimensions ($\mathscr{L}$ = length, $\mathscr{M}$ = mass, $\mathscr{T}$ = time).

Parameter	Dimension
δ, L, c	$\mathscr{L}$
D	$\mathscr{L}^2\mathscr{T}^{-1}$
C_0, Os^*	$\mathscr{M}\mathscr{L}^{-3}$
P	$\mathscr{M}^{-1}\mathscr{L}^4\mathscr{T}^{-1}$
N_0	$\mathscr{M}\mathscr{L}^{-2}\mathscr{T}^{-1}$
a	$\mathscr{L}^2$

The most difficult entry, that for the permeability P, is obtained from the following considerations. To measure P, one must set up a known sol*ute* concentration difference across a membrane and measure the volume of sol*vent* that crosses a given area in a given time. This volume is proportional to the concentration difference via the permeability constant, so that

$$\text{permeability} = \frac{\text{solvent volume/(area} \times \text{time)}}{\text{difference in solute mass per unit volume}},$$

giving the dimensions $\mathscr{M}^{-1}\mathscr{L}^4\mathscr{T}^{-1}$ of Table 8.1.

Since Os^*/C_0 is dimensionless, the value of the function on the right side of (2) must also be independent of units, so that f must depend on dimensionless combinations of its arguments. Since $\mathscr{M}$, $\mathscr{L}$, and $\mathscr{T}$ appear as powers, only powers of the arguments will permit their cancellation. Thus f must involve its variables in combinations of the form

$$\delta^{a_1}L^{a_2}c^{a_3}D^{a_4}C_0^{a_5}P^{a_6}N_0^{a_7}a^{a_8}, \tag{3a}$$

where the a_i are constants. This combination has dimensions

$$\mathscr{L}^{a_1}\mathscr{L}^{a_2}\mathscr{L}^{a_3}(\mathscr{L}^2\mathscr{T}^{-1})^{a_4}(\mathscr{M}\mathscr{L}^{-3})^{a_5}(\mathscr{M}^{-1}\mathscr{L}^4\mathscr{T}^{-1})^{a_6}(\mathscr{M}\mathscr{L}^{-2}\mathscr{T}^{-1})^{a_7}\mathscr{L}^{2a_8} \tag{3b}$$

or

$$\mathscr{L}^{a_1+a_2+a_3+2a_4-3a_5+4a_6-2a_7+2a_8}\mathscr{M}^{a_5-a_6+a_7}\mathscr{T}^{-a_4-a_6-a_7} \tag{3c}$$

and will be dimensionless if and only if

$$\begin{aligned} a_1 + a_2 + a_3 + 2a_4 - 3a_5 + 4a_6 - 2a_7 + 2a_8 &= 0, \\ a_5 - a_6 + a_7 &= 0, \\ -a_4 - a_6 - a_7 &= 0. \end{aligned} \tag{4a}$$

The solution to the above equations can be written (Exercise 1)

$$\begin{aligned} &a_6 = \alpha, \quad a_7 = \beta, \quad a_4 = -(\alpha+\beta), \quad a_5 = \alpha - \beta, \\ &a_2 = \gamma, \quad a_3 = \theta, \quad a_8 = \varepsilon, \quad a_1 = \alpha + \beta - \gamma - \theta - 2\varepsilon, \end{aligned} \tag{4b}$$

where α, β, γ, θ, and ε are arbitrary. With this, (3) becomes

$$\delta^{\alpha+\beta-\gamma-\theta-2\varepsilon}L^{-\gamma}c^{\theta}D^{-\alpha-\beta}C_0^{\alpha-\beta}P^{\alpha}N_0^{\beta}a^{\varepsilon}$$

or

$$\left(\frac{\delta P C_0}{D}\right)^{\alpha}\left(\frac{\delta N_0}{DC_0}\right)^{\beta}\left(\frac{L}{\delta}\right)^{\gamma}\left(\frac{c}{\delta}\right)^{\theta}\left(\frac{a}{\delta^2}\right)^{\varepsilon}. \tag{4c}$$

Thus, as discussed in Section 6.2, we see that the only dimensionless groupings of the eight parameters are

$$\pi_1 = \frac{\delta P C_0}{D}, \quad \pi_2 = \frac{\delta N_0}{DC_0}, \quad \pi_3 = \frac{L}{\delta}, \quad \pi_4 = \frac{c}{\delta}, \quad \pi_5 = \frac{a}{\delta^2}, \tag{5}$$

or some combination thereof. Consequently,

$$\frac{O_s^*}{C_0} = f(\pi_1, \pi_2, \pi_3, \pi_4, \pi_5).$$

From the assumption that the emergent osmolarity depends on the eight parameters of Table 8.1, we have deduced that the ratio of emergent to ambient osmolarity depends on the five dimensionless groups of (5). With further thought we can do better. It is possible to convince oneself that the permeability P must always occur in the combination cP, for with one-dimensional flow (or even just axisymmetric flow) only the permeability per unit length matters. Similarly, N and D must occur only in the combinations cN and aD. Moreover, the area a and the circumference c do not enter the problem except in the above combinations. (This last assertion is not easy to justify at the present stage, but let us accept it for the moment.)

With the definitions

$$P_1 = cP, \quad N_1 = cN_0, \quad D_1 = aD,$$

we have, for some function g,

$$\frac{O_s^*}{C_0} = g(\delta, L, D_1, C_0, P_1, N_1). \tag{6}$$

As above, we can therefore assert that the dimensionless emergent osmolarity must depend on dimensionless combinations of the form

$$\delta^{b_1}L^{b_2}D_1^{b_3}C_0^{b_4}P_1^{b_5}N_1^{b_6}. \tag{7}$$

This term has the dimensions

$$\mathscr{L}^{b_1+b_2+4b_3-3b_4+5b_5-b_6}\mathscr{M}^{b_4-b_5+b_6}\mathscr{T}^{-b_3-b_5-b_6},$$

which (Exercise 2) can be shown to be dimensionless if and only if, for arbitrary α, β, and γ,

$$b_1 = 2\alpha + 2\beta - \gamma, \quad b_2 = \gamma, \quad b_3 = -\alpha - \beta, \quad b_4 = \alpha - \beta, \quad b_5 = \alpha, \quad b_6 = \beta. \tag{8}$$

With this, (7) becomes

$$\left(\frac{\delta^2 C_0 P_1}{D_1}\right)^\alpha \left(\frac{\delta^2 N_1}{D_1 C_0}\right)^\beta \left(\frac{L}{\delta}\right)^\gamma.$$

Returning to our original notation, we deduce that for some function f_1,

$$\frac{O_s^*}{C_0} = f_1(\pi_6, \pi_7, \pi_8), \tag{9}$$

where

$$\pi_6 = \frac{c\delta^2 C_0 P}{aD}, \qquad \pi_7 = \frac{c\delta^2 N_0}{aDC_0}, \qquad \pi_8 = \frac{L}{\delta}. \tag{10}$$

Instead of the *eight* dimensional parameters listed on the right side of (2) we need only consider the *three* combinations of (9)!

Other combinations than those in (9) can be used. For example, a function of π_6, π_7, and π_8 is also a (different) function of π_7/π_6, $1/\pi_7$, and π_8. Making the abbreviations

$$\nu \equiv \frac{\pi_7}{\pi_6} = \frac{N_0}{PC_0^2}, \qquad \eta \equiv \frac{1}{\pi_7} = \frac{aC_0 D}{cN_0 \delta^2}, \qquad \lambda \equiv \pi_8 = \frac{L}{\delta}, \tag{11}$$

we can therefore write, for some function f_2,

$$\frac{O_s^*}{C_0} = f_2(\nu, \eta, \lambda). \tag{12}$$

The same basic result, that Os^*/C_0 depends only on three dimensionless parameters, will be deduced below from the governing differential equations of the problem. There we shall see some theoretical consequences that can be drawn from this result. Now let us examine some possible consequences for experimenters.

POSSIBILITY OF A SCALE MODEL FOR STANDING GRADIENT FLOW

Our dimensional analysis reveals the possibility of making scale models for standing gradient flows. In contrast with airplane and ship models, here the model would be much larger than the true system, for the tubes in which standing gradient flow would actually take place are so tiny that it is only with great difficulty and ingenuity that measurements can be made.

Suppose that it was deemed worthwhile to construct a model secreting tube that was 100 times as long as the genuine tube. Using primes to denote the parameters for the model system, this means that $L' = 100L$. If we also make the region of solute pumping 100 times as long ($\delta' = 100\delta$), then $\pi_8' \equiv L'/\delta' = L/\delta = \pi_8$, so that one of the three parameters governing the original system remains the same in the model system. If we make the area to circumference ratio 100 times as large ($a'/c' = 100a/c$) and retain the same values of the remaining dimensionless parameters ($C_0' = C_0$, $P' = P$, $N_0' = N_0$, $D' = D$), then the other two governing dimensionless parameters in (9) retain their values ($\pi_6 = \pi_6'$, $\pi_7 = \pi_7'$). *If* (6) *is valid*, then (9) is valid and the ratio of emergent to ambient osmolarity will be the same in the two systems ($Os^{*\prime}/C_0' = Os^*/C_0$). Measurements in the model system will yield information about behavior of the original system, but only if we have not omitted any important parameters. When dealing with living organisms, it is very difficult to be sure that all essential factors are included in a model.

Further progress requires formulation of a definite mathematical problem. To this we now turn.

EXERCISES

1. Use the theory of linear equations to verify that the solution to (4a) contains five arbitrary parameters. (Students often feel that systems of linear equations containing fewer unknowns than equations are "exceptional," but dimensional analysis is an area where such equations occur all the time.)
2. Verify that (7) is dimensionless if and only if (8) holds.

8.2 A Mathematical Model and Its Dimensional Analysis

In this section we formulate a mathematical model for standing gradient flow and then subject it to a dimensional analysis. The required equations could be obtained by specializing general results obtained elsewhere, but we shall proceed essentially from first principles, if only in order to make the present discussion nearly self-contained. There are two governing equations, one for conservation of fluid mass and one for conservation of solute mass.

CONSERVATION OF FLUID MASS

We can assume that water has a uniform density so that it will suffice to require conservation of fluid volume. Consider a test cylinder extending from x^* to $x^* + \Delta x^*$. Here Δx^* stands for a positive number, not necessarily a small one. Let $v^*(x^*)$ denote the fluid velocity at x^*, taken positive in the direction x^* increasing. Then the rate at which fluid *enters* the test cylinder through its ends is

$$a[v^*(x^*) - v^*(x^* + \Delta x^*)], \tag{1}$$

since a is the cross-section area of the cylinder. By the assumption of proportionality between rate of water flow through the lateral walls and concentration difference, the rate at which water *enters* the test cylinder through its lateral walls is

$$Pc \int_{x^*}^{x^*+\Delta x^*} [C^*(s) - C_0]\, ds. \tag{2}$$

[As a check on signs, if $C^* > C_0$, so solute concentration is higher inside the tube, then the term (2) is positive, corresponding correctly to the fact that water will *enter* the tube to dilute the relatively high solute concentration there.]

Since no water is created inside the test cylinder, the terms in (1) and (2) must add to zero. Thus, using the integral mean value theorem on the second term,

$$a[v^*(x^*) - v^*(x^* + \Delta x^*)] + Pc[C^*(x^* + \theta\, \Delta x^*) - C_0]\, \Delta x^* = 0,$$

where $0 < \theta < 1$. Dividing by Δx^* and taking the limit as $\Delta x^* \to 0$, we obtain the first equation:

$$\frac{dv^*}{dx^*} = Pca^{-1}(C^* - C_0). \tag{3}$$

CONSERVATION OF SOLUTE MASS

Let $N^*(x^*)$ denote the rate at which solute is actively transported inward across a unit area of lateral boundary located at x^*.† As above, let $F^*(x^*)$ be the amount of solute per unit time that crosses a unit area which is located at x^*, and which is oriented normally to the generators of the cylinder. Recall that F^* is positive if solute flows in the direction x^* increasing. Setting to zero the net rate at which solute flows into the test cylinder, we obtain

$$c \int_{x^*}^{x^*+\Delta x^*} N^*(s)\, ds + a[F(x^*) - F(x^* + \Delta x^*)] = 0. \tag{4}$$

† For the moment we consider the general case. Later we return to the special assumption that N^* is constant for $0 \leq x^* \leq \delta$ and zero elsewhere.

As in the previous equation, by using the integral mean value theorem, dividing by Δx^*, and then letting $\Delta x^* \to 0$, we obtain

$$cN^*(x^*) - a\frac{dF^*}{dx^*} = 0. \tag{5}$$

The solute flux per unit area F^* is the sum of two parts, (a) flux due to convection F^*_{conv} and (b) flux due to diffusion F^*_{diff}. First, let us ascertain the convective contribution, that due to the bodily transport of the solute by fluid motion. We observe that if the fluid at x^* has speed $v^*(x^*)$ then, per unit time, the volume of fluid which crosses a unit area at x^* (in the direction of increasing x^*) is $v^*(x^*)$. (Here the area is assumed to be oriented normally to the generators of the cylinder.) The mass of solute that crosses this unit area in a unit time is therefore $C^*(x^*)v^*(x^*)$, since C^* is the mass of solute per unit fluid volume. Hence

$$F^*_{\text{conv}} = C^*v^*. \tag{6}$$

The diffusive flux is given by

$$F^*_{\text{diff}} = -D\left(\frac{dC^*}{dx^*}\right), \tag{7}$$

where D, a constant, is the **solute diffusivity**. Equation (7) is sometimes called **Fick's law**; it states the solute flux is proportional to the spatial derivative in solute concentration, just as the Newton–Fourier Law of cooling says that heat flux (or temperature flux) is proportional to the spatial derivate of temperature. With the negative sign in (7) the constant D will be positive, since solute flows away from regions of relatively high concentration.

A justification of assumption (7) can be provided by an argument identical with that used to justify Equation (1.3.7). Or the analogy with heat flux can be accepted, in which case (7) is essentially the same as Equation (4.1.4). At any rate, combining (6) and (7) we have

$$F^* = v^*C^* - D\left(\frac{dC^*}{dx^*}\right). \tag{8}$$

BOUNDARY CONDITIONS

The differential equations necessary for our problem are supplied by (3), (5), and (8). It remains to state boundary conditions. These follow.

No fluid is to flow across the closed end of the channel, so that

$$\text{at } x^* = 0, \qquad v^* = 0. \tag{9}$$

No solute is to flow across the closed end of the channel so that $F^* = 0$ at $x^* = 0$. Using (8) and (9), we can write this condition as

$$\text{at } x^* = 0, \qquad \frac{dC^*}{dx^*} = 0. \tag{10}$$

As stated earlier, we assume that the solute concentration at the open end of the channel equals C_0, the assumed concentration of bathing fluid. Thus

$$\text{at } x^* = L, \qquad C^* = C_0. \tag{11}$$

To determine as simply as possible the effect of a solute "pumping" confined to the closed end of the tube, we take N^* to have the constant value N_0 near the closed end of the channel and to be zero elsewhere:

$$N^* = N_0 \quad \text{for } 0 \le x^* \le \delta; \qquad N^* = 0 \quad \text{for } \delta < x^* \le L. \tag{12}$$

Since N^* has been assumed to be discontinuous at $x^* = \delta$, our problem has now been effectively split up into two problems, one for $0 < x^* < \delta$, the other for $\delta < x^* < L$. We must assume further conditions to "link" these problems together appropriately. We shall derive the conditions from naïve physical considerations. Other derivations are requested in the exercises.

No solute is being created at $x^* = \delta$; therefore the solute that enters from the left must equal the solute leaving from the right. In symbols,

$$F^*(\delta^-) = F^*(\delta^+), \qquad \text{where } F^*(\delta^\pm) \equiv \lim_{\varepsilon \downarrow 0} F^*(\delta \pm \varepsilon).\dagger \tag{13}$$

Similarly, no fluid is being created at $x^* = \delta$, so the fluid volume entering from the left must equal the fluid volume leaving from the right. Thus

$$v^*(\delta^-) = v^*(\delta^+). \tag{14}$$

We adopt as an additional physical hypothesis, because its denial is contrary to our intuition concerning molecular behavior, that the solute concentration itself is a continuous function of x^* at $x^* = \delta$. If C^* and F^* are continuous at δ, then dC^*/dx^* must be continuous at $x^* = \delta$, or

$$\frac{dC^*}{dx^*}(\delta^-) = \frac{dC^*}{dx^*}(\delta^+). \tag{15}$$

With the form of active solute transport assumed in (12), the solute mass conservation equation (5) can easily be integrated once. We find

$$aF^* = cN_0 x^* + Q_1, \quad 0 \le x^* < \delta; \qquad aF^* = Q_2, \quad \delta < x^* \le L. \tag{16}$$

$F^*(0) = 0$ implies that $Q_1 = 0$. The continuity of F^* at $x^* = \delta$ implies that $Q_2 = cN_0\,\delta$. Thus

$$aF^* = cN_0 x^*, \quad 0 \le x^* < \delta; \qquad aF^* = cN_0\,\delta, \quad \delta < x^* \le L. \tag{17}$$

Since the lateral walls have been assumed impermeable to solute, (17) could almost have been written down at once. It merely states that solute flux increases linearly with x^* in the region of the channel where there is uniform active transport per unit length, and that the flux remains constant in the region of the channel where no further transport takes place.

† $\lim_{\varepsilon \downarrow 0} f(\varepsilon) \equiv \lim_{\substack{\varepsilon \to 0 \\ \varepsilon > 0}} f(\varepsilon)$.

Of central interest is the emergent osmolarity Os^*, which was defined in Equation (1.1). Summarizing our previous discussion, we see that Os^* is determined by the following mathematical problem:

$$\frac{dv^*}{dx^*} = Pca^{-1}(C^* - C_0) \quad \text{for } 0 < x^* < \delta \text{ and } \delta < x^* < L.$$

$$v^*C^* - D\frac{dC^*}{dx^*} = \begin{cases} a^{-1}cN_0 x^* & \text{for } 0 < x^* < \delta, \\ a^{-1}cN_0 \delta & \text{for } \delta < x^* < L. \end{cases} \tag{18}$$

$$v^*(0) = 0, \quad C^*(L) = C_0, \quad v^* \text{ and } C^* \text{ continuous at } x^* = \delta.$$

Determine

$$Os^* = \frac{F^*(L)}{v^*(L)} = \frac{cN_0\delta}{av^*(L)}.$$

[In obtaining the last line, we have used (17).]

In the above, we have eliminated the condition that dC^*/dx^* must vanish at $x^* = 0$ and the condition that dC^*/dx^* must be continuous at $x^* = \delta$. These are implied by the conditions already required. This conforms to our expectation that having integrated an equation once, fewer boundary conditions are necessary.

As a check on the above equations we can test for **dimensional homogeneity**: in a given equation each term must have the same dimension; otherwise, one could render an equation false by changing one of the fundamental units of measurement. (For a formal discussion of this matter, see Exercise 6.2.12.) To give an example, in the first equation the dimensions of the left side are those of velocity/length or, $\mathcal{T}^{-1}$. From Table 8.1 the dimensions of the right side are

$$(\mathcal{M}^{-1}\mathcal{L}^4\mathcal{T}^{-1})(\mathcal{L})(\mathcal{L}^2)^{-1}(\mathcal{M}\mathcal{L}^{-3}) = \mathcal{T}^{-1}$$

so that the equation is indeed dimensionally homogeneous.

The boxed equations (18) comprise the mathematical formulation of the standing gradient problem. We are faced with a system of two first order nonlinear differential equations for the two unknown functions $v^*(x^*)$ and $C^*(x^*)$. As expected, there are two boundary conditions. There is also a pair of continuity conditions, to "stitch together" the variables at the point $x^* = \delta$, where the specification of the problem changes.

INTRODUCTION OF DIMENSIONLESS VARIABLES

The next step is to switch to dimensionless variables. In doing so, we shall refer distance x^* to the length δ and concentration C^* to the reference concentration C_0. There is no obvious choice for a velocity scale. We can find one by trial and error, but the following procedure is a little more rational.

From (6) the convective flux density equals the product of velocity and concentration. Thus velocity must have the dimensions of flux density divided by concentration. From (17) the combination of parameters $cN_0\,\delta/a$ has the dimensions of flux density, for this is the total solute flux density at the open end of the tube. A typical concentration is C_0. Thus we take $cN_0\,\delta/(aC_0)$ as our reference velocity. To check, we consult Table 8.1, which shows that

$$\text{dimensions of } \frac{cN_0\,\delta}{aC_0} = \frac{(\mathscr{L})(\mathscr{M}\mathscr{L}^{-2}\mathscr{T}^{-1})\mathscr{L}}{\mathscr{L}^2(\mathscr{M}\mathscr{L}^{-3})} = \frac{\mathscr{L}}{\mathscr{T}}.$$

This is, of course, the correct dimension for a velocity.

We thus introduce the dimensionless variables x, v, and C, defined by

$$x = \frac{x^*}{\delta}, \qquad C = \frac{C^*}{C_0}, \qquad v = \frac{v^*}{cN_0\,\delta/(aC_0)}. \tag{19}$$

We obtain [Exercise 5(a)]

$$C - 1 = \nu v' \qquad \text{for } 0 < x < \lambda, \tag{20a}$$

$$vC - \eta C' = \begin{cases} x & \text{for } 0 < x < 1, \\ 1 & \text{for } 1 < x < \lambda, \end{cases} \tag{20b}$$

$$v(0) = 0, \qquad C(\lambda) = 1, \tag{20c, d}$$

$$C \text{ and } v \text{ continuous at } x = 1. \tag{20e}$$

The dimensionless emergent osmolarity Os is given by

$$Os \equiv \frac{Os^*}{C_0} = \frac{1}{v(\lambda)}, \tag{21}$$

Here

$$\nu = \frac{N_0}{PC_0^2}, \qquad \eta = \frac{aC_0\,D}{cN_0\,\delta^2}, \qquad \lambda = \frac{L}{\delta}, \tag{22}$$

just as in (1.11).

In the original *dimensional* problem (18), the *eight parameters* δ, L, c, D, C_0, P, N_0, a appear so that any feature of the "answer," e.g., the value of Os^*, can only depend on these eight parameters. In the *dimensionless* problem (20), only the *three parameters* ν, η, and λ appear. Any feature of the answer to the dimensionless problem, e.g., the value of $Os = Os^*/C_0 = 1/v(\lambda)$, can depend only on these three parameters.

Suppose that we had used L as a reference length rather than δ. What then? With $y = x^*/L$ replacing $x = x^*/\delta$, we would have obtained [Exercise 5(b)]

$$\bar{C} - 1 = \nu_1\,\bar{v}', \quad \text{for } 0 < \bar{y} < \lambda^{-1} \qquad \left(' = \frac{d}{dy}\right); \tag{23a}$$

$$\bar{v}\bar{C} - \eta_1 \bar{C}' = \begin{cases} \lambda y & \text{for } 0 < y < \lambda^{-1}, \\ 1 & \text{for } \lambda^{-1} < y < 1. \end{cases} \tag{23b}$$

$$\bar{v}(0) = 0, \qquad \bar{C}(1) = 1. \tag{23c, d}$$

$$\bar{C} \text{ and } \bar{v} \text{ continuous at } y = \lambda^{-1}, \qquad \overline{Os} = \frac{1}{v(1)}. \tag{23e}$$

Here a bar indicates that the dependent variables are a function of y, and

$$\eta_1 = \frac{aC_0 D}{cN_0 \delta L}, \qquad \nu_1 = \frac{N_0 \delta}{PC_0^2 L}.$$

Again, only three parameters are involved, this time η_1, ν_1, and λ. As expected, the new parameters are functions of the old, for $\eta_1 = \lambda^{-1}\eta$, $\nu_1 = \lambda^{-1}\nu$. This illustrates the fact that although the form of the equations and boundary conditions is changed when different nondimensionalizations are used (the dimensionless parameters are different and are located in different places), this inessential change does not alter the number of dimensionless parameters on which the answer depends.

COMPARISON OF PHYSICAL AND MATHEMATICAL APPROACHES TO DIMENSIONAL ANALYSIS

The same result, that only three dimensionless parameters govern the standing gradient problem, was obtained from a rather general physical formulation of the problem in Section 8.1, and a precise mathematical formulation here. (The same sort of result was obtained in Section 6.2, for the relatively trivial projectile problem.) Let us compare the two approaches.

It would appear that formulating the equations which govern a phenomenon has great value, even if the equations appear entirely intractable. Thus, murky depths of physical intuition must seemingly be plumbed if the parametric dependence of Os^* given in (1.2) is to be "intuited." Deriving the governing equations required a comparatively less difficult selection and simplification of known physical laws. And *once the equations are derived, the relevant dimensional parameters appear automatically*. Again, rather considerable physical intuition is required to be sure of our earlier statement [leading to (1.6)] that the parameters a, c, N_0, P, and D enter our standing gradient flow problem only in the combinations cP, cN_0, and aD. From the governing equations (18), one sees at once that N_0, P, and D are always coupled with a and c as in the just-mentioned combinations. The fact that a and ca appear in no other combinations is not obvious but becomes so if one replaces the speed v^* by the volume flux w^*, where $w^* = av^*$ (Exercise 4). Thus *introduction of dimensionless variables into a set of equations isolates the dimensionless parameters in a way that is computationally easy and does not require the somewhat deep justification of the pi theorem.*

Knowing dimensionless parameters on which the solution to a problem must depend, one can often obtain without further effort some significant aspects of the phenomenon under investigation. For example, we see from the parameter list in (22) that the shape of the cross section affects standing gradient flow only through the parameter a/c. It is natural to define an **effective radius** r by $a/c \equiv r/2$. If the cross section is circular, then the effective radius equals the actual radius. Since r and D occur only in the parameter η, we can be certain that doubling the effective radius (say by squeezing a channel of given circumference) has precisely the same effect as halving the diffusivity D. To give another example, the flow is unaffected if N_0, P, and r are all doubled.

EXERCISES

1. (a) Deduce (13) by integrating (5) from $x^* - \varepsilon$ to $x^* + \varepsilon$, using (12), and then letting ε approach zero.

(b) Using the same process, deduce (14) from (3). [As pointed out, e.g., by Friedman (1956, p. 176), for this formal process for finding jump conditions in the presence of discontinuities to be valid, it is necessary that the differential equations be written in a form that holds for an inhomogeneous medium, where all "parameters" are functions of the independent variables. For example, the heat equation must be written in the form (4.1.13) rather than (4.1.12). Another approach to the derivation of jump conditions, is given in Exercise 2.]

2. (a) Deduce from (4) that if N^* is bounded, then F^* is continuous.

(b) Similarly, deduce from the fact that (1) and (2) add to zero that v^* is continuous if C^* is bounded. (This use of the fundamental integral equations of conservation is the best way to derive jump conditions. For more complicated examples, see Section 12.3 and Exercises 14.4.6–9.)

3. Check all the equations of (18) for dimensional homogeneity.

4. Show that if $w^* = av^*$ is introduced into (18), then a and c appear only in the combinations cP, cN_0, and aD, thereby verifying the assumptions leading up to (1.6). Can you see why w^*, not v^*, is the appropriate dependent variable?

5. (a) Derive (20), the dimensionless version of (18).

(b) Use $y = x^*/L$, instead of $x = x^*/\delta$, to derive (23).

6. The purpose of this exercise is to guide the reader through certain calculations that led to the realization that a simple view of solute transport in membranes had to be abandoned. As pointed out at the beginning of Section 8.1, the standing gradient hypothesis is an attempt to remedy the situation.

For definiteness, let us focus our attention on frog skin. It is known that an isolated piece of skin, immersed in a salt solution of concentration C_b, can transport salt and water from outside to inside. Consider the steady state situation. Suppose that there is a uniform salt flux F_s through the skin, leading to a salt concentration C_0 on the outside of the skin and a concentration C_i on the inside. Fluid layers are assumed to exist on the outside and inside of the skin, of thicknesses δ_0 and δ_i, respectively. Let $C(x)$ denote the salt concentration in these layers, where x is a coordinate normal to the (parallel) bounding planes of the skin.

‡(a) It is asserted that C satisfies the following differential equation

$$D\frac{dC}{dx} - F_w C = -F_s. \tag{24}$$

Here the constant F_w is the water flux that arises due to the osmotic effect of salt gradients, and F_s is the salt flux. What is the constant D? What has been assumed in (24)? Why is it legitimate to assume that F_s and F_w are constants?

(b) Solve the differential equation in the inner layer; impose boundary conditions; and obtain the result

$$C_i = (C_b - F)\exp\left(\frac{-F_w\,\delta_i}{D}\right) + F, \qquad F \equiv \frac{F_s}{F_w}.$$

(c) Correspondingly obtain the following result for the outer layer:

$$C_0 = (C_b - F)\exp\left(\frac{F_w\,\delta_0}{D}\right) + F.$$

(d) Justify the assumption

$$F_w = L(C_i - C_0).$$

Use this assumption to obtain

$$F_w = \frac{F_s L(\delta_0 + \delta_i)}{D + C_b L(\delta_0 + \delta_i)}, \tag{25}$$

For the frog skin, the various parameters on the right side of (25) have been measured. The resulting prediction for F_w is one to two orders of magnitude below measured values. See "Unstirred Layers in Frog Skin" by J. Dainty and C. R. House, *J. Physiol. 182*, 66–78 (1966).

8.3 Obtaining the Final Scaled Dimensionless Form of the Mathematical Model

The next step in our analysis is to choose scaled dimensionless variables so that the various dimensionless parameters which appear will correctly give the relative orders of magnitudes of the various terms. This we shall do.

We then show that there exists a small parameter ν in the problem. It appears that straightforward application of the regular perturbation method can be made, but difficulty arises when this is attempted. To avoid the difficulty, the relation between a second dimensionless parameter, η, and ν must be established. When this is done (in the second half of the section), at last the problem is in final form.

SCALING

As was mentioned in the discussion of scaling in Section 6.3, we are faced with estimating the size of various terms that will only be known when the problem is solved. But we are not operating in a vacuum—we have computer calculations for some typical values of the various parameters (Figure 8.3).

Determining a scale for the concentration $C^*(x^*)$ is easy. The obvious guess that $C^*(x^*)$ is always about the magnitude of the concentration C_0 of bathing fluid is reinforced by the results displayed in Figure 8.3. We

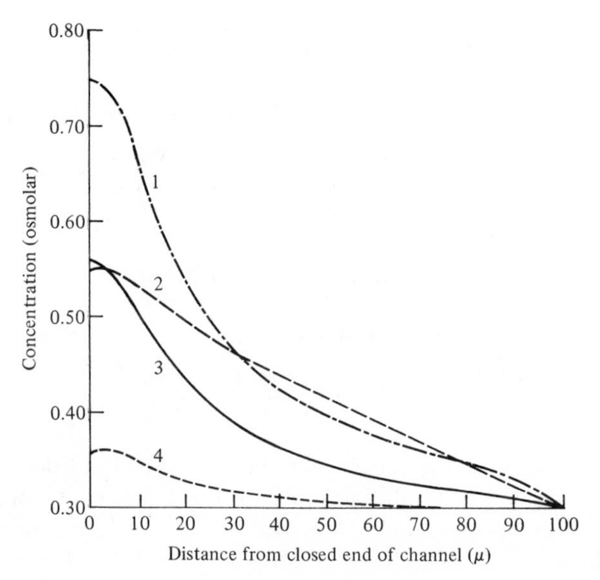

FIGURE 8.3. *Examples of concentration variation in a standing gradient flow system, from Figure* 3 *of DB (reproduced with permission). In the numerical calculations, whose results are given here, the parameters had the following values:* $L = 100\mu = 10^{-2}$ *cm*, $\delta = 10\mu$, $r = 0.05\mu$, $C_0 = 0.3$ mOsm/cm^3, $D = 10^{-5}$ cm^2/sec, $N_0 = 10^{-5}$ (*curve* 1), 5×10^{-6} (*curve* 3), *or* 10^{-6} (*curves* 2 *and* 4) mOsm/cm^2 sec, $P = 10^{-5}$ (*curves* 1, 3, *and* 4) *or* 10^{-6} (*curve* 2) cm^4/sec mOsm. *The calculated values of* Os^* *for curves* 1–4 *were* 0.342, 0.803, 0.318, *and* 0.304.

therefore take C_0 as the concentration scale and introduce a dimensionless concentration C by

$$C = \frac{C^*}{C_0}. \tag{1}$$

Direct estimate of the magnitude of v^* seems difficult. After some thought, however, an indirect approach suggests itself, based on the fact that it is concentration differences that drive the flow. We first note that the maximum velocity (which the scaling procedure requires us to estimate) is expected to be at the open end of the tube. This is because the cumulative amount of water drawn in by osmosis increases with x^*. We therefore focus our attention on $x^* = L$.

From (2.8), the flux density at $x^* = L$ satisfies

$$F^*(L) = v^*(L)C^*(L) - D\frac{dC^*}{dx^*}(L). \tag{2}$$

The two terms on the right side reflect the effects of convection and diffusion, respectively, in discharging solute from the tube. We expect that dC^*/dx^* will be negative at $x^* = L$ so that we can write

$$F^*(L) \geq v^*(L)C^*(L) \qquad \text{or} \qquad v^*(L) \leq \frac{F^*(L)}{C^*(L)}. \tag{3}$$

Using the boundary condition $C^*(L) = C_0$, and the expression in (2.17) for $F^*(L)$, we can write the last inequality as

$$v^*(L) \leq \frac{cN_0\,\delta}{aC_0}. \tag{4}$$

We anticipate that diffusion effects at their most intense will still not *greatly* outweigh convection in discharging solute.* The two sides of the inequalities in (3) should therefore not differ by an enormous amount. Consequently, the right side of (4) can serve as an estimate of the largest value of v^*, and this is precisely what a velocity scale should do.

In determining a scale for the independent variable x^*, we should focus our attention on the region where the dependent variables C^* and v^* are changing most rapidly. The scale should be a constant that estimates the distance over which a significant part of that change takes place. From Figure 8.3 it appears that the dependent variables change appreciably over distances considerably smaller than the channel length, and that δ is the appropriate length scale.

* Using (2.18), we see that $F^*(L)/[C_0 v^*(L)] = Os^*/C_0$. In the DB calculations this ratio of total to convective flux ranged from around unity in three cases to about $\frac{8}{3}$ in one case. This bears out our contention that diffusive flux will never greatly exceed convective flux.

To summarize, concentration, velocity, and length scales are estimated to be C_0, $cN_0\,\delta/aC_0$, and δ, respectively. Thus we should introduce the scaled dimensionless variables

$$C = \frac{C^*}{C_0}, \qquad v = \frac{v^*}{cN_0\,\delta/aC_0}, \qquad x = \frac{x^*}{\delta}. \tag{5}$$

But this choice of variables is one of those considered [in (2.19)] when our only concern was to put the problem in dimensionless form. The resulting equations are given in (2.20).

ESTIMATING THE SIZE OF THE DIMENSIONLESS PARAMETERS

The equations of (2.20) form a nonlinear system; there is no apparent way to solve them exactly. It is natural in such circumstances to look for a simplification. Because scaled variables were employed, the dimensionless parameters ν and η that appear in (2.20) should provide good estimates for the relative magnitudes of the various terms. If one of these parameters is relatively small, the corresponding term should be negligible to a first approximation. In such a case, perturbation theory might provide a method of treating the problem.

To estimate the parameters, we must turn to the physiologists for quantitative information. Using data compiled and analyzed by DB, we find the estimates for minimum, typical, and maximum values of the various dimensional and dimensionless parameters listed in Table 8.2. This table

TABLE 8.2. Estimated values of the various parameters.

Parameter	Units	Minimum value	Typical value	Maximum value
r	cm	10^{-6}	5×10^{-6}	10^{-4}
L	cm	4×10^{-4}	10^{-2}	2×10^{-2}
δ	cm	4×10^{-5}	10^{-3}	2×10^{-3}
D	cm^2/sec	10^{-6}	10^{-5}	5×10^{-5}
N_0	mOsm/cm^2 sec	10^{-10}	10^{-7}	10^{-5}
P	cm^4/sec mOsm	10^{-6}	2×10^{-5}	2×10^{-4}
C_0	mOsm/cm^3	—	3×10^{-1}	—
ν	Dimensionless	10^{-5}	5×10^{-2}	10^{2}
η	Dimensionless	4×10^{-3}	75	10^{10}
κ	Dimensionless*	4×10^{-3}	5	200

* κ will be defined below.

shows that ν is typically small and can be very small. (The maximum and minimum values of the dimensionless parameters are exaggerated because all the constituent dimensional parameters are unlikely to be of extreme size in a single system.)

AN UNSUCCESSFUL REGULAR PERTURBATION CALCULATION

Since we have employed scaled dimensionless variables, we can deduce from (2.20a) that $C(x) \approx 1$; i.e., $C^*(x^*) \approx C_0$. To obtain more information, we assume that the dependent variables can be expanded as a power series in the small parameter ν:

$$C(x) = C^{(0)}(x) + \nu C^{(1)}(x) + \nu^2 C^{(2)}(x) + \cdots, \tag{6a}$$

$$v(x) = v^{(0)}(x) + \nu v^{(1)}(x) + \nu^2 v^{(2)}(x) + \cdots. \tag{6b}$$

We now substitute into (2.20). From (2.20b), for example, we obtain

$$[v^{(0)} + \nu v^{(1)} + \cdots][C^{(0)} + \nu C^{(1)} + \cdots] - \eta[C^{(0)} + \nu C^{(1)} + \cdots] = \begin{pmatrix} x \\ 1 \end{pmatrix}.$$

(Here we introduce a two-element column vector, with the convention that the upper element is to be used for $0 \le x < 1$ and the lower for $1 < x \le \lambda$.) The next step is to equate to zero the coefficients of each power of ν. In doing so, we must decide on the role of η. A natural assumption is that η is independent of ν, in which case we find

$$\nu^0: C^{(0)} = 1, \qquad v^{(0)}C^{(0)} - \eta C^{(0)\prime} = \begin{pmatrix} x \\ 1 \end{pmatrix}. \tag{7a, b}$$

$$\nu: C^{(1)} = v^{(0)\prime}, \qquad v^{(1)}C^{(0)} + v^{(0)}C^{(1)} - \eta C^{(1)\prime} = 0. \tag{8a, b}$$

It is readily seen that the solutions to (7) and (8) which satisfy the boundary conditions (2.20c, d) are (Exercise 1)

$$C^{(0)} = 1, \quad v^{(0)} = \begin{pmatrix} x \\ 1 \end{pmatrix}, \quad C^{(1)} = \begin{pmatrix} 1 \\ 0 \end{pmatrix}, \quad v^{(1)} = -\begin{pmatrix} x \\ 0 \end{pmatrix}. \tag{9}$$

The functions $v^{(1)}$ and $C^{(1)}$ are not continuous at $x = 1$, which violates (2.20e). There appear to be no solutions of the type sought.

RELATION BETWEEN PARAMETERS

The difficulty arises from inadequate handling of the *two parameters* ν and η that appear in the differential equations of the problem.* (It turns out that the parameter λ in the boundary conditions does not play a crucial role.) Our basic simplification of (2.20a, b) for small ν was to omit the term $\nu v'$ in (2.20a). We retained all the terms in (2.20b). But we *may* set $\eta = \nu\eta_1$, where, from the definitions of η and ν in (2.22), $\eta_1 = \nu/\eta = cN_0^2\, d^2/aPC_0^3\, D$.

* In Segel's work, the assumption that η is independent of ν was rejected on the grounds that a discontinuous solution is unacceptable. But the discontinuity can be "fixed up" by the methods of singular perturbation theory (Exercise 9.2.9). This type of solution can be ruled out, however, on the grounds that no trace of such behavior is found in the numerical results provided by DB. That is, although a solution with a rapidly varying part near $x = 1$ is certainly possible in principle, it does not appear to be appropriate in the parameter domain of biological interest.

If it is assumed that η_1 remains fixed as ν approaches zero, then the term proportional to C' would be omitted in the basic simplification of (2.20b). Still another possibility is to write

$$\eta = \frac{\eta_2}{\nu}, \qquad \text{where} \qquad \eta_2 = \nu\eta = \frac{aD}{PC_0 cd^2}. \tag{10}$$

Suppose that η_2 remains fixed as ν goes to zero. Making the substitution $\eta = \eta_2/\nu$, multiplying by ν, and then letting $\nu \to 0$, we see that the leading term in (2.20b) is now $C' = 0$. It has become clear that the presence of two parameters allows contemplation of a number of different limits. Which one is appropriate to our problem?

From (9) we know that

$$\lim_{\substack{\nu\to 0\\ \eta,\,\lambda\text{ fixed}}} C(x; \nu, \eta, \lambda)$$

has the acceptable value unity. We run into trouble, however, in looking at the next approximation, since

$$\lim_{\substack{\nu\to 0\\ \eta,\,\lambda\text{ fixed}}} \nu^{-1}[C(x; \nu, \eta, \lambda) - 1] = C^{(1)} = \begin{pmatrix}0\\1\end{pmatrix} \tag{11}$$

is not continuous. It turns out, as we shall see, that the appropriate limit keeps fixed the quantity η_2 defined in (10). For

$$\lim_{\substack{\nu\to 0\\ \eta_2,\,\lambda\text{ fixed}}} C(x; \nu, \eta, \lambda) = 1, \quad \text{and} \quad \lim_{\substack{\nu\to 0\\ \eta_2,\,\lambda\text{ fixed}}} \nu^{-1}[C(x; \nu, \eta, \lambda) - 1] \tag{12}$$

is continuous.

Before proceeding, it is worth noting that the limit used in (12) can be written more fully as

$$\lim_{\substack{\nu\to 0,\ \eta\to\infty\\ \eta\nu\equiv\eta_2\\ x,\,\eta_2,\,\lambda\text{ fixed}}} C(x; \nu, \eta, \lambda). \tag{13}$$

An example of a function of η, λ, and ν that does not permit the limit (11) but does permit the limit (13) is $\lambda e^{\nu} \sin(1/\eta\nu)$. For

$$\lim_{\substack{\nu\to 0\\ \eta,\,\lambda\text{ fixed}}} \lambda e^{\nu} \sin\left(\frac{1}{\eta\nu}\right)$$

does not exist but

$$\lim_{\substack{\nu\to 0\\ \lambda,\,\eta\nu\equiv\eta_2\text{ fixed}}} \lambda e^{\nu} \sin\left(\frac{1}{\eta\nu}\right) = \lambda \sin\left(\frac{1}{\eta_2}\right).$$

There is no automatic procedure that will lead inevitably to success in problems, such as the present one, which involve several parameters. *Try various possibilities* is one bit of advice. Many a confident invitation to the reader to "consider the following dimensionless variables" leaves unmentioned the trial and error leading to the correct choice. Somewhat better advice is: *Use your physical intuition.* Because one of the authors provided the analysis of the present problem it can be stated that the correct approach here was originally arrived at (after a time) by a combination of trial and error and the following physical reasoning.

When v is small, we see from the volume conservation equation (2.20a) that the dimensionless solute concentration C is approximately unity. In dimensional terms, this means that the solute concentration C^* is approximately equal to the ambient solute concentration C_0. If $C = 1$, then of course $C' = 0$, but making the approximation $C' = 0$ and eliminating the term $\eta C'$ in (2.20b) is dangerous. One reason for this is that although C may remain nearly equal to unity, it may fluctuate rapidly around unity, so C' could be rather large. Even if C' is not too large, it is multiplied by a parameter which might very well be large (see Table 8.2) so that the term $\eta C'$ could be important.

On the other hand, to use the fact that $C \approx 1$ to justify the approximation $vC \approx v$ seems very sensible. The term vC represents the convection of solute. To assume that $vC \approx v$ is tantamount to replacing convection of solute at the actual spatially varying concentration $C(x)$ by convection of solute at the approximate (dimensionless) concentration unity. This should be appropriate either if the actual concentration $C^*(x^*)$ does not stray too far from C_0 or if solute convection is not too important.

FINAL FORMULATION

Just a little *directed* trial and error now leads to the final answer. If the substitution (10) is made, the lowest order approximation gives $C^{(0)} = 1$, while the next approximation gives an $O(v)$ correction $C^{(1)}$ to C and an $O(1)$ velocity $v^{(0)}$.* The quantities $C^{(1)}$ and $v^{(0)}$ satisfy a (linear) conservation of solute equation in which vC is replaced by $v^{(0)}$, i.e., in which fluid carries along solute of concentration unity.

One other consideration enters the final formulation of the problem, but it is a trivial one. Formulas look a little simpler if one writes the parameter η_2 of (10) in the form

$$\eta_2 = \frac{\lambda^2}{\kappa^2}. \tag{14}$$

Although the argument leading to it may have seemed circuitous, the procedure now is straightforward. Using (10) and (14), we replace the parameter η by means of

$$\eta = \frac{\lambda^2}{\nu\kappa^2} \quad \text{so} \quad \kappa = \left(\frac{cPC_0 L^2}{aD}\right)^{1/2}. \tag{15}$$

* It is encouraging at this point to note, using Table 8.2, that η_2, unlike η_1, is not typically large. It has a typical value of about 4.

The problem of (2.20) now becomes

$$C - 1 = \nu v', \qquad \nu\kappa^2 vC - \lambda^2 C' = \nu\kappa^2\binom{x}{1}. \tag{16a, b}$$

$$v(0) = 0, \quad C(\lambda) = 1, \qquad C \text{ and } v \text{ continuous at } x = 1. \tag{16c, d, e}$$

$$0 < \nu \ll 1, \qquad \kappa, \lambda = O(1).$$

Determine

$$Os = \frac{1}{v(\lambda)}.$$

This is the final scaled dimensionless form of the problem. It was quite a struggle to obtain the proper formulation. But it is a straightforward matter now, as we shall see in the next section, to solve the problem by the methods of regular perturbation theory.

We stated in Section 1.1 that the solution of a mathematical problem may not be the hardest job facing an applied mathematician. Our discussion here provides a good illustration of this point.

EXERCISES

1. Verify (7) and (8), and obtain the solutions of (9). Also, verify (16).

8.4 Solution and Interpretation

We are now ready to solve our problem by a regular perturbation. We shall find that the very first approximation to Os, the dimensionless quantity of primary interest, is adequate for our purposes. The analytic expressions obtained in our approximate solution are used to provide a deeper understanding of standing gradient flow. A noteworthy point here is the physical interpretation of the dimensionless parameters.

A FIRST APPROXIMATION TO THE SOLUTION

Upon substituting the series (3.6a) and (3.6b) into (3.16a, b), we obtain the following equations on collecting powers of ν (Exercise 1):

$$\nu^0: C^{(0)} = 1, \qquad \lambda^2 C^{(0)\prime} = 0; \tag{1a, b}$$

$$\nu^1: C^{(1)} = v^{(0)\prime}, \qquad \lambda^2 C^{(1)\prime} = \kappa^2\left[-\binom{x}{1} + v^{(0)}C^{(0)}\right]; \tag{2a, b}$$

$$\nu^2: C^{(2)} = v^{(1)\prime}, \qquad \lambda^2 C^{(2)\prime} = \kappa^2[v^{(0)}C^{(1)} + v^{(1)}C^{(0)}]. \tag{3a, b}$$

Equations (1a) and (1b) hold if and only if $C^{(0)} = 1$. Using this, on combining (2a) and (2b) we find that

$$\lambda^2 v^{(0)''} - \kappa^2 v^{(0)} = -\kappa^2 \binom{x}{1}, \tag{4}$$

The solution to this equation and the relevant boundary conditions can be found by standard methods (Exercise 2). Thus, for $0 \le x \le 1$,

$$v^{(0)} = x - K_1 \sinh \kappa\lambda^{-1}x, \qquad C^{(1)} = 1 - K_1\kappa\lambda^{-1} \cosh \kappa\lambda^{-1}x; \tag{5}$$

and for $1 \le x \le \lambda$,

$$v^{(0)} = 1 - K_2 \cosh \kappa(1 - \lambda^{-1}x), \qquad C^{(1)} = K_2\,\kappa\lambda^{-1} \sinh \kappa(1 - \lambda^{-1}x). \tag{6}$$

Here

$$K_1 = \frac{\lambda \cosh \kappa(1 - \lambda^{-1})}{\kappa \cosh \kappa}, \qquad K_2 = \frac{\lambda \sinh \kappa\lambda^{-1}}{\kappa \cosh \kappa}. \tag{7}$$

The first approximation to the emergent osmolarity is

$$Os = \frac{Os^*}{C_0} = [v(\lambda)]^{-1} \approx [v^{(0)}(\lambda)]^{-1} = (1 - K_2)^{-1}. \tag{8}$$

Since λ typically has the value 10, $\kappa\lambda^{-1}$ will often not exceed unity. In this case we can make the approximation $\sinh(\kappa\lambda^{-1}) \approx \kappa\lambda^{-1}$, which enables us to write the strikingly simple formula

$$Os \approx \frac{\cosh \kappa}{\cosh \kappa - 1}. \tag{9}$$

COMPARISON WITH NUMERICAL CALCULATIONS

Let us compare our approximate results with numerical calculations done for a number of particular cases by DB. Figure 8.4 gives a plot of Os as determined by (9). Almost all of DB's numerical results, displayed by them in five graphs, fall on this single curve. To show the extent of agreement and disagreement we have also plotted a number of points from their Figure 7, a graph of emergent osmolarity as a function of permeability in which $L = 100\ \mu\text{m}$, $\delta = 10\ \mu\text{m}$, $r = 0.05\ \mu\text{m}$, $C_0 = 0.3\ \text{mOsm/cm}^3$, $D = 10^{-5}\ \text{cm}^2/\text{sec}$, and $N_0 = 10^{-5}$(◐), 10^{-6}(○), 10^{-7}(▽), 10^{-8}(+), or 10^{-9}(□) mOsm/cm² sec. We have used a semilog plot of Os versus $5\kappa^2/6$ to facilitate the comparison. It will be seen that agreement ranges from good to excellent except for the largest values of N_0. For such values the parameter v is no longer small enough to permit neglect of the higher order terms in (6). The $O(v^2)$ terms in the series expansions are given in Exercise 3. As will be discussed further, these terms improve the agreement at large v.

A clearer presentation of our approximate results is obtained by plotting Os versus κ as in Figure 8.5. We use (9) and do not employ a logarithmic

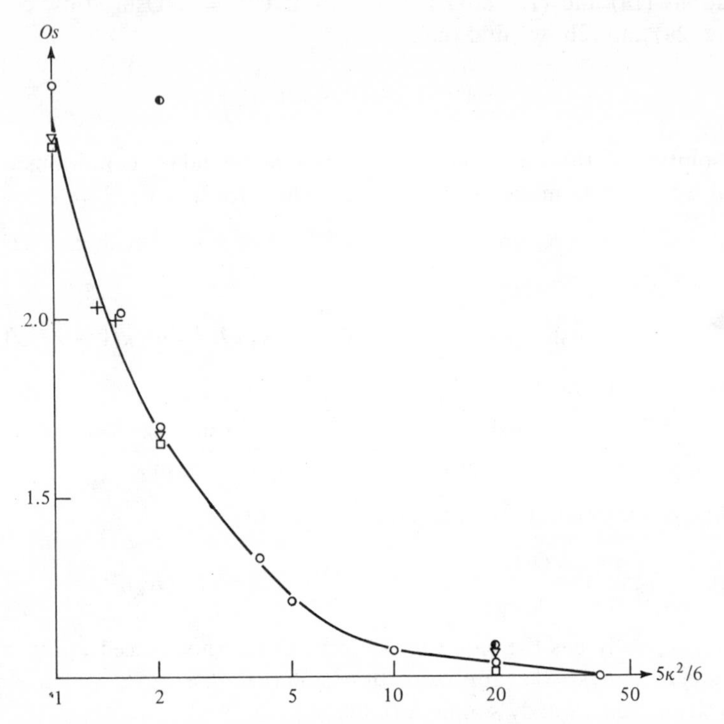

FIGURE 8.4. *A plot of Os, the ratio of the concentration of fluid emerging from the open end of the channel (Os^*) to the ambient concentration (C_0). According to* (9), *the approximate theoretical expression used, Os depends only on the parameter κ that was defined in* (3.15). *The various points were obtained numerically from the "exact" solution by DB, using various parameter values as detailed in the text.*

scale on the abscissa as we did in Figure 8.4. To see the effect of solute transport site, we also specialize (8) to plot Os versus κ when $\lambda = 2$, i.e., when solute is actively transported over the first half of the channel.

We shall discuss our results further in a moment. First, let us pause to make explicit what we have achieved from the point of view of perturbation theory. Our problem was a second-order nonlinear inhomogeneous system of ordinary differential equations. There was in addition one boundary condition at each end of the interval. The inhomogeneous terms were different in each of two different subintervals. Continuity conditions linked the solutions appropriate for each of the subintervals. Thus the problem was not a particularly simple one.

Diamond and Bossert treated the problem numerically for a number of cases. Their analysis involved setting an extra boundary condition at $x = 0$, "shooting" the solution toward $x = \lambda$, and adjusting the extra boundary

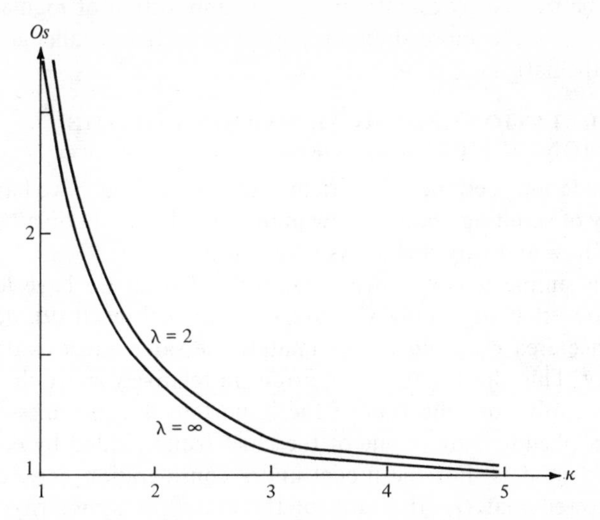

FIGURE 8.5. *Plots of the dimensionless emergent concentration Os versus κ. (The parameter κ measures the relative importance of convection and diffusion; it is independent of the rate at which the solute is actively pumped into the tube.) Bottom curve: theoretical prediction when $\lambda \to \infty$, where λ is the ratio of the channel length L to the length δ of the region of solute pumping. The results for $\lambda = 10$ are indistinguishable from those plotted here. Top curve: theoretical prediction for $\lambda = 2$.*

condition at $x = 0$ again and again (according to a certain scheme) until the solution behaved properly at $x = \lambda$.

By a regular perturbation method we were able to fit virtually all the results of DB with the simple expression (9). The straightforward calculations necessary to derive this expression could be repeated in an hour or two. But the reader should not lose sight of what was required to set up these calculations. In particular, he should recall the intuitive thinking (which was aided by perusal of some numerical solutions) that went into the process of scaling. By this process we were assured that the size of the dimensional parameters represented the relative magnitudes of the various terms. The reader should also remember that experimental information (borrowed from the compilation of DB) enabled us to assert that one of these parameters was small for many problems of physiological interest and that the other was rather large. Finally, we should keep in mind that physical intuition coupled with some trial and error was required to exploit the information on parameter size by means of the perturbation method.

Perhaps it appears discouraging at first that straightforward perturbation calculations must be supplemented by considerable additional reasoning in some problems. In the authors' view, although there are many problems that can in fact be treated by essentially routine application of regular perturbation theory, it is the more difficult applications that challenge a working applied mathematician.

INTERPRETATION: PHYSICAL MEANING OF THE DIMENSIONLESS PARAMETERS

To conclude this section, let us attempt to use the results we have obtained in our study of standing gradient flow in tiny tubules to obtain a better understanding of how and why such flows take place.

To give a simple but worthwhile example of what we have learned, our analysis shows that since only κ appears in the final result (9), doubling the circumference/area ratio has approximately the same effect as doubling the permeability. This effect may be important in relatively short channels.

More generally, identification of the important dimensionless parameters governing a phenomenon is one of the chief fruits yielded by consideration of simplified models. For the model under consideration here, dimensional analysis showed that Os, the ratio of the emergent osmolarity Os^* to the ambient osmolarity C_0, is determined by three dimensionless parameters, whatever the value of the eight dimensional parameters governing the problem. Approximate solution of the governing equations by means of perturbation theory led to the realization that Os is often only a function of the single parameter κ. In determining the meaning of this parameter, it is more convenient to consider $\frac{1}{2}\kappa^2$, which can be written

$$\frac{1}{2}\kappa^2 = \frac{[c(\bar{C} - C_0)PL]C_0}{aD(\bar{C} - C_0)/(2L)}. \tag{10}$$

In the above expression $\bar{C}$ is a typical concentration in the channel. (For definiteness, $\bar{C}$ can be regarded as the concentration halfway along the channel.) The square-bracketed factor in the numerator is thus an estimate of the total water flow per unit time into the channel across its boundaries. This of course equals the total volume of water that flows out of the open end of the channel per unit time. The entire numerator is therefore an estimate of aF^*_{conv}, the convective flux of solute out of the channel. The denominator is an estimate of aF^*_{diff}, the diffusive flux of solute out of the channel, since $(\bar{C} - C_0)/(L/2)$ should approximate the concentration gradient at the end of the channel.* Thus

$$\frac{1}{2}\kappa^2 \approx \frac{F^*_{\text{conv}}}{F^*_{\text{diff}}}. \tag{11}$$

To be compared with (11) is

$$Os \equiv \frac{Os^*}{C_0} = \frac{F^*_{\text{conv}} + F^*_{\text{diff}}}{F^*_{\text{conv}}} = 1 + \frac{F^*_{\text{diff}}}{F^*_{\text{conv}}}. \tag{12}$$

* Our estimate of the concentration gradient is obtained by dividing the concentration difference between mid-channel and the open end by the half-length of the channel.

This is an exact result, which can be obtained from the defining relationship for Os^* in (2.18). Combining (11) and (12), we obtain

$$Os \approx 1 + 2\kappa^{-2}. \tag{13}$$

Thus, when the solution (9) is expanded for small values of κ, we find that $Os \approx 2\kappa^{-2}$, in agreement with the corresponding expansion of (13).

We have seen that the parameter κ provides an estimate of the relative importance of convection and diffusion that is closely related to Os. When κ is small, diffusion is dominant so that the graph of concentration vs. distance very nearly represents the constant gradient, which is typical of diffusion problems. (Compare curve 2 of Figure 8.3.) It is no wonder that (13) is nearly exact in this case—for $C^*(L/2)$ is almost exactly the average concentration in the tube, and $[C^*(L/2) - C_0]/[L/2]$ is an excellent estimate for the concentration gradient.

When κ becomes large, then the interpretation (11) of $2\kappa^{-2}$ loses its quantitative accuracy, but still (13) provides a rough approximation. As we shall now show, the approximate interpretation (11) of Equation (10) for κ^2 can be exploited in order to understand a fundamental aspect of the phenomenon under investigation.

A striking feature of flow from the salt glands of birds and the kidney tubules of crocodiles is that the concentration of the secretion seems virtually unaffected by changes in the rate of active transport into the tubule. DB's numerical calculations showed that the standing gradient system is able to reproduce this feature. In present terms we are speaking of the fact that Os is independent of v unless v is large. How can this be? Formula (10) gives the answer. If $N_0\,\delta$ decreases, then the typical solute concentration $\bar{C}$ decreases, less water is drawn into the channel, and the importance of convection in sweeping out solute is lessened. According to (10), however, the decrease of $\bar{C}$ results in a decrease of the concentration gradient at the open end. This restores the balance between convection and diffusion, and thereby keeps the emergent osmolarity at its original value. It appears that only when the solute concentration in the fluid that issues from the active region is rather large do changes in this concentration affect the proportions of solute removed from the open end of the channel by convection and diffusion. Determination of the next term in the perturbation series (Exercise 3) gives the expected qualitative result, originally illustrated in DB's numerical work, that when v is large, (9) provides an underestimate of the emergent osmolarity

Note that *it is essentially the first term in the series that provides almost all the interesting information.* This occurs frequently.

To get an idea of what values of v signal the inappropriateness of the lowest order approximation (9) in physiological situations, we observe from Figure 8.5 that agreement with (9) is good for the point ◐ when $5\kappa^2/6 = 20$ but poor when $5\kappa^2/6 = 2$. In the former case $v = 5$ but in the latter $v = 50$. From Table

8.2, even values of ν as high as 5 would be unusual. We conclude that the lowest order approximation is expected to furnish an adequate description of standing gradient flows in all but highly exceptional physiological situations.

It is apparent from its definition (2.22), with N_0 in the numerator, that ν is a dimensionless measure of the importance of active transport. If desired, a more accurate interpretation of ν can be obtained by writing

$$\nu = \left[\frac{N_0 c\, \delta}{Pc\, \delta(C_2 - C_0)C(1)}\right]\left[\frac{C(1)}{C_0}\right]\left[\frac{C_2 - C_0}{C_0}\right], \tag{14}$$

where C_2 is a typical concentration toward the closed end of the tube. (Again it facilitates interpretation to introduce the same factor into the numerator and the denominator.) From (14) one obtains

$$\nu \approx \left[1 + \frac{F^*_{\text{diff}}}{F^*_{\text{conv}}}\right]\left[\frac{C(1)}{C_0}\right]\left[\frac{C_2 - C_0}{C_0}\right], \tag{15}$$

where F^*_{diff} and F^*_{conv} are evaluated at $x = 1$, the end of the active transport region.

The last of the three parameters governing Os is $\lambda \equiv L/\delta$, which obviously measures the degree of concentration of the actively transporting sites at the closed end of the tube. Also, λ does not enter the simplified formula (9), while Figure 8.6 shows that use of (9) makes very little change even when λ is as small as 2. (When λ is decreased, the pumping rate N_0 must be decreased if the total solute transported is to remain the same, but for the range of parameters of primary interest such a decrease in N_0 has no effect on Os.) This very weak dependence of Os on λ gives confidence in the model; it shows that the features of standing gradient flow will be found as long as active transport is confined "to a reasonable extent" toward the closed end of the tube. To give one more observation on the effect of λ, we note from Exercise 3 that the larger λ is, the higher ν must be before effects of high transport rate becomes evident. It is in accord with intuition that even high transport rates have little effect when the active site only occupies a very small fraction of the tube length.

FINAL REMARKS

In closing this chapter let us recall that the problem of standing gradient osmotic flow has served to illustrate a number of basic techniques in applied mathematics such as nondimensionalization, scaling, and perturbation theory. More important, the problem serves as a paradigm of how theory can be used to take a step forward in scientific understanding.

As we have seen, the standing gradient problem was originally formulated to make precise a suggested answer to a physiological riddle. The numerical calculations of DB showed that the suggestion was a feasible one: active solute pumping at the far end of a channel was demonstrated to be capable of doing the job required of it for reasonable values of the various parameters involved. The numerical calculations also helped in ascertaining the typical values of various terms, as was required in the present analytic treatment of the problem. The analytic treatment yielded a simple formula (9), which summed up the operation of the standing gradient model.

There is some value in being able to consult this formula rather than a number of tables or graphs for the "answer," given a particular set of parameters. A more important result of the analysis is the understanding of the flow process, for the model is doubtless oversimplified so that precise predictions cannot be trusted. The understanding resides (i) in the isolation, from eight dimensional parameters, of the dimensionless parameter κ as the governing parameter of the flow, (ii) in the physical interpretation of κ, (iii) in the prediction that the independence of emergent osmolarity Os^* on the solute pumping rate N_0 is to be expected in physiological systems, and (iv) in the analysis of why this prediction emerges. To see the virtue of theoretical work, compare what you now know of standing gradient flow to the ingenious but unsupported idea that started the whole train of reasoning: "secreting tissues may work by active pumping deep in their ubiquitous infoldings."*

EXERCISES

1. Derive (1), (2), and (3).

2. Solve (4), subject to appropriate boundary conditions. Thereby obtain (5) and (6).

3. (a) Use (5) and (6) to show that $v^{(1)}$ satisfies the following equations:

$$0 < x < 1: \left(\frac{\lambda}{\kappa}\right)^2 v^{(1)''} - v^{(1)} = \frac{1}{2} K_1^2 \kappa \sinh 2\kappa\lambda^{-1}x$$

$$- K_1 \lambda \sinh \kappa\lambda^{-1}x - K_1 \kappa x \cosh \kappa\lambda^{-1}x + \lambda x.$$

$$1 < x \le \lambda: \left(\frac{\lambda}{\kappa}\right)^2 v^{(1)''} - v^{(1)} = -\frac{1}{2} K_2^2 \sinh 2\kappa(1 - \lambda^{-1}x)$$

$$+ K_2 \kappa \sinh \kappa(1 - \lambda^{-1}x).$$

* Refinements and extensions of the standing gradient model have been made and analyzed. See, for example, S. Weinbaum and J. R. Goldgraben, "On the Movement of Water and Solute in Extracellular Channels with Filtration, Osmosis, and Active Transport," *J. Fluid Mech. 53*, 481–512 (1972). Weinbaum and Goldgraben stress that $C^*(L)$ can take values other than C_0, depending on conditions at the exit. They draw interesting conclusions from considering a one-parameter family of solutions wherein $C^*(L) = \tilde{C}$, $\tilde{C}$ arbitrary. Another important part of their paper concerns effects of an imposed pressure gradient.

(b) Show that solutions are

$$0 \le x \le 1: v^{(1)} = A_1 \sinh \kappa\lambda^{-1}x + B_1 \cosh \kappa\lambda^{-1}x + \tfrac{1}{6}K_1^2\kappa \sinh 2\kappa\lambda^{-1}x - \tfrac{1}{4}K_1\kappa x \cosh \kappa\lambda^{-1}x - \tfrac{1}{4}K_1\kappa^2\lambda^{-1} \sinh \kappa\lambda^{-1}x - \lambda x, \quad C^{(2)} = v^{(1)\prime};$$

$$1 \le x \le \lambda: v^{(1)} = A_2 \sinh \kappa(1 - \lambda^{-1}x) + B_2 \sinh \kappa(1 - \lambda^{-1}x) - \tfrac{1}{2}K_2 \kappa^2\lambda^{-1}x \cosh \kappa(1 - \lambda^{-1}x) - \tfrac{1}{6}K_2^2 \kappa \sinh 2\kappa(1 - \lambda^{-1}x), \quad C^{(2)} = v^{(1)\prime},$$

where the four constants A_1, B_1, A_2, and B_2 are to be determined by the requirements $v^{(1)}(0) = 0$, $v^{(1)\prime}(\lambda) = 0$, and $v^{(1)}$ and $v^{(1)\prime}$ continuous at $x = 1$.

(c) Simplify otherwise unwieldy expressions by neglecting terms of order unity compared to terms of order λ. Obtain

$$v^{(1)}(\lambda) \approx \frac{-5 \sinh 2\kappa - 4 \sinh \kappa - 6\kappa \cosh^2 \kappa}{12\lambda \cosh^3 \kappa}.$$

(d) Make a further simplification by expanding the hyperbolic functions in powers of κ and retain the first nonvanishing term. Obtain

$$v^{(1)}(\lambda) \approx -\frac{7\kappa^7}{120\lambda \cosh^3 \kappa}.$$

The correction is thus seen to decrease the dimensionless exit speed $v(\lambda)$ (at least when λ^{-1} and κ are not too large) and therefore to increase Os.

CHAPTER 9
Introduction to Singular Perturbation Theory

SINGULAR perturbation theory has grown out of attempts to deal with physical problems, the most important of which involves the boundary layer phenomenon in fluid mechanics. The classical instance of this phenomenon occurs in the flow past an object such as an airplane wing. As will be discussed with some care in Chapter 3 of II, the appropriately scaled equations in this case contain the effects of viscosity only in terms multiplied by an extremely small parameter. Yet if these terms are neglected, computations show that the air exerts no drag on the wing. This is D'Alembert's paradox (see Section 15.4). The paradox was resolved early in this century when Ludwig Prandtl showed experimentally and theoretically that however small the viscosity or "stickiness" of the fluid, viscous effects are important in a thin layer near the wing. Prandtl's discovery not only was a breakthrough in the understanding of aerodynamics, but it opened up the investigation, via singular perturbation theory, of how small causes (e.g., small fluid viscosity) can lead to large effects (appreciable drag).

It might seem that the practitioner of singular perturbation theory would find relatively few important problems to work on, but this is not at all the case. The abundance of such problems is at least partially explained by the fact that singular perturbation techniques seem to be appropriate for examining the behavior of solutions near their "worst" singularities. "Taking the bull by the horns" and examining functions near their worst* singularities is the best way to obtain information about qualitative behavior, and it is elucidation of this behavior that is very often the goal of a theoretician.† In this lies the point of a remark once made about a distinguished applied mathematician, "All he can do is singular perturbation problems. But of course he can turn all problems into singular perturbation problems!"

Our introduction to singular perturbation theory begins with a discussion of algebraic equations having a small parameter multiplying the term of highest degree. First, introductory concepts of singular perturbation theory are used to find the behavior of certain roots which are "missed" by regular perturbation theory. Then successive approximations procedures are applied to find better approximations to these roots, or to suggest appropriate series

* In the language of complex variable theory, the worst singularities are *essential*.

† See L. A. Segel, "The Importance of Asymptotic Analysis in Applied Mathematics," *Amer. Math. Monthly 73*, 7–14 (1966).

assumptions. After this we turn to an extensive treatment of a particular ordinary differential equation which serves as a *model* equation for singular perturbation phenomena. With it we illustrate a goodly amount of the analysis required in applying singular perturbation theory to simple boundary value problems.

We are using the word "model" here in a somewhat different sense than we have heretofore. Most of the time we use **descriptive models**, which are designed to caricature definite natural phenomena. Now we are speaking of a **conceptual model**, which is "constructed to elucidate delicate and difficult points of a theory. Not always is it easy (or even desirable) to draw a sharp dividing line between the two types, but the extremes on both sides are clearly recognizable."*

The model equation that we shall study is a linear second order equation with constant coefficients and a small parameter multiplying the highest derivative. We carefully examine the exact solution to this equation and its associated boundary conditions. This enables us to develop the main lines of a singular perturbation procedure that works for a large class of problems in which a small parameter multiplies the highest derivative in a differential equation. The variety of singular perturbation problems that one is likely to meet is so large as to preclude the hope of a comprehensive recipe. Certainly, none is offered here; indeed, none is available. Instead, stress is laid on the "spirit of singular perturbations" with the hope that the reader will be spurred to further study of this fascinating and valuable subject.

Opportunity for further study is offered in the next chapter, where singular perturbation theory is used to analyze enzyme-catalyzed chemical reactions. Then in Section 11.2 we employ the powerful "multiple-scale" approach to singular perturbation calculations on the now-familiar pendulum problem. All these illustrations concern problems that are governed by ordinary differential equations. In Chapter 3 of II we discuss the application of singular perturbation techniques to viscous flow at high Reynolds number. This is the boundary layer problem which, as mentioned above, provided the primary stimulus to the development of singular perturbation methods.

9.1 Roots of Polynomial Equations

A SIMPLE PROBLEM

A very simple problem which gives insight into singular perturbation theory is the quadratic equation

$$\varepsilon m^2 + 2m + 1 = 0, \qquad \text{where } 0 < \varepsilon \ll 1. \tag{1}$$

* See "Some Mathematical Models in Science" by M. Kac, *Science 166*, 695–99 (1969).

To obtain an approximate solution, we neglect the seemingly small term εm^2 and find

$$2m + 1 \approx 0, \qquad m \approx -\tfrac{1}{2}.$$

We can improve this "naïve" approximation by any of the three methods discussed in Chapter 7 (Exercise 1). For example, if we assume a series solution

$$m = -\tfrac{1}{2} + \varepsilon m_1 + \varepsilon^2 m_2 + \cdots, \tag{2}$$

we find that

$$m = -\tfrac{1}{2} - \tfrac{1}{8}\varepsilon - \tfrac{1}{16}\varepsilon^2 + \cdots. \tag{3}$$

We can thus obtain an arbitrarily accurate approximation to *one* solution of the quadratic (1). But according to the fundamental theorem of algebra, there are *two* such solutions. Here, then, is a perturbation problem in which a straightforward approach does not produce a completely satisfactory answer. This is the kind of problem in which the singular perturbation method can be used with advantage.

The singular nature of (1) is connected with the fact that the full equation is of second degree (and therefore has two roots), while the approximate equation $2m + 1 = 0$ has but a single root. By comparison, the equation $m^2 + \varepsilon m - 4 = 0$ (treated in Exercise 7.2.9) is second degree, and so is the approximate equation $m^2 - 4 = 0$; this is an example of a regular perturbation problem. *It is a characteristic feature of singular perturbations that the naïvely approximated problem has a different qualitative character from the original problem.*

In searching for an approximation to the second root $\bar{m}$ of the quadratic, we begin with the knowledge that neglect of the term $\varepsilon\bar{m}^2$ is not justified. We have already found the root to the first degree equation that results when this term is neglected.

If $\bar{m}$ were $O(1)$, i.e., if $\bar{m}$ approached a (finite) limit as $\varepsilon \to 0$, then the neglect of the term $\varepsilon\bar{m}^2$ presumably *would* be justified for small ε—because then this term would be small compared to the term 1. We are presumably dealing with a root that becomes very large when $\varepsilon \to 0$ (or, possibly, with a root that does not approach a limit when $\varepsilon \to 0$).

As $\varepsilon \to 0$, each of the terms $\varepsilon\bar{m}^2$, $2\bar{m}$, and 1 has a certain magnitude. If all three terms have the same magnitude, the task of obtaining $\bar{m}$ is not lightened by the fact that ε is small: all three terms in (1) are important and we are back to the original equation. Perhaps two of the terms are of the same magnitude and the third is relatively small. We have agreed that $\varepsilon\bar{m}^2$ must be retained, so the possibilities are $\varepsilon\bar{m}^2$ and 1 of the same magnitude, $2\bar{m}$ relatively small; $\varepsilon\bar{m}^2$ and $2\bar{m}$ of the same magnitude, 1 relatively small. If the first possibility obtains, then from (1) we see that $\varepsilon\bar{m}^2 + 1 \approx 0$ or $\bar{m} \approx i\varepsilon^{-1/2}$. In this case, however, $2\bar{m} \approx 2i\varepsilon^{-1/2}$ and the "relatively small" term is in fact of relatively large magnitude ($|2\overline{m}| \gg 1$). The second possibility is that $\varepsilon\overline{m}^2 + 2\overline{m} \approx 0$. If we discard the alternative $\overline{m} \approx 0$ for the moment, we find that $\overline{m} \approx 2\varepsilon^{-1}$. Here the term omitted is, indeed, relatively small compared to the terms retained:

$$1 \ll |\varepsilon\overline{m}^2| \approx 4\varepsilon^{-1}, \; 1 \ll |2\overline{m}| \approx 4\varepsilon^{-1}.$$

Thus our approximation appears to be consistent. The discarded alternative should be ignored, for if $\overline{m} \approx 0$, then the neglected term is *not* small compared to the terms retained, which is an inconsistency.

To improve on our first estimate of $\overline{m}$, let us employ the successive approximations procedure. As discussed in Section 7.2, the idea is to write the equation in the form $f(m) = g(m)$, where the initial approximation $m^{(0)}$ is provided by the solution to $f(m^{(0)}) = 0$ and higher approximations by the solutions to $f(m^{(i)}) = g(m^{(i-1)})$, $i = 1, 2, \ldots$.

It might appear that the way to do this is to choose $f(m) \equiv \varepsilon m^2 + 2m$, but then setting $f = 0$ yields the spurious first approximation $m^{(0)} \approx 0$ as well as the true first approximation $m^{(0)} \approx -2\varepsilon^{-1}$. To avoid this, we divide (1) by m. Thus a scheme to obtain successive approximations $m^{(i)}$ to $\overline{m}$ is given by $m^{(0)} = -2\varepsilon^{-1}$ and

$$\varepsilon m^{(i)} + 2 = -\frac{1}{m^{(i-1)}}$$

or

$$m^{(i)} = -\frac{2}{\varepsilon} - \frac{1}{\varepsilon m^{(i-1)}}, \qquad i = 1, 2, \ldots. \tag{4}$$

With this

$$m^{(1)} = -\frac{2}{\varepsilon} - \frac{1}{\varepsilon(-2\varepsilon^{-1})} = -\frac{2}{\varepsilon} + \frac{1}{2}$$

and

$$m^{(2)} = -\frac{2}{\varepsilon} + \frac{1}{2 - \frac{1}{2}\varepsilon}.$$

Note that if we expand the second term in powers of ε, we obtain

$$\overline{m} \approx m^{(2)} = -2\varepsilon^{-1} + \tfrac{1}{2} + \tfrac{1}{8}\varepsilon + \cdots.$$

Inspection of (4) shows that a similar series can be obtained for $m^{(3)}$, $m^{(4)}$, etc.

Here, as in other problems, it is probably best to abandon the successive approximations method once the form of the solution becomes clear. We have seen from the above calculations that the root $\overline{m}$ can be written in the form

$$\overline{m} = -2\varepsilon^{-1} + m_0 + m_1\varepsilon + m_2\varepsilon^2 + \cdots. \tag{5}$$

The coefficients m_i can be obtained with less labor by the series method, starting with the assumption (5), than by successive approximation [Exercise 2(a)]. Alternatively, one can begin with the assumption that εm can be expanded in powers of ε and use the parametric differentiation procedure [Exercise 2(b)].

Let us summarize. From (3) and (5) we see that the quadratic equation $\varepsilon m^2 + 2m + 1 = 0$ has the roots

$$m = -\tfrac{1}{2} + \cdots, \qquad \overline{m} = -2\varepsilon^{-1} + \cdots. \tag{6}$$

An arbitrarily accurate expression for the first root can be obtained by the methods of regular perturbation theory, since that root has a power series expansion. The second root becomes infinite as $\varepsilon \to 0$. The leading term in its approximate expression can be determined by finding and equating to zero a pair of terms with matching and relatively large magnitudes. One of these must be the term εm^2 whose neglect, on the "naïve" grounds that it has a small parameter as a factor, led to an approximation only for the $O(1)$ root $m = -\frac{1}{2} + \cdots$.

Our first example (1) is readily solved by the quadratic formula. The expressions for the two roots may be expanded to provide a check on our calculations (Exercise 3). It should already be apparent, however, that the approach just illustrated can be used on any algebraic equation with a small parameter multiplying the term of highest power.

A MORE COMPLICATED PROBLEM

Consider the quartic equation

$$\varepsilon x^4 + \varepsilon x^3 - x^2 + 2x - 1 = 0, \qquad 0 < \varepsilon \ll 1. \tag{7}$$

Roots that approach a nonzero limit as $\varepsilon \to 0$ can be approximated by neglecting the terms εx^4 and εx^3. This gives the quadratic $x^2 - 2x + 1 = 0$ with the double root of unity.

Example 1. Find higher approximations to the roots of (7) that are approximately equal to unity.

Partial Solution. We assume

$$x = 1 + x_1\varepsilon + O(\varepsilon^2).$$

Upon substitution into (7) we obtain

$$\varepsilon[2 + O(\varepsilon)] - x_1^2\varepsilon^2 + O(\varepsilon^3) = 0.$$

Collecting terms of $O(\varepsilon)$, we obtain $2 = 0$, a contradiction. The form we assumed for the solution must be incorrect. To find the correct form, we set up a successive approximations scheme. Putting the dominant portion of (7) on the left, we consider

$$(x - 1)^2 = \varepsilon x^4 + \varepsilon x^3.$$

This leads to the scheme

$$x_0 = 1; \; x_n = \sqrt{\varepsilon x_{n-1}^4 + \varepsilon x_{n-1}^3}, \qquad n = 1, 2, \ldots,$$

for one of the two roots and

$$x_0 = 1; \; x_n = -\sqrt{\varepsilon x_{n-1}^4 + \varepsilon x_{n-1}^3}, \qquad n = 1, 2, \ldots,$$

for the other. A couple of iterations indicate that the roots can be expanded in powers of $\varepsilon^{1/2}$. Let us introduce $a \equiv \varepsilon^{1/2}$ so that (7) becomes

$$a^2x^4 + a^2x^3 - (x-1)^2 = 0.$$

Then the problem reduces to a straightforward determination of the coefficients in the expansion

$$x = 1 + x_1 a + x_2 a^2 + \cdots.$$

The reader is asked to carry out the calculations [Exercise 4(a).]

To obtain first approximations for the two remaining roots, we again employ the very frequently useful **balancing procedure** which supposes a pair of terms to be of the same magnitude and the remaining terms to be negligible. We have already considered the case in which the first two terms of (7) are negligible. Thus the alternatives that remain for balancing terms are (a) $\varepsilon x^4 \approx -\varepsilon x^3$, (b) $\varepsilon x^4 \approx x^2$, (c) $\varepsilon x^4 \approx -2x$, (d) $\varepsilon x^4 \approx 1$, (e) $\varepsilon x^3 \approx x^2$, (f) $\varepsilon x^3 \approx -2x$, (g) $\varepsilon x^3 \approx 1$, with the remaining terms negligible.

The zero roots that arise from alternative (a) can be excluded, as above. The remaining root $x \approx -1$ also leads to an inconsistency, for the "large" terms εx^4 and $-\varepsilon x^3$ are in fact $O(\varepsilon)$.

Alternative (b) gives the nonzero roots

$$p^{(0)} = \frac{1}{\sqrt{\varepsilon}}, \qquad n^{(0)} = \frac{-1}{\sqrt{\varepsilon}}. \tag{8}$$

[In (8) we have denoted the positive and negative roots by p and n.] Using the approximations (8) to x, we see that εx^4 and x^2 are $O(\varepsilon^{-2})$, while the remaining terms behave as follows:

$$\varepsilon x^3 = O(\varepsilon^{-1/2}), \qquad 2x = O(\varepsilon^{-1/2}), \qquad 1 = O(1). \tag{9}$$

These are indeed relatively small so that (8) gives initial approximations to the two roots we are investigating. It can be shown, as expected, that alternatives (c)–(g) are all inconsistent [Exercise 4(b)].

To obtain more accurate expressions for the $O(\varepsilon^{-1/2})$ roots by successive approximations, we mimic our earlier procedure and, dividing through by x^2, rewrite (7) as

$$\varepsilon x^2 - 1 = x^{-2} - 2x^{-1} - \varepsilon x,$$

where the left side is the binomial whose two roots are the zeroth approximations (8). The iterates therefore satisfy

$$\varepsilon[x^{(0)}]^2 - 1 = 0, \quad \varepsilon[x^{(k)}]^2 - 1 = [x^{(k-1)}]^{-2} - 2[x^{(k-1)}]^{-1} - \varepsilon x^{(k-1)}; \qquad k = 1, 2, \ldots.$$

Solving for $x^{(k)}$, we obtain the final iteration scheme

$$p^{(0)} = \frac{1}{\sqrt{\varepsilon}}, \qquad p^{(k)} = \sqrt{\varepsilon^{-1} + \varepsilon^{-1}[x^{(k-1)}]^{-2} - 2\varepsilon^{-1}[x^{(k-1)}]^{-1} - x^{(k-1)}}, \tag{10a}$$

$$n^{(0)} = \frac{-1}{\sqrt{\varepsilon}}, \qquad n^{(k)} = -\sqrt{\varepsilon^{-1} + \varepsilon^{-1}[x^{(k-1)}]^{-2} - 2\varepsilon^{-1}[x^{(k-1)}]^{-1} - x^{(k-1)}}. \tag{10b}$$

Here $p^{(k)}$ denotes the kth approximation to the positive root, p, of magnitude $\varepsilon^{-1/2}$ and $n^{(k)}$ is the kth approximation to the negative root, n, of this magnitude. Note that each of the two roots p and n is associated with a uniquely determined sequence of iterates. Beginners often make the mistake of sprinkling $\pm$ signs throughout formulas that correspond to (10). This results in a proliferation of "approximations" that cannot be correct, since the total number of roots must, of course, equal the degree of the equation. For each root there must be an iteration scheme that yields a unique answer at every stage.

From (10) we find

$$p^{(1)} = \sqrt{\varepsilon^{-1} - 3\varepsilon^{-1/2} + 1}, \qquad n^{(1)} = -\sqrt{\varepsilon^{-1} + 3\varepsilon^{-1/2} + 1}. \tag{11}$$

We can expand these expressions about $\varepsilon = 0$. As is common in successive approximation schemes, only the leading term in the correction is trustworthy, so we obtain finally

$$p^{(1)} = \frac{1}{\sqrt{\varepsilon}} - \frac{3}{2} + \cdots, \qquad n^{(1)} = \frac{-1}{\sqrt{\varepsilon}} - \frac{3}{2} + \cdots. \tag{12}$$

Determination of the next approximation confirms what is already indicated by (11) and (12): p and n have series expansions in powers of $\sqrt{\varepsilon}$.

The reader should fill in the details of the above example and carry out some exercises to gain facility in exploiting the presence of a small parameter in algebraic equations. Our discussion has illustrated the only relatively difficult features: (i) the need to be careful of multivalued roots, and (ii) the possible occurrence of fractional powers of ε in series expansions for the roots. The appearance of fractional powers can be anticipated by application of the successive approximations procedure.

THE USE OF SCALING

Those who are familiar with the scaling ideas of Section 6.3 might find it worthwhile to apply those ideas here. Thus, in (1) only one of the two roots is $O(1)$, and only for this root is the εm^2 term expected to have a small effect on the answer. The other root is $O(\varepsilon^{-1})$, so that the change of variable $m = \varepsilon^{-1}\hat{m}$ is necessary to obtain a problem with a correctly scaled $O(1)$ variable $\hat{m}$. In terms of $\hat{m}$, (1) becomes

$$\varepsilon(\varepsilon^{-1}\hat{m})^2 + 2\varepsilon^{-1}\hat{m} + 1 = 0 \qquad \text{or} \qquad \hat{m}^2 + 2\hat{m} + \varepsilon = 0,$$

which yields the approximation $\hat{m} \approx -2$ in accord with earlier results. Note that in this example, and in the other example of this section, *no single scale can be used to characterize the complete solution of the problem.* This will be found to be a distinctive property of singular perturbation problems.

EXERCISES

1. (a) Assume a solution to (1) of the form (2) and thereby verify (3).
(b) Obtain (3) by the parametric differentiation procedure.
(c) Obtain (3) by successive approximations.

2. (a) Find a series expansion to the second root $\overline{m}$ of (1) by assuming (5). At the least, find m_0 and m_1.
‡(b) Verify the above calculations by using the parametric differentiation method.

‡3. Solve (1) by the quadratic formula; expand the expressions for the two roots in series; and compare with the results of Exercises 1 and 2.

4. (a) To a first approximation, as the text showed, (7) has a pair of equal roots. Find a three-term approximation to these roots by completing Example 1. In particular, show that the roots are not equal.
(b) Show that alternatives (c)–(g) that were proposed in the discussion of (7) are inconsistent.
(c) Graph the four roots of (7) as a function of ε, for small ε.
(d) Verify (12) and find one more term in the series using successive approximations.
(e) Obtain the result of (d) by using series.
‡(f) Obtain the result of (d) by a *modified* parametric differentiation approach.

5. Find a first approximation, valid for small positive ε, to all roots of the following equations.
(a) $\varepsilon x^4 - x^2 + 3x - 2 = 0$.
(b) $\varepsilon x^5 + x^3 - 1 = 0$.
(c) $x + \varepsilon x^3 = 2$.
(d) $\varepsilon x^3 + \varepsilon x^2 + x - 1 = 0$.

6. Find a second approximation to the roots estimated in Exercise 4.

7. Discuss in detail the character of the roots of

$$\varepsilon p_n(x) + p_m(x) = 0, \qquad n > m,$$

where p_i denotes a polynomial of degree i.

8. This problem outlines a more deductive approach to the estimate of the root $\overline{m}$ of (1). Let $\overline{m}(\varepsilon) = E(\varepsilon) + R(\varepsilon)$. Because we wish the remainder R to be small compared to the estimate E, we require that $\lim_{\varepsilon\to 0}(R/E) = 0$. From (1),

$$\varepsilon(E + R)^2 + 2(E + R) + 1 = 0. \tag{13}$$

‡(a) Suppose that $|\varepsilon E| \to \infty$. Obtain a contradiction.
‡(b) Suppose that $|\varepsilon E| \to 0$. Obtain a contradiction.
(c) From (a) and (b), if εE has a limit, then it must approach a nonzero constant. Determine this constant.

9.2 Boundary Value Problems for Ordinary Differential Equations

In this section we shall study the use of singular perturbation methods to obtain approximate solutions for second order ordinary differential equations with a small parameter multiplying the highest derivative. Most of the discussion concerns a simple "model" problem. Some of the results found in the previous section will prove useful here.

EXAMINATION OF THE EXACT SOLUTION TO A MODEL PROBLEM

Consider the following equation for $y(x)$:

$$\varepsilon\frac{d^2y}{dx^2} + 2\frac{dy}{dx} + y = 0; \qquad 0 < x < 1, \quad 0 < \varepsilon \ll 1, \tag{1}$$

with the boundary conditions

$$y(0) = 0, \qquad y(1) = 1. \tag{2a, b}$$

Note that ε is positive.

Since ε is very small, we naïvely suppose that the first term in (1) may be neglected. This gives the equation

$$2\frac{dy}{dx} + y = 0 \tag{3}$$

with the general solution

$$y = K \exp\left(-\tfrac{1}{2}x\right),$$

where K is an arbitrary constant. A dilemma at once appears, for if K is determined by (2a) the "approximation"

$$y \equiv 0 \tag{4}$$

results, while if K is determined by (2b), the "approximation" is

$$y = \exp\left(\tfrac{1}{2} - \tfrac{1}{2}x\right). \tag{5}$$

Neither "approximation" can satisfy both boundary conditions, so that as $\varepsilon \to 0$, neither can approach the true solution for all x in the closed interval $[0, 1]$. Is either of the two "approximations" ever a good one? If so, in what part of the interval? How does the solution behave in the remainder of the interval? Can one obtain an approximation that is good throughout the interval?

To answer these questions, we shall solve (1) and (2) exactly and then approximate the answer for small positive ε. Then we shall reconsider the problem, pretending that we do not know the exact solution but keeping in mind its qualitative behavior. We shall be able to develop an approximation procedure that will work for a large class of problems whose exact solutions are impossible to attain.

Before embarking on this program, let us note that the present problem possesses "singular" features that are similar to those of the algebraic equations discussed above. For the second degree polynomial equation (1.1), the approximate equation was of first degree and had a single root. Here the approximate equation (3) is first order, so the two boundary conditions (2) cannot both be satisfied. In Equation (1.1) the term εm^2 could not be negligible for all roots. Here the term $\varepsilon d^2y/dx^2$ must be important at least for portions of $(0, 1)$. It must be that d^2y/dx^2 is large in those portions; otherwise $\varepsilon(d^2y/dx^2)$ would indeed be negligible for small enough ε.

Returning to the analysis of (1), we first observe that since this equation is linear and has constant coefficients, it has the general solution

$$y = C_1 \exp(m_1 x) + C_2 \exp(m_2 x), \tag{6}$$

where m_1 and m_2 are the two roots of

$$\varepsilon m^2 + 2m + 1 = 0. \tag{7}$$

When the boundary conditions of (2) are imposed, we obtain the final solution,

$$y = \frac{\exp(m_1 x) - \exp(m_2 x)}{\exp(m_1) - \exp(m_2)}. \tag{8}$$

Equation (7) is the same as Equation (1.1), so that when ε is small, we may employ our earlier analysis, as summarized in (1.6), and approximate m_1 and m_2 by

$$m_1 = -\frac{1}{2}, \qquad m_2 = -\frac{2}{\varepsilon}. \tag{9}$$

Using (9), *since ε is positive*, we see that we can neglect $\exp(m_2)$ compared with $\exp(m_1)$ in the denominator of (8). We thus write the approximate solution to our problem as

$$y(x, \varepsilon) \approx e^{1/2}[e^{-x/2} - e^{-2x/\varepsilon}], \qquad 0 < \varepsilon \ll 1. \tag{10}$$

What is the qualitative behavior of the approximate solution (10)? As required by boundary condition (2a), this approximation has the value zero at $x = 0$, since the second term in the square bracket has exactly the same magnitude as the first and is of opposite sign. As x increases, however, the second term decreases rapidly. When $x = \varepsilon$ this term is only $-e^{-2}$, about one-seventh of its value at $x = 0$. When $x = 2\varepsilon$ it is less than 2 per cent of this value. Thus, except when x is within a few times ε of $x = 0$, the second term is negligible and (10) may be further approximated to yield

$$y = e^{1/2}e^{-x/2} = e^{(1/2)(1-x)}. \tag{11}$$

The graph of (10) may then be drawn, as in the solid line of Figure 9.1. We see that the solution changes rapidly in a layer near $x = 0$ whose thickness is $O(\varepsilon)$. Such a region of rapid change near a boundary is called a **boundary layer**.

Notice that (11) is the same as (5). Thus, for small positive ε, the solution to the differential equation (1) and the boundary conditions (2) rises rapidly from its assigned value of zero at $x = 0$ until it merges with (5), the solution

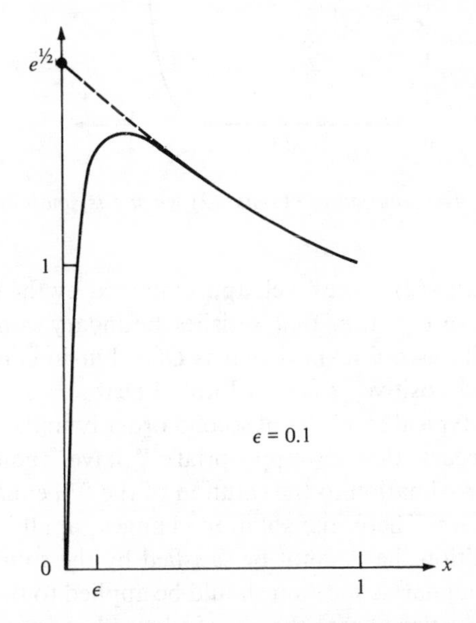

FIGURE 9.1. *The graph of* (10), *a good approximation to the solution of* (1) *and* (2) *when ε is small and positive* (*solid line*). *The "outer approximation"* [(11), (14), *or* (5)] *is depicted by a dashed line.*

to the naïve equation (3) that satisfies boundary condition (2b) at $x = 1$. This "naïve" solution is depicted by a dashed line in Figure 9.1.

It is instructive to consider the case when ε is small and *negative*. Approximate roots of (7) are still given by (9), but now it is the first term of the denominator of the exact solution (8) that is negligible. We have

$$y(x, \varepsilon) \approx -e^{2/\varepsilon}[e^{-x/2} - e^{-2x/\varepsilon}], \qquad |\varepsilon| \ll 1, \varepsilon < 0. \tag{12}$$

Moreover, the first term of this expression is exponentially small for all x in $[0, 1]$ so that we can write, with negligible loss of accuracy,

$$y(x, \varepsilon) \approx e^{(2/\varepsilon)(1-x)}, \qquad |\varepsilon| \ll 1, \varepsilon < 0. \tag{13}$$

The sole term remaining in (13) is virtually zero except in a boundary layer where x is within a few times ε of $x = 1$. (See Figure 9.2.) For small negative ε,

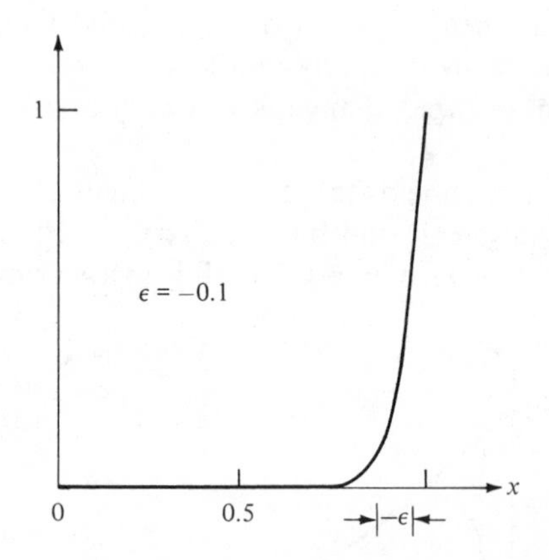

FIGURE 9.2. *The solution to* (1) *and* (2) *when* ε *is small and negative.*

the solution of (1) and (2) is thus well approximated by the (identically zero) solution of the naïve equation that satisfies boundary condition (2a). The boundary layer still has a thickness that is $O(\varepsilon)$, but in contrast to the case where ε is small and positive, it is now located near $x = 1$.

If this example is typical of a class of second order boundary value problems (and it is), it appears that an appropriate "naïve" equation serves to furnish a good approximation to the solution of the full equation except near one of the boundaries. There, the solution changes rapidly to conform with the boundary condition that cannot be satisfied by the solution to the naïve equation. Which boundary condition should be applied to the naïve equation, and which left to be satisfied with the aid of a boundary layer, depends on the relative signs of the various terms. (It is not merely the sign of ε that determines this, for the equation can always be multiplied by -1 if necessary, to make the small parameter positive.) If we assume that the boundary layer will either be near $x = 0$ or $x = 1$, its location can presumably be ascertained fairly efficiently by trial and error.

So that the reader will not entertain false notions, we must warn him that although the generalizations of the previous paragraph apply to a large class of important problems, they do not, for example, apply to all boundary value problems for linear second order ordinary differential equations with a small parameter multiplying the highest derivative. We shall return to the matter of delineating the problems for which singular perturbation methods are appropriate. For the present, a mood of cautious optimism is appropriate.

It is profitable to examine more carefully the relation between the solution to the approximate equation (3) and the solution to the full problem (1). Because the solution to (3) and the appropriate boundary condition provides an approximation that is valid *outside* a boundary layer, it is customary to associate the word **outer** with that approximation. *Until further notice we shall consider the case* $\varepsilon > 0$; therefore the appropriate outer approximation will be that of (11). We use O for outer and write

$$y_O(x) \equiv \exp\left[\tfrac{1}{2}(1 - x)\right]. \tag{14}$$

We know that $y_O(x)$ is a good approximation to the exact solution $y(x, \varepsilon)$ except near $x = 0$. Indeed, it is not difficult to show [Exercise 1(a)] that

$$\lim_{\varepsilon \downarrow 0} y(x, \varepsilon) = y_O(x) \qquad \text{for } x \text{ in } (0, 1].^* \tag{15}$$

The limit is *uniform* for $a \leq x \leq 1$, a positive. That is, given any $E > 0$, there exists a positive number D that depends on E but *not* on x, such that $|y(x, \varepsilon) - y_O(x)| < E$ whenever $0 < \varepsilon < D$ for all x in $[a, 1]$.

Equation (15) contains the first of several limits in this section that involve a function of more than one variable. *Whenever we take a limit with respect to one of the stated variables, it will be tacitly understood that the remaining variables are fixed during the limiting process.* Thus, in (15) the dependence of y on x and ε is explicitly stated. In the limit, x is considered to be fixed as ε decreases toward zero.

Returning to our discussion of $y_O(x)$, we note that, using (12) and (13),

$$\lim_{x \downarrow 0} \left[\lim_{\varepsilon \downarrow 0} y(x, \varepsilon)\right] = y_O(0) = e^{1/2}. \tag{16}$$

But

$$\lim_{\varepsilon \downarrow 0} \left[\lim_{x \downarrow 0} y(x, \varepsilon)\right] = \lim_{\varepsilon \downarrow 0} 0 = 0, \tag{17}$$

so that the operations $\lim_{\varepsilon \downarrow 0}$ and $\lim_{x \downarrow 0}$ cannot be interchanged. Such an interchange would be permitted if the limit of (15) were uniform over $(0, b)$

* The downward arrow in (15) signifies that ε approaches zero through positive values.

for some positive b.* Thus $y(x, \varepsilon)$ does not approach $y_O(x)$ uniformly on such an interval, as can also be deduced directly.

To sum up, the outer solution $y_O(x)$ does not satisfy the boundary condition at $x = 0$. However, $y_O(x)$ *is* the limit as $\varepsilon \downarrow 0$ of $y(x, \varepsilon)$ in (0, 1], but the limit is not uniform. We say that the approximation $y_O(x)$ is **not uniformly valid** in (0, 1].

However close a *fixed* x is to the origin, that x becomes farther and farther outside the boundary layer as $\varepsilon \downarrow 0$. Thus, from (15) the outer solution $y_O(x)$ is aptly named, for it approximates the true solution only outside the boundary layer. But what is an appropriate approximation for small ε *inside* the boundary layer? We now turn to this question.

It turns out that there is little to be gained in arbitrarily pinpointing a precise "edge" to the boundary layer. On the other hand, it seems clear, for example, that $x = 0.1\varepsilon$ is well inside the boundary layer, while $x = 4\varepsilon$ is far toward the edge of the boundary layer. This is true *regardless of the value of* ε, as long as ε is small. To find a typical value of the solution well inside and far toward the edge of the boundary layer, then, one could consider, respectively,

$$\lim_{\varepsilon \downarrow 0} \left[y(x, \varepsilon)\big|_{x=0.1\varepsilon} \right] \quad \text{and} \quad \lim_{\varepsilon \downarrow 0} \left[y(x, \varepsilon)\big|_{x=4\varepsilon} \right].$$

More generally,

$$\lim_{\varepsilon \downarrow 0} \left[y(x, \varepsilon)\big|_{x=\xi\varepsilon} \right] \equiv y_I(\xi), \qquad \xi \text{ fixed}, \tag{18}$$

will provide an approximation $y_I(\xi)$ to the solution at an unchanging proportional distance ξ into a boundary layer whose absolute thickness shrinks continually as $\varepsilon \downarrow 0$. We call this the **inner** or boundary layer approximation, where I stands for "inner." Using alternative notation, we define the function Y by

$$Y(\xi, \varepsilon) \equiv y(\xi\varepsilon, \varepsilon) \tag{19}$$

with which

$$y_I(\xi) \equiv \lim_{\varepsilon \downarrow 0} Y(\xi, \varepsilon). \tag{20}$$

We emphasize that ξ is held fixed in the limits of (18) and (20).

In our present example, from (8),

$$y(x, \varepsilon) = K[\exp(m_1 x) - \exp(m_2 x)], \qquad K \equiv \frac{1}{\exp(m_1) - \exp(m_2)},$$

so

$$Y(\xi, \varepsilon) = K[\exp(m_1 \varepsilon\xi) - \exp(m_2 \varepsilon\xi)].$$

* See, e.g., Franklin (1964), p. 395.

From (9), or more fully, from our earlier discussion concerning the roots of the quadratic equation (7),

$$\lim_{\varepsilon \downarrow 0} \varepsilon m_1(\varepsilon) = 0, \qquad \lim_{\varepsilon \downarrow 0} \varepsilon m_2(\varepsilon) = -2.$$

Hence

$$y_I(\xi) \equiv \lim_{\varepsilon \downarrow 0} Y(\xi, \varepsilon) = e^{1/2}(1 - e^{-2\xi}). \tag{21}$$

Figure 9.3 shows that $y_I(\xi)$ does indeed provide a good approximation to the exact solution within the boundary layer.

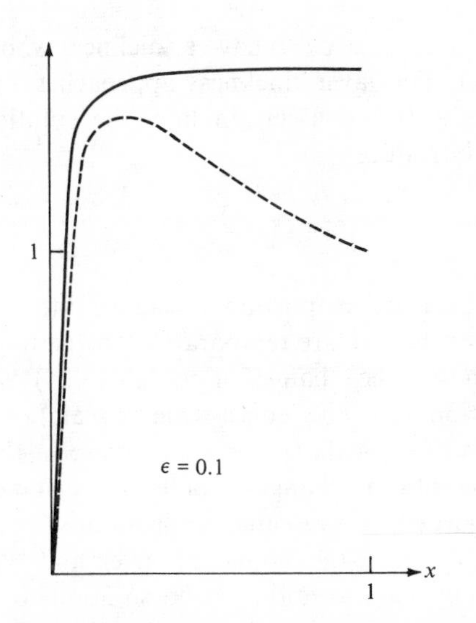

FIGURE 9.3. *Comparison between the exact solution to* (1) *and* (2) *for small positive* ε *(dashed line) and the "inner" boundary layer approximation* (21) *(solid line). Agreement is good near* $x = 0$.

FINDING AN APPROXIMATE SOLUTION BY SINGULAR PERTURBATION METHODS

Let us return to the original problem of (1) and (2):

$$\varepsilon \frac{d^2y}{dx^2} + 2\frac{dy}{dx} + y = 0; \qquad y(0) = 0, \quad y(1) = 1, \quad 0 < \varepsilon \ll 1. \tag{22}$$

We shall now derive an approximate solution to this problem, pretending that we do not know its exact solution. We shall take from our previous discussion the supposition that the solution to the outer equation well approximates the exact solution except in a thin layer near one of the boundaries. For the moment we shall also take as given that the boundary layer is located near $x = 0$.

The outer approximation does not vary rapidly, so it should be permissible to neglect the term $\varepsilon d^2y/dx^2$. Using the advance knowledge that the boundary layer is near $x = 0$, we require the outer approximation y_O to satisfy the boundary condition at $x = 1$. Thus

$$2\frac{dy_O}{dx} + y_O = 0, \qquad y_O(1) = 1, \tag{23a, b}$$

so

$$y_O(x) = \exp\left[\tfrac{1}{2}(1 - x)\right], \tag{24}$$

in agreement with (14).

Let the boundary layer near $x = 0$ have a thickness whose order of magnitude is given by $\delta(\varepsilon)$. The layer thickness approaches zero as $\varepsilon \downarrow 0$, so we require that $\delta \downarrow 0$ as $\varepsilon \downarrow 0$. To ascertain how the solution behaves in the boundary layer, we introduce

$$\xi = \frac{x}{\delta}, \tag{25}$$

which permits us to measure proportional distance into the layer. [We know from (18) that $\delta(\varepsilon) = \varepsilon$, but we are temporarily "forgetting" this.]

Strong motivation for the change of variable (25) is provided by the scaling concept (Section 6.3). The length scale (x scale) of the outer solution (24) is of order unity. This scale is not appropriate in the boundary layer, however, for there the solution changes rapidly in a distance of magnitude δ. Thus (25) can be regarded as a **rescaling** appropriate for the boundary layer region. Two scales, δ inside the boundary layer and unity outside, must thus be used to specify the variation of the solution to (22). Such double scales are a telltale sign of a singular perturbation problem, as we have mentioned in the discussion of "unorthodoxy" in Section 6.3.

Let us introduce the change of variable (25) into the differential equation of (22). Replacing the function y by Y as in (19), we obtain

$$\frac{\varepsilon}{\delta^2}\frac{d^2Y}{d\xi^2} + \frac{2}{\delta}\frac{dY}{d\xi} + Y = 0. \tag{26}$$

We now balance terms pairwise, just as we did in determining the magnitude of the second root $\overline{m}$ of the quadratic equation (1.1). Since δ is the length scale in the boundary layer, the expressions $d^2Y/d\xi^2$, $dY/d\xi$, and Y are all $O(1)$ there. The orders of magnitudes of the various terms are

$$\frac{\varepsilon}{\delta^2}, \qquad \frac{1}{\delta}, \qquad 1. \tag{27}$$

Hopefully, two of these balance and the third is negligible (for then the problem can be simplified by neglecting a term). One term of the balancing pair must be ε/δ^2, since we have already seen that an approximation which is valid throughout [0, 1] cannot be obtained if the second derivative term is never taken into account.

If the first expression in (27) balances the *third*, then $\delta = \varepsilon^{1/2}$. Hence $1/\delta$ is *large* compared to the other expressions ($1/\delta = \varepsilon^{-1/2} \gg 1$), not negligible. If the first expression in (27) balances the *second*, then

$$\frac{\varepsilon}{\delta^2} = \frac{1}{\delta} \quad \text{or} \quad \delta = \varepsilon \qquad \left(\text{so } \xi = \frac{x}{\varepsilon}\right). \tag{28}$$

Now the remaining expression is negligible ($1 \ll \varepsilon^{-1}$), so (28) provides the scaling that we seek. With (28), (26) becomes

$$\frac{d^2 Y}{d\xi^2} + \frac{2dY}{d\xi} + \varepsilon Y = 0. \tag{29}$$

This is the **inner equation**; it has been obtained with a scaling appropriate to the inner or boundary layer. As a first approximation, we neglect the last term and take as an approximate equation describing behavior in the boundary layer

$$\frac{d^2 y_I}{d\xi^2} + \frac{2dy_I}{d\xi} = 0. \tag{30}$$

[In passing from (29) to (30), we have let $\varepsilon \downarrow 0$ with ξ fixed so that the replacement of Y by y_I is called for, by (20).] This second order equation requires two boundary conditions. One stems from $y(0) = 0$. In the present variables this is

$$y_I(0) = 0. \tag{31}$$

The second boundary condition on y, $y(1) = 1$, will *not* supply a second restriction on y_I, since $x = 1$ is far *outside* the boundary layer, while all the above discussion has been aimed at approximating a function that describes the solution *inside* this layer of rapid change. For the moment, then, let us put aside the question of the second boundary condition and solve (30) and (31). We obtain

$$y_I(\xi) = C(1 - e^{-2\xi}), \tag{32}$$

where C is a constant.

MATCHING

It turns out, not surprisingly with hindsight, that to specify C we must match the inner solution at its "farthest extremity" with the outer solution at its "nearest extremity," for both these extremities correspond to the same "edge" of the boundary layer. One hopes for the existence in some sense of such an "edge" (or, better, an *overlap* or **intermediate region**), where *both* the inner and outer solutions are reasonable approximations to the true solution. The two approximations may reasonably be required to agree in an overlap region.

The matching is a subtler matter than it might seem at first and is an object of continuing intensive study at this writing. It appears that the best way to proceed begins by recognizing that when $x = O(\delta)$, then x is within the boundary layer, whereas when $x = O(1)$, then x is outside the boundary layer. For x to be in the intermediate region, we should let $x = O[\Theta(\varepsilon)]$ where $O(\Theta)$ lies between $O(\delta)$ and $O(1)$; i.e.,

$$\lim_{\varepsilon \downarrow 0} \frac{\Theta}{\delta} = \infty, \qquad \lim_{\varepsilon \downarrow 0} \Theta = 0. \tag{33a, b}$$

Now, when x was within the boundary layer $[x = O(\delta)]$, it was appropriate to introduce the variable $\xi = x/\delta$ and to let $\varepsilon \downarrow 0$ with ξ fixed. Analogously, when x is in the intermediate region it is appropriate to introduce η, where

$$\eta = \frac{x}{\Theta} \qquad \left(x = \eta\Theta, \quad \xi = \frac{\eta\Theta}{\delta}\right) \tag{34}$$

and to let $\varepsilon \downarrow 0$ with η fixed. Thus the inner and outer approximations are said to **match** if they have a common limit when the new variable η is introduced and held fixed as ε tends to zero. Hence *matching requires that*

$$\lim_{\varepsilon \downarrow 0} [y_O(x)|_{x=\eta\Theta}] = \lim_{\varepsilon \downarrow 0} [y_I(\xi)|_{\xi=\eta\Theta/\delta}], \qquad \eta \text{ fixed.} \tag{35}$$

In the present case, the left side of (35) is, from (24) and (33b),

$$\lim_{\varepsilon \downarrow 0} [e^{(1/2)(1-x)}|_{x=\eta\Theta}] = \lim_{\varepsilon \downarrow 0} e^{(1/2)(1-\eta\Theta)} = e^{1/2}. \tag{36}$$

Since $\delta(\varepsilon) \equiv \varepsilon$, the right side of (35) is, from (32) and (33a),

$$\lim_{\varepsilon \downarrow 0} C(1 - e^{-2\eta\Theta/\delta}) = C. \tag{37}$$

Thus (35) requires that $C = e^{1/2}$ so, from (32),

$$y_I(\xi) = e^{1/2}(1 - e^{-2\xi}), \tag{38}$$

and this agrees with (21).

For our particular problem, both from the exact solution and from a fairly comprehensive-seeming approximation method, we have derived an inner approximation $y_I(\xi)$ and an outer approximation $y_O(x)$. It would be useful to have a **uniform approximation** y_U, which would be valid throughout [0, 1]. This can be obtained by adding y_I and y_O and subtracting their common part. Thus

$$y_U(x) = y_O(x) + y_I\left(\frac{x}{\delta}\right) - \lim_{\varepsilon \downarrow 0} y_O(\eta\Theta). \tag{39}$$

In the inner region the sum of the first and third terms on the right is negligible, leaving the inner approximation. In the outer region the sum of the second and third terms is negligible, as can be seen from (35), which leaves the outer approximation. Thus y_U reduces to the appropriate approximation in each region. For the example under study we have, from (24), (38), and (36),

$$y_U(x) = e^{(1/2)(1-x)} + e^{1/2}(1 - e^{-2x/\varepsilon}) - e^{1/2} = e^{1/2}(e^{-x/2} - e^{-2x/\varepsilon}),$$

which agrees with (10), our initial approximation to the exact solution.

FURTHER EXAMPLES

In our study of the illustrative problem (22) we took as given that the boundary layer was located near $x = 0$. We know that this is the case when $\varepsilon \downarrow 0$, but the boundary layer is near $x = 1$ when $\varepsilon \uparrow 0$. Fortunately, if we make an incorrect assumption concerning the location of the boundary layer, it shows up, since it leads to a breakdown in our procedure. We illustrate this and, at the same time, provide more familiarity with the singular perturbation procedure in the following examples.

Example 1. Show that the boundary layer procedure breaks down when $\varepsilon \uparrow 0$ if a boundary layer is assumed to exist near $x = 0$.

Solution. All steps of the analysis go through just as before except that at the very last stage, (37) is replaced by

$$\lim_{\varepsilon \uparrow 0} C(1 - e^{-2\eta\Theta/\delta}) = \infty$$

and no matching is possible.

Example 2. Carry through the calculations assuming that a boundary layer is present near $x = 1$.

Solution. We should now require the outer approximation to satisfy the boundary condition at $x = 0$. The outer approximation still satisfies (23a) then, but (23b) is replaced by $y_O(0) = 0$, giving $y_O(x) \equiv 0$.

For x to be within the boundary layer, it is necessary that $1 - x$ be $O(\delta)$ so that the appropriate boundary layer variable is $(1 - x)/\delta \equiv \xi$. (Note that ξ is positive.) Introducing ξ, we find that

$$\frac{\varepsilon}{\delta^2}\frac{d^2Y}{d\xi^2} - \frac{2}{\delta}\frac{dY}{d\xi} + Y = 0. \tag{40}$$

Since (40) differs from (26) only in a sign change, term balancing is accomplished just as before and again $\delta(\varepsilon) = \varepsilon$. Neglecting the last term in (40) and requiring Y to have the value unity at $\xi = 0$ (corresponding to $x = 1$), we find

$$y_I(\xi) = C(e^{2\xi} - 1) + 1.$$

To specify the constant C, we introduce the intermediate variable $\bar{\eta} = (1 - x)/\Theta$, where Θ is subject to the restrictions (33) as before. We then impose the matching requirement (35) (with appropriate small modifications):

$$\lim_{\varepsilon \downarrow 0} y_I\left(\frac{\bar{\eta}\Theta}{\delta}\right) = \lim_{\varepsilon \downarrow 0} y_O(1 - \Theta\bar{\eta}) = 0.$$

If ε approaches zero through positive values, the limit of y_I does not exist (as expected), but if it approaches zero through negative values, the limit is $-C+1$. In the latter case $C=1$, $y_I(\xi)=\exp(2\xi)$, and

$$y_U(x) = 0 + e^{2(1-x)/\varepsilon} - 0 = e^{2(1-x)/\varepsilon},$$

in agreement with (13).

Generalizing, we propose the following basic procedure for singular perturbation boundary value problems for second order ordinary differential equations. *This procedure will not always work.* However, it will work for a large class of problems, and will work for a still larger class of problems with modifications that should be within reach of the reader.

Consider the boundary value problem

$$\varepsilon y'' + f(x, y, y') = 0, \quad y(a) = A, \quad y(b) = B; \qquad ' \equiv \frac{d}{dx}. \tag{41}$$

Suppose (with no loss of generality) that ε is positive. Suppose also that the problem has a unique solution* with a boundary layer at $x = a$. This supposition does not commit us to whether the boundary layer is to the right or to the left, for we have not specified which of a or b is the larger.

(a) Determine the outer approximation y_O by solving

$$f(x, y_O, y_O') = 0, \qquad y_O(b) = B. \tag{42}$$

(b) Introduce into (41) the boundary layer variable

$$\xi = \pm \frac{x-a}{\delta(\varepsilon)}, \tag{43}$$

where the sign is chosen to make ξ positive in the interior of the interval over which the differential equation holds. (Note that $x = a \pm \delta\xi$.) With the notation $Y(\xi, \varepsilon) = y(a \pm \delta\xi, \varepsilon)$, (41) can be written

$$\frac{\varepsilon}{\delta^2}\frac{d^2 Y}{d\xi^2} + f\left(a \pm \delta\xi, Y, \pm\frac{1}{\delta}\frac{dY}{d\xi}\right) = 0. \tag{44}$$

Suppose that for small ε (and fixed ξ) the dominant contribution to the second term in (44) has the form

$$\delta^s F\left(\xi, Y, \frac{dY}{d\xi}\right) \tag{45}$$

* The reader should be aware of the fact that even if (41) is linear, it may have an infinite number of solutions or it may have no solutions at all. See, for example, Section 1.5 of I. Stakgold's *Boundary Value Problems of Mathematical Physics* (N.Y.: Macmillan, 1967).

for some function F and some constant s. [Note that $F(\xi, Y, dY/d\xi)$ is independent of δ.] Balance the second derivative term with (45) by choosing

$$\varepsilon\delta^{-2} = \delta^s \qquad \text{or} \qquad \delta = \varepsilon^{1/(2+s)}. \tag{46}$$

The problem must be such that $s > -2$ if δ is to approach zero with ε as required.

(c) Determine $y_I(\xi)$ by solving

$$\frac{d^2y_I}{d\xi^2} + F\left(\xi, y_I, \frac{dy_I}{d\xi}\right) = 0, \qquad y_I(0) = A. \tag{47}$$

(d) To find the arbitrary constant in the solution to (47), introduce the intermediate variable

$$\eta \equiv \pm\frac{x-a}{\Theta(\varepsilon)}, \tag{48}$$

where Θ is a function that is unspecified except for the requirements

$$\lim_{\varepsilon\downarrow 0} \Theta(\varepsilon) = 0 \qquad \text{and} \qquad \lim_{\varepsilon\downarrow 0} \frac{\Theta(\varepsilon)}{\delta(\varepsilon)} = \infty. \tag{49}$$

Impose the **matching requirement**

$$\lim_{\varepsilon\downarrow 0} \left[y_O(x)\big|_{x=a\pm\Theta\eta}\right] = \lim_{\varepsilon\downarrow 0} \left[y_I(\xi)\big|_{\xi=\eta\Theta/\delta}\right], \qquad \eta \text{ fixed.} \tag{50}$$

REMARKS. (a) Further restrictions on Θ may be imposed to obtain matching. (b) The above procedure should not be slavishly "plugged into" but should rather be regarded as a summary and a guide. This will result in a more efficient attack on easier problems and will increase the reader's chance of developing the "feeling" for singular perturbations which is necessary for generalizing the method to harder problems. (c) Nonlinear problems frequently require considerable adaptation of the above procedure. Typically, the nonlinear reduced equation (42) will have more than one solution. Different solutions serve as the outer approximation in different regions. The choice of outer solution is governed by the possibility of matching with "boundary layers" located near the boundaries *and* in the interior. [See Cole (1968), Chap. II.]

Example 3. Find outer and inner approximations for small positive ε to the solution of

$$\varepsilon\frac{d^2y}{dx^2} + \alpha(x)\frac{dy}{dx} + \beta(x)y = 0; \qquad y(0) = 0, \quad y(1) = 1, \quad \alpha(0) \equiv a_0 > 0, \quad \beta(0) \text{ finite.}$$

To save trial and error, the information that the boundary layer is near $x = 0$ is given.

Solution. The outer approximation satisfies $\alpha y' + \beta y = 0$, $y(1) = 1$, and so is given by

$$y_O(x) = \exp\left[\int_x^1 \frac{\beta(s)}{\alpha(s)}\right] ds.$$

We introduce $\xi \equiv x/\delta$, where $\delta(\varepsilon) \to 0$ as $\varepsilon \downarrow 0$. We obtain

$$\frac{\varepsilon}{\delta^2}\frac{d^2Y}{d\xi^2} + \frac{\alpha(\delta\xi)}{\delta}\frac{dY}{d\xi} + \beta(\delta\xi)Y = 0.$$

Since $\alpha(\delta\xi) \to a_0$ when $\varepsilon \downarrow 0$ for fixed ξ, the magnitudes of the various terms again are as in (27); hence, by the same argument as was used there, $\delta = \varepsilon$. The inner approximation satisfies $d^2y_I/d\xi^2 + a_0\, dy_I/d\xi = 0$, $y_I(0) = 0$, so $y_I(\xi) = C(1 - e^{-a_0\xi})$. Matching gives $C = \exp\left\{\int_0^1 [\beta(s)/\alpha(s)]\, ds\right\}$.

Other interesting features of singular perturbation problems are illustrated in the exercises, e.g., problems

(a) of higher order than second (Exercise 7),
(b) where it is not true that $\delta(\varepsilon) = \varepsilon$ [Exercise 2(c)], and
(c) where the outer solution in a second order equation does not satisfy either boundary condition (Exercise 6).

As we have mentioned, singular perturbation theory is very much an art at present. The procedures have been rigorously justified only on a fraction of the problems upon which they have been used. As is illustrated in the books of Van Dyke (1964), Cole (1968), and Nayfeh (1973), ingenuity in applying singular perturbation techniques, particularly to nonlinear ordinary differential equations and to partial differential equations, leads to elegant and satisfying results. Traps threaten the unwary, but as has been remarked,* "the inalienable right to think while using any technique provides a very real degree of protection against the acceptance of ... spurious constructions."

EXERCISES

1. (a) Verify (15).
(b) Verify that the limit of (15) is uniform in $[a, 1]$ as stated in the text.
(c) Show directly that the limit is not uniform in $(0, 1]$.

2. Find inner, outer, and uniform approximations to the solutions of the following problems. Assume that ε is small and positive and that there is a boundary layer at $x = 0$. Proceed from first principles in each case; do not "plug" into formulas. ($' \equiv d/dx$.)
(a) $\varepsilon y'' + (1 + x)y' + y = 0$; $y(0) = 0$, $y(1) = 1$.
(b) $\varepsilon y'' + y' + y^2 = 0$; $y(0) = \frac{1}{4}$, $y(1) = \frac{1}{2}$.
‡(c) $\varepsilon y'' + x^{1/3}y' + y = 0$; $y(0) = 0$, $y(1) = e^{-3/2}$.

* G. F. Carrier, "Singular Perturbation Theory and Geophysics," *SIAM Rev. 12*, 175–93 (1970).

3. Show that contradictions result in the problems of Exercise 2 if the boundary layer is assumed to be near $x = 1$.

4. A "small" mass m hangs from a weightless spring with internal damping proportional to speed. A vertical impulse I (instantaneous momentum change) is imparted to the mass by striking it with a hammer. Initial conditions on the vertical deflection y^* at time t^* can thus be taken as

$$\text{at } t^* = 0: \qquad y^* = 0, \quad m\frac{dy^*}{dt^*} = I.$$

The governing equation is

$$m\left(\frac{d^2y^*}{dt^{*2}}\right) + \mu\left(\frac{dy^*}{dt^*}\right) + ky^* = 0,$$

where μ and k are the damping and spring constants, respectively. Since the mass is small, we have a strongly "overdamped" situation wherein the mass will quickly return to rest after the impulse is expended in stretching the spring. This exercise requests a perturbation analysis of the situation.

(a) Show that a certain choice of dimensionless variables reduces the problem to

$$\varepsilon y'' + y' + y = 0, \qquad y(0) = 0, \quad \varepsilon y'(0) = 1. \tag{51}$$

(b) Solve the problem by singular perturbation techniques, using the following hints. Do not impose an initial condition on the outer approximation. Find an inner approximation that satisfies both initial conditions. Then complete determination of the outer approximation.

‡(c) Is your answer plausible? Discuss briefly. In particular, explain why (51) is correctly scaled, provided that t is not too small.

(d) Find a composite solution valid for the entire range $t > 0$.

(e) Check your answer by examining the exact solution when $\varepsilon \ll 1$.

‡5. Use singular perturbation theory to obtain outer, inner, and composite expansions to the solution of the problem (Carrier, op. cit.)

$$\varepsilon u'' - (2 - x^2)u = -1, \qquad u(-1) = u(1) = 0.$$

REMARK. It is sufficient to solve the differential equation on (0, 1) subject to the boundary conditions $u'(0) = 0$, $u(1) = 0$. Why?

6. (a) Use singular perturbation techniques to approximate the solution to

$$\varepsilon^2 y'' - y = 0, \qquad y(0) = 1, \quad y(1) = 2, \quad 0 < \varepsilon \ll 1.$$

(b) Show that singular perturbation techniques fail on the problem

$$\varepsilon^2 y'' + y = 0; \qquad y(0) = 1, \quad y(1) = 2, \quad 0 < \varepsilon \ll 1.$$

Use the exact solution to show why things go wrong.

7. For a slightly stiff string with fixed ends, the modes of vibration have a shape $y(x)$ given by the eigenvalue problem

$$\varepsilon y^{(iv)} - y'' = \lambda y, \qquad y(0) = y'(0) = y(1) = y'(1) = 0.$$

Here ε is a measure of the stiffness; $0 < \varepsilon \ll 1$. The eigenvalue λ is a dimensionless frequency of vibration. Find outer and inner approximations, assuming that λ is $O(1)$, and conclude that the stiffness has no effect on the eigenvalues to lowest order. [For a discussion of corrections due to stiffness see Sec. 3 of R. E. O'Malley's "Topics in Singular Perturbations," *Advan. Math. 2*, 365–40 (1968). Reprinted in *Lectures on Ordinary Differential Equations* (New York: Academic, 1970).]

‡8. Find the exact solution to

$$y'' - \varepsilon^2 y = 0; \qquad y(0) = 1, \quad y'(x) \to 0 \text{ as } x \to \infty,$$

and take its limit as $\varepsilon \downarrow 0$. Show that there are three different types of limits, depending on the range of x under consideration. Thereby demonstrate that this problem has a singular perturbation character for small ε in spite of the fact that the order of the equation does not change when ε is set equal to zero.*

9. The object of this exercise is to show how the "unacceptable" discontinuity that appeared in (8.3.9) can be "fixed" with the aid of singular perturbation theory.

(a) In (8.4.5), (8.4.6), and (8.4.7), reintroduce the parameter η, using $\kappa \equiv \lambda(\eta\nu)^{-1/2}$. Show that when $\eta\nu \ll 1$,

$$\frac{C-1}{\nu} \begin{cases} \approx 1 - \frac{1}{2}e^{(x-1)/(\eta\nu)^{1/2}}, & x \le 1, \\ \approx \frac{1}{2}e^{-(x-1)/(\eta\nu)^{1/2}}, & x \ge 1. \end{cases}$$

Sketch. Also sketch the discontinuous function

$$\lim_{\substack{\nu \to 0 \\ \eta \text{ fixed}}} \frac{C-1}{\nu}.$$

(b) It should now be apparent that (8.3.6) provides an outer expansion for the problem of (8.2.20). Pretend that you do not know the results of part (a) and "fix" the discontinuity obtained in (8.3.9) with a boundary layer.

10. In studying viscous fluid flow past an infinite plane, one is led to the following problem for $y(x; \varepsilon)$ on $0 \le x < \infty$, $0 < \varepsilon$:

$$\varepsilon(y''' + yy'') + 1 - (y')^2 = 0, \qquad y(0) = y'(0) = 0, \qquad \lim_{x \to \infty} y'(x) = 1.$$

* This exercise is taken from Sec. 1 of the O'Malley paper cited in Exercise 7.

We are interested in the case $\varepsilon \downarrow 0$, and you may assume that there is a boundary layer at $x = 0$. Determine the first term in the outer solution. Also determine the boundary layer equation, boundary conditions, and matching conditions. Do not solve. [To do this, you will have to introduce boundary layer variables for both x and y, and also make use of the requirement that in passing from the inner to the outer region $y'(x)$ must be continuous as $\varepsilon \downarrow 0$.]

CHAPTER 10
Singular Perturbation Theory Applied to a Problem in Biochemical Kinetics

AS AN example of a relatively simple but important application of the techniques introduced in Chapter 9, we shall now examine a problem in biochemical kinetics. We shall be led to consider singular perturbation theory for an *initial* value problem involving a certain system of ordinary differential equations. Some details of the discussion of boundary value problems in Chapter 9 are inapplicable, but the arguments are the same in spirit. The reader's understanding should thus be reinforced.

In Section 10.1 we formulate a mathematical problem that describes the course of a chemical reaction catalyzed by an enzyme. We shall consider the frequently occurring case when the initial enzyme concentration is small (in a sense to be made more precise). The problem can then be profitably analyzed with the techniques of singular perturbation theory. This we do in Section 10.2. In particular, we make a technical advance over Chapter 9 by indicating how higher approximations can be obtained.

10.1 Formulation of an Initial Value Problem for a One Enzyme–One Substrate Chemical Reaction

THE LAW OF MASS ACTION

We shall be mainly interested in the time course of reactions that are catalyzed by enzymes, but first we must consider some simple cases of the *law of mass action*, which forms the foundation for quantitative studies of chemical reactions.

For definiteness let us start by considering a situation wherein molecules a and b combine to give molecule c. If a and b are to react, they must undergo an effective collision. Not every collision may be effective because juxtaposition of active or important parts of the molecules is probably required for the reaction (Figure 10.1).

If we hold external factors (such as temperature) fixed, and also keep the concentration of b fixed, then it is natural to suppose that the frequency of effective collisions (and hence the rate of reaction) is proportional to the concentration of a. For if we double the amount of a involved, there should be twice as many effective collisions per unit time. (This assumes that the concentration of a is not too high; otherwise doubling this concentration

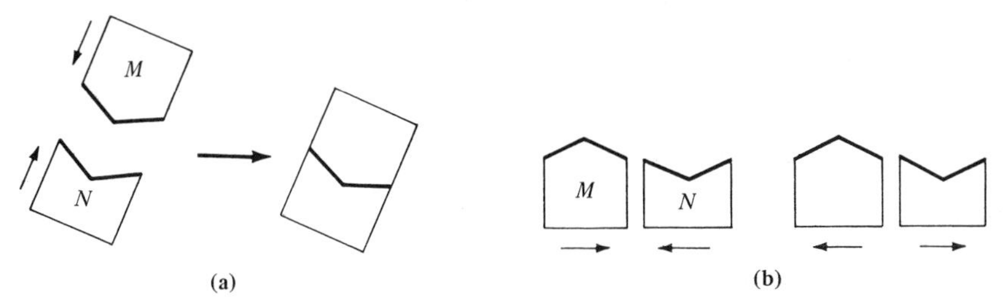

FIGURE 10.1. *Schematic illustration of molecular collisions. The heavily bordered portions of the "molecules" M and N must join for a successful reaction.* (a) *An effective collision.* (b) *An ineffective collision.*

may lead to interference effects.) Using the same symbols for the chemicals and for their concentrations (number of molecules per unit volume), we may write $dc/dt \sim a$. By the same token, $dc/dt \sim b$. This leads us to the equation

$$\frac{dc}{dt} = k_1 ab,$$

where k_1 is the **rate constant**. Chemists use the symbolism

$$a + b \xrightarrow{k_1} c$$

to denote the reaction in question. As should be apparent, the equations governing the concentrations of a and b in this case are

$$\frac{da}{dt} = -k_1 ab, \qquad \frac{db}{dt} = -k_1 ab.$$

An important special case is the situation where a and b are identical, and combine to form the two element chain or **dimer** a_2. We then have

$$a + a \xrightarrow{k_1} a_2, \qquad \frac{da_2}{dt} = k_1 a^2.$$

This illustrates the fact that when just two molecules of a substance are involved in a reaction, the rate is proportional to the square of the corresponding concentration.

The above discussion, while plausible, is speculative. The acceptability of the ideas we have put forth dates from 1867 when their validity was experimentally established by the Norwegians Guldberg and Waage.

It is only a slight extension of our basic ideas to consider a reversible reaction

$$a + b \underset{k_{-1}}{\overset{k_1}{\rightleftarrows}} c, \tag{1}$$

wherein the chemical c may spontaneously break up into its original components a and b. The reader should satisfy himself that the governing equations now are

$$\frac{da}{dt} = -k_1ab + k_{-1}c, \qquad \frac{db}{dt} = -k_1ab + k_{-1}c, \qquad \frac{dc}{dt} = k_1ab - k_{-1}c.$$

From these equations we can at once deduce that

$$\frac{d(a+c)}{dt} = 0, \qquad \frac{d(b+c)}{dt} = 0,$$

so that $a + c = \text{constant}$ and $b + c = \text{constant}$. This is to be expected, because each new molecule of c results from a combination of a and b, and hence from a disappearance of a molecule of a and a molecule of b.

ENZYME CATALYSIS

Certain chemicals have been found to be of vital importance in speeding up and controlling biochemical processes. These **enzymes**† are now recognized to be large protein molecules that somehow can catalyze chemical reactions, most often a specific reaction for each enzyme. The precise details of enzyme action remain a topic of intensive research. But, as was suggested in the late nineteenth century by the Swedish chemist Svante Arrhenius, the essence of the matter is that the enzyme (concentration e^*) combines with the "input" molecule or **substrate** (concentration s^*). This produces a combined enzyme–substrate molecule, or **complex** (concentration c^*). Bound to the enzyme molecule in this complex, the substrate molecule is "activated" and far more likely to form the "output" or **product** (concentration p^*) of the reaction. When this happens (in the simple case under consideration), it is presumed that the complex breaks up into the product molecule and the original "free" enzyme molecule. One must also allow for the possibility that the complex breaks apart without the reaction having occurred. Schematically, then, one can write

$$s^* + e^* \underset{k_{-1}}{\overset{k_1}{\rightleftharpoons}} c^* \overset{k_2}{\longrightarrow} p^* + e^*. \tag{2}$$

Let us assume that originally only the substrate and enzyme are present, at concentrations $\bar{s}$ and $\bar{e}$, respectively. Then the concentration of chemicals at time t^* is determined by the following equations and initial conditions:

$$\frac{ds^*}{dt^*} = -k_1 s^* e^* + k_{-1}c^*, \tag{3a}$$

$$\frac{de^*}{dt^*} = -k_1 s^* e^* + k_{-1}c^* + k_2 c^*, \tag{3b}$$

† A classical reference, which includes theoretical discussions, is *The Enzymes* by M. Dixon and E. C. Webb (New York: Academic, 2nd ed., 1964).

$$\frac{dc^*}{dt^*} = k_1 s^* e^* - k_{-1} c^* - k_2 c^*, \tag{3c}$$

$$\frac{dp^*}{dt^*} = k_2 c^*, \tag{3d}$$

$$\text{at } t^* = 0: \quad s^* = \bar{s}, \quad e^* = \bar{e}, \quad c^* = 0, \quad p^* = 0. \tag{3e}$$

Adding (3b) and (3c), and using (3e), we conclude that

$$e^*(t^*) + c^*(t^*) = \text{constant} = \bar{e}. \tag{4}$$

We can thus eliminate e^* and consider the problem

$$\frac{ds^*}{dt^*} = -k_1 \bar{e} s^* + (k_1 s^* + k_{-1}) c^*, \tag{5a}$$

$$\frac{dc^*}{dt^*} = k_1 \bar{e} s^* - (k_1 s^* + k_{-1} + k_2) c^*, \tag{5b}$$

$$s^*(0) = \bar{s}, \qquad c^*(0) = 0. \tag{5c}$$

Once $s^*(t^*)$ and $c^*(t^*)$ have been determined by the above equations, $e^*(t^*)$ and $p^*(t^*)$ can be determined from the easily deduced equation

$$s^* + c^* + p^* = \bar{s} + \bar{e}. \tag{6}$$

Setting up the equations for the time course or **kinetics** of the reaction is not an idle exercise, for one cannot merely take a snapshot of the molecular interaction to check on whether or not Arrhenius' speculations are correct. But experimenters have been able to follow the time course of reactions, for example by taking advantage of certain changes in light absorption properties. Comparison of such observations with theoretical kinetic results form the classical evidence for the existence of the enzyme-complex mechanism.

Certain simplifying assumptions have enabled clear comparison with experiment. A number of different people are given credit here. The names most often mentioned are those of the Berlin scientists Leonor Michaelis and Maude Menten, but others prefer to cite the Frenchman Henri or the Englishmen Briggs and Haldane.† Be that as it may, the main idea is as follows.

It often occurs, or can be arranged, that the initial concentration of substrate is much larger than that of enzyme. ("Concentration" means number of molecules per unit volume.) In such a case there will be a short initial

† Later analyses from a singular perturbation point of view occur in articles by J. R. Bowen, A. Acrivos, and A. K. Oppenheim, "Singular Perturbation Refinement to Quasi-steady State Approximation in Chemical Kinetics," *Chem. Eng. Sci. 18*, 177–88 (1963); and F. G. Heineken, H. M. Tsuchiya, and R. Aris, "On the Mathematical Status of the Pseudo-steady State Hypothesis of Biochemical Kinetics," *Math. Biosci. 1*, 95–113 (1967).

period when the enzyme will quickly "load up" with substrate so that the concentration of complex increases rapidly. During most of the reaction time, however, the concentration of complex will remain approximately constant. The reason for this is that as long as there are many substrate molecules around for each enzyme molecule, the speed of the reaction will only be limited by how fast the enzyme can work, not by the availability of substrate to work on. The time for an enzyme molecule that becomes "free" to bind to a new substrate molecule is always about the same. Of course, when most of the substrate is used up, this time will become longer and longer. In the final stage of the reaction, therefore, complex concentration should noticeably decrease. Ultimately, all the substrate will have been converted to product by reaction (2), the enzyme will be entirely in its "free" form once again, and no complex will be present.

In comparing their observations with theory, biochemists therefore make the hypothesis that complex concentration can be regarded as constant to first approximation. Equation (5b) with $dc^*/dt^* = 0$ can then be used to express c^* in terms of s^*. Upon substitution of this result into (5a), the problem reduces to a single ordinary differential equation for ds^*/dt^* in terms of s^*. This gives the "Michaelis–Menten kinetics." In this chapter we shall give a detailed treatment of the problem by singular perturbation methods. In particular, we shall show that the Michaelis–Menten kinetic equation corresponds to the "outer" equation in the terminology of the previous chapter. Our analysis follows rather closely that of Heineken et al. (op. cit.).

SCALING AND FINAL FORMULATION

The first step is to introduce dimensionless variables in such a way that the maximum magnitude of each term is correctly estimated by the parameter that precedes it (scaling). In doing this, we focus our attention on the situation after the initial "loading-up" phase and before the final stage. For the period in question, $\bar{s}$ is an excellent estimate for s^*, since our entire approach is based on the fact that s^* will not decrease much from its initial value $\bar{s}$. Further, in the presence of an abundance of substrate, most of the enzyme initially present should soon join with substrate to form complex. We thus take $\bar{e}$ as an estimate for c^*. The choice of time scale is perhaps not immediately obvious; therefore, let us simply denote it by $\bar{t}$. We thus introduce dimensionless substrate and complex concentrations and a dimensionless time by

$$s = \frac{s^*}{\bar{s}}, \qquad c = \frac{c^*}{\bar{e}}, \qquad t = \frac{t^*}{\bar{t}}. \tag{7}$$

Upon substituting into (5a), we obtain

$$\frac{\bar{s}}{\bar{t}}\frac{ds}{dt} = -k_1\bar{e}\bar{s}s(1 - c) + k_{-1}\bar{e}c. \tag{8}$$

The first term on the right side of (8) gives the rate of decrease of substrate concentration caused by conversion into complex. The second term on the right gives the rate of increase of substrate concentration caused by spontaneous breakdown of complex. If the latter effect were large, there would be considerable formation of complex which splits apart without accomplishing anything. Such dominance of the "back reaction" apparently occurs on occasion and performs a controlling function. But we shall consider more usual circumstances in which there is a rough balance between the terms $(\bar{s}/\bar{t})\, ds/dt$ and $-k_1\bar{e}\bar{s}s(1-c)$ of (8). This requires that the time scale be

$$\bar{t} = \frac{1}{k_1\bar{e}}. \tag{9}$$

Introducing (7) and (9) into (5), we see that the dimensionless scaled equations are

$$\boxed{\begin{aligned} \dot{s} &= -s + (s + \kappa - \lambda)c, \\ \varepsilon\dot{c} &= s - (s+\kappa)c, \\ s(0) &= 1, \qquad c(0) = 0, \end{aligned}} \qquad \left(\cdot \equiv \frac{d}{dt}\right) \qquad \begin{matrix}(10a)\\(10b)\\(10c)\end{matrix}$$

where

$$\varepsilon = \frac{\bar{e}}{\bar{s}}, \qquad \kappa = \frac{k_{-1} + k_2}{k_1\bar{s}}, \qquad \lambda = \frac{k_2}{k_1\bar{s}},$$

are parameters.

The boxed equations (10) comprise the mathematical formulation of our problem. We are faced with an initial value problem for a pair of first order nonlinear ordinary differential equations. Three parameters occur, of which one is small. If the small parameter is set equal to zero, we can reduce the problem to a single first order differential equation, but now at least one of the initial conditions cannot be satisfied. Singular perturbation methods seem called for. These are the subject of the next section.

EXERCISES

1. Verify (4), (5), and (6).

2. Suppose that the second reaction in (2) is reversible, with rate constant k_{-2}.

(a) How is (3) changed?

(b) How are (4), (5), and (6) changed?

(c) Begin to trace the consequences of this reversibility in the remaining calculations of this chapter.

10.2 Approximate Solution by Singular Perturbation Methods

In this section we use singular perturbation methods to obtain an approximate solution to the problem posed in (1.10).

MICHAELIS–MENTEN KINETICS AS AN OUTER SOLUTION

In (1.10) the dimensionless parameter ε is small by hypothesis. Since we have introduced scaled dimensionless variables, when the scaling is valid the term multiplied by ε may be neglected, yielding the following equations for the approximations s_0 and c_0:

$$\dot{s}_0 = -s_0 + (s_0 + \kappa - \lambda)c_0, \tag{1a}$$

$$s_0 - (s_0 + \kappa)c_0 = 0 \qquad \text{or} \qquad c_0 = \frac{s_0}{s_0 + \kappa}. \tag{1b}$$

Substituting (1b) into (1a), we obtain the single first order equation

$$\dot{s}_0 = -\frac{\lambda s_0}{s_0 + \kappa}. \tag{2}$$

We shall refer to the solutions of (1) and (2) as the **Michaelis–Menten approximation**. With this approximation, data from chemical experiments can be organized to permit ready determination of the parameters κ and λ (Exercise 8).

It should be evident that the Michaelis–Menten approximation can be regarded as the outer approximation of a singular perturbation problem. The first order equation (2) cannot be expected to have a solution that satisfies both initial conditions of (1.10c). This is no surprise, for we have reduced the order of the system by neglecting $\varepsilon\dot{c}$. Furthermore, our scaling was made for a period after the initial moments when the complex concentration builds up rapidly from zero. Solving (2), we find easily [Exercise 1(a)] that

$$s_0 + \kappa \ln s_0 = -\lambda t + Q, \tag{3}$$

where Q is a constant. This equation and (1b) provide the outer solution, where Q must be determined by matching.

INNER SOLUTION

To find the inner equations, we postulate the existence of an inner or **initial layer** of "width" $\delta(\varepsilon)$. On introducing the appropriate change of variable $\tau = t/\delta(\varepsilon)$, we see that the inner solutions $S(\tau, \varepsilon) = s(\delta\tau, \varepsilon)$ and $C(\tau, \varepsilon) = c(\delta\tau, \varepsilon)$ satisfy

$$\delta^{-1}\frac{dS}{d\tau} = -S + (S + \kappa - \lambda)C, \qquad \varepsilon\,\delta^{-1}\frac{dC}{d\tau} = S - (S + \kappa)C. \tag{4}$$

In order to retain the term involving $dC/d\tau$, we must take

$$\delta(\varepsilon) = \varepsilon \qquad \text{so } \tau = t/\varepsilon.^* \tag{5}$$

Thus the inner equations are

$$S' = \varepsilon[-S + (S + \kappa - \lambda)C], \quad C' = S - (S + \kappa)C, \qquad ' \equiv \frac{d}{d\tau}. \tag{6}$$

Because we want the inner solutions to be appropriate to the first few moments of the reaction, the original initial conditions must be satisfied. Thus

$$S(0, \varepsilon) = 1, \qquad C(0, \varepsilon) = 0. \tag{7}$$

Let $S_0(\tau)$ and $C_0(\tau)$ denote the first approximation to the inner solutions. From (6) and (7) these satisfy

$$S_0' = 0, \quad C_0' = S_0 - (S_0 + \kappa)C_0; \qquad S_0(0) = 1, \quad C_0(0) = 0.$$

Hence [Exercise 1(b)]

$$S_0(\tau) \equiv 1, \qquad C_0(\tau) = (\kappa + 1)^{-1}[1 - e^{-(\kappa+1)\tau}], \tag{8a, b}$$

and the inner approximations are completely determined.

To match, we introduce an intermediate variable τ_i such that

$$\tau_i \equiv \frac{t}{\Psi(\varepsilon)}, \qquad \lim_{\varepsilon \downarrow 0} \Psi(\varepsilon) = 0, \qquad \lim_{\varepsilon \downarrow 0} \frac{\Psi(\varepsilon)}{\delta(\varepsilon)} = \infty, \tag{9a, b, c}$$

in parallel with (9.2.49). Recall from (5) that $\delta(\varepsilon) = \varepsilon$. (Note also that $\tau = \tau_i \psi/\delta$.) In parallel with (9.2.50), the matching condition on the substrate concentration is that, for fixed Ψ,

$$\lim_{\varepsilon \downarrow 0} [s_0(t)\big|_{t=\Psi\tau_i}] = \lim_{\varepsilon \downarrow 0} [S_0(\tau)\big|_{\tau=\tau_i\Psi/\delta}] \qquad \text{or} \qquad s_0(0) = 1, \tag{10}$$

since $S_0 \equiv 1$. Hence $Q = 1$ in (3), so that

$$s_0 + \kappa \ln s_0 = -\lambda t + 1. \tag{11}$$

This completes the (implicit) determination of the outer approximation to the substrate concentration.

The corresponding matching condition on the concentration of the complex requires that

$$c_0(0) = \lim_{\varepsilon \downarrow 0} \{(\kappa + 1)^{-1}[1 - e^{-(\kappa+1)\tau_i\psi/\delta}]\} = (\kappa + 1)^{-1}, \tag{12}$$

* We can regard the introduction of τ rather than t as a required rescaling in an initial layer where, far from being negligible compared to s and $(s + \kappa)c$, $\dot{c}$ is comparable to these terms as complex concentration rapidly builds. Here, then, is another instance of an unorthodox function whose proper description requires the introduction of more than one scale. (Compare Section 6.3.)

using (8b). There are no remaining constants which can be chosen so that this condition will be satisfied. From (1b) and (10), however, we have

$$c_0(0) = \frac{s_0(0)}{s_0(0) + \kappa} = \frac{1}{1 + \kappa} \tag{13}$$

so that the condition is automatically satisfied. This state of affairs is not a coincidence. It comes about because of our correct approach to the problem.

A UNIFORM APPROXIMATION

The **procedure for obtaining a uniform approximation** is always the same: *add the inner and outer approximations and subtract their common part.* The justification given for this procedure after its first introduction in (9.2.39), is not restricted to the particular context in which it appeared. In the present case the initial term $s_0^{(u)}$ in the uniform approximation $s^{(u)}$ to s is given by

$$s_0^{(u)}(t) = s_0(t) + S_0\left(\frac{t}{\varepsilon}\right) - 1 \qquad \text{or} \qquad s_0^{(u)}(t) = s_0(t). \tag{14}$$

Here we have used (10) and the fact that $S_0 \equiv 1$. Correspondingly, for the complex concentration we have, using (12),

$$c_0^{(u)}(t) = c_0(t) + C_0\left(\frac{t}{\varepsilon}\right) - \frac{1}{\kappa + 1}, \qquad \text{so } c_0^{(u)}(t) = \frac{s_0(t)}{s_0(t) + \kappa} - \frac{e^{-(\kappa+1)t/\varepsilon}}{\kappa + 1}. \tag{15}$$

All our various results are expressed in terms of $s_0(t)$, which is given implicitly by the transcendental equation (11). For fixed values of the parameters κ and λ, s_0 can be plotted as a function of t using various numerical methods [e.g., see Ralston (1965)]. Furthermore, as we shall now show, useful analytic approximations can be obtained for small and large values of t.

For small t a power series expansion is appropriate. To obtain this expansion, we require the derivatives of $s_0(t)$ evaluated at $t = 0$. But the initial condition on s_0 is $s_0(0) = 0$. Also, from the differential equation (1.10a) and the initial condition (13), we find at once that

$$\dot{s}_0(0) = -1 + \frac{1 + \kappa - \lambda}{1 + \kappa} = -(\kappa + 1)^{-1}\lambda. \tag{16}$$

Higher order terms can be found either by successively differentiating (1.10a) or by substituting a power series into (11) and equating coefficients (Exercise 2).

For our purposes it will be sufficient to record the result

$$s_0(t) = 1 - (\kappa + 1)^{-1}\lambda t + O(t^2). \tag{17}$$

When $t \gg \lambda^{-1}$, (11) can be approximated by

$$s_0 + \kappa \ln s_0 \approx -\lambda t. \tag{18}$$

Since s_0 is positive, the right side equals a large negative number when t is large. The only way the left side of (18) can yield such a number is for s_0 to approach zero as $t \to \infty$.* When s_0 is small and positive, we have $|\ln s_0| \gg s_0$. Hence (18) can be approximated by $\kappa \ln s_0 \approx -\lambda t$, giving

$$s_0(t) \approx \exp\left(\frac{-\lambda t}{\kappa}\right), \qquad t \text{ large.} \tag{19}$$

Using the above result, our approximation $|\ln s_0| \gg s_0$ seems consistent when $\lambda t/\kappa \gg \exp(-\lambda t/\kappa)$. Since $\exp(-2) \approx 0.14$, we can probably regard this requirement as fulfilled when $\lambda t/\kappa > 2$. Another requirement, made at the beginning, is that $t \gg \lambda^{-1}$. Since λ and κ are typically of order unity, we can thus expect the exponential substrate decay of (19) after a few units of dimensionless time t, i.e., when the dimensional time t^* is a few multiples of $(k_1\bar{e})^{-1}$.

COMMENTS ON RESULTS SO FAR

Until fairly recently, biochemists have been content with an informal derivation of the Michaelis–Menten relation (1b) between complex and substrate. Interpreted as in Exercise 8, this relation has proved very useful in understanding the action of enzymes in biological systems. As one would expect, however, eventually problems arise in which deeper analytically based understanding is required in order that the fundamental ideas can be applied to more complicated situations. One example that shows this is some work of S. Rubinow and J. Lebowitz† concerning the interaction with a substrate and an inhibitor of the enzyme L-asparagine amidohydrolase. (This chemical shows promise in cancer therapy.)

From our detailed analysis we can obtain a quantitative estimate of how long an initial interval must pass before the Michaelis–Menten approximation can be used. From (15) we see that the approximation $s_0/(s_0 + \kappa)$ to c_0 [see (1b)] will be accurate only after an initial interval of dimensionless duration equal to a few multiples of $\varepsilon(\kappa + 1)^{-1}$. During this interval, the complex concentration $c_0(t)$ increases rapidly from its initial value of zero to a value approximately equal to $(\kappa + 1)^{-1}$. In practice the duration of the initial interval is typically a second or less.

HIGHER APPROXIMATIONS

Until now we have found only the simplest kinds of inner, outer, and uniform approximations to the solutions of singular perturbation problems. It is possible to improve on these approximations in a systematic manner, as we now illustrate. The main idea is to posit series expansions for the

* We certainly expect on physical grounds that the substrate will all be used up eventually.

† "Time-dependent Michaelis–Menten Kinetics for an Enzyme–Inhibitor–Substrate System," *J. Amer. Chem. Soc.* *92*, 3888–3893 (1970).

solutions to the inner and outer equations. (The approximations obtained thus far are the first terms in these expansions.) Additional conditions necessary for the unique specification of higher approximations can be obtained by introducing an intermediate variable as before, but requiring a more accurate match.

To illustrate the procedure for finding higher approximations, we return to the enzyme kinetics problem. We now deal with the inner and outer equations by assuming series expansions. For the inner equations (6) this requires the assumption

$$S(\tau, \varepsilon) = S_0(\tau) + \varepsilon S_1(\tau) + \cdots, \qquad C(\tau, \varepsilon) = C_0(\tau) + \varepsilon C_1(\tau) + \cdots. \tag{20}$$

Upon the usual substitution and collection of terms (Exercise 3), we find, of course, that the lowest approximations S_0 and C_0 are given as before by (8). The next higher order approximations satisfy

$$S_1' = -S_0 + (\kappa - \lambda + S_0)C_0, \qquad C_1' = -(\kappa + S_0)C_1 + (1 - C_0)S_1. \tag{21}$$

Upon substitution of the series into the boundary conditions (7), we find that the higher approximations satisfy the homogeneous boundary conditions

$$S_1(0) = 0, \qquad C_1(0) = 0. \tag{22}$$

By elementary but somewhat lengthy calculations, the solutions of (21) and (22) are found to be

$$\begin{aligned} S_1 &= -(\kappa + 1)^{-2}[\lambda\tau_1 + (1 + \kappa - \lambda)(1 - e^{-\tau_1})], \qquad \tau_1 \equiv (\kappa + 1)\tau, \\ C_1 &= -(\kappa + 1)^{-4}\{\lambda\kappa\tau_1 + \kappa(1 + \kappa - 2\lambda) + (1 + \kappa - \lambda)e^{-2\tau_1} \\ &\quad + [\tfrac{1}{2}\lambda\tau_1^2 + (1 + \kappa - \lambda)(1 - \kappa)\tau_1 - (1 + 2\kappa + \kappa^2 - \lambda - 2\kappa\lambda)]e^{-\tau_1}\}. \end{aligned} \tag{23}$$

For the outer solutions we also assume expansions in powers of ε:

$$s(t, \varepsilon) = s_0(t) + \varepsilon s_1(t) + \cdots, \qquad c(t, \varepsilon) = c_0(t) + \varepsilon c_1(t) + \cdots. \tag{24}$$

Note that here the independent variable is the outer variable t. In (20) we used the inner variable τ. Again, the lowest order approximations s_0 and c_0 are given, as before, by (1b) and (11). Exercise 4(a) shows that the next approximations satisfy the equations

$$\dot{s}_1 = (c_0 - 1)s_1 + (\kappa - \lambda + s_0)c_1, \qquad \dot{c}_0 = s_1(1 - c_0) - (\kappa + s_0)c_1. \tag{25a, b}$$

We lack initial conditions. These must be determined by matching.

For higher order matching, the reader will do well if he introduces the intermediate variable and matches "as accurately as is reasonable." To know what is "reasonable" requires experience (some of which will now be provided) and varying degrees of ingenuity.

In the present initial value problem the inner solution is completely determined. Let us construct an intermediate expansion for the inner substrate concentration. To do this, we express the inner variable in terms of the intermediate variable $\tau_i = t/\Psi$ of (9). We find from the expressions for $S_0(\tau)$ and $S_1(\tau)$ in (8) and (23) that

$$S\left(\frac{\tau_i \Psi}{\varepsilon}\right) = S_0\left(\frac{\tau_i \Psi}{\varepsilon}\right) + \varepsilon S_1\left(\frac{\tau_i \Psi}{\varepsilon}\right) + \cdots$$

$$= 1 - \varepsilon(\kappa + 1)^{-2}\left[\lambda(\kappa + 1)\left(\frac{\tau_i \Psi}{\varepsilon}\right) + (1 + \kappa - \lambda) + \text{TST}\right] + \cdots$$

$$= 1 - (\kappa + 1)^{-1}\lambda\tau_i \Psi - (\kappa + 1)^{-2}(1 + \kappa - \lambda)\varepsilon + \text{TST} + \cdots. \quad (26)$$

Following Cole (1968), we use the abbreviation TST for *transcendentally small terms*. In $S_1(\tau_i \Psi/\varepsilon)$ such a term is

$$(\kappa + 1)^{-2}(1 + \kappa - \lambda) \exp\left(-\frac{\tau_i \Psi}{\varepsilon}\right),$$

which decreases exponentially to zero as $\varepsilon \downarrow 0$ with τ_i fixed, since $\Psi/\varepsilon \to +\infty$ by (9c). This is to be contrasted with the algebraic decrease of a term such as $-(\kappa + 1)^{-1}\lambda\tau_i \Psi$, which is explicitly written in (26).

The above expression must be matched with the outer solution $s(t)$ when t is expressed in terms of the intermediate variable τ_i. But how can the term proportional to Ψ in (26) be matched? With the insight provided by the early masters of singular perturbation theory, we observe that at lowest order a match was afforded by proper choice of $s_0(0)$, the "innermost" value of the outer approximation $s_0(t)$. Now we must examine this approximation more closely near $t = 0$. In fact we utilize not just $s_0(0)$ but the first *two* terms of the expansion of $s_0(t)$ about $t = 0$. From (17)

$$s_0(t) = 1 - (\kappa + 1)^{-1}\lambda t + O(t^2) \qquad \text{so } s_0(\tau_i \Psi) = 1 - (\kappa + 1)^{-1}\lambda\tau_i \Psi + O(\Psi^2).$$

Upon introduction of the intermediate variable, the $O(t)$ term in $s_0(t)$ provides just what is necessary to match the $O(\Psi)$ term in (26).

Not only must our more accurate match take account of *two terms* in s_0, but it must also introduce *one term* from s_1. Thus we should match (26) with

$$s_0(\tau_i \Psi) + \varepsilon s_1(\tau_i \Psi) + O(\varepsilon^2) = 1 - (\kappa + 1)^{-1}\lambda\tau_i \Psi + O(\Psi^2) + \varepsilon s_1(0) + O(\varepsilon\Psi) + O(\varepsilon^2). \quad (27)$$

This requires that

$$s_1(0) = -(\kappa + 1)^{-2}(1 + \kappa - \lambda), \quad (28)$$

which is the desired initial condition for s_1.

Let us discuss the matching a little further. We ask: What will match the $O(\Psi^2)$ term in (27)—which came from the $O(t^2)$ term in $s_0(t)$? The matching term must come from the neglected $\varepsilon^2 S_2$ term in (26). Just as a term proportional to $\varepsilon\tau$ in $\varepsilon S_1(\tau)$ gave rise to an $O(\Psi)$ term in the intermediate region, so we expect a term proportional to $\varepsilon^2\tau^2$ in $\varepsilon^2 S_2(\tau)$ to give rise to an $O(\Psi^2)$ term in the intermediate region. In a very satisfactory way, then, the leading terms that are neglected when the approximation is carried to a certain point will be taken into consideration when the approximation is carried one more step.

Having made the specification (28), we expect from (26) and (27) that

$$[S_0(\tau) + \varepsilon S_1(\tau)]_{\tau=\tau_i\Psi/\varepsilon} - [s_0(0) + ts_0'(0) + \varepsilon s_1(0)]_{\tau=\Psi\tau_i}$$
$$= O(\Psi^2) + O(\varepsilon\Psi) + O(\varepsilon^2).$$

We see in retrospect that we can impose the **two-term matching condition**

$$\lim_{\varepsilon\downarrow 0} \frac{1}{\varepsilon} \{[S_0(\tau) + \varepsilon S_1(\tau)]_{\tau=\tau_i\Psi/\varepsilon} - [s_0(0) + ts_0'(0) + \varepsilon s_1(0)]_{t=\Psi\tau_i}\} = 0. \quad (29)$$

Equation (29) holds if $\Psi^2/\varepsilon \to 0$ as $\varepsilon \downarrow 0$. This is a restriction on Ψ in addition to the earlier requirement that $\Psi/\varepsilon \to \infty$ as $\varepsilon \downarrow 0$. It means that our matching is imposed in a layer of $O(\Psi)$ thickness, where $\varepsilon \ll \Psi \ll \sqrt{\varepsilon}$.

Example. By appropriate matching, find the initial condition that $c_1(t)$ must satisfy. Verify that this condition is implied by (28).

Solution. Using (8b) and (23), we find that

$$C_0\left(\frac{\tau_i\Psi}{\varepsilon}\right) + \varepsilon C_1\left(\frac{\tau_i\Psi}{\varepsilon}\right) = (\kappa+1)^{-1} - (\kappa+1)^{-3}\lambda\kappa\tau_i\Psi$$
$$- (\kappa+1)^{-4}\kappa(1+\kappa-2\lambda)\varepsilon + \text{TST}.$$

From (1b) and (17)

$$c_0(t) = \frac{s_0(t)}{s_0(t)+\kappa} = \frac{1-(\kappa+1)^{-1}\lambda t + O(t^2)}{\kappa+1-(\kappa+1)^{-1}\lambda t + O(t^2)}$$
$$= (\kappa+1)^{-1} - (\kappa+1)^{-3}\lambda\kappa t + O(t^2)$$

so that

$$[c_0(t) + \varepsilon c_1(t)]_{t=\tau_i\Psi} = (\kappa+1)^{-1} - (\kappa+1)^{-3}\lambda\kappa\tau_i\Psi + O(\Psi^2) + \varepsilon c_1(0) + O(\varepsilon\Psi).$$

Matching requires the initial condition

$$c_1(0) = -(\kappa+1)^{-4}\kappa(1+\kappa-2\lambda). \quad (30)$$

But from (25b)

$$c_1 = -\frac{\dot{c}_0 + s_1c_0 - s_1}{\kappa + s_0}. \quad (31)$$

Differentiating (1b) and using (2), we find that

$$\dot{c}_0 = \kappa\dot{s}_0(s_0+\kappa)^{-2}. \quad (32)$$

Using this, (2), and (28), we recover (30).

We can now complete the determination of the second approximation. From (25b) we can derive (31), which, upon substitution into (25a), gives an equation for $s_1(t)$. After some computation (Exercise 5), it is seen that this equation, together with initial condition (28), has the solution

$$s_1 = \frac{s_0}{s_0 + \kappa}\left[\frac{\kappa - \lambda}{\kappa}\ln\left(\frac{\kappa + s_0}{(1+\kappa)s_0}\right) - \frac{\kappa - \lambda + s_0}{\kappa + s_0}\right] \tag{33}$$

so that from (31) and (32)

$$c_1 = \frac{1}{(\kappa + s_0)^3}\left[\frac{2\kappa s_0 \lambda}{\kappa + s_0} - \kappa s_0 + s_0(\kappa - \lambda)\ln\left(\frac{\kappa + s_0}{(1+\kappa)s_0}\right)\right]. \tag{34}$$

As in the case of the lowest approximation, the uniform solution $s^{(u)}$ equals the inner solution plus the outer solution minus the common part, but now the solutions are taken to higher order. For example, let $s_1^{(u)}$ be the approximation to $s^{(u)}$ in which $O(\varepsilon)$ terms are retained. Then

$$\begin{aligned} s_1^{(u)}(t,\varepsilon) &= [s_0(t) + \varepsilon s_1(t)] + \left[S_0\left(\frac{t}{\varepsilon}\right) + \varepsilon S_1\left(\frac{t}{\varepsilon}\right)\right] \\ &\quad - [1 - (\kappa+1)^{-1}\lambda t - \varepsilon(\kappa+1)^{-2}(1+\kappa-\lambda)] \\ &= s_0(t) + \frac{\varepsilon s_0(t)}{s_0(t)+\kappa}\left[\frac{\kappa-\lambda}{\kappa}\ln\left(\frac{\kappa + s_0(t)}{(\kappa+1)s_0(t)}\right) - \frac{\kappa - \lambda + s_0(t)}{\kappa + s_0(t)}\right] \\ &\quad + \frac{\varepsilon(1+\kappa-\lambda)e^{-(\kappa+1)t/\varepsilon}}{(\kappa+1)^2}. \end{aligned} \tag{35}$$

FURTHER ANALYSIS FOR LARGE TIMES

We have investigated a problem in enzyme kinetics by developing and explaining principles that are of wide application in singular perturbation theory. The present problem, however, is one of the widening (but by no means all-inclusive) class that can be treated by a rigorously based "prescription." As applied by Heineken et al. (op. cit.), the prescription comes from work by the Soviet mathematician A. B. Vasil'eva.* This work proves, for example, that the above expression for $s_1^{(u)}(t, \varepsilon)$ is an asymptotic approximation to the exact solution $s^{(e)}(t, \varepsilon)$ in the sense that

$$|s^{(e)}(t,\varepsilon) - s_1^{(u)}(t,\varepsilon)| < a\varepsilon^2 \qquad \text{when } 0 \le t \le T, \tag{36}$$

for sufficient small ε. The quantity a is independent of ε and t but it may depend on the constant T. In other words the error is uniformly $O(\varepsilon^2)$ throughout a time interval of fixed length T.

In contemplating a result such as the one just stated, one naturally is curious as to whether the restriction to a fixed time interval is made because

* *Russ. Math. Surv. 18*, 13 (1963).

of technical difficulties in constructing the proof or whether the restriction is required by the nature of the problem. That the former is likely is seen by calculating the ratio $\varepsilon s_1(t)/s_0(t)$, using the "large time" approximation $s_0(t) \approx \exp(-\lambda t/\kappa)$ of (19). We find

$$\frac{s_1(t)}{s_0(t)} \approx \frac{1}{\kappa}\left[\frac{\kappa-\lambda}{\kappa}\ln\frac{\kappa}{(1+\kappa)\exp(-\lambda\tau/\kappa)} - \frac{\kappa-\lambda}{\kappa}\right] \approx \frac{\lambda(\kappa-\lambda)t}{\kappa^3}. \tag{37}$$

For our approximation to be apparently consistent (in the terminology of Section 6.1), the correction $\varepsilon s_1(t)$ should be small compared with $s_0(t)$. From the above calculation we see that this is the case (if $\lambda \neq \kappa$) unless $\varepsilon t = O(1)$, $t = O(1/\varepsilon)$. But when $t = O(1/\varepsilon)$ then, by (19), $s_0 \approx \exp(-\lambda/\kappa\varepsilon)$, a transcendentally small quantity. Thus the inconsistency affects the solution only when it is completely negligible. Indeed, direct calculation of the solution for very large times produces results that are the same as those of (19) and (1b). (See Exercise 7.)

Our remarks lead to the conjecture that inequality (36) holds for *all* nonnegative t. Proofs of this type of conjecture have in fact been given. [See F. C. Hoppensteadt, "Singular Perturbations on the Infinite Interval," *Trans. Amer. Math. Soc. 123*, 521–35 (1966).]

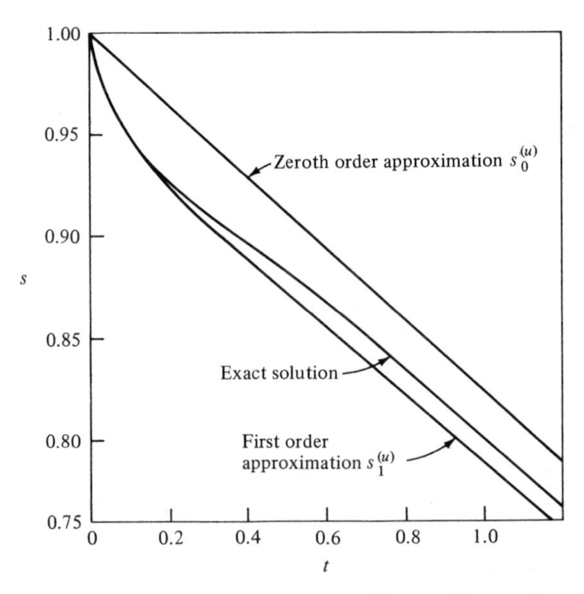

FIGURE 10.2. *Approximations to the dimensionless substrate concentration when* $\varepsilon = 0.1$. *The "exact" solution was obtained numerically. [Redrawn with permission, from F. G. Heineken, H. M. Tsuchiya, and R. Aris, "On the Mathematical Status of the Pseudo-steady State Hypothesis of Biochemical Kinetics,"* Mathematical Biosciences **1**, 95–113 (1967).]

FURTHER DISCUSSION OF THE APPROXIMATE SOLUTIONS

To gauge the accuracy of the approximations they obtained, Heineken et al. (op. cit.) found by numerical calculations some "exact" solutions to the governing equations (1.10). Employing data relevant to the hydrolysis of benzoyl-L-arginine ethyl ester catalyzed by the enzyme trypsin, they took $\kappa = 1$, $\lambda = 0.375$. Their results are shown in Figures 10.2–10.4. These figures show excellent agreement with $\varepsilon = 0.1$, "a value that the biochemist would regard as quite large," and good agreement even when $\varepsilon = 1.0$.

The more analytic information that is available concerning the time course of the chemical reaction, the easier it is to deduce the values of the various parameters from concentration measurements. Analytic solutions lend themselves readily to investigations of parameter variation. But perhaps the largest recompense for an investigation such as the one just concluded is the thorough understanding it gives of the scientific phenomenon.

From the point of view of presenting important ideas in applied mathematics, the present example has served to illustrate the technique for finding higher order approximations in singular perturbation theory. The reader

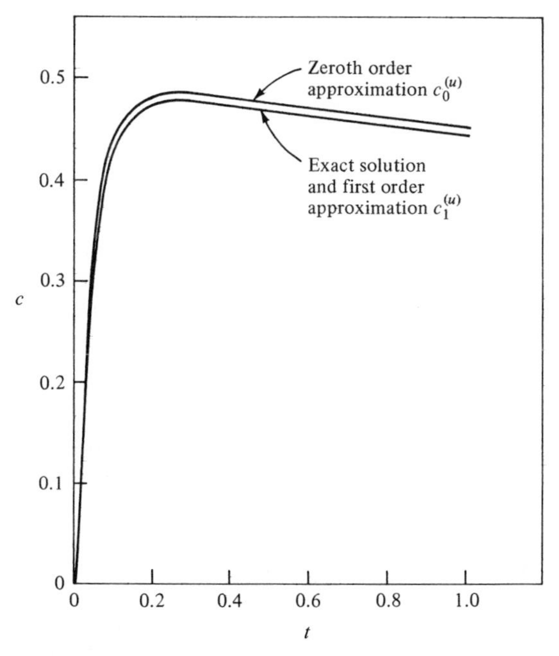

FIGURE 10.3. *Approximations to the dimensionless complex concentration when* $\varepsilon = 0.1$. [*Redrawn, with permission, from F. G. Heineken, H. M. Tsuchiya, and R. Aris, "On the Mathematical Status of the Pseudo-steady State Hypothesis of Biochemical Kinetics,"* Mathematical Biosciences **1**, 95–113 (1967).]

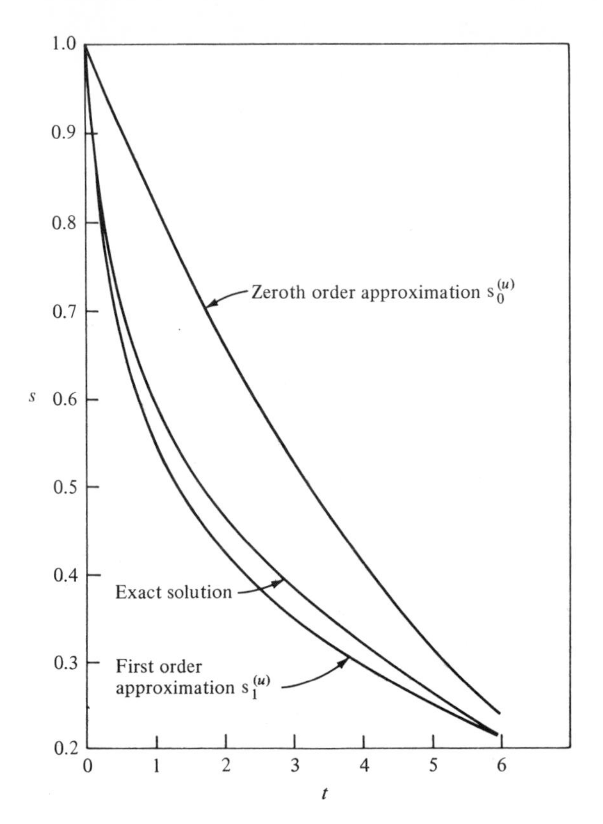

FIGURE 10.4. *Approximations to the dimensionless substrate concentration when* $\varepsilon = 1.0$. [*Redrawn, with permission, from F. G. Heineken, H. M. Tsuchiya, and R. Aris,* "*On the Mathematical Status of the Pseudo-steady State Hypothesis of Biochemical Kinetics,*" Mathematical Biosciences **1**, 95–113 (1967).]

should obtain a good initial grasp of this matter if he fills in the details of the present discussion, particularly as in Exercises 3, 4, and 6, and also completes Exercise 9, which involves higher approximations for a boundary value problem.

EXERCISES

1. (a) Verify (3).
(b) Verify that (8) gives the outer approximations S_0 and C_0.

2. (a) Find $\ddot{s}_0(0)$ and $\dddot{s}_0(0)$ by successively differentiating (1.10a).
(b) Assume that $s_0(t) = 1 + \sigma_1 t + \sigma_2 t^2 + \sigma_3 t^3 + \cdots$; find σ_1, σ_2, and σ_3 by substituting into (11) and collecting terms. Check your answer by comparing with part (a).

3. (a) Show that S_1 and C_1, the $O(\varepsilon)$ coefficients in the outer expansions, satisfy (21) and (22).
(b) Find the equations that are satisfied by S_2 and C_2.
‡(c) Obtain S_1 and C_1, as in (23).

4. (a) Show that s_1 and c_1, the $O(\varepsilon)$ coefficients in the inner expansions, satisfy (25).
(b) Find the equations that are satisfied by s_2 and c_2.

5. ‡(a) Use (31) to obtain a nonhomogeneous first order differential equation for s_1. Show that this equation and the initial condition (28) has the solution (33).
(b) Find the solution for c_1, and thereby verify (34).

6. Find in a manner similar to that used in (35), the uniform approximation $c_1^{(u)}$.

7. For very large times we expect that $s = s^*/ \ll 1$ and $c = c^*/\bar{c} \ll 1$. Subject to this expectation, find an approximate solution to (10). Show that retention of the dominant term in the solution permits a match with (19) and (1b), and thereby show that s_0 and c_0 continue to give first approximations to s and c even for very large times.

8. The object of this exercise is to indicate the way that biochemists utilize the Michaelis–Menten approximation.
(a) Define the rate of product formation V by $V = dp^*/dt^*$. Let $V_{\max}$ be the maximum rate of product formation at any time for any initial substrate concentration s_0. Use (1.3e) to show that $V_{\max} = k_2 \bar{e}$.
(b) Using the Michaelis–Menten equation (2), show that when $s = \kappa$, then $V = \frac{1}{2}V_{\max}$. Equivalently, show that $V = \frac{1}{2}V_{\max}$ when $s^* = K_m$, where $K_m = (k_{-1} + k_2)/k_1$. (This gives an interpretation of κ or, as biochemists prefer, of K_m, the **Michaelis constant**.)
(c) Show that the results can be presented as shown in Figure 10.5. This double-reciprocal plot is deservedly popular with biochemists because (if applicable) one can fit data with a straight line.

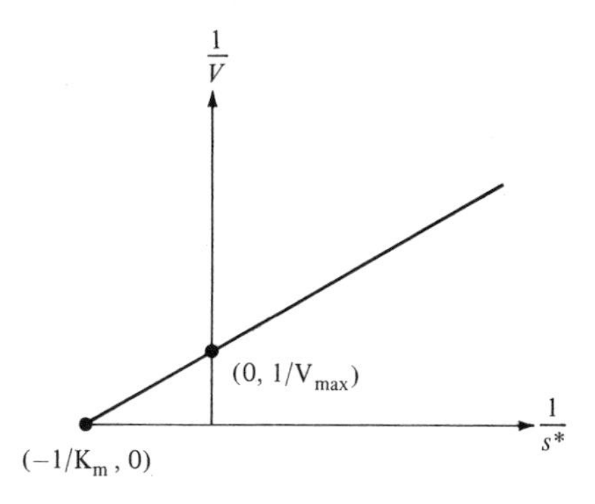

FIGURE 10.5. *Lineweaver–Burk double reciprocal plot relating velocity V (rate of product formation) with substrate concentration s^*.*

‡(d) Part (c) gives an interpretation of κ. Rather than use the other dimensionless constant, λ, it is perhaps preferable to present the results in terms of $\beta \equiv k_{-1}/k_2$ $[\lambda = \kappa/(1+\beta)]$. The quantity β is the rate at which the complex dissociates back into substrate and enzyme, divided by the rate at which it yields enzyme plus product. But λ, or other combinations of κ and λ, are also directly interpretable. What is the interpretation of λ? Of some other combinations?

9. Consider

$$\varepsilon y'' + (1+x)y' + y = 0, \qquad y(0) = 0, \quad y(1) = 1, \quad 0 < \varepsilon \ll 1.$$

This is a special case of an illustrative problem discussed by Cole (1968).

(a) Show that the outer approximation is $y_0(x) = 2/(1+x)$ and the inner approximation is $Y_I(\xi) = 2 - 2\exp(-\xi)$, where $\xi = x/\varepsilon$. Show that the uniform approximation is $2/(1+x) - 2\exp(-x/\varepsilon)$.

(b) Show that the $O(\varepsilon)$ contribution to the outer solution is

$$y_1(x) = -\tfrac{1}{2}(1+x)^{-1} + 2(1+x)^{-3},$$

while the $O(\varepsilon)$ contribution to the inner solution is

$$Y_1(\xi) = C(1 - e^{-\xi}) - 2(e^{-\xi} + \xi - 1 - \tfrac{1}{2}(\xi^2 e^{-\xi}).$$

(c) Verify that to determine the constant C, one should set $\eta(\varepsilon) = x/\Theta(\varepsilon)$ and match

$$y_0(\Theta\eta) + \varepsilon y_1(\Theta\eta) \equiv 2 - 2\Theta\eta + O(\Theta^2\eta^2) + \frac{3\varepsilon}{2} + O(\varepsilon\Theta\eta)$$

and

$$Y_0\left(\frac{\Theta\eta}{\varepsilon}\right) + \varepsilon Y_1\left(\frac{\Theta\eta}{\varepsilon}\right) = 2 - 2\Theta\eta + O(\Theta^2\eta^2) + 2\varepsilon + C\varepsilon + \text{TST},$$

giving $C = -\frac{1}{2}$. (Recall that TST indicates transcendentally small terms.)

(d) Find the uniform approximation that retains terms through $O(\varepsilon)$.

CHAPTER 11
Three Techniques Applied to the Simple Pendulum

THIS chapter contains some brief discussions of certain useful techniques and concepts. All the discussions are centered on the pendulum problem. We have already treated this problem in some detail, but nevertheless, as will be seen, much can be learned from further investigation. From this we draw the general moral that a surprising amount of scientific understanding can be gleaned from an intensive study of a properly selected special problem.

In Section 11.1, we investigate the fate of small perturbations to the two possible equilibrium positions of the pendulum, bob upward and bob downward. This provides a simple illustration of the concept of stability, which was briefly introduced in our earlier investigation of ameba aggregation (Section 1.3) and which is taken up again in more detail when we study the stability of density-stratified fluids (Section 15.2).

Section 11.2 illustrates the power of the "multiple scale" perturbation method. For the pendulum this method recognizes that there are two time scales, one during which the pendulum executes an oscillation, and a much longer one during which the period of the oscillation changes significantly. The multiple scale or "two-timing" method has recently found application in many contexts.

Section 11.3 shows how qualitative understanding of all possible motions of a simple pendulum can be obtained by an analysis of the displacement–angular velocity plane. The damped and undamped pendulum are contrasted. Another method of wide applicability is thereby illustrated, phase plane analysis.

The three sections are mutually independent and can be read in any order.

11.1 Stability of Normal and Inverted Equilibrium of the Pendulum

Very often, in subjects ranging from mechanics to economics, the equations that govern a phenomenon have time-independent **steady state** or **equilibrium** solutions. A simple pendulum, for example, has two such solutions—one with the bob downward (**normal equilibrium**) and one with the bob upward (**inverted equilibrium**).

Why is it that we never see a pendulum in inverted equilibrium? It is because the equilibrium is unstable, in the sense that the slightest disturbance will result in a radically different situation. This rather simple problem will be examined here, to provide an elementary example of a general technique that will be developed further in Section 15.2.

DETERMINING STABILITY OF EQUILIBRIUM

The equation governing the motion of the pendulum contains the time variable t only in t derivatives. This occurs in many problems, indeed in all problems in which the choice of a time origin is immaterial. Using the pendulum as an example for such time-translation-invariant situations, we shall outline a procedure for investigating the stability of equilibrium. We shall restrict ourselves to small departures from equilibrium or **perturbations**. As a consequence we need only consider linear equations. Nevertheless, much useful information can be obtained.

STEP A. *Determine all equilibrium solutions.* For the pendulum it is convenient to write the equation for the angular displacement θ^* in the form

$$\frac{d^2\theta^*}{dt^2} + \sin\theta^* = 0. \tag{1}$$

[Here t is time, measured in units $(L/g)^{1/2}$. The derivation of (1) was given in Section 2.2.]

By definition an equilibrium state does not change with time; hence we search for a constant Θ such that $\theta^* = \Theta$. Thus $\sin\Theta = 0$ and we have two cases.

Case (*i*): $\Theta = 0$, normal equilibrium.

Case (*ii*): $\Theta = \pi$, inverted equilibrium.

STEP B. *Introduce a variable that measures departure from equilibrium.* For the pendulum, such a variable is θ', where

$$\theta' = \theta^* - \Theta. \tag{2}$$

In terms of θ', the governing equation is

$$\frac{d^2\theta'}{dt^2} + \sin(\Theta + \theta') = 0. \tag{3}$$

STEP C. *Assume that the departure from equilibrium is just a small perturbation and retain only linear terms in the governing equations.* To obtain an approximate equation for θ', we use the Taylor formula

$$f(\theta') = f(0) + f'(0)\theta' + f''(\xi)\frac{(\theta')^2}{2!},$$

where ξ is between 0 and θ'. We neglect the remainder term, since it is proportional to $(\theta')^2$, a nonlinear expression composed of the square of a small quantity. This gives the *linear* approximation

$$\sin(\Theta + \theta') = \sin\Theta + \theta'\cos\Theta.$$

Substituting for Θ, we find in case (i) that

$$\frac{d^2\theta'}{dt^2} + \theta' = 0, \tag{4a}$$

and in case (ii) that

$$\frac{d^2\theta'}{dt^2} - \theta' = 0. \tag{4b}$$

STEP D. *Determine the behavior of the perturbations as functions of time.* We find that (4a) has the solution

$$\theta' = \theta_0 \cos t + \theta_1 \sin t, \tag{5}$$

where the constants θ_0 and θ_1 are determined by the initial conditions

$$\theta'(0) = \theta_0, \qquad \frac{d\theta'}{dt}(0) = \theta_1. \tag{6}$$

According to (5), if one slightly perturbs a pendulum hanging in normal equilibrium, small oscillations result. The smaller the initial perturbation displacement θ_0 and velocity θ_1, the smaller the amplitude of the resulting oscillations.

To investigate the stability of the inverted pendulum, we solve (4b) and find, using the initial conditions (6),

$$\theta' = \tfrac{1}{2}(\theta_0 + \theta_1)e^t + \tfrac{1}{2}(\theta_0 - \theta_1)e^{-t}, \tag{7a}$$

or equivalently,

$$\theta' = \theta_0 \cosh t + \theta_1 \sinh t. \tag{7b}$$

DISCUSSION OF RESULTS

The inverted pendulum is an example of an equilibrium state wherein small perturbations almost certainly increase with time. Since slight disturbances are inevitable, such unstable equilibrium points should not be observed in practice.*

An exceptional situation occurs when $\theta_1 = -\theta_0$. Here (7a) shows that $\theta'(t)$ decays exponentially to zero. Evidently, for an initial angular displacement from equilibrium θ_0, an initial angular velocity of $-\theta_0$ is just exactly enough to restore the pendulum, finally, to the inverted equilibrium position. While theoretically interesting, this exceptional situation is of little practical significance.

* The pendulum *can* persist in an inverted position if the pivot oscillates appropriately. See, for example, J. P. Den Hartog, *Mechanical Vibrations* (New York: McGraw-Hill, 4th ed., 1956), pp. 348–50.

Equilibrium with bob upward is said to be **unstable** because there exists a small perturbation that grows with time. This definition takes into account the fact that only unusual initial conditions would be such as to exclude this growing solution.

Equilibrium with bob downward is termed **neutrally stable** to the class of perturbations considered, because these small disturbances neither grow nor decay. Another term is also used to describe this situation, **Liapunov stable**. This term is applicable if (as here) the perturbed solution will remain arbitrarily close to the equilibrium solution, provided that the initial disturbance is sufficiently small.

The effect of damping is considered in Section 11.3. There we see (from a somewhat wider point of view) that small perturbations from normal equilibrium eventually die out completely, so that the equilibrium is again attained. In such a case we speak of **asymptotic stability**. As a matter of fact the pendulum eventually comes to rest in the bob-down position, whatever the initial condition, when damping is present. This position is therefore said to be **globally asymptotically stable**. The large number of definitions, and there are more, give a hint of the subtleties involved in stability theory.

We remind the reader that we have deleted all nonlinear terms, a procedure that is reasonable in a study of small perturbations, but one that has not been rigorously justified. On this matter we comment that our results are expected to be correct except perhaps for the borderline case of neutral stability, where a tiny change in the problem could nudge the state of the system either way.

A more extensive study of stability theory will be found in Section 15.2. There, the stability of a layer of fluid whose density varies is analyzed. As one would expect, there are close parallels between the behavior of a top-heavy pendulum and a top-heavy fluid layer. But the latter situation is considerably more complicated; its description involves partial differential equations.

EXERCISES

1. In biological applications the population P of certain organisms at time t is sometimes assumed to obey the **logistic** or **Verhulst–Pearl equation**

$$\frac{dP}{dt} = aP(1 - PE^{-1}), \qquad t = \text{time}, E \text{ a constant}. \tag{8}$$

(a) Determine the equilibrium population levels.
(b) Examine their stability.
(c) With the information obtained from (a) and (b), discuss the qualitative behavior of the population. Reinforce your assertions by examining the levels of P at which P is decreasing and increasing, respectively.
(d) Solve (8) exactly and compare the results with (a), (b), and (c).

2. Another population equation in common use, associated with the name of Gompertz, is

$$\frac{dP}{dt} = aP(\ln E - \ln P).$$

Carry out the instructions of Exercise 1(a), (b), (c) for this model.

3. Still another population model is

$$\frac{dP}{dt} = aP(E_1 P^{-c} - E_2),$$

where the "competition constant" c satisfies $0 < c \leq 1$. Carry out the instructions of Exercise 1(a), (b), (c) for this model.

11.2 A Multiple Scale Expansion

The stability analysis of Section 11.1 revealed important distinctions between equilibrium points, and, by virtue of linearization, this was easily accomplished. But our approach is of limited validity. The limitations are most manifest in our treatment of the inverted pendulum. The solution (1.7) predicts a displacement that continues to increase exponentially, but if we tap an inverted pendulum we do not expect it to whirl with ever-increasing speed. After a time, however, (1.7) cannot be expected to hold, because there is an inconsistency between the large predicted values of θ' and "small perturbation theory." There are no apparent inconsistencies connected with (1.5) but, as we have discussed in Section 7.1, this solution, too, has deficiencies that become more and more pronounced as time passes. In Chapter 2 we have briefly indicated one method of remedying the deficiency. This is Poincaré's introduction of a perturbed independent variable designed to take into account the fact that small oscillations have a period that depends mildly on the amplitude of oscillation. We now discuss another approach.

In our discussion of singular perturbation theory we pointed out that the solutions to a number of problems possessed more than one scale. For example, in the illustrative boundary value problem (Equation 9.2.1)

$$\varepsilon \frac{d^2y}{dx^2} + 2\frac{dy}{dx} + y = 0; \qquad y(0) = 0, \quad y(1) = 1, \quad 0 < \varepsilon \ll 1,$$

the solution varied rapidly with scale $1/\varepsilon$ in the boundary layer near $x = 0$ and then varied more slowly with an $O(1)$ scale in the rest of the interval. Similar variation occurred in the enzyme kinetics problem of Chapter 10. In the pendulum problem, we saw in Section 7.1 that the regular perturbation solution fails because on a long $O(1/a^2)$ time scale there is a significant deviation in the pendulum's position compared to the prediction made by linearized theory, with its slightly erroneous amplitude-independent $O(1)$ period.

It has become increasingly evident in recent years that explicit formal recognition of the presence of multiple scales leads to approximation methods of great power. For example, consider the simple problem

$$\ddot{y} + 2\varepsilon\dot{y} + (1 + \varepsilon^2)y = 0, \quad y(0) = 0, \quad \dot{y}(0) = 1 \qquad \left(\cdot = \frac{d}{dt}\right). \tag{1}$$

The exact solution is

$$y(t, \varepsilon) = \exp(-\varepsilon t) \sin t. \tag{2}$$

Suppose that one applied regular perturbation theory to obtain an expansion in powers of ε, assuming that

$$y(t, \varepsilon) = y_0(t) + \varepsilon y_1(t) + \varepsilon^2 y_2(t) + \cdots . \tag{3}$$

One would then obtain the following expansion of (2):

$$y(t, \varepsilon) = \sin t - \varepsilon t \sin t + \tfrac{1}{2}\varepsilon^2 t^2 \sin t + \cdots . \tag{4}$$

As expected, terms containing powers of t appear and limit the usefulness of the expansion. On the other hand, suppose that a solution of the form $y = f(t, \varepsilon t, \varepsilon)$ were sought, where

$$f(t, \tau, \varepsilon) = f^{(0)}(t, \tau) + \varepsilon f^{(1)}(t, \tau) + \varepsilon^2 f^{(2)}(t, \tau) + \cdots . \tag{5}$$

Here the existence of two scales is explicitly recognized. By comparing with the exact answer, we know that $f^{(i)}(t, \tau) \equiv 0$, $i \geq 1$, and that

$$f^{(0)}(t, \tau) = e^{-\tau} \sin t, \qquad \text{so } f(t, \varepsilon t, \varepsilon) = e^{-\varepsilon t} \sin t. \tag{6}$$

The first term of the expansion would yield the exact solution in this specially designed problem.

We shall now illustrate the multiple scale method on the pendulum problem written in the scaled form of (7.1.4):

$$\frac{d^2\Theta}{dt^2} + \frac{\sin(a\Theta)}{a} = 0; \qquad \text{at } t = 0, \quad \Theta = 1, \quad \frac{d\Theta}{dt} = 0. \tag{7a}$$

We shall show that if an expansion like (5) is assumed, then the successive terms in the series can be systematically obtained. Approximations result that are valid for successively longer time intervals.

The solution Θ of (7a) is actually a function of a^2 (see Section 7.1). To bring this out, we introduce $a^2 = \varepsilon$ and expand $\sin(a\Theta) = \sin(\varepsilon^{1/2}\Theta)$ in a Taylor series. Equation (7a) is thus replaced by

$$\frac{d^2\Theta}{dt^2} + \Theta - \tfrac{1}{6}\varepsilon\Theta^3 + \cdots = 0. \tag{7b}$$

SUBSTITUTION OF A TWO-SCALE SERIES INTO THE PENDULUM EQUATION

To obtain an approximate solution to (7b) we assume that $\Theta(t, \varepsilon) = f(t, \varepsilon t, \varepsilon)$ and that an expansion of the form (5) is valid in some sense. In computing derivatives, we must use the chain rule to find, for example, that if $f^{(0)}(t, \tau) \equiv \exp(-\tau) \sin t$, then

$$\frac{d}{dt} f^{(0)}(t, \varepsilon t) = \frac{\partial}{\partial t}(e^{-\tau} \sin t)\bigg|_{\tau=\varepsilon t} \frac{dt}{dt} + \frac{\partial}{\partial \tau}(e^{-\tau} \sin t)\bigg|_{\tau=\varepsilon t} \frac{d\tau}{dt}$$

$$= e^{-\varepsilon t} \cos t - \varepsilon e^{-\varepsilon t} \sin t.$$

More generally, it is convenient to write

$$\frac{d}{dt} f^{(0)}(t, \varepsilon t) = f_1^{(0)} \frac{dt}{dt} + f_2^{(0)} \frac{d\tau}{dt} = f_1^{(0)} + \varepsilon f_2^{(0)},$$

where the subscript i indicates partial differentiation with respect to the ith argument. Thus if we assume that

$$f(t, \varepsilon t, \varepsilon) = f^{(0)}(t, \varepsilon t) + \varepsilon f^{(1)}(t, \varepsilon t) + \varepsilon^2 f^{(2)}(t, \varepsilon t) + \cdots, \tag{8}$$

then

$$\frac{d}{dt} f(t, \varepsilon t, \varepsilon) = f_1^{(0)} + \varepsilon[f_2^{(0)} + f_1^{(1)}] + \varepsilon^2[f_2^{(1)} + f_1^{(2)}] + \cdots \tag{9}$$

and

$$\frac{d^2}{dt^2} f(t, \varepsilon t, \varepsilon) = f_{11}^{(0)} + \varepsilon[f_{12}^{(0)} + f_{21}^{(0)} + f_{11}^{(1)}]$$
$$+ \varepsilon^2[f_{22}^{(0)} + f_{12}^{(1)} + f_{21}^{(1)} + f_{11}^{(2)}] + \cdots, \tag{10}$$

where all functions on the right side of (9) and (10) are to be evaluated at $(t, \varepsilon t)$. Assuming continuity of second partial derivatives, $f_{21}^{(i)} = f_{12}^{(i)}$, so that we can write (10) as

$$\frac{d^2 f(t, \varepsilon t, \varepsilon)}{dt^2} = f_{11}^{(0)} + \varepsilon[2f_{12}^{(0)} + f_{11}^{(1)}] + \varepsilon^2[f_{22}^{(0)} + 2f_{12}^{(1)} + f_{11}^{(2)}] + \cdots. \tag{11}$$

We are now ready for the *standard step* of *substituting* the *assumed series into the governing equation*. Using (10), we thus find that if Θ is given by the series f of (8), then (7b) implies that

$$0 = [f_{11}^{(0)}(t, \varepsilon t) + f^{(0)}(t, \varepsilon t)]$$
$$+ \varepsilon[f_{11}^{(1)}(t, \varepsilon t) + f^{(1)}(t, \varepsilon t) + 2f_{12}^{(0)}(t, \varepsilon t) - \tfrac{1}{6}(f^{(0)}(t, \varepsilon t))^3] + \cdots.$$

Since the terms in the square brackets depend on ε, we have here a situation which is *not* that of a power series identically equal to zero. It is not *necessary* to set equal to zero the factors multiplying successive powers of ε. It is, however, sufficient and convenient, so we impose the conditions

$$f_{11}^{(0)} + f^{(0)} = 0, \tag{12}$$

$$f_{11}^{(1)} + f^{(1)} = \tfrac{1}{6}[f^{(0)}]^3 - 2f_{12}^{(0)}. \tag{13}$$

The initial conditions of (7) require, using (9), that

$$f^{(0)}(0, 0) = 1, \qquad f_1^{(0)}(0, 0) = 0; \tag{14}$$

$$f^{(1)}(0, 0) = 0, \qquad f_1^{(1)}(0, 0) = -f_2^{(0)}(0, 0). \tag{15}$$

SOLVING LOWEST ORDER EQUATIONS

We are faced with the *partial* differential equation (12). If we regard f as a function of the variables t and τ, this equation can be written

$$\frac{\partial^2 f^{(0)}}{\partial t^2} + f = 0.$$

We see that τ appears only as a parameter; so the partial differential equation is, in fact, no more difficult to solve than the simple ordinary differential equation one meets in finding an approximate solution to the pendulum problem by regular perturbation theory. Thus

$$f^{(0)}(t, \tau) = A(\tau) \cos t + B(\tau) \sin t, \tag{16}$$

where the "arbitrary constants" A and B must be regarded as functions of τ. The initial conditions (14) lead to the requirement

$$A(0) = 1, \qquad B(0) = 0. \tag{17}$$

At this stage we have reached essentially the same conclusions as we did in our regular perturbation calculation, where the lowest approximation was

$$\Theta \approx 1 \cdot \cos t + 0 \cdot \sin t. \tag{18}$$

We do not know what $A(\tau)$ and $B(\tau)$ are, but we do know that they are initially unity and zero, respectively. We also know that it is $A(\varepsilon t)$ and $B(\varepsilon t)$ which appear in the solution so that these quantities will remain close to their initial values until t becomes comparable with $1/\varepsilon$. Equation (18) gives no fuller information than (16) and (17), for we know that (18) becomes a poor approximation just when $t = O(1/\varepsilon)$.

HIGHER APPROXIMATIONS; REMOVING RESONANT TERMS

To proceed with the calculation, we substitute the results we have obtained so far into (13), obtaining

$$\frac{\partial^2 f^{(1)}}{\partial t^2} + f^{(1)} = \frac{1}{6}(A \cos t + B \sin t)^3 - 2(-A' \sin t + B' \cos t),$$

where $' = d/d\tau$. Using trigonometric identities [as in Exercise 2(a)], we can rewrite this equation in the form

$$\frac{\partial^2 f^{(1)}}{\partial t^2} + f^{(1)} = [2B' + \frac{1}{8}(A^3 + AB^2)] \cos t$$
$$+ [-2A' + \tfrac{1}{8}(B^3 + A^2B)] \sin t$$
$$+ \tfrac{1}{24}(A^3 - 3AB^2) \cos 3t - \tfrac{1}{24}(B^3 - 3A^2B) \sin 3t. \qquad (19)$$

Since (19) is "essentially" an ordinary differential equation, we can employ the method of undetermined coefficients. According to this method, since $\cos t$ and $\sin t$ are solutions of the homogeneous equation, the corresponding particular solutions will contain terms proportional to $t \cos t$ and $t \sin t$, respectively.

The presence of $\cos t$ and $\sin t$ on the right side of (19) will thus force the term $\varepsilon f^{(1)}$ in the series (8) to contain terms proportional to $\varepsilon t \cos t$ and $\varepsilon t \sin t$. However small ε is, these terms will eventually cease to become negligible, compared to the lowest order approximation $f^{(0)}$. We wish to avoid this. Indeed we shall require that the successive approximations $\sum_{i=0}^{N} \varepsilon^i f^{(i)}$ are of order ε^N uniformly in time. To this end *we require that each* $f^{(i)}(t, \varepsilon t)$ *is bounded for* $0 \le t \le \infty$. *Methodical application of this boundedness requirement forms the heart of multiple scale methods.*

In the present instance the boundedness requirement demands the deletion of the "resonant" forcing terms proportional to $\cos t$ and $\sin t$ in (19); therefore we must have

$$2B' + \tfrac{1}{8}(A^3 + AB^2) = 0, \qquad -2A' + \tfrac{1}{8}(B^3 + A^2B) = 0. \qquad (20a, b)$$

At first this nonlinear system might seem formidable, but soon we notice that since $A(\tau) \neq 0$, at least near $\tau = 0$ [since $A(0) = 1$], we can use (20a) to find $A^2 + B^2 = -16B'/A$. Substitution into (20b) then gives

$$-2A' - \frac{2BB'}{A} = 0 \qquad \text{or} \qquad (A^2 + B^2)' = 0.$$

Using the initial condition (17), we at once infer that $A^2 + B^2 = 1$. Substituting back into (20), we find that $16B' + A = 0$, $-16A' + B = 0$. Again employing (17), we find that [Exercise 2(b)]

$$A(\tau) = \cos \frac{\tau}{16}, \qquad B(\tau) = -\sin \frac{\tau}{16}. \qquad (21)$$

We use (16) and recall that $\varepsilon = a^2$, and so we conclude that the pendulum amplitude can be approximated to lowest order by

$$f^{(0)}(t, a^2 t) = \cos \frac{a^2 t}{16} \cos t - \sin \frac{a^2 t}{16} \sin t$$
$$= \cos \frac{1 + a^2}{16} t, \qquad (22)$$

in agreement with the result of (2.2.35) and (2.2.37).

With (20), (19) becomes [Exercise 2(c)]

$$\frac{\partial^2 f^{(1)}}{\partial t^2} + f^{(1)} = \frac{1}{24}(A^3 - 3AB^2)\cos 3t - \tfrac{1}{24}(B^3 - 3A^2B)\sin 3t. \qquad (23)$$

This has the solution [Exercise 2(d)]

$$f^{(1)} = C(\tau)\cos t + D(\tau)\sin t - \tfrac{1}{192}(A^3 - 3AB^2)\cos 3t + \tfrac{1}{192}(B^3 - 3A^2B)\sin 3t + O(\varepsilon). \qquad (24)$$

At this stage, all that is known about C and D are the initial conditions required by (15). The calculations can in principle be continued. We shall not carry them further, however, for the main ideas have already been illustrated.

We note the appearance in (24) of higher frequency harmonics proportional to $\cos 3t$ and $\sin 3t$. It should be apparent that other harmonics will continue to appear if the calculations are continued. This harmonic generation is a characteristic nonlinear phenomenon.

The simple linearized theory of the pendulum makes an $O(a^2)$ error in its prediction of the period. This means, as we have seen, that predictions become unreliable when $t = O(a^{-2})$. The improved result (22) provides an $O(a^2)$ correction to the period so that predictions are expected to be reliable until t becomes $O(a^{-4})$. Exercises 4 and 5 show that the next approximation should hold until $t = O(a^{-6})$. It is typical of a class of solution methods for nonlinear problems defined on unbounded intervals that successive approximations adequately cover successively larger portions of these intervals.

SUMMARY AND DISCUSSION

Let us sum up the basic elements of the procedure. Consider an ordinary differential equation for $f(t, \varepsilon)$, where ε is a small parameter. Suppose that failure of a regular perturbation expansion to be uniformly valid, or some other reason, leads to the supposition that f varies on two scales, one $O(1)$ and the other $O(\varepsilon)$. Calculations can then proceed as follows.

(a) Assume the expansion

$$f(t, \varepsilon) = f^{(0)}(t, \tau, \varepsilon) + \varepsilon f^{(1)}(t, \tau, \varepsilon) + \varepsilon^2 f^{(2)}(t, \tau, \varepsilon) + \cdots, \qquad \tau \equiv \varepsilon t.$$

(b) Substitute into the original equation and obtain a sequence of equations for the $f^{(i)}$ by equating to zero the coefficients of successively higher powers of ε.

(c) Remove indeterminancy by requiring that the $f^{(i)}$ remain bounded for all time.

(d) Solve the equations for the $f^{(i)}$, successively.

REMARKS. (i) Higher approximations often require that the variables t and $\tau \equiv \varepsilon t$ be supplemented by $\tau_2 \equiv \varepsilon^2 t$, $\tau_3 \equiv \varepsilon^3 t$, etc. (See Exercises 5 and 6.) (ii) The equations for the $f^{(i)}$ are *partial* differential equations, but they involve both variables only in the nonhomogeneous terms. Thus there is little added difficulty on this score, since "extra" variables in many respects behave like parameters.

We have provided a guide to the use of multiple scale techniques in a class of problems, but this is certainly not a universal recipe. As usual, a flexible attitude will help the reader to widen the class of problems with which he can deal. So will practice, as encouraged by the exercises.

We can now usefully characterize, more precisely than we have heretofore, problems for which singular perturbation methods are appropriate. These problems have the property that an expansion in powers of a suitable small parameter, with coefficients depending only on the independent variable, is not uniformly valid for all relevant values of the independent variable. We have seen three types of singular perturbation methods that can be used advantageously on such problems. These involve (i) introduction of a slightly distorted independent variable (Chapter 2), (ii) use of differently scaled variables in different domains, with the separate expansions joined by the matching technique (Section 9.2), and (iii) application of multiple scale assumptions (this section). For a given simple problem it is often possible to decide which of these methods is "best." More often than not, however, even on simple problems this decision involves value judgments such as whether or not it is "better" to sacrifice some brevity in calculation to gain a more automatic approach.

EXERCISES

1. Assume a solution of (1) of the form (5). Verify that (6) is obtained by substituting into the equation.

2. (a) Verify that $f^{(1)}$ satisfies (19).
(b) Verify (21).
(c) Verify (23).
(d) Verify (24).

‡3. Consider

$$\ddot{y} + \varepsilon\dot{y}^3 + y = 0; \qquad y(0) = 0, \quad \dot{y}(0) = 1.$$

Assume that

$$y(t, \varepsilon) = f^{(0)}(t, \tau) + \varepsilon f^{(1)}(t, \tau) + \cdots, \qquad \text{where } \tau = \varepsilon t,$$

and obtain the approximate solution

$$y \approx \left(1 + \frac{3\varepsilon t}{4}\right)^{-1/2} \sin t + O(\varepsilon).$$

Give a physical interpretation of the problem and of the solution.

4. Poincaré noticed that introduction of a slightly distorted time scale τ, where

$$t = \tau(1 + h_2 a^2 + h_4 a^4 + \cdots), \tag{25}$$

will lead to a uniformly valid perturbation solution of the pendulum problem. (Do not confuse this τ with the abbreviation for εt used elsewhere.) In Section 2.2 the successive coefficients were chosen to remove "resonant" terms in the various differential equations. M. Pritulo* suggested that in such problems, one should carry out a regular perturbation calculation and then introduce the new time scale chosen in such a way that secular (growing) terms are eliminated. This exercise applies Pritulo's approach to the pendulum problem.

(a) To find a first correction to the period in the pendulum problem, replace t by $\tau(1 + h_2 a^2 + \cdots)$ in (7.1.24) and show that no secular terms remain when $h_2 = \frac{1}{16}$. In doing so, prove and use the approximation $\cos(\tau + h_2 a^2) \approx \cos\tau - h_2 \tau a^2 \sin\tau$.

(b) As a start toward finding the coefficient h_4, show that if (25) holds, then

$$\cos t = \cos\tau - (h_2 \tau \sin\tau)a^2 - \tfrac{1}{2}(h_2^2 \tau^2 \cos\tau + 2h_4 \tau \sin\tau)a^4 + \cdots .$$

(c) Given (25), show that the secular terms in the regular perturbation solution $\Theta_0(t) + a^2\Theta_2(t) + a^4\Theta_4(t)$ of (7.1.24) and Exercise 7.1.3 are

$$-\left(h_2 - \frac{1}{16}\right) a^2\tau \sin\tau - \left(\frac{h_2^2}{2} - \frac{h_2}{16} + \frac{1}{512}\right)a^4\tau^2 \cos\tau - \left(h_4 + \frac{h_2}{192} - \frac{h_2}{16}\right)a^4\tau \sin\tau + \left(\frac{h_2}{64} - \frac{1}{1024}\right)a^4\tau \sin 3\tau.$$

Show that these terms vanish if

$$t = \tau[1 + \tfrac{1}{16}a^2 + \tfrac{11}{3072}a^4 + \cdots].$$

(That a single choice of h_2 makes three expressions vanish at first seems miraculous but is merely a consequence of the fact that we have picked the correct form for the solution, so that everything *must* work out.)

(d) With (c) the solution

$$\cos\tau + \left(\frac{a^2}{192}\right)(\cos\tau - \cos 3\tau) + \cdots$$

has period 2π in τ and

$$2\pi[1 + \tfrac{1}{16}a^2 + \tfrac{11}{3072}a^4 + \cdots]$$

in t. Verify this answer by expanding the integrand of (2.2.19) in powers of a^2.

* *J. Appl. Math. Mech. 26*, 661–67 (1962).

REMARK. Pritulo's approach offers an easy way to get the first correction. It seems to offer no advantage at higher orders, however. Then, in contrast to Poincaré's original method, one must actually compute the secular terms, which are later discarded.

5. From Exercise 4 it should be apparent that determination of higher corrections to the period of the pendulum by the multiple scale method would require an assumption of the form

$$\Theta = f^{(0)}(t, \varepsilon t, \varepsilon^2 t) + \varepsilon f^{(1)}(t, \varepsilon t, \varepsilon^2 t) + \varepsilon^2 f^{(2)}(t, \varepsilon t, \varepsilon^2 t) + O(\varepsilon^3) \quad (\varepsilon \equiv a^2).$$

It turns out that the calculations are very formidable.

(a) To get some idea of how things go, show that the method fails if (8) is assumed and the calculations are carried to $O(\varepsilon^2)$.

(b) Assume that the $f^{(i)}$ are also a function of $\varepsilon^2 t$ and carry the calculations at least through $O(\varepsilon)$.

6. For a relatively simple calculation involving three scales consider

$$\ddot{y} + 2\varepsilon\dot{y} + y = 0; \qquad y(0) = 0, \quad \dot{y}(0) = 1.$$

Assume that

$$y = f^{(0)}(t, \tau_1, \tau_2) + \varepsilon f^{(1)}(t, \tau_1, \tau_2) + \varepsilon^2 f^{(2)}(t, \tau_1, \tau_2) + O(\varepsilon^3),$$

where $\tau_1 = \varepsilon t, \quad \tau_2 = \varepsilon^2 t.$

(a) Show that

$$f_{00}^{(0)} + f^{(0)} = 0,$$
$$f_{00}^{(1)} + f^{(1)} = -2f_{01}^{(0)} - 2f_0^{(0)},$$
$$f_{00}^{(2)} + f^{(2)} = -2f_{02}^{(0)} - f_{11}^{(0)} - 2f_1^{(0)} - 2f_{01}^{(1)} - 2f_0^{(1)},$$

where subscripts 0, 1, 2 denote partial derivatives with respect to t, τ_1 and τ_2. What are the initial conditions?

(b) Show that $f^{(0)} = A(\tau_1, \tau_2) \cos t + B(\tau_1, \tau_2) \sin t$.

(c) Show that

$$A(\tau_1, \tau_2) = e^{-\tau_1}\alpha(\tau_2), \qquad B(\tau_1, \tau_2) = e^{-\tau_1}\beta(\tau_2),$$
$$f^{(1)} = C(\tau_1, \tau_2) \cos t + D(\tau_1, \tau_2) \sin t.$$

(d) Find $\alpha(\tau_2)$ and $\beta(\tau_2)$ and compare $f^{(0)}$ with the exact solution.

7. To show that boundary layer problems can be treated by the multiple scale method, consider again the example [Equation (9.2.1)]

$$\varepsilon \frac{d^2y}{dx^2} + 2\frac{dy}{dx} + y = 0; \qquad y(0) = 0, \quad y(1) = 1, \quad 0 < \varepsilon \ll 1.$$

Assume that

$$y = f^{(0)}(x, X) + \varepsilon f^{(1)}(x, X) + \cdots + \text{transcendentally small terms},$$

where $X = x/\varepsilon$.

(a) Show that on substituting the series into the differential equation, the vanishing of $O(\varepsilon^{-1})$ terms leads to

$$f^{(0)}(x, X) = C_1(x) + C_2(x) \exp(-2X).$$

‡(b) Show that the vanishing of $O(1)$ terms requires

$$f_{XX}^{(1)} + 2f_X^{(1)} = -[2C_1' + C_1] + [2C_2' - C_2] \exp(-2X).$$

Why must the terms in the square brackets be required to vanish? Why is it not necessary to require the vanishing of C_2? Show that

$$C_1(x) = \exp[\tfrac{1}{2}(1 - x)],$$

while $C_2(x)$ is unspecified except for the condition $C_2(0) = -e^{1/2}$. Compare your answer with that obtained by other approaches to the problem.

(c) What happens when ε is small and negative?

11.3 The Phase Plane

Our previous discussions of the pendulum have all been concerned with what happens when it is slightly deflected from an equilibrium position. All the perturbation methods we employed in these discussions are useful in a wide variety of problems, e.g., problems governed by systems of nonlinear partial differential equations. (The methods must be extended, but their principles remain the same.)

The pendulum problem can, in fact, be solved in closed form with the aid of elliptic integrals. We have not followed this approach because it is rarely the case that solutions can be expressed exactly in terms of known functions, so that discussion of the exact solution would not lead to ideas with the desired breadth of applicability. We now take a point of view midway between those of perturbation theory and exact solution. This will enable us to illustrate the important *phase plane* approach to nonlinear problems.

We wish to consider in one "glance" all possible initial conditions; therefore, precise scaling is not helpful. Nondimensionalizing merely to reduce the number of parameters that appear, we thus start with the pendulum equation in the standard dimensional form [Equation (7.1.1)]:

$$\frac{d^2\theta^*}{d(t^*)^2} + \frac{g}{L}\sin\theta^* = 0, \qquad \theta^*(0) = a, \qquad \frac{d\theta^*}{dt^*}(0) = \Omega. \tag{1}$$

We introduce the variables

$$\theta = \theta^*, \qquad t = t^*\left(\frac{L}{g}\right)^{-1/2}. \tag{2}$$

Here time is referred to the scale $(L/g)^{1/2}$ and angular displacement to the radian. With (2) our problem becomes

$$\ddot{\theta} + \sin\theta = 0, \qquad \theta(0) = a, \qquad \dot{\theta}(0) = \Omega\left(\frac{L}{g}\right)^{1/2} \equiv b. \tag{3}$$

We now take advantage of the fact that (3) is a second order differential equation in which the independent variable does not appear explicitly. Using a "trick" that has become part of the elementary literature of differential equations (Boyce and DiPrima, 1969, p. 89), we make the abbreviation $\omega \equiv \dot{\theta}$ and regard ω as a function of θ. Thus

$$\frac{d^2\theta}{dt^2} = \frac{d\omega}{dt} = \frac{d\omega}{d\theta}\frac{d\theta}{dt} = \omega\frac{d\omega}{d\theta},$$

and our equation becomes

$$\omega\omega' + \sin\theta = 0, \qquad \omega' \equiv \frac{d\omega}{d\theta}. \tag{4}$$

This can be integrated once, yielding

$$\tfrac{1}{2}\omega^2 - \cos\theta = \text{constant} = \tfrac{1}{2}b^2 - \cos a \tag{5}$$

or

$$\omega^2 = 2\cos\theta - 2\cos a + b^2. \tag{6}$$

We have used the fact that $\omega = b$ when $\theta = a$, from (3).

THE PHASE PORTRAIT OF AN UNDAMPED SIMPLE PENDULUM

Let us sketch the family of curves given by (6) when all possible values of a and b are considered. To do this, we make the following observations. (i) Both ω and θ can be replaced by their negatives without altering (6) so that the curves are symmetric in both the θ and ω axes. It is therefore sufficient to restrict our attention to the first quadrant. (ii) The right side of (6) has period 2π in θ so that we can further restrict attention to the strip $\omega \geq 0$, $0 \leq \theta \leq 2\pi$. (iii) If ω is positive, it follows from (4) that $\omega' = 0$ at $\theta = 0$ (horizontal tangent), $\omega' < 0$ for $0 < \theta < \pi$ (ω decreasing), $\omega' = 0$ at $\theta = \pi$ (horizontal tangent), and $\omega' > 0$ for $\pi < \theta < 2\pi$ (ω increasing). (iv) If $\theta \neq 0, \pi$, then $|\omega'| \to \infty$ as $\omega \to 0$ (vertical tangent on θ axis). We have deduced the qualitative behavior shown in Figure 11.1.

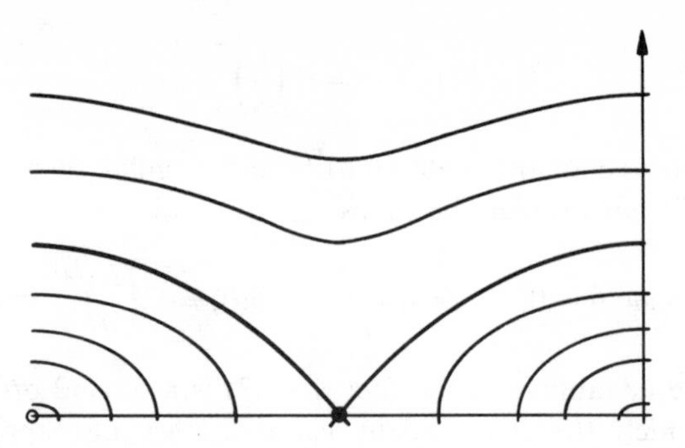

FIGURE 11.1. *Portion of the phase plane for the undamped pendulum: schematic drawing.*

Of particular interest are the points (0, 0) and (π, 0). Here a tangential slope ω' is undefined by (4). But for θ near zero, (4) gives approximately

$$\omega\, d\omega + \theta\, d\theta = 0, \tag{7}$$

so that $\omega^2 + \theta^2 =$ constant, and in the neighborhood of (0, 0) the curves are circles. For θ near π we write $\theta = \pi + \bar{\theta}$ and assume that $\bar{\theta}$ is small. We obtain

$$\omega\, d\omega - \bar{\theta}\, d\bar{\theta} = 0, \qquad \omega^2 - \bar{\theta}^2 = \text{constant}. \tag{8}$$

Near (π, 0) the curves are thus hyperbolas, of which the degenerate hyperbola

$$\omega = \pm\bar{\theta}, \qquad \text{i.e., } \omega = \pm(\theta - \pi),$$

passes through (π, 0).

One more observation will be useful. From $\omega = \dot{\theta}$ it follows that θ increases with time if and only if ω is positive. This enables us to indicate with an arrow how the curves are traced as time passes. Combining all this, we obtain Figure 11.2.

To each initial condition $\theta(0) = a$, $\dot{\theta}(0) = b$, there corresponds a point (a, b) in Figure 11.2. With the exception of the points $(n\pi, 0)$, $n = 0, \pm 1, \ldots,$ from each initial point one can follow in the direction of the arrow a unique curve that traces out the subsequent angular displacement θ and angular speed $\dot{\theta} \equiv \omega$ of the pendulum. These curves are called **trajectories**, and the plane itself is called the **phase plane**. The collection of trajectories in the phase plane is aptly said to comprise the **phase portrait** of the system, for this collection depicts the behavior of which the system is capable.

It must be kept in mind that values of θ that differ by 2π correspond to the same position. One way to take account of this fact is to imagine that the phase plane strip $0 \le \theta \le 2\pi$ is wound on a right circular cylinder with $\theta = 0$ identified with $\theta = 2\pi$. Thus all trajectories correspond to periodic motions.

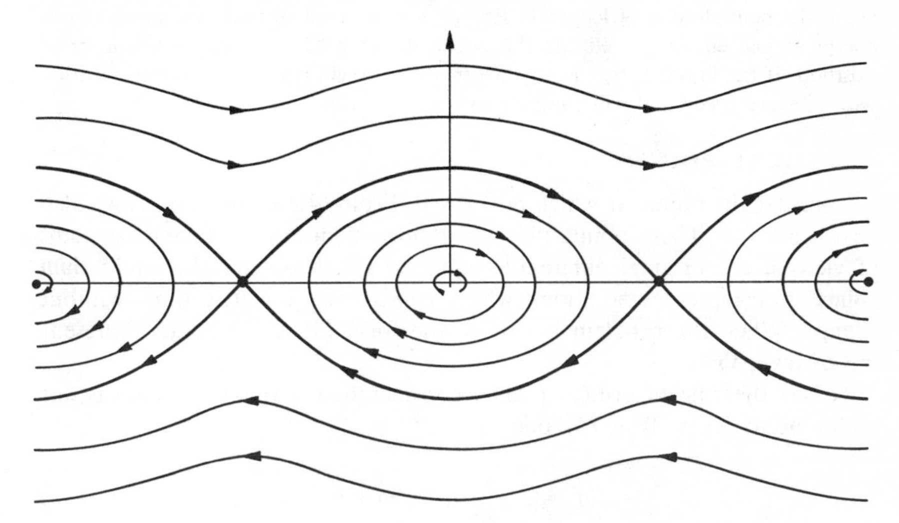

FIGURE 11.2. *Schematic phase portrait for the undamped pendulum. The separatrices S_1 and S_2 (heavy lines) bound a domain consisting of closed trajectories about the center* (0, 0). *These trajectories represent periodic oscillations of bounded extent—i.e., the pendulum never overturns. A trajectory such as T represents a continually overturning motion.*

SEPARATRICES

Trajectories that start near (0, 0) satisfy $|\theta| < \pi$ (pendulum never overturns). By contrast, on all trajectories starting at (π, b) for nonzero b, the angle θ periodically passes through π. (An inverted pendulum, given a nonzero initial speed, will continually circle its support.)

Consider now the special trajectories passing through odd integer multiples of π. These separate the trajectories that correspond to paths circling the support from trajectories that correspond to nonoverturning paths. Such a special curve, which separates regions of different qualitative behavior, is called a **separatrix**.

Let the separatrix passing through $(\pi, 0)$ intersect the ω axis at $(0, b_0)$. Then $\omega^2 = 2\cos\theta - 2 + b_0^2$, from (6). In order that $\omega \to 0$ as $\theta \to \pi$, we must have $b_0^2 = 4$ or $b_0 = \pm 2$. Hence a pendulum at bob-down equilibrium will circle its support if and only if its initial angular speed has magnitude greater than 2.

Example. Give a physical explanation of the condition just obtained.

Solution. An initial dimensionless angular speed of 2 corresponds by (3) to an initial dimensional angular speed of $\Omega = 2(L/g)^{-1/2}$. Let the bob have mass m, and take the zero of potential energy at the bob-down position. Then the initial energy is all kinetic and has the magnitude $\frac{1}{2}m(\Omega L)^2 = 2mLg$, since the bob has linear speed ΩL if the pendulum is of length L. Energy is conserved so that this initial kinetic energy is just enough to elevate the bob a distance $2L$ to a motionless inverted position. If the initial energy is any larger, the bob will reach the inverted position with nonzero speed and will continue to rotate.

CRITICAL POINTS

Consider the points at which $\omega = \dot{\theta} = 0$. From (4), at such points we also have $\sin \theta = 0$. These points play an important dual role. From the point of view of the original differential equation (3), these are the **equilibrium points** that represent solutions which can persist through time without change. When the problem is cast in the form (4), these points represent the **critical points**.

To see the role of critical points, consider first a point P which is *not* such a point. Then, from (4), one has

$$\omega' = -\frac{\sin \theta}{\omega}, \qquad \omega \neq 0.$$

This gives a unique value to the slope of the trajectory $\omega = \omega(\theta)$ that passes through P. By contrast, the above equation cannot be used as it stands at a critical point (ω_0, θ_0). The only way to obtain a slope at such a point is to consider

$$\lim_{\substack{\omega \to \omega_0 \\ \theta \to \theta_0}} -\frac{\sin \theta}{\omega},$$

and this limit generally has different values depending on how the approach to (ω_0, θ_0) is made. (See Exercise 1.)

Because trajectories have a unique tangent except at critical points, it follows that *curves composed of trajectories cannot cross except at critical points.* Consequently, examination of trajectories in the neighborhood of critical points is particularly revealing. But since critical points and equilibrium points are identical, such an examination is equivalent to a study of the stability of equilibrium. Thus the small perturbation solutions given in (1.5) and (1.7) describe behavior in the neighborhood of normal and inverted equilibrium. Equivalent descriptions are provided by (7) and (8). In particular, the exceptional perturbation of the inverted pendulum that decays to zero [see the material below (1.7)] is represented in the phase plane as a perturbation along a separatrix. A solution starting precisely along the appropriate separatrix will follow that trajectory into equilibrium at $(\pi, 0)$.

The wide applicability of our approach can be appreciated if it is realized that for any dynamical system whose position can be described by a single coordinate, say x, Newton's second law takes the form $\ddot{x} = g(x, \dot{x}, t)$. Most important are **autonomous** systems in which time does not explicitly appear; i.e., $g = g(x, \dot{x})$. For these, if we introduce the speed $v \equiv \dot{x}$, we have $v(dv/dx) = g(x, v)$, and consideration of the v-x phase plane is again appropriate.

More generally, one can consider the first order equation for $y(x)$:

$$f(x, y)\frac{dy}{dx} = g(x, y). \tag{9}$$

The slope dy/dx is uniquely defined except at critical points where $f(x, y) = 0$, $g(x, y) = 0$. Let trajectories in the x-y (phase) plane be parametrized by t, so $dv/dx = (dv/dt)/(dx/dt)$. Then the trajectory through (x_0, y_0) can be described by $x = x(t)$, $y = y(t)$, where

$$\frac{dx}{dt} = f(x, y), \quad \frac{dy}{dt} = g(x, y); \qquad x(t_0) = x_0, \quad y(t_0) = y_0. \tag{10}$$

LIMIT CYCLES

By the relevant existence and uniqueness theorems (assuming sufficient smoothness—see Chapter 2), precisely one trajectory passes through each point. The further fact that the slope is uniquely defined except at critical points makes plausible the following classification of possible trajectories emerging from a point (x_0, y_0).*

(i) The trajectory through (x_0, y_0) consists entirely of the point (x_0, y_0). Then this point is an equilibrium point (i.e., a critical point).

(ii) The trajectory through (x_0, y_0) returns to (x_0, y_0). The resulting simple closed curve is called a **limit cycle** and is the phase plane path of a periodic solution.

(iii) The trajectory through (x_0, y_0) neither remains at this point nor returns to it, but either passes to infinity, or approaches some other equilibrium point, or approaches a closed curve. Although the closed curve could pass through several critical points, in practice it is usually a limit cycle (and therefore passes through no critical points). The location of limit cycles is generally not a simple matter and will not be discussed further. It is frequently easy, however, and is always valuable, to determine behavior in the neighborhood of critical points.

BEHAVIOR OF TRAJECTORIES NEAR CRITICAL POINTS

If (X, Y) is a critical point of (9) or (10), then by definition

$$f(X, Y) = 0, \qquad g(X, Y) = 0. \tag{11}$$

Introducing deviations from equilibrium $\bar{x}$ and $\bar{y}$ by

$$x(t) = X + \bar{x}(t), \qquad y(t) = Y + \bar{y}(t),$$

* S. Lefschetz, *Differential Equations: Geometric Theory* (New York: Wiley–Interscience, 1957), Chap. 10.

we have by formal application of the two-variable Taylor formula

$$\frac{d\bar{x}}{dt} = a\bar{x} + b\bar{y}, \qquad \frac{d\bar{y}}{dt} = c\bar{x} + d\bar{y}, \tag{12}$$

where the constants a, b, c, and d are given by

$$a = f_x(X, Y), \qquad b = f_y(X, Y), \qquad c = g_x(X, Y), \qquad d = g_y(X, Y),$$

and where remainder terms have been neglected. The linear system (12) has solutions of the form

$$\begin{bmatrix} \bar{x} \\ \bar{y} \end{bmatrix} = \begin{bmatrix} \hat{x} \\ \hat{y} \end{bmatrix} e^{mt}$$

if

$$\begin{bmatrix} a - m & b \\ c & d - m \end{bmatrix} \begin{bmatrix} \hat{x} \\ \hat{y} \end{bmatrix} = \begin{bmatrix} 0 \\ 0 \end{bmatrix}.$$

For a nontrivial solution the determinant of the coefficient matrix must vanish so that

$$m^2 + \beta m + \gamma = 0 \qquad \text{where} \quad \beta = -(a + d), \quad \gamma = ad - bc. \tag{13}$$

We assume that $\gamma \neq 0$ and $\beta^2 - 4\gamma \neq 0$. There are then these possibilities.

$$\text{I}: \beta^2 - 4\gamma > 0.$$

Equation (13) has the real roots

$$m_1 = -\beta + \sqrt{\beta^2 - 4\gamma}, \qquad m_2 = -\beta - \sqrt{\beta^2 - 4\gamma}.$$

I(a): $\gamma < 0$, so $m_1 > 0$, $m_2 < 0$. Equilibrium is unstable and trajectories near (X, Y) appear like the level lines near a mountain pass. Hence (X, Y) is called a **saddle point** or **col**. An example is the equilibrium point $(\pi, 0)$ in Figure 11.2.

I(b): $\gamma > 0$. Both roots are positive (negative) and equilibrium is unstable (asymptotically stable) if β is negative (positive). The trajectories can be shown to form a **node** as illustrated in Figure 11.3.

$$\text{II}: \beta^2 - 4\gamma < 0.$$

Now (13) has the complex conjugate roots $-\beta \pm i\sqrt{4\gamma - \beta^2}$ and (12) has solutions proportional to

$$e^{-\beta t} \cos(t\sqrt{4\gamma - \beta^2}), \qquad e^{-\beta t} \sin(t\sqrt{4\gamma - \beta^2}).$$

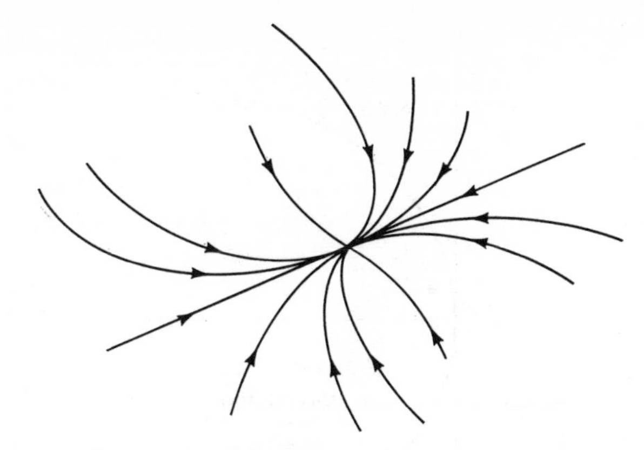

FIGURE 11.3. *Some trajectories near a typical stable node. The arrows are reversed when the node is unstable.*

II(a): $\beta = 0$. Perturbations are periodic and trajectories near the equilibrium point are thus closed. Such an equilibrium point is called a **center.** Equilibrium is Liapunov stable and neutrally stable. An example of a center is the point (0, 0) in Figure 11.2.

II(b): $\beta \neq 0$. The equilibrium point is a **focus** for spirals. If $\beta < 0$, as time passes the phase point spirals farther and farther from the focus, which is unstable. If $\beta > 0$, on the other hand, the focus is asymptotically stable.

The different types of qualitative behavior exhibited by solutions to (12) in the various situations can be quickly ascertained from Figure 11.4. But is this behavior appropriate for the nonlinear equations (10) in the neighborhood of the equilibrium point (X, Y)? That is, can one justify the linearization leading to (12)? It turns out that a sufficient condition for an affirmative answer is the continuity of F and G and their first partial derivatives in a neighborhood of (X, Y). An *exception,* not surprisingly, is the center, for the slightest inaccuracy can prevent the perfect return of the state point to a given position that is required for a periodic solution.

The case of the center, and the cases $\gamma = 0$, $\beta^2 - 4\gamma = 0$, which were not included in the possibilities discussed above, certainly require more discussion. This is not within our scope here.

As a worked example, we now consider the *damped* pendulum. The device used above to obtain an explicit expression for the trajectories no longer can be employed. Thus general phase plane methods are essential in this case.

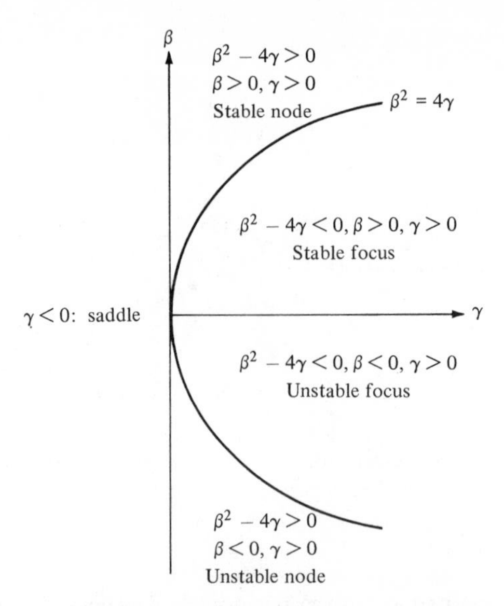

FIGURE 11.4. *Qualitative behavior of the solutions to* (12): $d\bar{x}/dt = a\bar{x} + b\bar{y}$, $d\bar{y}/dt = c\bar{x} + d\bar{y}$. *Abbreviations:* $\beta = -(a + d)$, $\gamma = ad - bc$.

Example. Analyze phase plane behavior for the equation

$$\ddot{\theta} + \nu\dot{\theta} + \sin\theta = 0, \tag{14}$$

governing a damped pendulum.

Solution. We write

$$x = \theta, \qquad \dot{x} = y,$$

so that

$$\dot{x} = y, \qquad \dot{y} = -\nu y - \sin x.$$

Linearizing about the equilibrium point (0, 0), we have

$$\dot{x} = y, \qquad \dot{y} = -\nu y - x,$$

so that, from (12) and (13),

$$a = 0, \qquad b = 1, \qquad c = -1, \qquad d = -\nu;$$

$$\beta = \nu, \qquad \gamma = 1, \qquad \beta^2 - 4\gamma = \nu^2 - 4.$$

Therefore $(0, 0)$ is a stable node if $\nu^2 - 4 > 0$ and a stable focus if $\nu^2 - 4 < 0$. The two possibilities are identical with the over- and under-damped situations encountered in elementary treatments of the (linearized) pendulum problem.

Consider the equilibrium point $x = \pi$, $y = 0$. Introducing perturbations $\bar{x}$ and $\bar{y}$ by

$$x = \pi + \bar{x}, \qquad y = \bar{y},$$

and linearizing, we obtain

$$\dot{\bar{x}} = \bar{y}, \qquad \dot{\bar{y}} = -\nu\bar{y} + \bar{x}.$$

Hence $\beta = \nu$, $\gamma = -1$, $\beta^2 - 4\gamma = \nu^2 + 4$, and $(\pi, 0)$ is a saddle point (as it was for the undamped pendulum). Physical intuition rules out limit cycles, for there can be no periodic solutions with no supply of energy to replace that which is dissipated through damping. With slight damping, trajectories like T in Figure 11.2 now spiral slowly into an equilibrium point that represents a motionless pendulum with bob down. It can be seen that in the underdamped case $0 < \nu < 2$, the phase portrait is as given in Figure 11.5.

Our treatment of the phase plane provides the means to obtain some useful results. Still, it only scratches the surface. For an exposition that is simple and brief but a good deal more comprehensive than that given here, see

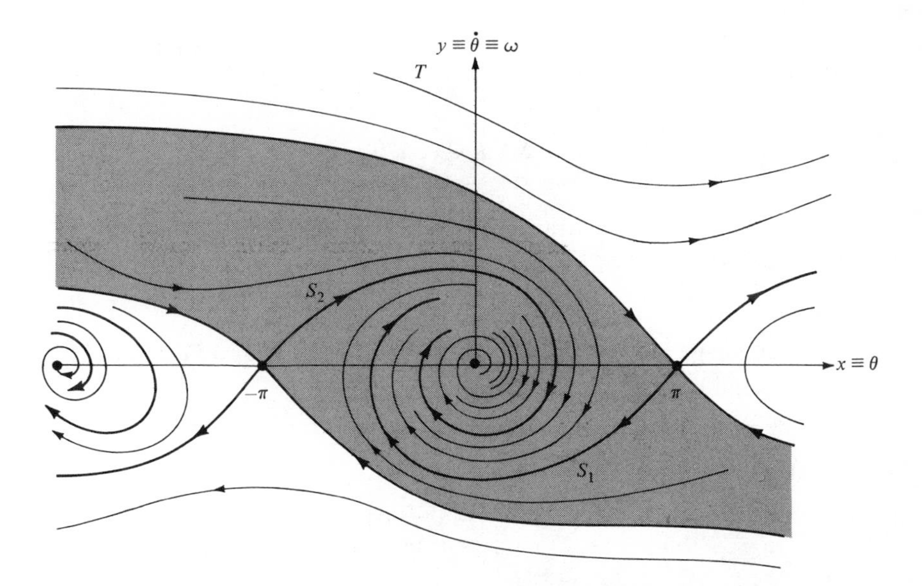

FIGURE 11.5. *Phase portrait for the slightly damped pendulum. The "domain of attraction" of* (0, 0) *is shaded, and heavy lines denote separatrices. Along trajectory T, the angle increases by several multiples of* 2π *(several overturnings) before the trajectory finally approaches an asymptotically stable focus. Just above the separatrix* S_2, *such "overturning" trajectories are heading for the stable focus at* (0, 0). *Just below* S_2, *the trajectories represent decaying oscillations of bounded extent. The reader should try to convince himself that the qualitative behavior depicted here follows from the nature of the singular points (alternating saddles and foci), from the absence of limit cycles and the fact that trajectories never cross but terminate at singular points, and from some physical considerations. Compare Figure* 11.2.

Chapter 9 of Boyce and DiPrima (1969). Much richer treatment with an eye to applications can be found, for example, in books by J. Stoker* and by A. Andronov, A. Vitt, and S. Khaikin.† For theoretical results, start with the classical text of Coddington and Levinson (1955).

The phase plane concept is capable of wide generalization. For example, consider a system of N particles. Initial conditions require specification of three position components and three velocity components for each particle. Instead of the phase plane, we must consider a phase space of $6N$ dimensions, half for the position coordinates and half for the velocity coordinates. One speaks of a **system of 6*N* degrees of freedom**. In continuum mechanics, particles lose their identity, but it has proved useful to introduce infinite dimensional phase spaces, with a single dimension being identified with a Fourier component of some variable.

Observe that what is of interest in phase plane analysis is the qualitative behavior of trajectories, not their precise course. Modern research on differential equations is largely concerned with elucidating such qualitative behavior, and this is just what often concerns the applied mathematician.

EXERCISES

1. (a) Use (7) to show that if one approaches the origin along the line $\omega = K\theta$, then the trajectories one crosses have the slope $-K^{-1}$. (Thus the limiting slope depends on the mode of approach.) Interpret geometrically.

(b) Harmonize (1.5) and (1.7) with (7) and (8).

2. (a) Eliminate e^* and c^* from (10.1.3) to obtain a pair of equations for s^* and p^*.

(b) Show that there is only one physically significant equilibrium point, and determine its stability.

(c) Find the four regions in the first quadrant of the s^*-p^* plane where ds^*/dt^* and dp^*/dt^* are both positive, both negative, or have the two combinations of opposite signs. Use this information to sketch the possible trajectories.

(d) Compare the present approach to the problem with that of Chapter 10. [See I. G. Darvey and R. F. Matlak, "An Investigation of a Basic Assumption in Enzyme Kinetics Using Results of the Geometric Theory of Differential Equations," *Bull. Math. Biophys. 29*, 335–41 (1967).]

3. (a) The equations

$$\frac{dx}{dt} = (a - bx - cy)x, \qquad \frac{dy}{dt} = (e - fx - gy)y,$$

* *Nonlinear Vibrations* (New York: Wiley–Interscience, 1950).

† *Theory of Oscillations* (Elmsford, N.Y.: Pergamon, 1966).

are a simple model for the competition between two species of organisms. (Here a, b, c, e, f, and g are constants.) Write a brief essay on what is assumed in this model, and on what some of its limitations are expected to be.

(b) By examining the phase plane, show that if $(a/c) > (e/g)$ and $(a/b) > (e/f)$, then species x wins. Is this reasonable?

(c) Discuss the case $(a/c) < (e/g)$, $(e/f) < (a/b)$. (See Section 3D of J. M. Smith, *Mathematical Ideas in Biology*, Cambridge: Cambridge U.P., 1968.)

4. Use the phase plane approach on Exercise 15.2.7 (which is self-contained).

PART C

Introduction to Theories of Continuous Fields

CHAPTER 12
Longitudinal Motion of a Bar

WE NOW begin the third part of the present volume, a part devoted to the study of basic continuum mechanics. This chapter will be devoted to a discussion of the one-dimensional longitudinal motion of a bar, a physical problem that arises in many different contexts. As we shall see, it is possible to elucidate a number of significant aspects of the phenomena involved, with relatively elementary mathematical techniques. To give one example, an incoming elastic wave doubles its force at a plane where it is reflected (Exercise 12.2.9). There could be dire consequences if this were not anticipated.

We have based the development of the basic equations on the same approach that is used in deriving similar results for the three-dimensional continuum. We feel that the general approach is clearer if it is first presented without the complications of three-dimensional geometry. Thus we recommend that at least Section 12.1 be studied before the general material, which begins in Chapter 13. (On the other hand, some will prefer to regard the present chapter as providing a particular illustration of the general theory of continuum mechanics, and this is certainly a feasible procedure.) Many of the specific questions that will be considered are prototypes of similar problems which will be tackled later with more complicated geometries. Such items as wave propagation, reflection and transmission, dispersion, propagation of discontinuities, and free vibrations arise in a natural manner. The mathematical techniques used in studying these can be generalized, although with difficulties at times, so that they apply in two and three dimensions.

Our general strategy will be to adopt an optimistic attitude toward obtaining a simple but meaningful one-dimensional formulation of elastic* motion. Certain subtleties that a deep discussion would require are not treated here.

12.1 Derivation of the Governing Equations

GEOMETRY

The bar shown in Figure 12.1 is assumed to be slender. That is, the average diameter of a cross section must be small when compared with the length of the bar. Although the cross section may be variable, we shall require that its

* "The property of recovery of an original size and shape is the property that is termed elasticity" (Love, 1944).

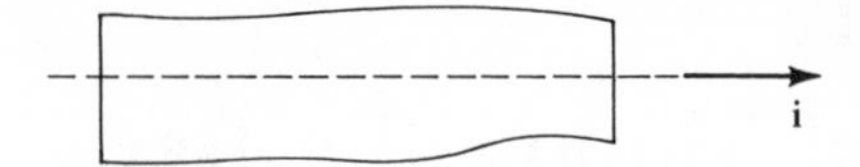

FIGURE 12.1. *A bar that probably can be regarded as "slender." Cross sections normal to the unit vector* **i** *are assumed to remain plane and to retain their shape as they move.*

variation along the bar be small. Because of these geometrical restrictions it is not unreasonable to *assume* (as we shall) *that each cross section remains plane* throughout the motion and *moves longitudinally without distortion.* It is this assumption, that the only variation is longitudinal, which gives the problem its one-dimensional character.

Throughout this chapter we assume that the bar can be regarded as composed of continuously distributed matter in smooth motion (with possible exceptions at special locations). This is perfectly natural, but some discussions of the deeper implications of the continuum assumption will be found in Section 13.1.

We shall switch back and forth between two independent variables. The **spatial coordinate** x picks out a particular location in space, while the **material coordinate** A picks out a particular cross section. We shall always specify a cross section by stating its distance A from a reference point (the origin). This measurement will be made at some fixed reference time (usually time $t = 0$). Distance to the right (left) of the origin will be taken positive (negative), as usual.

In our one-dimensional problem, complete information about the motion of the bar requires knowledge of the position of every section at all times. We shall write the required functional description of the **cross-section paths** as follows:

$$x = x(A, t), \qquad x(A, 0) = A. \tag{1a, b}$$

In words, these equations state "x is the position now (at time t) of the section that was initially (at time zero) located at A; A is the position at time zero of the section initially located at A."

We shall assume that (1) is uniquely invertible so that we can write

$$A = A(x, t) \qquad [\text{with } A(A, 0) = A]. \tag{2}$$

In words, "A is the initial position of the particle now at x." Using common language, (2) is obtained from (1) by solving for A in terms of x. By definition, this implies the identities

$$x[A(x, t), t] = x, \qquad A[x(A, t), t] = A. \tag{3a, b}$$

Example 1. If $x = A + t$, then $A = x - t$ and, indeed, $(x - t) + t = x$, $(A + t) - t = A$.

As stated, sometimes we use A as the independent variable [in (1)] and sometimes x [in (2)]. But note that we also sometimes use A as a *dependent* variable [in (2)] and we sometimes use x in this role [in (1)].

Usually, we shall use **lowercase** letters for functions of the **spatial** variable and **uppercase** letters for functions of the **material** variable. Thus we shall write

$$f[x(A, t), t] = F(A, t), \qquad f(x, t) = F[A(x, t), t]. \tag{4a, b}$$

Example 2. Let $\theta(x, t)$ denote the temperature at time t at position x. Let $\Theta(A, t)$ denote the temperature at time t of the section which was initially located at A. Then

$$\theta[x(A, t), t] = \Theta(A, t), \qquad \theta(x, t) = \Theta[A(x, t), t]. \tag{5a, b}$$

Note that (5a) can be converted into (5b) by substituting for A in terms of x and t and then using (3a). Similarly, (3b) can be used to convert (5b) into (5a).

We shall frequently have to convert from spatial to material variables. Rather than use the somewhat bulky notation of (4a), or the even bulkier

$$f(x, t)\big|_{x=x(A,\, t)},$$

we shall sometimes make the abbreviation

$$f[x(A, t), t] \equiv f(x, t)\big|_A \qquad \text{or} \qquad f[x(A, t), t] \equiv f\big|_A. \tag{6}$$

This should be read "f evaluated at A." Similarly, we write

$$F[A(x, t), t] \equiv F\big|_x. \tag{7}$$

To see this notation displayed in useful formulas, consider the differentiation of (4a) by means of the chain rule.* We find [Exercise 1(a)]

$$\frac{\partial F}{\partial A} = \frac{\partial f}{\partial x}\bigg|_A \frac{\partial x}{\partial A}, \qquad \frac{\partial F}{\partial t} = \frac{\partial f}{\partial x}\bigg|_A \frac{\partial x}{\partial t} + \frac{\partial f}{\partial t}\bigg|_A. \tag{8a, b}$$

Focus attention on a section that is initially located at A and now, at time t, is to be found at position x. The vector joining the initial position to the current position is called the **displacement vector** and is denoted by

$$\mathbf{U}(A, t) = U(A, t)\mathbf{i}.$$

* A detailed discussion of the chain rule will be found in Appendix 13.1. We note here only that it might be helpful to think of x and t as spatial variables, and A and t' as material variables, with $t' = t$. Then chain rule applied to (4a) gives

$$\frac{\partial F}{\partial A} = \frac{\partial f}{\partial x}\bigg|_A \frac{\partial x}{\partial A} + \frac{\partial f}{\partial t}\bigg|_A \frac{\partial t}{\partial A}, \quad \frac{\partial F}{\partial t'} = \frac{\partial f}{\partial x}\bigg|_A \frac{\partial x}{\partial t'} + \frac{\partial f}{\partial t}\bigg|_A \frac{\partial t}{\partial t'}.$$

These equations reduce to (8) when t' is identified with t.

Here **i** is a unit vector in the positive A direction. The component function $U(A, t)$ will be known as the **displacement** (in material coordinates). We see from Figure 12.2 that

$$x(A, t) = A + U(A, t). \tag{9}$$

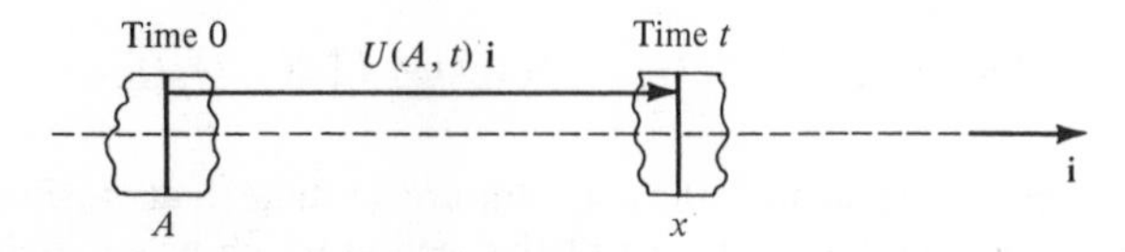

FIGURE 12.2. *The displacement $U(A, t)$ takes place when a section moves from initial position A to final position $x(A, t) = A + U(A, t)$.*

We define the displacement in spatial coordinates, $u(x, t)$, by

$$u(x, t) = U[A(x, t), t]. \tag{10}$$

If we substitute for A in terms of x, we find from (9) that

$$x = A(x, t) + u(x, t). \tag{11}$$

This has the interpretation "the initial position of the section now at x plus that section's displacement equals its present position."

A section's longitudinal velocity component, or **velocity** for short, is defined in material coordinates by

$$V(A, t) = \frac{\partial x(A, t)}{\partial t}. \tag{12}$$

From (9) we also have

$$V(A, t) = \frac{\partial U(A, t)}{\partial t}. \tag{13}$$

The spatial velocity component v is related to V by

$$v(x, t) = V[A(x, t), t], \qquad v[x(A, t), t] = V(A, t). \tag{14a, b}$$

The velocity vector is then given by $v(x, t)\mathbf{i} = \mathbf{v}(x, t)$ and $V(A, t)\mathbf{i} = \mathbf{V}(A, t)$ in spatial and material coordinates, respectively.

THE MATERIAL DERIVATIVE AND THE JACOBIAN

In present notation the chain rule result (8b) can be written

$$\frac{\partial F}{\partial t} = \left.\frac{\partial f}{\partial t}\right|_A + V\left.\frac{\partial f}{\partial x}\right|_A. \tag{15}$$

We shall define the **material derivative** Df/Dt of the function $f(x, t)$ by the important equation

$$\boxed{\frac{Df(x, t)}{Dt} = \left.\frac{\partial F(A, t)}{\partial t}\right|_{A=A(x,\, t)}.} \tag{16}$$

By making the substitution $A = A(x, t)$ in (15), we thereupon find

$$\boxed{\frac{Df(x, t)}{Dt} = \frac{\partial f(x, t)}{\partial t} + v(x, t)\frac{\partial f(x, t)}{\partial x}.} \tag{17}$$

The material derivative $Df(x, t)/Dt$ represents the rate of change of the function f following a material section. It can be thought of as giving the change with time of f, as seen by an observer who is riding with the section now located at x.

Example 3. Consider a *motionless* bar ($v \equiv 0$) of uneven temperature $\theta(x, t)$. Here

$$\frac{D\theta(x, t)}{Dt} = \frac{\partial\theta(x, t)}{\partial t},$$

for the only reason that the temperature can change at x is because of *local* heating or cooling processes which occur at that fixed point. But if the bar moves, we have

$$\frac{D\theta}{Dt} = \frac{\partial\theta}{\partial t} + v\frac{\partial\theta}{\partial x}. \tag{18}$$

There is an additional term because material is being bodily carried or *convected* past the position x at velocity v.

Example 4. Consider the case of an infinite bar that *steadily* moves through a roller which thins it. (See Figure 12.3.) Let some time be designated $t = 0$, and label with the coordinate A some section far to the left of the roller. The velocity of this section will be approximately constant until the section reaches the roller. Then it will speed up to the faster velocity that is required to move the thinner bar out of the way. Thus $V(A, t)$ will have a graph like that in Figure 12.4. On the other hand, if one fixes one's gaze at a particular point x, the velocity there will never

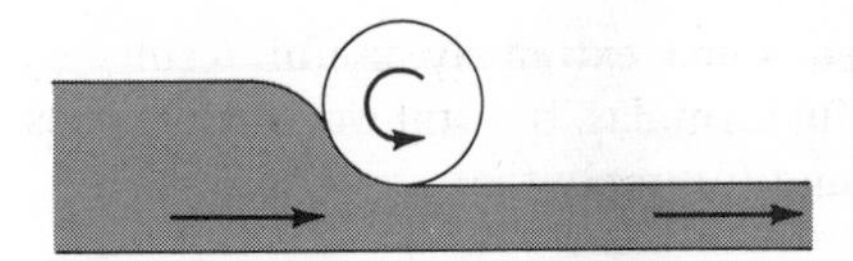

FIGURE 12.3. *A very long bar moving steadily through a roller. The thinner material moves faster than the thicker.*

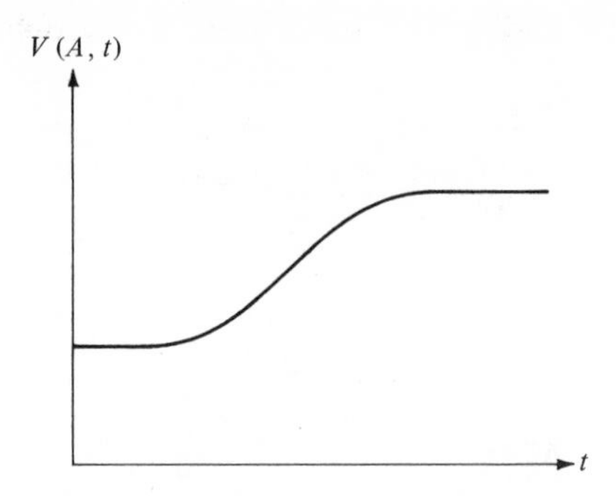

FIGURE 12.4. *The velocity V of a fixed section of the bar of Figure* 12.3 *increases as the bar is stretched because of passage under the roller.*

change, for the process has been assumed to proceed steadily. Thus at any fixed value of x, $v(x, t)$ is constant; i.e., $\partial v/\partial t = 0$.

For the velocity, (17) generally takes the form

$$\frac{Dv(x, t)}{Dt} = \frac{\partial v(x, t)}{\partial t} + v(x, t)\,\frac{\partial v(x, t)}{\partial x}. \tag{19}$$

The left side of this equation is the rate of change with time of the velocity of a certain cross section, namely, that cross section now located at x. In other words, (19) is an expression for the acceleration of the cross section now located at x. As we have seen, in the present example the first term on the right side of (19) is zero, for the velocity at a given position is always the same. But we expect from Figure 12.4 that the acceleration of any given section is nonzero. And indeed, the second term in (19) gives a nonzero contribution to the acceleration. This arises from the fact that the section which is at x at time t is not the section that was at x at time $t - \Delta t$.

It is hard for some people to believe that $\partial v(x, t)/\partial t$ is not the acceleration of the section located at x. But note that

$$\frac{\partial v(x, t)}{\partial t} = \lim_{\Delta t \to 0} \frac{v(x, t + \Delta t) - v(x, t)}{\Delta t}.$$

The numerator gives the difference in velocities of sections located at x at times $t + \Delta t$ and t, respectively. Generally, these are *different sections*, but to find the acceleration we must compute the time rate of change of a *single* section's velocity.

We have derived some extremely useful results by manipulating (8b). Somewhat less useful formulas, but still worthwhile ones, can be derived from (8a). Indeed, (8a) and (9) imply that

$$\frac{\partial F}{\partial A} = \left(1 + \frac{\partial U}{\partial A}\right) \frac{\partial f}{\partial x}\bigg|_A$$

or

$$\left.\frac{\partial f}{\partial x}\right|_A = \frac{\partial F/\partial A}{1 + \partial U/\partial A}. \tag{20}$$

Moreover [Exercise 1(b)], one can show that

$$\left.\frac{\partial F}{\partial A}\right|_x = \frac{\partial f/\partial x}{1 - \partial u/\partial x}. \tag{21}$$

The derivative $\partial x/\partial A$ that appears in the preceding calculations is the Jacobian of the one-dimensional transformation between the x coordinates and the A coordinates. With this in mind, let us write $\partial x/\partial A = J(A, t)$. We shall have need in the near future for a simple expression for $\partial J/\partial t$. Applying the chain rule, we see that

$$\frac{\partial J}{\partial t} = \frac{\partial}{\partial t}\left(\frac{\partial x}{\partial A}\right) = \frac{\partial}{\partial A}\left(\frac{\partial x}{\partial t}\right) = \frac{\partial V}{\partial A} = \left.\frac{\partial v}{\partial x}\right|_A \frac{\partial x}{\partial A}.$$

We thus obtain the **Euler expansion formula** in material coordinates,

$$\boxed{\frac{\partial J(A, t)}{\partial t} = J(A, t)\left.\frac{\partial v(x, t)}{\partial x}\right|_A.} \tag{22}$$

In spatial coordinates, (22) becomes

$$\frac{DJ[A(x, t), t]}{Dt} = J[A(x, t), t] \cdot \frac{\partial v(x, t)}{\partial x}. \tag{23}$$

The extension of this expansion formula to higher dimensions is found in Section 13.4, where it will be seen that the results are essentially the same as (22) and (23), but the calculations are more elaborate.

CONSERVATION OF MASS

As will be the case for the three-dimensional continuum, we shall take as basic assumptions generalizations of the various fundamental laws that are postulated in the study of the mechanics of systems of particles.

We shall begin by considering the **conservation of mass**, by which we mean that the mass of an arbitrary material portion of the bar is unchanged throughout the motion. In order to see the analytic implications of this statement, let us designate the mass per unit volume of the bar at a section occupying the position x at time t by $\rho(x, t)$. The **mass density**, or simply **density**, $\delta(A, t)$ in material coordinates is therefore given by

$$\delta(A, t) = \rho[x(A, t), t]. \tag{24}$$

Similarly, the **area of a cross section** occupying the position x at the time t will be denoted by $\sigma(x, t)$. In material coordinates the cross-section area is denoted by S, where

$$S(A, t) = \sigma[x(A, t), t]. \tag{25}$$

Recall that we have assumed that each cross section moves as a whole in the longitudinal direction only. Consequently, the area of a given cross section will not change in time; i.e.,

$$S(A, t) = S(A, 0). \tag{26}$$

Thus

$$\sigma(x, t) = S[A(x, t), 0]. \tag{27}$$

Furthermore, since $A(x, t) = x - u(x, t)$, we may write

$$\sigma(x, t) = S[x - u(x, t), 0]. \tag{28}$$

Since the distribution of area at time $t = 0$, $S(A, 0)$, is assumed known, we observe that $\sigma(x, t)$ is a known function of $x - u(x, t)$, although the displacement $u(x, t)$ must itself be found.

We shall follow an arbitrary portion of the bar in time. At $t = 0$, this portion is designated by the interval

$$M \leq A \leq N,$$

while at t it is given by

$$m(t) \leq x \leq n(t),$$

where

$$x(M, t) \equiv m(t), \qquad x(N, t) \equiv n(t). \tag{29}$$

The assumption of mass conservation can now be written in the form

$$\frac{d}{dt}\int_{m(t)}^{n(t)} \rho(x, t)\sigma(x, t)\, dx = 0. \tag{30}$$

In order to avoid differentiating the limits of integration, we change the variable of integration to material coordinates with the result

$$\frac{d}{dt}\int_{m}^{n} \rho(x, t)\sigma(x, t)\, dx = \frac{d}{dt}\int_{M}^{N} \delta(A, t)S(A, t)\frac{\partial x}{\partial A}\, dA.$$

Taking the derivative of the integrand, and replacing $\partial x/\partial A$ by J, we obtain

$$\int_{M}^{N}\left[\frac{\partial}{\partial t}(\delta S)J + \delta S\frac{\partial}{\partial t}J\right] dA. \tag{31}$$

Using the Euler expansion formula (22), we may write the last integral as

$$\int_M^N \left[\frac{\partial}{\partial t}(\delta S) + \delta S \left.\frac{\partial v}{\partial x}\right|_A\right] J\, dA.$$

Returning this last result to spatial coordinates, we find that the conservation of mass requirement takes the form

$$\int_m^n \left[\frac{D(\rho\sigma)}{Dt} + \rho\sigma \frac{\partial v}{\partial x}\right] dx = 0. \tag{32}$$

Since the interval $m \le x \le n$ is taken over an arbitrary portion of the bar, the integrand in the above equation must itself vanish if (as we assume) it is continuous.* We thus obtain the **differential equation of mass conservation**

$$\boxed{\frac{D(\rho\sigma)}{Dt} + \rho\sigma \frac{\partial v}{\partial x} = 0.} \tag{33}$$

This equation is frequently known as the **continuity equation**. A form that may look more familiar to some can be written if we define the mass per unit length $\bar{\rho} \equiv \rho\sigma$ with which (33) becomes

$$\frac{D\bar{\rho}}{Dt} + \bar{\rho} \frac{\partial v}{\partial x} = 0. \tag{34}$$

The continuity equation can be simplified in the one-dimensional situation we have been considering. To see this, recall that $S(A, t) = S(A, 0)$. Thus

$$\frac{\partial S}{\partial t} = 0 \qquad \text{so } \frac{D\sigma}{Dt} = 0 \tag{35}$$

and (33) becomes

$$\sigma \frac{D\rho}{Dt} + \rho\sigma \frac{\partial v}{\partial x} = 0. \tag{36}$$

In the event that the material is **incompressible,**

$$\delta(A, t) = \delta(A, 0). \tag{37}$$

This implies that

$$\frac{\partial \delta}{\partial t} = 0, \qquad \text{so } \frac{D\rho}{Dt} = 0, \tag{38}$$

* Formally, we have used the Dubois–Reymond lemma. [See the remarks following Equation (4.1.15).] It might be preferable to use this lemma before making the final change to spatial coordinates, for then the arbitrary nature of the integration interval is clearer.

and the continuity equation reduces to

$$\frac{\partial v}{\partial x} = 0. \tag{39}$$

Consequently, the bar does not deform as it moves (Exercise 11).

FORCE AND STRESS

In order to consider the hypothesis of balance of linear momentum, we shall have to say a word or two about the types of forces that will be considered. These will be divided into two classes, body forces and surface tractions. **Body forces** (like gravity) will be defined as forces acting on a portion of the bar which are functions of the volume or the mass of that portion. **Surface forces** are forces that are transmitted across a cross section from one portion of the bar to the other. It is assumed that they depend only on the position of the section, and whether the portion of the bar being acted upon lies on the left or the right of the section. (For the corresponding analysis in three dimensions, the reader is referred to Section 14.2.)

The forces and other vectors we shall be considering are all parallel to the **i** direction, and hence the distinction between the vector and its one component is not too significant. Nevertheless, it will be useful for the moment at least to preserve this difference. The **body force per unit mass** will be denoted in spatial coordinates by $\mathbf{f}(x, t)$. In material coordinates the body force is $\mathbf{F}(A, t)$, where

$$\mathbf{F}[A(x, t)t] = \mathbf{f}(x, t).$$

We shall also write

$$\mathbf{f}(x, t) = f(x, t)\mathbf{i}.$$

The **surface traction** or **stress vector** will be denoted by $\mathbf{t}(x, t; +)$. This designates the force per unit area across a section located at x at time t. The force is exerted *by* material on the positive side *on* material on the negative side.* Similarly, $\mathbf{t}(x, t; -)$ is the force per unit area across a section at x at time t exerted *by* material on the negative side acting *on* material on the positive side.

We emphasize that if we consider a section of area σ at the location x at time t [Figure 12.5(a)], we cannot specify the traction until we agree which portion of the bar is considered to exert the stress. Thus, in Figure 12.5(b), we consider the stress produced by the material to the right of the section. The force so produced is $\sigma(x, t)\mathbf{t}(x, t; +)$. Similarly, in Figure 12.5(c) we have

* Remember, by convention the positive side is the right side. Another notation, which is more cumbersome but which is required in Section 14.2, uses $\mathbf{t}(x, t; \mathbf{i})$ instead of $\mathbf{t}(x, t; +)$. Here $\mathbf{i}$ is the unit normal pointing outward from the material on which the stress is acting. Similarly, $\mathbf{t}(x, t, -\mathbf{i})$ is used instead of $\mathbf{t}(x, t, -)$. Again, the indicated vector ($-\mathbf{i}$ in this case) points outward from the material on which the stress is acting.

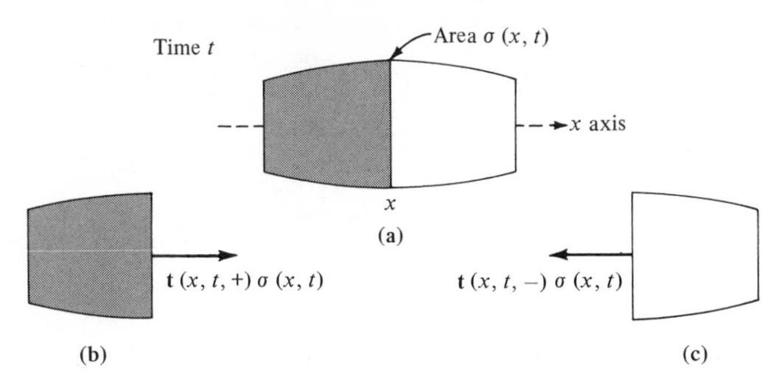

FIGURE 12.5. (a) *The stress in the bar at the section x is undefined.* (b) *The product of the stress vector $\mathbf{t}(x, t, +)$ and the cross-section area $\sigma(x, t)$ gives the surface force exerted by the unshaded portion of the bar on the shaded portion.* (c) *In contrast to* (b), *here the surface force represents the effect exerted by the shaded portion on the unshaded.*

indicated the force $\sigma(x, t)\mathbf{t}(x, t; -)$ produced by the material to the left of the section on the material on the right. We have drawn the vectors $\mathbf{t}(x, t; +)$ and $\mathbf{t}(x, t; -)$ oppositely directed in anticipation of some type of law of "action and reaction." Such a law will be proved below.

We must point out that it is not known at this stage whether the force in Figure 12.5(b) points to the right (as drawn) or to the left, i.e., whether the material to the right of the section pulls or pushes on the remainder of the bar. When a problem is completely analyzed, this question can be settled. In the absence of this analysis the direction of the arrow is arbitrarily chosen, and the lack of significance of this direction must not be overlooked.

By contrast, if we know that a portion of the bar is in a state of tension, then the forces applied at the end sections will be as shown in Figure 12.6(a). If the section of the bar is in a state of compression, the forces are as shown in Figure 12.6(b).

A state of tension is sometimes referred to as one of "positive" stress, whereas a state of compression is considered to be one of "negative" stress. If we use this nomenclature, we see that there are two sign conventions at work. Considering tension as positive means that we are using a sign convention intrinsic to the body with positive stresses pointing outward and

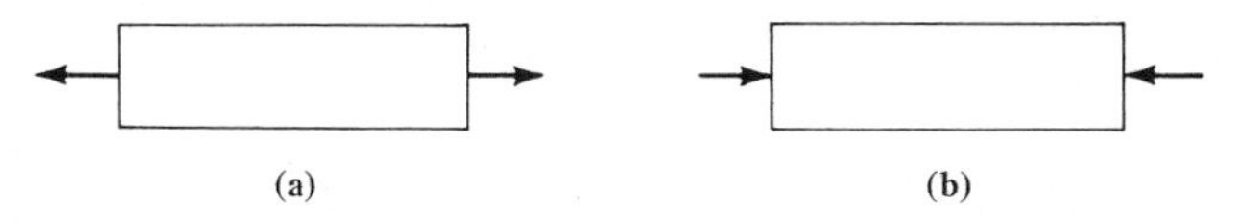

FIGURE 12.6. (a) *A bar under tension (positive stress).* (b) *A bar under compression (negative stress).*

negative stresses pointing inward. On the other hand, there is a sign convention for the components induced by the orientation of the A axis. Thus we see from Figure 12.6(a) that in a state of tension, of "positive" stress, the stress $\mathbf{t}(m, t; -)$ has a negative component, whereas $\mathbf{t}(n, t; +)$ has a positive component. The opposite is, of course, true for a state of compression. Both types of sign conventions are useful and can be effective, provided that one is careful to state which is being employed.

BALANCE OF LINEAR MOMENTUM

As in our discussion of the conservation of mass, let us follow an arbitrary portion of the bar described in spatial coordinates by $m(t) \le x \le n(t)$. In mechanics, (linear) momentum is the product of mass and velocity. Thus the **linear momentum** of this section is defined by the integral

$$\int_{m(t)}^{n(t)} \rho\sigma\mathbf{v}\, dx. \tag{40}$$

The hypothesis of **balance of linear momentum*** states that the time rate of change of linear momentum of any portion of the bar is equal to the sum of all the external forces acting on it. Thus

$$\frac{d}{dt}\int_m^n \rho\sigma\mathbf{v}\, dx = \int_m^n \rho\sigma\mathbf{f}\, dx + \mathbf{t}(m, t; -)\sigma(m, t) + \mathbf{t}(n, t; +)\sigma(n, t). \tag{41}$$

The calculation of the left-hand side of (41) can be handled in exactly the same manner as that used in our discussion of mass conservation. One shifts to material coordinates, carries out the differentiation using the Euler expansion formula, and then returns to spatial coordinates. After observing that certain terms vanish because of the mass conservation formula (33), one obtains [Exercise 1(c)] the result

$$\frac{d}{dt}\int_m^n \rho\sigma\mathbf{v}\, dx = \int_m^n \rho\sigma\frac{D\mathbf{v}}{Dt}\, dx. \tag{42}$$

With this, our expression (41) for the balance of linear momentum becomes

$$\int_m^n \left[\rho\sigma\left(\frac{D\mathbf{v}}{Dt} - \mathbf{f}\right)\right] dx = \mathbf{t}(m, t; -)\sigma(m, t) + \mathbf{t}(n, t; +)\sigma(n, t). \tag{43}$$

The right-hand side of (43) is in somewhat awkward form, since it does not appear as an integral and thus precludes use of the Dubois–Reymond lemma. This difficulty can be alleviated by the following procedure. Let us consider a generic section $x = x_0$, in the arbitrary portion of the bar, which divides it into two pieces I and II, satisfying

$$\text{I}: m \le x \le x_0, \qquad \text{II}: x_0 \le x \le n.$$

* If the linear momentum of the bar were truly *conserved*, then the time derivative of (40) would be zero.

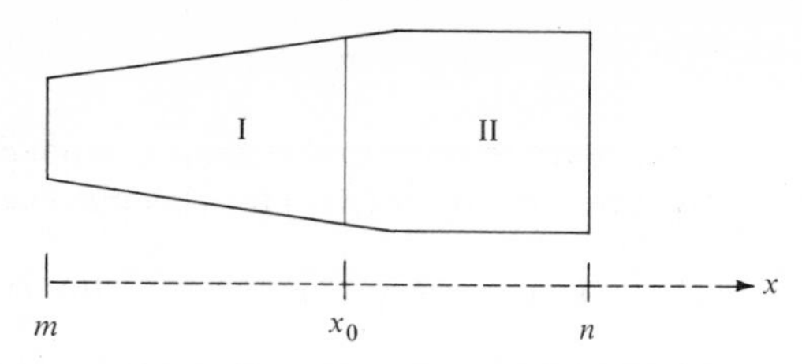

FIGURE 12.7. *Section* $x = x_0$ *divides the bar into pieces* I *and* II.

(See Figure 12.7.) We shall now apply formula (43) to each of these parts, with the results

$$\int_m^{x_0} \rho\sigma\left(\frac{D\mathbf{v}}{Dt} - \mathbf{f}\right) dx = \mathbf{t}(x_0, t; +)\sigma(x_0, t) + \mathbf{t}(m, t; -)\sigma(m, t), \qquad (44a)$$

$$\int_{x_0}^{n} \rho\sigma\left(\frac{D\mathbf{v}}{Dt} - \mathbf{f}\right) dx = \mathbf{t}(x_0, t; -)\sigma(x_0, t) + \mathbf{t}(n, t; +)\sigma(n, t). \qquad (44b)$$

Adding (44a) and (44b) and subtracting (43), we find that

$$\mathbf{t}(x_0, t; +)\sigma(x_0, t) + \mathbf{t}(x_0, t; -)\sigma(x_0, t) = 0.$$

Since $x = x_0$ is a generic section, we may then conclude* for any section

$$\boxed{\mathbf{t}(x, t; -) = -\,\mathbf{t}(x, t; +).} \qquad (45)$$

It should be noted that the result derived in (45) can be considered as a form of Newton's third law in ordinary particle mechanics (action equals reaction), although in this case it is found as a consequence of the conservation of linear momentum. Equation (45) also tells us that we predicted correctly in drawing the oppositely directed vectors $\mathbf{t}(x, t, +)$ and $\mathbf{t}(x, t, -)$ in Figures 12.5(b) and 12.5(c).

We are now in a position to simplify (43) through the use of the action–reaction result (45). To this end, we introduce the following definition: the stress component, $T(x, t)$, is defined† by

$$\boxed{\mathbf{t}(x, t; +) = T(x, t)\mathbf{i}.} \qquad (46)$$

* The derivation just completed, save for notation, is precisely the same as one that can be applied in three dimensions. See Exercise 14.2.1.

† The reader will note that we have been inconsistent in our notation in using a capital letter to refer to a quantity evaluated in spatial coordinates. We shall find it convenient, however, to make this deviation to match more closely with the symbolism conventionally used in three dimensions.

We thus see from this definition and (45) that

$$\mathbf{t}(x, t; -) = -T(x, t)\mathbf{i}. \tag{47}$$

In a few instances, we shall want to refer to the stress component in material coordinates and shall use the notation $\mathscr{T}(A, t)$ for this purpose; i.e.,

$$\mathscr{T}(A, t) = T[x(A, t), t] \qquad \text{or} \qquad T(x, t) = \mathscr{T}[A(x, t), t].$$

We are now in a position to write the right-hand side of (43) in the form of an integral, for

$$\begin{aligned}\mathbf{t}(m, t; -)\sigma(m, t) + \mathbf{t}(n, t; +)\sigma(n, t) &= \mathbf{i}[T(n, t)\sigma(n, t) - T(m, t)\sigma(m, t)] \\ &= \mathbf{i}\int_m^n \frac{\partial(T\sigma)}{\partial x}\, dx. \end{aligned} \tag{48}$$

Dropping the vector form of (43), since there is only one component present, we can then write the balance of linear momentum as

$$\int_m^n \left[\rho\sigma\left(\frac{Dv}{Dt} - f\right) - \frac{\partial(T\sigma)}{\partial x}\right] dx = 0. \tag{49}$$

Since the portion of the bar has been arbitrarily chosen, if the above integrand is continuous, then the Dubois–Reymond lemma allows us to conclude the following final form for the law of balance of linear momentum:

$$\boxed{\rho\sigma \frac{Dv}{Dt} = \rho\sigma f + \frac{\partial(T\sigma)}{\partial x}\cdot} \tag{50}$$

STRAIN AND STRESS–STRAIN RELATIONS

In analyzing the motion of a bar, we normally consider as known (i) the cross-section area σ, (ii) the body force f, and (iii) appropriate initial data on all the other variables. We have three unknown functions, the density ρ, the velocity v (or equally well the displacement $u = Dv/Dt$), and the stress component T. On the other hand, we have developed but two basic differential equations (33) and (50). Thus it would appear that our mathematical description of the motion is incomplete.

At the same time, our discussion of the physics of the situation is also deficient. We have derived **field equations** that *describe the universal physical requirements* of mass conservation and momentum balance. But we have in no way incorporated any properties of the particular material from which the bar has been made. Common experience shows us that two bars which are identical in geometry and subject to the same loadings can vary considerably in their subsequent motion if they are composed of different materials. This deficiency in our analysis can be repaired through the introduction of a **constitutive relation**, which in this case *describes the particular nature of the material* by relating a measure of the distortion of the bar to the amount of stress required to produce this distortion.

We must first assign a precise meaning to the notion of "distortion." To do this, consider two sections that originally (at $t = 0$) were located at the nearby locations A and $A + \Delta A$. At time t these sections occupy the positions $x(A, t)$ and $x(A + \Delta A, t)$. The sections in question originally enclosed an element of material having length ΔA. At time t this material has a *new length*

$$x(A + \Delta A, t) - x(A, t),$$

which can be approximated by

$$\frac{\partial x}{\partial A} \Delta A = \left(1 + \frac{\partial U}{\partial A}\right) \Delta A.$$

[We have used the relationship $x(A, t) = A + U(A, t)$ of (9).]

A reasonable measure of distortion is

$$\frac{\text{new length} - \text{original length}}{\text{original length}}. \tag{51}$$

In the present instance, to lowest approximation this measure is

$$\varepsilon \equiv \frac{(1 + \partial U/\partial A)\, \Delta A - \Delta A}{\Delta A} = \frac{\partial U}{\partial A}.$$

Thus the axial strain, or simply the **strain** (as ε is called), is given by

$$\varepsilon(A, t) = \frac{\partial U(A, t)}{\partial A}. \tag{52}$$

Since higher powers of ΔA have been neglected in approximating our measure of distortion, one often speaks of the strain as a change in infinitesimal length per unit original infinitesimal length. It will be observed later that the task of defining an appropriate measure of strain in three dimensions is more elaborate.

In forming a constitutive equation, we shall assume that the stress component $\mathcal{T}(A, t)$ is a function of the strain $\varepsilon(A, t)$. This makes concrete our feeling that the more one wants the bar to elongate, the harder one has to pull it. We shall limit ourselves to situations in which the initial state of the bar is free of stress. Thus the graph of stress component versus strain might have the appearance of the heavy line in Figure 12.8. In particular, the graph must go through the origin.

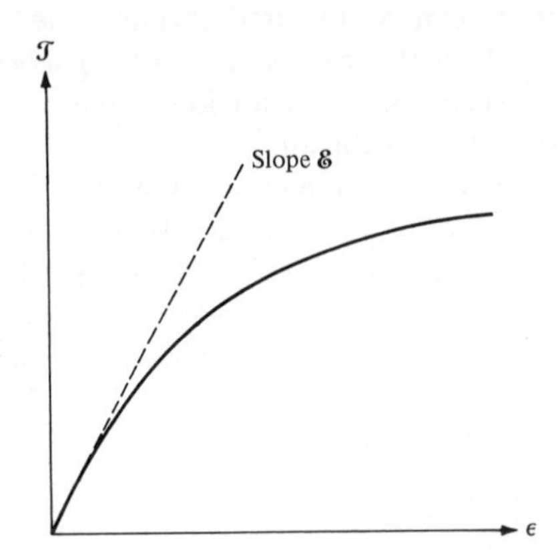

FIGURE 12.8. *Plot of a "typical" relation between stress* $\mathscr{T}$ *and strain* ε.

We can make some progress without reference to experiments.* We reason that for sufficiently small strain ε, the *graph* of Figure 12.8 can be *approximated by its tangent line* at the origin. If the slope of this line is denoted by $\mathscr{E}$, we thereby hypothesize that

$$\mathscr{T}(A, t) = \mathscr{E}(A, t) \cdot \varepsilon(A, t). \tag{53a}$$

More simply, we find that

$$\boxed{\mathscr{T} = \mathscr{E}\varepsilon = \mathscr{E}\frac{\partial U}{\partial A}.} \tag{53b}$$

In **Hooke's law** (53b), the proportionality factor $\mathscr{E}$ is called **Young's modulus**. By the way, Robert Hooke (a seventeenth-century contemporary of Newton) postulated the proportionality of stress and strain only for large scale or global situations. In our pointwise or local version of Hooke's law, Young's modulus may be a function of A and t—for the nature of the bar may vary with these variables. We shall assume, however, that

$$\mathscr{E}(A, t) = \mathscr{E}(A, 0), \tag{54}$$

where the initial composition of the bar, and hence the initial distribution of the stiffness factor $\mathscr{E}$, is known.†

* This is a dangerous procedure, but it is sometimes a necessary one. Perhaps only by guessing a reasonable constitutive equation and then making some plausible and interesting theoretical deductions can one convince an experimenter that it is worth his while to investigate certain phenomena.

† Aging effects could be included by allowing $\mathscr{E}$ to be a known function of t as well as of A.

In spatial variables, Young's modulus is designated by E. That is, $E(x, t) = \mathscr{E}[A(x, t), t]$. Typically, $E = 3 \times 10^7$ lb/in.^{2} (2×10^{11} dynes/cm^2) for steel and 10^8 lb/in.^{2} (7×10^{12} dynes/cm^2) for aluminum.

We note that, by (21) with $F \equiv U$ and $f \equiv u$, in spatial coordinates our constitutive assumption is

$$T(x, t) = E(x, t)\frac{\partial u/\partial x}{1 - \partial u/\partial x}. \tag{55}$$

There is no doubt that it is a good approximation to replace the actual stress–strain curve by its tangent line at the origin, provided that strains are sufficiently small. The question is: Are there significant problems in which such small strains are found? It turns out that there is a wide range of such problems. To give one indication of this, consider the change in length of a standard 39-ft railroad track. During the temperature change from winter to summer an expansion of $\frac{1}{4}$ in. could easily occur. The corresponding strain is 0.00053. [If no provision were made by a gap between rails to allow for this expansion, the resultant axial stress (approximately 16,000 lb/in.^{2}) would be more than sufficient to cause the rail to buckle out of its straight position.]

The above remarks should provide sufficient motivation to pursue the consequences of our constitutive assumption. Using Hooke's law, we can write the momentum balance requirement (50) as

$$\rho\sigma\frac{Dv}{Dt} = \rho\sigma f + \frac{\partial}{\partial x}\left(\mathscr{E}S\frac{\partial U}{\partial A}\bigg|_x\right). \tag{56}$$

Alternatively, using (55) and the result $v = Du/Dt$, we have

$$\boxed{\rho\sigma\frac{D^2u}{Dt^2} = \rho\sigma f + \frac{\partial}{\partial x}\frac{E\sigma\, \partial u/\partial x}{1 - \partial u/\partial x}.} \tag{57}$$

At this point, there are no obvious omissions in our basic equations. Let us therefore review the situation to see whether the number of equations appears sufficient.

In the momentum equation (57) we regard σ, E, and f as known, while u and ρ are unknown. Thus from the initial area distribution, one can determine σ as a function of u from (28). Similarly, from (54), Young's modulus is a function of u determined from the original distribution of $\mathscr{E}$:

$$E(x, t) = \mathscr{E}[x - u(x, t), 0]. \tag{58}$$

Finally, the body force $F(A, t)$ (typically gravity) will be regarded as equal to its given initial distribution $F(A, 0)$ so that

$$f(x, t) = F[x - u(x, t), 0] \tag{59}$$

can be regarded as known.

Again recalling that $v = Du/Dt$, we see that the mass conservation equation (33),

$$\frac{D(\rho\sigma)}{Dt} + \rho\sigma \frac{\partial v}{\partial x} = 0,$$

also contains the two unknown functions u and ρ. We are thus faced with two equations in as many unknowns. Even for linear algebraic equations, one can draw no definitive conclusions from the fact that there are as many equations as unknowns. Nevertheless, we can draw the *tentative* conclusion that we have done a satisfactory job in formulating the governing differential equations.

INITIAL AND BOUNDARY CONDITIONS

One expects that a properly formulated problem will have a unique solution. But in the simplest problems of mechanics, and here, too, uniqueness cannot be expected unless initial conditions are prescribed. Different displacements and density variations must certainly be expected from bars that have different initial configurations. Thus we make the stipulation

$$u(x, 0) = u_0(x), \qquad v(x, 0) = v_0(x), \qquad \rho(x, 0) = \rho_0(x), \tag{60}$$

where u_0, v_0, and ρ_0 are given functions.

We must also make some statement about what happens at the ends of the bar. This is certainly reasonable in the case of a bar of finite length or at the finite end of a semiinfinite bar. The situation is perhaps not as clear in the case of an infinite bar; but we have an intuitive feeling that if nothing is said about what happens at "infinity," there may be serious difficulties in trying to find a unique solution. We shall make no attempt to discuss this uniqueness question at this time. Rather, the case of the finite end will be examined in a few examples. We shall formulate boundary conditions in material coordinates and shall leave to the reader the task of translation into spatial coordinates.

Let us assume that the bar has a finite terminus at $A = L$. First consider the case in which the bar is built in at the end, so that it is constrained from movement. In such a case we refer to the boundary as a **fixed end**. The boundary condition can then be stated as

$$U(L, t) = 0. \tag{61}$$

[It is understood that (61) must hold for all positive t.] In contrast with the case just treated, we can have the situation where no constraint whatsoever is applied, a **free end**. Under these conditions, no force is applied at the end to restrict the motion. At the end $A = L$, this requirement can be written, using Hooke's law (53), as

$$\mathscr{E}(L, t)S(L, t)U_A(L, t) = 0, \tag{62a}$$

where we have used subscript A to denote partial differentiation.

In most cases neither $\mathscr{E}$ nor S will vanish at $A = L$, and (62a) can be replaced by the simpler form

$$U_A(L, t) = 0. \tag{62b}$$

But S will vanish if the bar comes to a point at $A = L$. It must not be concluded, however, that (62a) is vacuous in such an instance. We find frequently, when S vanishes at $A = L$, that possible solutions of the equation of motion can have singularities at that point. Equation (62a) then tells us that the singularity cannot be so strong that the indicated product will not vanish. For example, if S behaves like $(A - L)$ near $A = L$, a purported solution U that behaves like $\log (A - L)$ is ruled out by (62a).

The fixed end and the free end are, by all odds, the most common end conditions met in practice. There are, however, a number of others that do arise; and we shall consider two of them. These boundary conditions are a bit more subtle in derivation, primarily because one must be careful with the algebraic signs.

Let us consider a rigid mass fixed to the free end of the bar at $A = L$, as shown in Figure 12.9. The mass M will undergo the same displacement $U(L, t)$

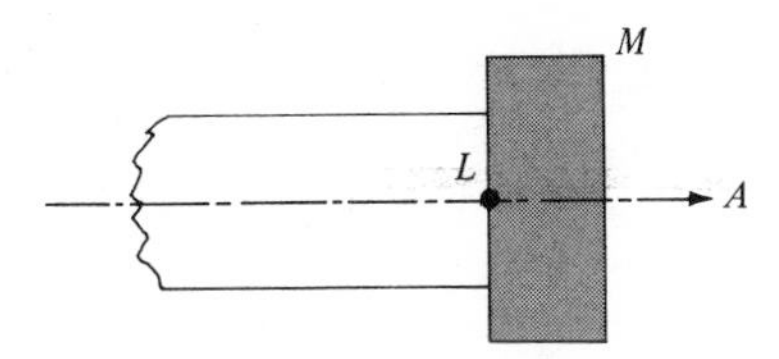

FIGURE 12.9. *A semiinfinite bar joined to a mass M.*

as the end of the bar. Now let us isolate the mass itself. If we ignore the effect of gravity, the force acting *on* the mass due to the bar is $S(L, t)\mathbf{t}[x(L, t), t; -]$. In terms of the stress component,

$$\mathbf{t}[x(L, t), t; -] = -\mathbf{i}T[x(L, t), t].$$

Applying Newton's second law to the mass M and replacing the stress component through Hooke's law, we obtain the following boundary condition:

$$-S(L, t)\mathscr{E}(L, t)U_A(L, t) = MU_{tt}(L, t). \tag{63}$$

As it should be, (62a) is recovered in the limit $M \to 0$.

As our final example, we shall consider a linear spring attached to the free end of the bar on one side and rigidly fixed at the other end (Figure 12.10). As the bar undergoes various displacements, the spring will follow. But there will be a lag, due to the inertia of the spring. In general, one would be faced with a separate, difficult, problem for the spring. One would have to find a

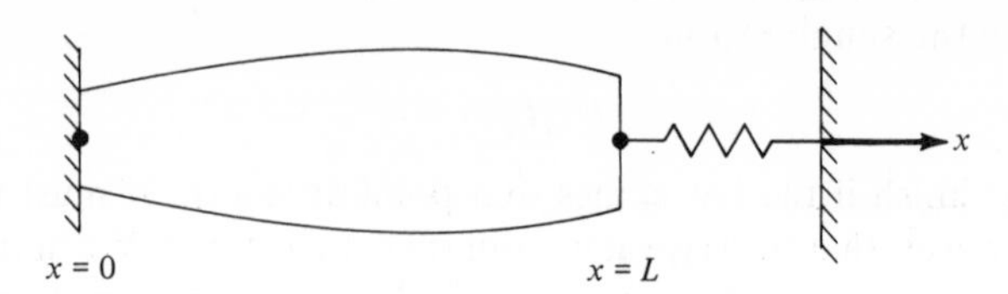

FIGURE 12.10. *Bar with one end built in and one end constrained by a linear spring.*

solution that matches the solution of the bar at the bar–spring junction. To avoid this difficulty, we shall assume that the spring is so light that its inertia is negligible. Under such circumstances, a displacement of the bar will instantly cause a corresponding uniform stretching of the spring.

Consider a slice at the end of the bar, corresponding to initial coordinates between $L - \Delta$ and L. There is a surface stress $\mathbf{t}[x(L - \Delta, t), t; -]$ acting on the left face of this slice. In obtaining the force on the right face, we assume that the spring is initially in an unstretched position. We denote the spring constant by k^2. Then the *magnitude* of the spring force is $|k^2 U(L, t)|$. We assert that the actual force on the bar due to the spring is $-k^2 U(L, t)\mathbf{i}$. Indeed, if $U(L, t)$ is positive, this force is leftward—which is the appropriate direction for the force due to a compressed spring. On the other hand, if $U(L, t)$ is negative, the force on the end of the bar is rightward—which is appropriate when the spring is extended.

The final step is to equate the rate of change of the slice's momentum with the sum of the surface forces just discussed and the body force. Upon taking the limit as $\Delta \to 0$, one sees that the inertia terms and the body force are negligible. Thus the boundary condition

$$S(L, t)\mathscr{E}(L, t)U_A(L, t) = -k^2 U(L, t) \tag{64}$$

emerges as the required balance between surface stress and spring force (Exercise 15). Boundary conditions, such as (64), that involve a linear combination of the unknown function and its first derivative are frequently referred to as **impedance boundary conditions**.

LINEARIZATION

A well-formulated problem seems to be offered by the governing equations (57) and (33), the initial conditions (60), and a pair of boundary conditions of the type we have just derived. But the equations are nonlinear, so that, in general, numerical or advanced analytical techniques are required to obtain useful exact or approximate solutions. Such techniques are out of place in this chapter, which is supposed to serve as a relatively simple introduction to continuum mechanics. Without further ado, therefore, we shall linearize the equations and boundary conditions.

The basic fact which makes linearization worthwhile is that in metallic materials, even a very large stress will produce but a small strain. We shall therefore systematically simplify all expressions under the assumption that the displacement u and its partial derivatives are small compared to unity. We shall also assume that the density at a particular location x does not vary much from its initial value.

Consider $\sigma(x, t)$, the cross-section area of the section now (at time t) located at position x. Recall that we assumed that the area of any given section does not change with time; i.e., $S(A, t) = S(A, 0)$. From this we deduced in (28) that

$$\sigma(x, t) = S[x - u(x, t), 0].$$

Since u has been assumed to be small, we should be able to make the approximation

$$\sigma(x, t) \approx S(x, 0). \tag{65}$$

The effect of this approximation is to replace the area of the section that is *now* x units from the origin by the area of the section which was *initially* x units from the origin. When displacements are small, such an approximation indeed seems reasonable.

A glance at the difference between the exact and approximate cross-sectional areas clarifies the nature of our approximation. By the mean value theorem we can write

$$|S(x, 0) - S(x - u, 0)| = |uS_x(\xi, 0)|, \tag{66}$$

where ξ is between $x - u$ and x. Assuming that the derivative S_x is bounded in the neighborhood of x, the right-hand side of (66) can be made as small as desired by taking u sufficiently small. The bigger the derivative, the smaller one has to take u to achieve a given level of approximation. This is to be expected; the more rapidly the cross-section area changes, the bigger is the error made in misestimating the correct location to measure this area.

With exact and approximate values of the same quantity both present, some notational difficulty is experienced. It seems best to use the function of a single variable $\sigma(x)$ to denote our approximation to $\sigma(x, t)$. A "hat" will be employed to signify the difference between the exact and approximate values. Thus we shall write

$$\sigma(x) \equiv S(x, 0), \qquad \hat{\sigma}(x, t) \equiv \sigma(x, t) - \sigma(x). \tag{67a, b}$$

With this notation, our approximation (65) is written

$$\sigma(x, t) \approx \sigma(x) \qquad [\text{i.e., } |\hat{\sigma}(x, t)| \ll |\sigma(x)|]. \tag{68}$$

Remember—$\sigma(x)$ is a known function, the initial area of a section x units from the origin.

In (58) and (59) we have exact analogues of (28) so that we can treat the Young's modulus $E(x, t)$ and the body force $f(x, t)$ in exactly the same way as we treated $\sigma(x, t)$. We *approximate the present value of each quantity at x by the known initial value of that quantity at x.** Thus we define

$$E(x) \equiv \mathscr{E}(x, 0), \qquad \hat{E}(x, t) \equiv E(x, t) - E(x); \tag{69}$$

$$f(x) \equiv F(x, 0), \qquad \hat{f}(x, t) \equiv f(x, t) - f(x). \tag{70}$$

We assume that

$$|\hat{E}(x, t)| \ll |E(x, t)|, \qquad |\hat{f}(x, t)| \ll |f(x, t)|, \tag{71}$$

so that

$$E(x, t) \approx E(x), \qquad f(x, t) \approx f(x). \tag{72}$$

We have said all along that $\sigma(x, t)$, $E(x, t)$, and $f(x, t)$ were regarded as known functions, although (before approximation) the argument of the functions contained the unknown displacement u. Previously, $\rho(x, t)$ was regarded as unknown, but this function, too, must be approximated by its known initial value. Otherwise, the term $\rho(x, t)\sigma(x, t)\, D^2u/Dt^2$ in the momentum equation (57) will remain nonlinear. Approximation of the density requires extra care, because the assumption

$$\delta(A, t) = \delta(A, 0)$$

is inappropriate. If this assumption holds, the material is *incompressible* and the only one-dimensional motions are uninteresting rigid motions (Exercise 11).

Although $\delta(A, t) \neq \delta(A, 0)$, we still write

$$\rho(x) \equiv \delta(x, 0), \qquad \hat{\rho}(x, t) \equiv \rho(x, t) - \rho(x), \tag{73}$$

and assume that

$$|\hat{\rho}(x, t)| \ll |\rho(x)| \qquad \text{so } \rho(x, t) \approx \rho(x). \tag{74}$$

If we approximate the difference $\hat{\rho}$ by the lowest order terms in the appropriate Taylor series, we find [Exercise 18(a)] that

$$|\hat{\rho}(x, t)| \approx |u\delta_x + t\delta_t|. \tag{75}$$

Knowledge that the displacement u is small, then, is not sufficient to guarantee that $|\hat{\rho}|$ is small. It could be that the action of small displacements produces continual compression or rarefaction that would eventually lead to a significant

* Roughly speaking, we ignore the difference between material and spatial coordinates.

deviation of the density of a section from its initial value. As we see from (75), this will not happen if the motion has been proceeding for a sufficiently short time (t sufficiently small).

As the final link in our chain of approximations, we recall that the velocity v is given by

$$v = \frac{Du}{Dt} = u_t + vu_x .$$

Solving for v, we find that

$$v = \frac{u_t}{1 - u_x}. \tag{76}$$

Since we are assuming that u and its derivatives are small compared to unity, we can make the approximation

$$v(x, t) \approx \frac{\partial u(x, t)}{\partial t}. \tag{77}$$

Note that the assumed smallness of u_t implies

$$|v(x, t)| \ll 1. \tag{78}$$

As is easily verified (Exercise 19), if we retain only lowest order terms we obtain the following linearized version of the momentum balance equation (57):

$$\boxed{\rho(x)\sigma(x)\frac{\partial^2 u}{\partial t^2} = \rho(x)\sigma(x)f(x) + \frac{\partial}{\partial x}\left[E(x)\sigma(x)\frac{\partial u}{\partial x}\right].} \tag{79}$$

The continuity equation (33) can be written in the form

$$\frac{\partial}{\partial t}[\rho(x, t)\sigma(x, t)] + v\frac{\partial}{\partial x}[\rho(x, t)\sigma(x, t)] + \rho(x, t)\sigma(x, t)\frac{\partial v}{\partial x} = 0.$$

The first term can be rewritten as follows:

$$\frac{\partial}{\partial t}[\rho(x, t)\sigma(x, t)] = \frac{\partial}{\partial t}\{[\rho(x) + \hat{\rho}(x, t)][\sigma(x) + \hat{\sigma}(x, t)]\}.$$

Again retaining the lowest order terms, we obtain the following linearized version of the continuity equation:

$$\frac{\partial}{\partial t}[\sigma(x)\hat{\rho}(x, t) + \rho(x)\hat{\sigma}(x, t)] + v\frac{\partial}{\partial x}[\rho(x)\sigma(x)] + \rho(x)\sigma(x)\frac{\partial v}{\partial x} = 0. \tag{80}$$

Equation (79) contains but a single unknown function, the displacement $u(x, t)$. Appropriate boundary conditions are linearized versions of (61)–(64). These versions are as follows [Exercise 16(b)]:

Fixed end:	$u = 0.$	(81a)
Free end:	$\sigma E u_x = 0.$	(81b)
Mass-loaded end:	$M u_{tt} + \sigma E u_x = 0.$	(81c)
End constrained by light spring:	$\sigma E u_x + k^2 u = 0.$	(81d)

Each boundary condition is to be imposed at the (known) initial position of the appropriate end.

NOTE. In the above equations, and in the remaining equations of this chapter, if σ, E, and f are written without arguments, the known functions $\sigma(x)$, $E(x)$, and $f(x)$ are implied.

It would appear that a self-contained linearized mathematical model for one-dimensional deflection of a bar is composed of the differential equation (79), a boundary condition such as those of (81) at each end of the bar, and the initial conditions

$$\boxed{u(x, 0) = u_0(x), \qquad u_t(x, 0) = v_0(x).} \tag{82}$$

What, then, is the role of the continuity equation (80)? The answer to this question is a little easier to see if we introduce the mass per unit length $\bar{\rho}(x, t)$ and its approximation $\rho(x)$:

$$\bar{\rho}(x, t) \equiv \rho(x, t)\sigma(x, t), \qquad \bar{\rho}(x) \equiv \rho(x)\sigma(x). \tag{83}$$

To lowest order

$$\begin{aligned}\bar{\rho}(x, t) - \bar{\rho}(x) &\equiv [\rho(x) + \hat{\rho}(x, t)][\sigma(x) + \hat{\sigma}(x, t)] - \rho(x)\sigma(x) \\ &\approx \rho(x)\hat{\sigma}(x, t) + \hat{\rho}(x, t)\sigma(x)\end{aligned}$$

so that the linearized continuity equation (80) can be regarded as stating

$$\frac{\partial}{\partial t}[\bar{\rho}(x, t) - \bar{\rho}(x)] \approx -\frac{\partial}{\partial x}[v(x, t)\rho(x)\sigma(x)]. \tag{84}$$

We see, then, that in linear theory the continuity equation provides a first correction to the difference between the actual and initial values of the mass per unit length. This correction is easily computed once the velocity $v \approx u_t$ is found. The correction is not of major interest in most applications, however, and the continuity equation is usually ignored in linear theory. Nevertheless, a careful worker would at least check to see that the predicted correction is indeed small, so that the approximate analysis has the appearance of consistency.

EXERCISES

1. (a) Verify (8).
(b) Verify (21).
(c) Verify (42).

2. Consider a motion described by

$$x = A + t(A - \tfrac{1}{2}), \qquad 0 \le A \le 1, \quad 0 \le t \le t_1.$$

(a) Describe the motion in qualitative terms.
(b) Find $U(a, t)$. Find $u(x, t)$ from the formula $u = x - A$. Verify that $u|_A = U$.
(c) Find

$$V(A, t), \qquad v(x, t), \qquad \frac{\partial V(A, t)}{\partial t}, \qquad \frac{\partial v(x, t)}{\partial t}.$$

(d) Make the substitution $x = x(A, t)$ in the second and fourth expressions. Compare the first pair; the second pair. Discuss fully in terms of the material derivative.

3. Consider a motion described by

$$x = \tfrac{1}{2}A + \tfrac{1}{2}Ae^t, \qquad 0 \le t \le t_1.$$

(a) Find the material variable A as a function of x and t.
(b) Calculate $U(A, t)$ and $u(x, t)$.
(c) Find the velocity $V(A, t)$ in material coordinates and $v(x, t)$ in spatial coordinates.
(d) Check the Euler expansion formula for this case by direct computation.

4. Establish the following differentiation formulas:
(a) $D(f \pm g)/Dt = Df/Dt \pm Dg/Dt$.
(b) $D(fg)/Dt = f(Dg/Dt) + g(Df/Dt)$.
(c) Does the analogue of the usual formula for the derivative of a quotient hold?

5. Consider a bar in the form of a truncated right circular cone. It is of length L and has radius r_1 at one end and r_2 at the other. The bar is assumed to undergo the motion described in Exercise 2.
(a) Find the cross-section area $S(A, 0)$ at time $t = 0$.
(b) Let $\sigma(x, t)$ denote the cross-section area at the position x, at time t. Find $\sigma(x, t)$ and calculate $D\sigma/Dt$. How does this compare with $\partial S(A, t)/\partial t$?

‡6. Derive the equation for mass conservation by a direct attack on the equation

$$\frac{d}{dt}\int_{x(M,t)}^{x(N,t)} \bar{\rho}(x, t)\, dt, \qquad \bar{\rho} \equiv \rho\sigma.$$

7. (a) Show by use of the procedure and notation of the text's discussion of mass conservation that

$$\frac{d}{dt}\int_{m(t)}^{n(t)} G\,dx = \int_{m(t)}^{n(t)} \left[\frac{DG}{Dt} + G\frac{\partial v}{\partial x}\right] dx.$$

(b) Use integration by parts to derive the one-dimensional version of the **Reynolds transport theorem**:

$$\frac{d}{dt}\int_{m(t)}^{n(t)} G\,dx = \int_{m(t)}^{n(t)} \frac{\partial G}{\partial t}\,dx + Gv\Big|_{m(t)}^{n(t)}.$$

(c) Interpret the result of (b). (In three dimensions the interpretation is the same; see Section 14.1.)

8. If $v(x, t)$ is independent of t, the motion is called *steady*.

(a) If you are familiar with first order partial differential equations, show that the most general steady motion has a displacement that can be written in the form $u(x, t) = x - F[t - g(x)]$ for some functions F and g.

(b) What can be said about the other quantities of interest?

9. Show that the mass conservation equation in material coordinates can be quickly obtained in the form

$$\frac{\partial}{\partial t}[\delta(A, t)S(A, t)J(A, t)] = 0.$$

Deduce the alternative form

$$\delta(A, t)S(A, t)J(A, t) = \delta(A, 0)S(A, 0).$$

[Since $S(A, t) \equiv S(A, 0)$ the cross-section area factor can be canceled.]

10. In the text's discussion of mass conservation, it is M and N, not $m(t)$ and $n(t)$, which are arbitrary. Nevertheless, (33) can be concluded directly from (32). Show this.

11. Show that (39) implies that $x(A, t) = A + f(t)$ for some function f. Interpret this result. Is it reasonable?

‡12. Find the "action–reaction" equation by the "thin-slice" approach. Consider a section of the bar of width Δx with the end forces as indicated in Figure 12.11. Apply the momentum conservation equation and then let $\Delta x \to 0$.

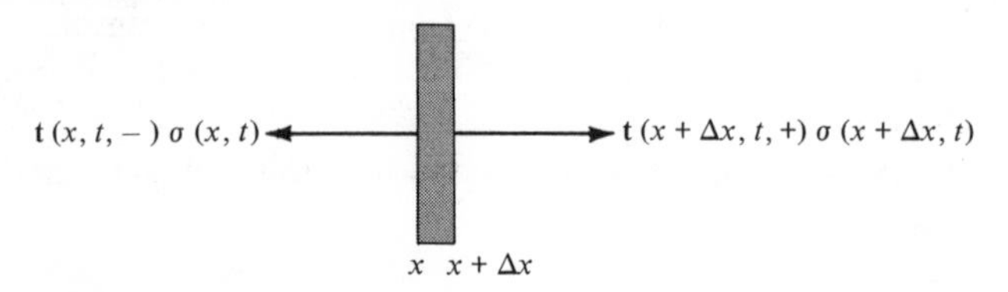

FIGURE 12.11. *Surface forces on a thin slice of bar with cross-section area σ.*

13. If we wish to be cavalier in our derivation of the approximate equation of motion (79), we can use a "thin-slice" approach and linearize as we go along. Carry out this procedure in the following manner. The forces acting on a small slice of the bar of width Δx are as shown in Figure 12.12.
(a) Apply Newton's second law of motion to the slice.
(b) Find the limiting form of the equation when $\Delta x \to 0$.
(c) Use Hooke's law.

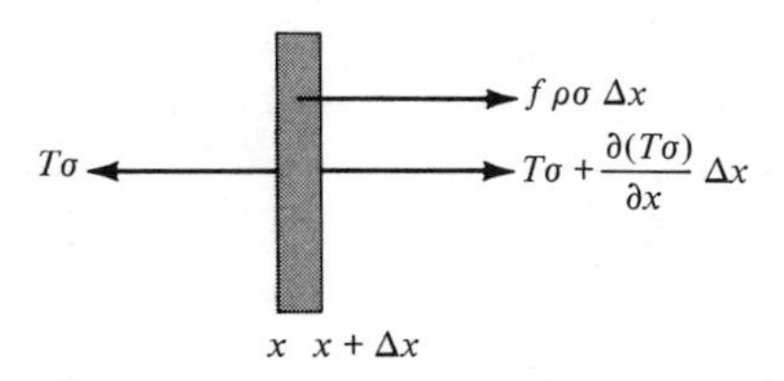

FIGURE 12.12. *Body and surface forces on a thin slice of bar.*

14. The **Eulerian strain*** is defined as the lowest order approximation to the quotient

$$\frac{\text{present length} - \text{original length}}{\text{present length}}.$$

Find an expression for the Eulerian strain in spatial coordinates. Use two different approaches.
(a) Consider a section of bar that is now (at time t) located between x and $x + \Delta x$.
(b) Consider a section of bar that was initially located between A and $A + \Delta A$. Approximate the quotient; then change to spatial coordinates.

15. Complete the derivation of (64).

16. (a) Transform the boundary conditions (62a), (63), and (64) into conditions on u.
(b) Linearize to obtain (81b–d).

17. Gravity was neglected in obtaining boundary condition (63). Under what circumstances is this justified?

18. (a) Verify (75).
(b) What is the relation between the function $\rho_0(x)$ of (60) and the function $\rho(x)$ of (73)?

* Euler's name is associated with spatial coordinates, Lagrange's with material. Consequently, Euler's name is invoked when the change in length is compared to the present length, and Lagrange's when comparison is made with the original length.

19. Show that our various assumptions give (79) as a first approximation to the momentum balance equation (57).

20. Show that (84) has a very reasonable physical interpretation. You may find this easier if you integrate between fixed limits $x = a$ and $x = b$.

21. Show that a $\frac{1}{4}$-in. elongation of a 39-ft steel rail is equivalent to a strain of 0.00053. Show that the corresponding stress is about 16,000 lb/in.^{2} (This verifies a statement in the text.)

22. For a certain spatial region, and in a certain time interval $(0, t_1)$, the one-dimensional motion of a bar is described by $x = (A - t)/(1 + At)$.

(a) Find $A(x, t)$ and verify that $A[x(A, t), t) = A$.

(b) Find a formula for the velocity, as a function of time, of the cross section that was initially located 3 units to the right of the origin.

(c) The density is given in material coordinates by the formula $\delta(A, t) = A^2$. Find an expression for the density that would be measured at a fixed point located one unit to the right of the origin.

23. Let us employ $J(A, t)$ and $j(x, t)$ to denote the Jacobian in material and spatial coordinates, respectively. Recall that

$$\frac{\partial J(A, t)}{\partial t} = J\left[\frac{\partial v}{\partial x}\right]_A, \qquad \frac{Dj}{Dt} = j\left(\frac{\partial v}{\partial x}\right).$$

The goal here is to find a formula for D^2j/Dt^2. Do this in two ways, and show that the answers are the same.

(a) Proceed in spatial coordinates and use the "natural" formula for the material derivative of a product.

(b) Find $\partial^2 J/\partial t^2$ in material coordinates and then switch to spatial coordinates at the end. Use very careful and explicit notation.

12.2 One-dimensional Elastic Wave Propagation

In this section and the following, we shall exclusively treat the linearized momentum balance equation (1.79). We shall consider, primarily, the case of constant physical properties. Consequently, the governing equation is the *one-dimensional wave equation.* Many of our conclusions concerning this special problem will be generalized to more complicated situations throughout the course of the text. It should be recalled that the linearization process told us essentially that to the order of magnitude that has been retained, there is no distinction between material coordinates and spatial coordinates. Furthermore, the density ρ, cross-section area σ, and Young's modulus E can be taken as known functions of position only. For reference, we repeat the linearized momentum equation (1.79):

$$[\sigma(x)E(x)u_x(x, t)]_x - \rho(x)\sigma(x)u_{tt}(x, t) + \rho(x)\sigma(x)f(x, t) = 0. \tag{1}$$

THE WAVE EQUATION

We shall begin by considering the important case that occurs when the material and geometric properties, A, E, and ρ, are independent of x and the body force, F, vanishes. Equation (1) then reduces to

$$u_{xx} - \frac{\rho}{E} u_{tt} = 0. \tag{2a}$$

It is frequently convenient to introduce the constant c, **the speed of sound**, given by

$$c^2 = \frac{E}{\rho}.$$

Equation (2a) can then be written as

$$u_{xx} - c^{-2} u_{tt} = 0 \qquad \text{or} \qquad u_{tt} = c^2 u_{xx}, \tag{2b}$$

a differential equation that is normally called the **wave equation**. We shall see shortly that these descriptions of c and (2a) or (2b) are indeed apt.

GENERAL SOLUTION OF THE WAVE EQUATION

In order to find the solutions of (2b), we shall try to ascertain an appropriate change of variables. Let us examine what we can do with what is perhaps the simplest such change, namely, a linear transformation of x and t. We shall set

$$\xi = \alpha x + \beta t \qquad \text{and} \qquad \eta = \gamma x + \delta t,$$

where the constants, α, β, γ, and δ will be determined so as to simplify the resulting differential equation. Direct application of the chain rule of partial differentiation then yields (Exercise 1)

$$(\alpha^2 - \beta^2 c^{-2}) u_{\xi\xi} + 2(\alpha\gamma - \delta\beta c^{-2}) u_{\xi\eta} + (\gamma^2 - \delta^2 c^{-2}) u_{\eta\eta} = 0. \tag{3}$$

As we shall see, the most helpful selection of the constants in the linear transformation is one for which the coefficients of the terms containing $u_{\xi\xi}$ and $u_{\eta\eta}$ vanish. One such selection is to set

$$\alpha = \beta c^{-1} \qquad \text{and} \qquad \gamma = -\delta c^{-1}.$$

(There are other choices that will accomplish the same end, but the examination of them is left as an exercise.) The transformation of variables now becomes

$$\xi = \alpha(x + ct), \qquad \eta = \gamma(x - ct).$$

We note that neither α nor γ can vanish if the transformation is to be non-degenerate. The coefficient of the $u_{\xi\eta}$ term can now be simplified, since

$$\alpha\gamma - \delta\beta c^{-2} = \alpha\gamma + \frac{\alpha\gamma c^2}{c^2} = 2\alpha\gamma.$$

Thus (3) can be reduced to

$$4\alpha\gamma u_{\xi\eta} = 0,$$

or, more simply, to*

$$u_{\xi\eta} = 0. \tag{4}$$

The general solution to (4) can now be readily obtained, for the equation states that u_ξ is independent of η. Consequently, we may integrate directly with respect to η, with the result

$$u_\xi = F(\xi),$$

where F is an arbitrary function. Integrating once more, we find

$$u = \int F(\xi)\, d\xi + G(\eta),$$

where G is an arbitrary function.

We have tacitly assumed that F is integrable. In fact, for future purposes the arbitrary functions that appear in the solution should be twice differentiable but otherwise arbitrary. Since F can be chosen at liberty, subject to these restrictions, we can equally well write

$$u = H(\xi) + G(\eta). \tag{5}$$

In terms of the original variables, x and t, (5) becomes

$$u(x, t) = H[\alpha(x + ct)] + G[\gamma(x - ct)].$$

Since H and G are arbitrary, we may write finally

$$u(x, t) = f(x - ct) + g(x + ct), \tag{6}$$

where f and g are arbitrary functions. Equation (6) is the general solution of (2b).

PHYSICAL SIGNIFICANCE OF THE SOLUTION

Before considering special forms of f and g, let us examine the physical significance of (6), concentrating on the first term $f(x - ct)$. Figure 12.13 is a typical plot of $f(x - ct)$ as a function of x with t considered as a parameter. The solid curve corresponds to $t = t_0$ and the dashed curve to $t = t_1$. (Only a portion of the curves are graphed.) We see that the second curve is obtained from the first simply by translation to the right through a distance $c(t_1 - t_0)$. Thus f can be considered as a fixed pattern or **wave** that moves to the right a distance $c(t_1 - t_0)$ in a time interval $t_1 - t_0$. In other words, the wave moves to the right with the *fixed speed c*.

* The reader who has some acquaintance with the theory of partial differential equations will recognize that what we have just done is to reduce the original equation to canonical form by introducing the characteristic coordinates ξ and η.

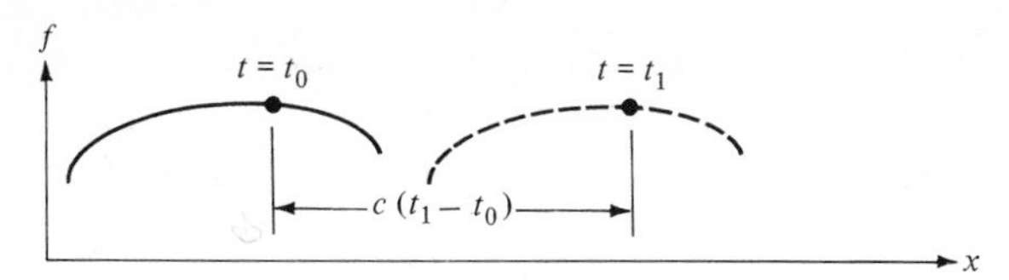

FIGURE 12.13 *A wave traveling to the right with speed c.*

An important and useful special case occurs when the functions f and g are taken to be sine and cosine functions. With such a choice, a solution could be written in the form

$$u(x, t) = A \sin \alpha(x - ct) + B \cos \alpha(x - ct) + E \sin \alpha(x + ct) + F \cos \alpha(x + ct). \quad (7)$$

SOLUTIONS IN COMPLEX FORM

It is frequently more convenient to replace the trigonometric expression by the complex exponential through the use of the relation

$$\exp (i\tau) = \cos \tau + i \sin \tau,$$

where $i = (-1)^{1/2}$. Equation (7) can then be rewritten as

$$u(x, t) = \mathscr{A} \exp [i\alpha(x - ct)] + \mathscr{B} \exp [i\alpha(x + ct)], \quad (8)$$

where the coefficients $\mathscr{A}$ and $\mathscr{B}$ are generally complex (Exercise 4).

We shall digress for a moment to discuss complex solutions. First of all, note that any linear partial differential equation in x and t, say

$$L(u) = a_0 + a_1 \frac{\partial u}{\partial x} + a_2 \frac{\partial u}{\partial t} + a_3 \frac{\partial^2 u}{\partial x^2} + \cdots = 0 \qquad (a_i \text{ are real constants})$$

has solutions of the form

$$u = f(x, t), \qquad f(x, t) \equiv \mathscr{A} \exp (\alpha x) \exp (\beta t).$$

(Here any of the constants $\mathscr{A}$ and α and β may be complex.) For

$$L\{f\} = \mathscr{A} \exp (\alpha x) \exp (\beta t)[a_0 + a_1\alpha + a_2 \beta + a_3 \alpha^2 + \cdots]$$

and a solution will result if the polynomial in the square brackets vanishes. Second, since L is linear and has real coefficients,

$$L(\text{Re} f) = \text{Re}\, L(f) = 0, \qquad L(\text{Im} f) = \text{Im}\, L(f) = 0.$$

(In the above equation, Re denotes "real part of" and Im "imaginary part of," as usual.) We therefore see that both Re f and Im f provide *real-valued* solutions of $L(u) = 0$.

Not all constant coefficient linear equations have solutions that are purely sines or purely cosines, but all have solutions of exponential form. One can always work with linear combinations of sines and cosines (or sinh and cosh) instead of exponentials, but the latter often provide more compact calculations.

ANALYSIS OF SINUSOIDAL WAVES

Let us consider one of the terms in the solution (7) in some detail. In particular, let

$$u(x, t) = a \cos \frac{2\pi(x - ct)}{\lambda}, \qquad a > 0. \tag{9}$$

Since the cosine is bounded between $+1$ and -1, the coefficient a is the maximum of the absolute value of $u(x, t)$ and is called the **amplitude.** Furthermore,

$$\begin{aligned} u(x + \lambda, t) &= a \cos \frac{2\pi(x + \lambda - ct)}{\lambda} \\ &= a \cos \left\{\frac{2\pi(x - ct)}{\lambda} + 2\pi\right\} = a \cos \frac{2\pi(x - ct)}{\lambda} \\ &= u(x, t). \end{aligned}$$

We can interpret this result in the following manner. If we regard t as fixed, then $u(x, t)$ as a function of x repeats itself in a distance λ. In other words, $u(x, t)$ is a periodic function of x of period λ. This "spatial period," λ, is called the **wavelength**; and its reciprocal, $k = \lambda^{-1}$, measures the number of waves in a unit distance and is known as the **wave number**.

Similarly, we can regard x as fixed and ask for the time τ at which the wave will repeat itself. This requires that

$$u(x, t + \tau) = a \cos \left[\frac{2\pi}{\lambda}(x - ct) - \frac{2\pi c\tau}{\lambda}\right] = u(x, t);$$

from which it follows that

$$\frac{2\pi c\tau}{\lambda} = 2\pi$$

or

$$\tau = \frac{\lambda}{c}. \tag{10}$$

We call τ the **period** and its reciprocal n, $n = c/\lambda$, the **frequency**. The frequency thus measures the number of waves passing a given point x in a unit time.

Now let us consider two solutions, u_1 and u_2, given by

$$u_1 = a \cos 2\pi(kx - nt),$$
$$u_2 = a \cos [2\pi(kx - nt) + \varepsilon]$$
$$= a \cos 2\pi \left[k\left(x + \frac{\varepsilon}{2\pi k}\right) - nt\right].$$

It is seen that u_2 is the same as u_1 except for a shift in x in the amount of $\varepsilon/2\pi k = \varepsilon\lambda/2\pi$. The parameter ε is known as the **phase of u_2 relative to u_1**. In the particular case in which $\varepsilon = \pi$, $u_2 = -u_1$; we refer to this situation by saying that u_1 and u_2 are π radians (or 180 degrees) **out of phase.**

We would like to point out finally that solutions of the form

$$u(x, t) = \exp [2\pi i(kx - nt)]$$
$$= \exp (2\pi ikx) \exp (-2\pi int)$$

is not as particular as it might first appear. More general time dependence can be synthesized as a series or integral through Fourier methods. For example, consider the sum of sinusoidal waves which are congruent but which travel in opposite directions

$$\sin 2\pi(kx - nt) + \sin 2\pi(kx + nt) = 2 \sin (2\pi kx) \cos (2\pi nt).$$

The result is a pattern that rises and falls between fixed nodes at $x = n/k$, $n = 0, \pm 1, \ldots$. This is a **standing wave**. Rather general functions of x can be similarly expressed in terms of complex exponential functions. Although the general line of attack is rather straightforward, there are numerous situations where the resulting mathematical problems can be both intriguing and challenging. See Chapter 4.

EFFECTS OF A DISCONTINUITY IN PROPERTIES

Let us now consider a very simple example that makes use only of exponential-type solutions and yet has some interesting conclusions. We take as our configuration two semiinfinite bars, I ($-\infty < x < 0$) and II ($0 < x < \infty$). Both have constant Young's modulus and density, the first with values E_1 and ρ_1 and the second with values E_2 and ρ_2. Both parts have the same shape and cross-section area. The velocity of sound in I will be denoted by c_1 and in II by c_2. The two semiinfinite bars are assumed to be welded together at $x = 0$ (Figure 12.14).

Considering a right-moving **incident wave**

$$\exp [i(x - c_1 t)]$$

in bar I, we shall try to determine both the ensuing **transmitted wave** in bar II and any additional **reflected wave** that might be produced by the presence of the interface.

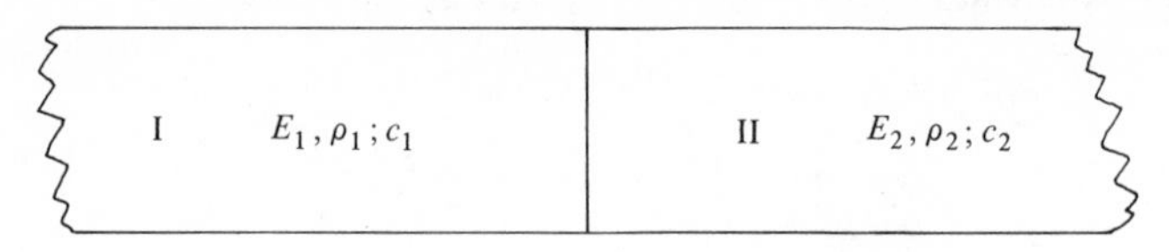

FIGURE 12.14. *Two bars welded together. The bars differ in their values of Young's modulus E and density ρ, and therefore in their sound speed* $c = (E/\rho)^{1/2}$.

In order to carry out the required calculations, we must first translate the conditions at the interface into analytic terms. Let us denote the displacement in I by $u_1(x, t)$ and in II by $u_2(x, t)$. Since the two bars are welded together, the displacement must be continuous across $x = 0$. Thus

$$u_1(0-, t) = u_2(0+, t), \tag{11}$$

where $0-$ implies a limit where x approaches the origin but always remains to its left, whereas $0+$ implies a limit where x approaches the origin but always remains to the right.

In addition, the force must be continuous across the interface (Exercise 13). Noting that the cross-section area of the two portions are the same, this requirement implies that

$$E_1\left[\frac{\partial u_1(0-, t)}{\partial x}\right] = E_2\left[\frac{\partial u_2(0+, t)}{\partial x}\right]. \tag{12}$$

We shall look for a solution in which u_1 is composed of the incoming wave together with a reflected left-moving wave and u_2 is a right-moving wave.* In other words, we try for solutions

$$u_1 = \exp[i(x - c_1 t)] + A \exp[i\alpha(x + c_1 t)],$$
$$u_2 = B \exp[i\beta(x - c_2 t)].$$

The constants A, B, α, and β will now be determined from the interface conditions (11) and (12).

We derive from (11) the requirement that

$$\exp(-ic_1 t) + A \exp(i\alpha c_1 t) = B \exp(-i\beta c_2 t)$$

for all values of t. In order for this to be true, the arguments of all the exponentials must be the same. Thus

$$\alpha = -1 \quad \text{and} \quad \beta = \frac{c_1}{c_2}.$$

* An effective way of solving simple problems is, as here, to guess the form of the answer. Expertise in this solution method, as with others, comes with practice.

In addition, the coefficients must satisfy

$$1 + A = B. \tag{13a}$$

Taking into account the conditions on the exponentials, we may write

$$u_1 = \exp[i(x - c_1 t)] + A \exp[-i(x + c_1 t)],$$

$$u_2 = B \exp\left[\frac{ic_1(x - c_2 t)}{c_2}\right].$$

Introducing these results into boundary condition (12), we find that

$$E_1 - E_1 A = \frac{E_2 c_1 B}{c_2}. \tag{13b}$$

Equations (13a) and (13b) can now be solved for the amplitudes with the result [Exercise 5(a)]:

$$A = \frac{1 - R}{1 + R}, \qquad B = \frac{2}{1 + R}, \qquad R \equiv \frac{E_2 c_1}{E_1 c_2} = \left(\frac{\rho_2 E_2}{\rho_1 E_1}\right)^{1/2}. \tag{14}$$

The expression for the amplitude A of the reflected wave shows us that there will be a phase shift in this wave of π if R is greater than unity.* As should be, the case when $R = 1$ (uniform bar), the transmitted wave is identical with the incident and there is no reflection.

The contemplation of limiting cases almost always increases one's understanding of a solution's implications. Often one can view the general case as some sort of compromise between comparatively understandable extremes.

In the present instance we note first that as $R \to \infty$, $A \to -1$ and $B \to 0$. For $R \gg 1$, the situation is that of a wave in a relatively light flexible bar impinging on a relatively heavy stiff bar. In the limit, the latter bar remains motionless. Taking the real part of the limiting incident solution

$$u_1 = \exp[i(x - c_1 t)] - \exp[-i(x + c_1 t)], \tag{15}$$

we obtain

$$\cos(x - c_1 t) - \cos(x + c_1 t) = 2 \sin x \sin c_1 t. \tag{16}$$

This is a standing wave. The displacement vanishes at the discontinuity $x = 0$ just as if the light flexible bar were rigidly fixed at that point. Indeed, the limiting solution for $x < 0$ is identical to that for a wave in a semiinfinite bar

* Since A and B are always real, there is no advantage in using complex notation. Cosine solutions or sine solutions could have been assumed from the outset. See Exercise 16 for an instance when this is not the case.

which is built in at one end [Exercise 7(a)]. This is no surprise, at least in retrospect.

Note that the imaginary part of (15) is

$$\sin(x - c_1 t) + \sin(x + c_1 t) = 2\sin x \cos c_1 t, \tag{17}$$

which differs from (16) only in phase.

In the other limiting case $R \to 0$, the reflected wave is the same as would be obtained if the bar along $x < 0$ had a free end at $x = 0$. The transmitted wave has the maximum possible amplitude. Details of the discussion are left to the reader [Exercise 7(c)].

It is illuminating to consider the distribution of energy in this problem. The required generalizations of energy concepts from particle mechanics are provided in Section 12.4. With these one can compute the energy per unit length in the various waves. Since no dissipative mechanisms are present, one expects that the energy in the incoming wave will just balance the sum of the energies in the transmitted and reflected waves. This is, in fact, the case (Exercise 4.6).

One more observation is required. In Section 12.1 we referred several times to a mathematical problem in which the governing equation (or equations) was supplemented by appropriate initial conditions (at time $t = 0$) and boundary conditions (at the two ends of the bar). The problem we have just solved, however, was not of this initial-boundary type. Rather, it was of the **scattering** type, which is commonly encountered in wave motion problems. Here an incoming signal is prescribed for all time, and the resultant effect is sought. To make sure that only the prescribed incoming signal is "causing" the phenomenon, one must (as here) require a **radiation condition**. Roughly speaking, this is

total solution minus prescribed incoming solution is entirely outgoing. (18)

EXERCISES

1. Carry out the detailed calculations required to derive (3).

2. In the discussion below (3), we made one selection for the parameters α and γ. Consider all the other choices that would make the appropriate coefficients in (3) vanish and determine which ones will lead to meaningful solutions. Do any of these additional choices yield essentially new solutions?

3. In the text we analyzed the geometrical significance of the solution $f(x - ct)$. Carry out a similar analysis for $g(x + ct)$. What would have happened if we had chosen the solution $f(ct - x)$?

4. (a) Writing $\mathscr{A} = M + iN$, $\mathscr{B} = P + iQ$, determine M, N, P, and Q in terms of A, B, C, and D so that (7) and (8) are equivalent.

(b) Also show that a solution of the form

$$u(x, t) = A \sin \alpha(x - ct) + B \cos \alpha(x - ct)$$

can be written in the form

$$u(x, t) = H \sin \alpha(x - ct + \varepsilon)$$

as well as

$$u(x, t) = K \cos \alpha(x - ct + \delta).$$

What is the corresponding exponential form?

5. (a) Derive the expressions for the coefficients A and B given in (14).
(b) Redo the problem, using dimensionless variables from the outset.

6. In imposing the displacement continuity requirement (11), we asserted that all the exponentials must have the same argument if the appropriate equation was to be valid for all t. Supply a formal proof for this assertion.

7. (a) Suppose that a semiinfinite uniform bar occupying $-\infty < x \le 0$ is subject to the incident wave $\exp [i(x - ct)]$. Find the reflected wave when the bar has a fixed end at $x = 0$.
(b) Repeat part (a) in the case when the bar has a free end at $x = 0$.
(c) Provide a full discussion of the solution given by (14) when $R \ll 1$.

8. In the case of the bar composed of two media, suppose that an incoming wave is given by the known function

$$f(c_1 t - x).$$

Find the reflected wave $g(c_1 t + x)$ and the transmitted wave $h(c_2 t - x)$. How does the behavior depend on the parameter R?

9. Consider a semiinfinite uniform bar occupying $-\infty < x \le 0$, carrying a mass M at the finite end $x = 0$. It is subject to a right-moving incident wave $f(ct - x)$, which has the property that

$$f(z) = 0 \qquad \text{whenever } z \le 0.$$

Writing the displacement in the form

$$u(x, t) = f(ct - x) + g(ct + x),$$

find the reflected wave $g(ct + x)$. After obtaining the general solution, examine the special cases that occur when M is allowed to approach zero and when M is allowed to approach infinity. In the latter case, calculate the force at $x = 0$ that would be produced by the incident wave $f(ct - x)$ alone, and compare it with the force produced at this location by the complete solution. (You will find that the force is doubled. Failure to anticipate this effect could lead to structural failure.)

10. If a bar of uniform properties is immersed in a viscous medium that imparts a resisting force proportional to the velocity u_t, the equation of longitudinal motion of the bar is

$$u_{xx} = \frac{1}{c^2}(u_{tt} + ku_t). \tag{19}$$

Derive (19), using the approach of Exercise 1.13. Explain the constant k and determine its units.

11. Show that if we look for a solution of (19) of the form

$$u(x, t) = \exp(-\tfrac{1}{2}kt)v(x, t),$$

then $v(x, t)$ will satisfy

$$v_{xx} = \frac{1}{c^2}(v_{tt} - \tfrac{1}{4}k^2 v). \tag{20}$$

Furthermore, if $k^2 \ll 1$ in (20), approximate solutions can be found by dropping the last term. Show that such approximate solutions can be written in the form

$$u(x, t) = \exp(-\tfrac{1}{2}kt)[f(x - ct) + g(x + ct)]. \tag{21}$$

Solution (21) is known as an **attenuated wave**.

12. Study the appropriateness of the approximation described in Exercise 11 by solving the following specific initial value problem

$$u(x, 0) = 0, \qquad u_t(x, 0) = \cos\frac{x}{c},$$

first using (20) as it stands and, second, approximating (20) by

$$v_{xx} = \frac{1}{c^2} v_{tt}. \tag{22}$$

How do the two solutions compare? Proceed by using separation of variables (Sections 4.1 and 15.4); i.e., write

$$v(x, t) = b(x)m(t),$$

from which

$$\frac{c^2 h''}{h} = \frac{m'' - \tfrac{1}{4}k^2 m}{m} = -\lambda^2,$$

where λ is a constant. Use the same technique on (22).

13. The derivation of boundary condition (12) was omitted. Supply an argument for this condition based on a "thin-slice" analysis similar to that used in the Exercise 1.12.

14. Consider the longitudinal motion of a uniform bar governed by

$$u_{tt} - c^2 u_{xx} = 0,$$

$$u(x, 0) = f(x), \qquad u_t(x, 0) = g(x), \qquad 0 < x < L,$$

$$u(0, t) = u(L, t) = 0, \qquad t > 0.$$

(a) Describe the physical meaning of these initial and boundary conditions.

‡(b) Derive the solution

$$u(x, t) = \frac{1}{2}\,[f(x + ct) + f(x - ct)] + \left(\frac{1}{2c}\right) \int_{x-ct}^{x+ct} g(s)\, ds$$

of the wave equation subject to these boundary conditions.

(c) The functions $f(x)$ and $g(x)$ are defined only for $0 < x < L$. However, using f and g in the solution written above requires that f and g be defined for all values of their arguments. Use the boundary conditions to argue that f and g should be extended as odd functions of period $2L$.

15. (a) Use the results of Exercise 14 to examine the following case. The initial velocity $g(x) = 0$ and the initial displacement is given by the graph shown in Figure 12.15. That is, $f = 0$ except for a small neighborhood about x_0. Examine how this "bump" propagates. Draw a sketch in the (t, x) plane of the regions where the bump in f has an effect. For example, what is the significance of the line $x - ct = x_0$?

(b) Perform a similar analysis when the material in a small region around some point x_0 is given an initial velocity, but no point of the bar is given an initial displacement.

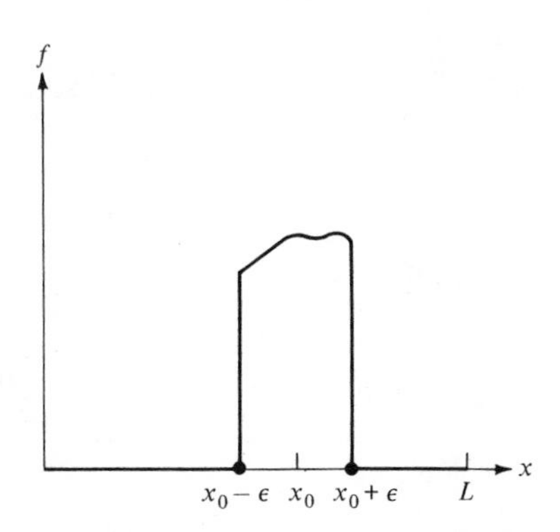

FIGURE 12.15. *Graph of a bump in displacement.*

16. Consider a semiinfinite uniform bar occupying $-\infty < x \le 0$. Suppose that a light spring, with spring constant k^2, attaches the bar to $x = 0$. We wish to examine the reflection of an incoming sinusoidal wave and thereby to explore the possible advantages of complex notation. We note that our problem is to determine a solution of the wave equation which satisfies

$$u' + \kappa u = 0 \quad \text{at } x = 0, \qquad \kappa = \frac{k^2}{\sigma E}.$$

‡(a) Look for a solution of the form

$$u = e^{i(ct-x)} + Qe^{i(ct+x)}$$

and determine Q. Take the real part of the final answer, and write it in the two ways

$$u = \cos(ct - x) + \alpha \cos(ct + x) + \beta \sin(ct + x) \tag{23}$$

and

$$u = \cos(ct - x) + B \cos(ct + x + \delta).$$

(b) Look for a solution of the form (23) and compare the amount of work with that required in (a) to obtain such a solution indirectly.

17. Show that a particular solution of the forced wave equation

$$u_{xx} - \left(\frac{1}{c^2}\right) u_{tt} = F(x, t)$$

is given by

$$u(x, t) = \frac{1}{2c} \int_0^t \int_{x-c(t-\bar{t})}^{x+c(t-\bar{t})} F(\bar{x}, \bar{t})\, d\bar{x}\, d\bar{t}.$$

18. (a) Use the concepts discussed in Chapter 6 to provide a more careful treatment of Exercise 11.

(b) Use perturbation theory to compute a correction to the approximate solution found in Exercise 11.

12.3 Discontinuous Solutions*

We have tacitly assumed in our discussion of particular examples in the preceding section that the solutions under study had all the analytic properties—continuity, differentiability, and the like—that one might desire. We shall now briefly consider what can be said if some of these restrictions

* Sections 12.3 and 12.4 can be read in either order.

are removed. In particular, we shall look at "mild" discontinuities, which will be characterized more explicitly in the course of the discussion. It should be noted at the very outset that the displacement, $u(x, t)$, will be assumed to be continuous, for we shall rule out motions that characterize tearing or folding of the material. Discontinuities of certain types will be allowed in the derivatives, however.

Investigations of discontinuous solutions arise naturally in the study of the motion of a bar. For example, one can ask what happens when the bar is suddenly struck at one end. This will no doubt produce a jump in the stress in the bar and very likely a jump in the velocity. Will an impulse imparted at one section be felt over the entire length of the bar? If so, will this occur instantaneously or will it take a finite length of time for the section at a particular location to "hear" the disturbance? Can the disturbance propagate at any speed and how will the properties of the bar affect the disturbance as it moves along? We shall see that it is possible to answer some of these questions with rather elementary methods.

The reader who is familiar with the elements of the theory of second order partial differential equations will recognize that what we shall be doing is, in disguised form, analyzing the characteristics and the equations that hold along them. The language to be used, however, is somewhat different from that normally employed in purely mathematical discussions.

It should be noted that we have already studied one example that can be thought of as having a form of discontinuous solution. We were concerned in Section 12.2 with two semiinfinite bars, with different properties, welded together at a fixed position, $x = 0$. This situation may also be interpreted as one infinite bar having a jump in section properties at $x = 0$. Boundary condition (2.12) states that the stress is continuous across the fixed material discontinuity. Since E jumps from E_1 to E_2 across the section $x = 0$, it follows that the displacement $u(x, t)$ has a jump in the derivative $u_x(x, t)$. Consequently, this problem may be considered as one in which the solution displays some form of discontinuity. We shall be concerned in the following, however, with somewhat more general situations in which the location of the discontinuity is not fixed in space and the material properties can vary from point to point.

One final remark should be made before proceeding with the specific analysis. We have blithely talked about waves and discontinuities propagating along the bar; and yet we have taken, as a paramount assumption, that the motions of the sections, or particles, are small. There is no inconsistency, however, for it is not the particles constituting the cross sections that propagate along the bar but rather certain properties, or functions of space and time, which do so. For example, if we note the distribution of stress in the bar at one instant of time and then at a later instant, we may find that the graph at the later instant appears to be more or less the same as that at the earlier time save for a shift in position. Thus we would conclude that the stress distribution has propagated along the bar with perhaps some distortion in shape. Similar descriptions could be made for the velocity as well as other properties of the bar. At the same time the particles that compose the bar have only moved within a small range.

MOTION OF THE DISCONTINUITY SURFACE

We shall now consider a bar in which the section properties $\sigma(x)$, $E(x)$, and $\rho(x)$ are twice continuously differentiable. We shall assume that the longitudinal displacement is also twice continuously differentiable and satisfies the equation of motion (2.1) with body forces neglected,

$$(\sigma E u_x)_x = \rho \sigma u_{tt}. \tag{1}$$

However, (1) is not required to hold across a section that may move along the bar in time. This section is a **moving discontinuity surface**. The displacement $u(x, t)$ is assumed to be continuous across the surface, but the first derivatives (hence the stress and the velocity) may undergo finite jumps.

We shall assume that the jumps in u_x and u_t are themselves continuous functions along the discontinuity surface. This means that if we consider the jump in u_x which occurs when the discontinuity surface occupies the position x_1 at time t_1 and the jump when the discontinuity surface occupies the position x_2 at time t_2, then

$$\lim_{\substack{x_2 \to x_1 \\ t_2 \to t_1}} \text{jump in } u_x \text{ at } (x_2, t_2) = \text{jump in } u_x \text{ at } (x_1, t_1),$$

where the limiting operation follows the motion of the discontinuity surface.

The purpose of our analysis will be to find equations from which we can determine the location of the discontinuity surface and the magnitude of the jumps for $t > t_0$ if we are given these quantities at an initial time t_0. We shall begin by finding an expression for the speed at which the discontinuity surface propagates.

The position x of the moving discontinuity surface is assumed to be given implicitly by

$$\phi(x, t) = 0 \tag{2}$$

or, explicitly, either by

$$x = x(t) \qquad \text{or} \qquad t = \psi(x).$$

Differentiating (2), we find that

$$\phi_x \frac{dx}{dt} + \phi_t = 0$$

or

$$\frac{dx}{dt} = -\frac{\phi_t}{\phi_x}. \tag{3}$$

Equation (3) relates the speed of the discontinuity surface, dx/dt, to the partial derivatives of the function $\phi(x, t)$. One of the questions that we shall study is whether there are any restrictions on this speed.

We shall denote the jump in a function $f(x, t)$ across such a discontinuity surface by the operator $[\![f(x, t)]\!]$, which is defined by

$$[\![f(x, t)]\!] = f(x-, t) - f(x+, t),$$

where

$$f(x-, t) = \lim_{\xi \to x} f(\xi, t), \qquad \xi < x,$$

and

$$f(x+, t) = \lim_{\xi \to x} f(\xi, t), \qquad \xi > x.$$

Recalling that the displacement $u(x, t)$ has been assumed to be continuous across the discontinuity surface, we may write

$$[\![u]\!] = 0. \tag{4}$$

Rather than regard the discontinuity surface as a moving section in one-dimensional x space, it will be more convenient to consider the surface as being represented by a fixed curve in an (x, t) space, as indicated in Figure 12.16. We shall designate the arc length along the curve $\phi(x, t) = 0$ by s. Integrating along the curve between two arbitrary points (x_0, t_0) and (x_1, t_1), we have

$$\int_{(x_0-, t_0)}^{(x_1-, t_1)} u_s(x-, t)\, ds = u(x_1-, t_1) - u(x_0-, t_0).$$

Similarly,

$$\int_{(x_0+, t_0)}^{(x_1+, t_1)} u_s(x+, t)\, ds = u(x_1+, t_1) - u(x_0+, t_0).$$

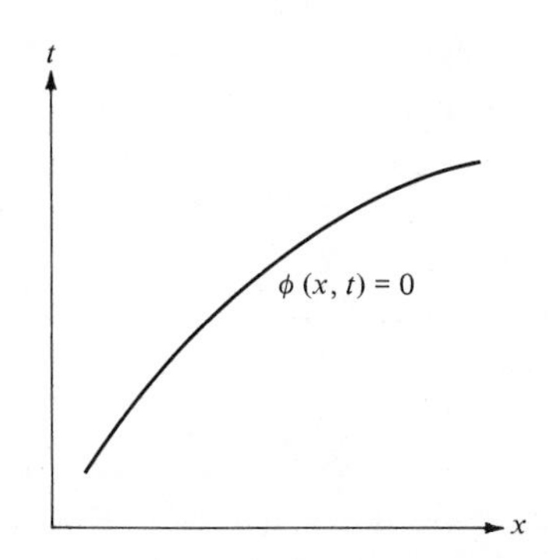

FIGURE 12.16. *Path of the discontinuity surface plotted in the xt plane.*

Subtracting the two integrals and using (4), we can then conclude that

$$\int_{(x_0, t_0)}^{(x_1, t_1)} [\![u_s(x, t)]\!] \, ds = 0. \tag{5}$$

From the chain rule

$$u_s(x, t) = u_x(x, t)\frac{dx}{ds} + u_t(x, t)\frac{dt}{ds},$$

and

$$[\![u_s(x, t)]\!] = [\![u_x(x, t)]\!]\frac{dx}{ds} + [\![u_t(x, t)]\!]\frac{dt}{ds}.$$

Thus $[\![u_s(x, t)]\!]$ is continuous along the discontinuity surface. Since the points (x_0, t_0) and (x_1, t_1) have been arbitrarily chosen along the curve, the Dubois–Reymond lemma may be applied to (5), yielding

$$[\![u_s(x, t)]\!] = 0. \tag{6}$$

Equation (6) states that the tangential derivative of u is also continuous across the discontinuity surface.

We may exploit (6) by rewriting it in terms of the derivatives u_x and u_t, with the result that

$$[\![u_x]\!]\frac{dx}{dt} + [\![u_t]\!] = 0.$$

Therefore,

$$\frac{dx}{dt} = -\frac{\phi_t}{\phi_x} = -\frac{[\![u_t]\!]}{[\![u_x]\!]},$$

or

$$[\![u_x]\!] = [\![u_t]\!]\frac{\phi_x}{\phi_t}. \tag{7}$$

Equation (7) shows that the jumps in the first derivatives are related to one another, and to the shape of the discontinuity surface.

We have yet to examine the implications that u is the solution of (1) on either side of the discontinuity surface. Let C be an arbitrary closed curve in the (x, t) plane bounding an interior G. The unit exterior normal $\mathbf{n}$ will be written in component form as

$$\mathbf{n} = n_1\mathbf{i} + n_2\mathbf{j},$$

where the unit vectors $\mathbf{i}$ and $\mathbf{j}$ are shown in Figure 12.17. The boundary curve C ranges from $t = t_1$ to $t = t_2$, the right-hand side of C is denoted by $x = n(t)$,

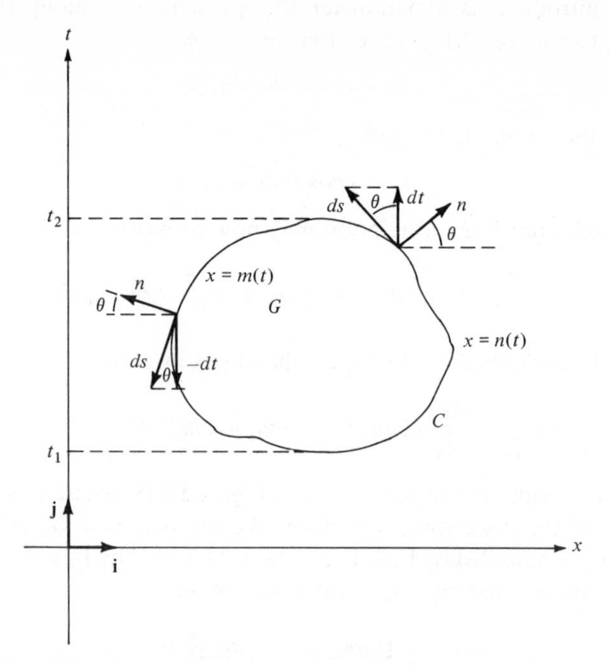

FIGURE 12.17. *The curve C is the boundary of a region G with unit exterior normal* **n**.

and the left-hand side by $x = m(t)$, as indicated. For a given value of t, let us consider a portion of the bar given by

$$m(t) \leq x \leq n(t).$$

We apply the integrated balance of linear momentum equation, in linearized form, to this portion of the bar. Combining (1.43), (1.46), and (1.47), we obtain

$$\int_{m(t)}^{n(t)} \rho\sigma u_{tt}\, dx = T(n, t)\sigma(n, t) - T(m, t)\sigma(m, t).$$

Integrating this relation over $t_1 \leq t \leq t_2$, we have

$$\begin{aligned}\int_{t_1}^{t_2}\int_{m(t)}^{n(t)} \rho\sigma u_{tt}\, dx\, dt &= \int_{t_1}^{t_2} T(n, t)\sigma(n, t)\, dt - \int_{t_1}^{t_2} T(m, t)\sigma(m, t)\, dt \\ &= \int_{t_1}^{t_2} T(n, t)\sigma(n, t)\, dt + \int_{t_2}^{t_1} T(m, t)\sigma(m, t)\, dt.\end{aligned}$$

It should be noted that we have assumed that the integrals involving the acceleration, u_{tt}, and the stress, T, exist.

Let us introduce as a parameter the arc length, s, along the curve C. Referring to Figure 12.17, we see that on $x = n(t)$

$$dt = \cos\theta\, ds = n_1\, ds.$$

Similarly, on $x = m(t)$, we have

$$dt = -\cos\theta\, ds = m_1\, ds.$$

Consequently, our integral relation may now be written as

$$\iint_G \rho\sigma u_{tt}\, dx\, dt = \oint_C T\sigma n_1\, ds = \oint_C E\sigma u_x\, n_1\, ds.$$

Applying Green's theorem to the left-hand side, we obtain

$$\oint_C (E\sigma u_x\, n_1 - \rho\sigma u_t\, n_2)\, ds = 0. \tag{8}$$

We now consider the region shown in Figure 12.18, where Γ is an arbitrary trajectory of the discontinuity surface. We see that $G_1 \cup G_2$ is bounded by $\Gamma_1 \cup \Gamma_2$, G_1 is bounded by $\Gamma_1 \cup \Gamma$, and G_2 is bounded by $\Gamma_2 \cup \Gamma$. Applying (8) to each of these three regions in turn, we obtain

$$\int_{\Gamma_1 \cup \Gamma_2} [E\sigma u_x n_1 - \rho\sigma u_t n_2]\, ds = 0,$$

$$\int_{\Gamma_1 \cup \Gamma(-)} [E\sigma u_x n_1 - \rho\sigma u_t n_2]\, ds = 0,$$

and

$$\int_{\Gamma_2} (E\sigma u_x n_1 - \rho\sigma u_t n_2)\, ds - \int_{\Gamma(+)} (E\sigma u_x n_1 - \rho\sigma u_t n_2)\, ds = 0,$$

where $\Gamma(-)$ indicates that the discontinuous functions are evaluated as limits from the G_1 side of Γ. Similarly, $\Gamma(+)$ indicates that the limit is taken

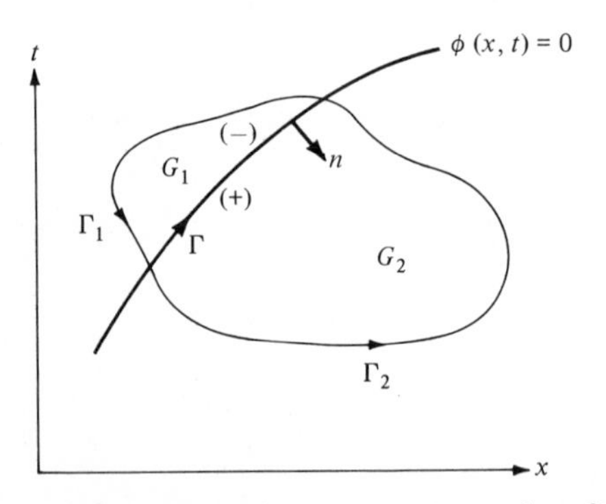

FIGURE 12.18. *Trajectory Γ of the discontinuity surface divides an arbitrary region into two portions, G_1 and G_2.*

from the G_2 side of Γ. Furthermore, the normal $\mathbf{n} = n_1\mathbf{i} + n_2\mathbf{j}$ appearing in the integrals over Γ is taken as directed from G_1 into G_2. Consequently, the second integral in the last equation must be prefixed by a minus sign, as indicated.

Subtracting the first integral from the sum of the remaining integrals, we then obtain

$$\int_{\Gamma(-)} (E\sigma u_x n_1 - \rho\sigma u_t n_2)\, ds - \int_{\Gamma(+)} (E\sigma u_x n_1 - \rho\sigma u_t n_2)\, ds = 0.$$

Since E, σ, and ρ are continuous across Γ, the last equation may be written in the form

$$\int_{\Gamma} (E\sigma[\![u_x]\!]n_1 - \rho\sigma[\![u_t]\!]n_2)\, ds = 0. \tag{9}$$

We recall that the *jumps* $[\![u_x]\!]$ and $[\![u_t]\!]$ have been assumed to be *continuous along the curve* Γ. Since the arc Γ has been arbitrarily chosen along $\phi(x, t) = 0$, the Dubois–Reymond lemma may be applied to (9) with the result

$$E\sigma[\![u_x]\!]n_1 - \rho\sigma[\![u_t]\!]n_2 = 0$$

along $\phi(x, t) = 0$. Now the unit normal vector $\mathbf{n}$ is proportional to $\phi_x\mathbf{i} + \phi_t\mathbf{j}$. Consequently, we may conclude that

$$E[\![u_x]\!]\phi_x - \rho[\![u_t]\!]\phi_t = 0. \tag{10}$$

Equations (7) and (10) constitute a pair of linear, algebraic, homogeneous equations in the unknowns $[\![u_x]\!]$ and $[\![u_t]\!]$. These will possess a nontrivial solution provided the determinant of the coefficients vanishes. Thus

$$-E\phi_x^2 + \rho\phi_t^2 = 0,$$

or

$$\left(\frac{\phi_x}{\phi_t}\right)^2 = \frac{\rho}{E}.$$

Applying (3), we may then conclude that

$$\left(\frac{dx}{dt}\right)^2 = \frac{E}{\rho},$$

or

$$\frac{dx}{dt} = \pm\left(\frac{E}{\rho}\right)^{1/2}. \tag{11}$$

Thus we see that the discontinuity surface or *wave front* moves either to the left or to the right at a determined speed. We recall that in the case of constant properties this speed, $(E/\rho)^{1/2}$, was designated by c. We shall use the same notation for variable properties, noting that c now depends upon x. Thus (11) can be written as

$$\frac{dx}{dt} = \pm c(x).$$

We see that weak discontinuities propagate at the local speed of sound. (The discontinuities must be small or "weak"; otherwise the basic linearization would be invalid.) This result is a generalization of that found in the previous section, where we showed that in a *uniform* bar all smooth small longitudinal disturbances propagate with the speed of sound. The generalization might have been anticipated if we had realized that a discontinuity can be regarded as a local disturbance which can only sense the local properties of the bar.

NOTE. Propagation of disturbances at the local sound speed is also discussed in the treatment of one-dimensional gas dynamics found in Section 15.3. In contrast to a solid, however, a gas generally possesses bulk convective motion. Consequently, disturbance propagation must be measured with respect to the bulk motion.

BEHAVIOR OF THE DISCONTINUITY

Now that we have determined the behavior of the moving discontinuity surface, let us see whether something can be said about the growth of the jump itself as time progresses. We have written the equation of the discontinuity surface in implicit form, $\phi(x, t) = 0$. Solving this equation for t, we can write

$$t = \psi(x),$$

i.e.,

$$\phi(x, t) \equiv \psi(x) - t = 0.$$

The function $\psi(x)$ is known as the **wave function**. Since

$$\phi_x \frac{dx}{dt} + \phi_t = 0,$$

it is seen that

$$\psi' = \frac{dt}{dx},$$

or

$$\psi'^2 = \frac{1}{c^2}, \tag{12}$$

which is known as the **eiconal equation**. There is a three-dimensional counterpart to (12), but the corresponding derivation is much more tedious.

Let us now return to the question of how the discontinuity itself behaves as the wave front propagates. Knowing the jump in stress or velocity at a section x_0 at time t_0, we would like to find out whether one can determine, from the properties of the bar, the subsequent jumps that will occur when the wave front has reached the position x at the time t. In particular, we shall consider a wave front propagating to the right with the material to the right of the front being undisturbed. The case in which the material to the right of the front is not quiescent will be left as an exercise.

We shall designate the velocity by

$$u_t(x, t) \equiv v(x, t).$$

It is convenient to introduce a special notation for the jumps in displacement and velocity at the discontinuity. (Of course, the first of these is identically zero.) Thus we make the definitions

$$U(x) \equiv [\![u(x, t)]\!]_{t=\psi(x)}$$

and

$$V(x) \equiv [\![v(x, t)]\!]_{t=\psi(x)}.$$

Taking into account the quiescent state to the right of the wave front, we have

$$U(x) = u(x, \psi)x)) \tag{13}$$

and

$$V(x) = v(x, \psi(x)). \tag{14}$$

Since the displacement is assumed to be continuous across the front,

$$0 = U(x). \tag{15}$$

Differentiating (15), we then find that

$$0 = U'(x) = u_x[x, \psi(x)] + u_t[x, \psi(x)]\frac{dt}{dx},$$

which can be rewritten through the use of (14) and the wave function ψ as

$$0 = u_x[x, \psi(x)] + V(x)\psi'(x).$$

Thus

$$u_x[x, \psi(x)] = -V(x)\psi'(x). \tag{16}$$

Since $U''(x)$ also vanishes, we can carry out a similar calculation with the result

$$u_{xx}[x, \psi(x)] = -V_x\psi' - (V\psi')'. \tag{17}$$

Performing a similar computation on $V(x)$, we find

$$V'(x) = v_x + v_t\psi'(x),$$

which can be transformed to yield

$$v_t = c^2(V'\psi' - v_x\psi'). \tag{18}$$

Finally, we evaluate the equation of motion (1) on the left-hand side of the front. Thus

$$\rho\sigma v_t = (E\sigma u_x)_x = (E\sigma)'u_x + E\sigma u_{xx},$$

or

$$v_t = \frac{u_x(E\sigma)'}{\rho\sigma} + c^2 u_{xx}. \tag{19}$$

Before continuing with the calculations, let us review the situation briefly. Our aim is to find an equation for $V = [\![u_t]\!]$ and for $[\![u_x]\!]$ in terms of the material properties of the bar. Equations (16), (17), and (18) express the pertinent derivatives in terms of V; these expressions can then be used in (19). The only possible complication appears to be terms containing v_x, which will have to be eliminated to carry out the program.

Introducing (18) into (19), we can show that

$$V'\psi' - v_x\psi' = \frac{(E\sigma)'}{E\sigma} u_x + u_{xx}.$$

Inserting (16) and (17), we find that

$$V'\psi' - v_x\psi' = \frac{(E\sigma)'}{E\sigma} V\psi' - v_x\psi' - (V\psi')'.$$

We observe that the terms containing v_x cancel and the resulting expression can be reduced to

$$\frac{V'}{V} = -\frac{1}{2}\frac{(E\sigma\psi')'}{E\sigma\psi'}. \tag{20}$$

Let us denote by x_0 the location of the discontinuity surface at time $t = t_0$. Then (20) may be integrated to yield

$$\frac{V(x)}{V(x_0)} = \frac{[E(x_0)\sigma(x_0)\psi'(x_0)]^{1/2}}{[E(x)\sigma(x)\psi'(x)]^{1/2}}.$$

Since $\psi' = 1/c$ and $E = c^2\rho$, we may rewrite this as

$$V(x) = V(x_0)\frac{[\rho(x_0)\sigma(x_0)c(x_0)]^{1/2}}{[\rho(x)\sigma(x)c(x)]^{1/2}}. \tag{21}$$

Equation (21) then states that the jump in velocity satisfies

$$[\![v(x, t)]\!]_{t=\psi(x)} = [\![v(x_0, t_0)]\!]_{t_0=\psi(x_0)}\left[\frac{\rho(x_0)\sigma(x_0)c(x_0)}{\rho(x)\sigma(x)c(x)}\right]^{1/2}. \tag{22}$$

Since $T = Eu_x$, (16) yields

$$[\![T(x, t)]\!]_{t=\psi(x)} = -E(x)V(x)\psi'(x).$$

Applying (21), we can then show that

$$[\![T(x, t)]\!]_{t=\psi(x)} = [\![T(x_0, t_0)]\!]_{t_0=\psi(x_0)} \left[\frac{\rho(x)c(x)\sigma(x_0)}{\rho(x_0)c(x_0)\sigma(x)}\right]^{1/2}. \tag{23}$$

Equations (22) and (23) are the propagation equations we have been seeking.

Consider, for example, a bar of uniform material; i.e., $\rho(x) \equiv \rho(x_0)$, $c(x) \equiv c(x_0)$. Suppose that the bar has the shape of a truncated wedge of unit thickness, and of half-height a at $x = 0$ and half-height b at $x = L$. Since $c(x) = c(x_0) = c$, a constant, we reach the conclusion that the eiconal equation (12)

$$\psi' = \frac{1}{c}$$

can be immediately integrated to yield

$$t = \psi(x) = \frac{x}{c},$$

where we have agreed that $x = 0$ when $t = 0$. In addition,

$$\sigma(x) = 2\left[a - \frac{(a-b)x}{L}\right].$$

Thus the jump in velocity will satisfy

$$[\![v(x, t)]\!]_{t=x/c} = [\![v(0, 0)]\!]\left[\frac{1}{1-(1-b/a)x/L}\right]^{1/2}$$

and the jump in stress obeys

$$[\![T(x, t)]\!]_{t=x/c} = [\![T(0, 0)]\!]\left[\frac{1}{1-(1-b/a)x/L}\right]^{1/2}.$$

We observe that in this special case, the jump in velocity increases as the area decreases and the jump in stress behaves similarly. This is not unreasonable, since one would expect the stress to increase as the cross-section area diminishes and the corresponding jump in velocity to increase as well.* For varying density ρ and local sound speed c, the situation is somewhat more complicated. We note from (22) and (23) that the jump in velocity will increase if $\rho(x)c(x)\sigma(x) > \rho(x_0)c(x_0)\sigma(x_0)$; whereas the jump in stress will increase if $\rho(x)c(x)/\sigma(x) > \rho(x_0)c(x_0)/\sigma(x_0)$.

* The fact that the intensity of the jump in stress produced by a hammer blow would increase as the area diminished could be of practical importance, for a high stress level can lead to failure of the material.

EXERCISES

1. In the analysis just concluded, we have assumed that the jumps were continuous along the discontinuity surface. Reexamine the arguments to determine what could be said if this requirement were lightened to piecewise continuity.
2. Carry out an analysis of the propagation of a discontinuity for the following equations:
   (a) The heat equation, $u_{xx} = cu_t$, where c is constant.
   (b) Laplace's equation, $u_{xx} + u_{tt} = 0$.*
3. The preceding analysis has been carried out under the hypothesis that the material properties E, σ, and ρ are continuous. A different form of discontinuity was considered in the previous section. The purpose of this exercise will be to show that the approach just used can be applied to the previously considered problem. Assume that we have a discontinuity surface in the bar which is *fixed for all time*. Furthermore, E, σ, and ρ may have jump discontinuities across this fixed surface. Apply the method of analysis of this section to the new problem. In particular, show that the result equivalent to (7) states that the velocity is continuous across the discontinuity surface, and that the equivalent to (10) yields the fact that the *force* is also continuous across the surface.
4. The detailed calculations that yield (17) through (23) have not been included in the text. Verify these equations.
5. Carry out the same analysis as that of the example at the end of the chapter, with the bar reversed.
6. Modify the discussion under (23) so that it applies to a truncated cone of basic radii a and b (see Figure 12.19). What is the difference in the behavior of the jumps?
7. Define an appropriate "difference displacement" so as to remove the text's restriction to wave fronts moving into undisturbed material.

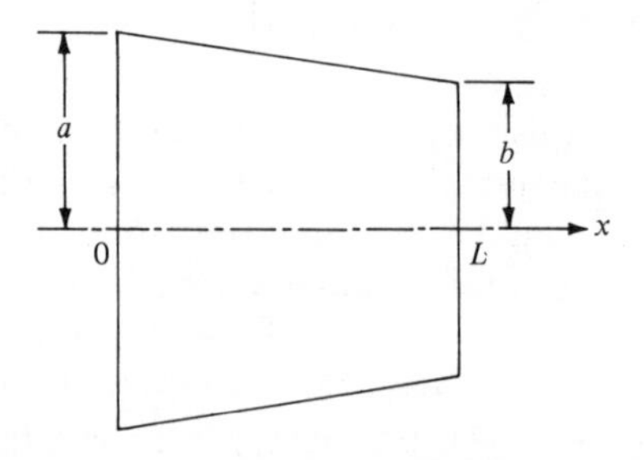

FIGURE 12.19. *Side view of a truncated cone whose ends are circular disks of radii a and b.*

* Imaginary propagation velocities should be rejected, as they have no direct physical meaning. Yet extensions into the complex domain have proved of considerable value. See Garabedian (1964), especially Chap. 16.

12.4 Work, Energy, and Vibrations

We shall consider briefly in this section the relation between work and energy. In addition, we shall examine some special solutions of the wave equation known as free vibrations. In the course of our discussion, we shall see how the balance between work and energy plays a role in the vibration problem that is both physically interesting and computationally valuable.

WORK AND ENERGY

Let us consider a bar of length L which has varying sectional properties and which is subject to end forces $\mathbf{F}(L, t)$ and $\mathbf{F}(0, t)$ only. (See Figure 12.20.) Using Hooke's law, we observe that

$$\mathbf{F}(L, t) = E(L)\sigma(L)u_x(L, t)\mathbf{i}$$

and

$$\mathbf{F}(0, t) = -E(0)\sigma(0)u_x(0, t)\mathbf{i}.$$

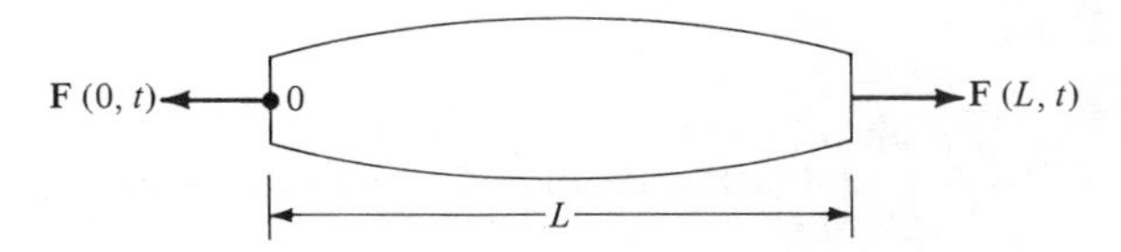

FIGURE 12.20. *Bar subject to end forces.*

In analogy with ordinary particle mechanics, we define the **kinetic energy density** (per unit length) of the bar as

$$\tfrac{1}{2}\rho(x)\sigma(x)u_t^2(x, t) \tag{1}$$

and the **total kinetic energy** of the bar at time t by

$$K(t) = \tfrac{1}{2}\int_0^L \rho(x)\sigma(x)u_t^2(x, t)\, dx. \tag{2}$$

Assuming that there are no distributed forces along the length of the bar, we define the **work of the external forces,** W_e, in the time interval $0 \le t \le t_1$ as*

$$W_e = \int_0^{t_1} [\mathbf{F}(L, t)\cdot u_t(L, t)\mathbf{i} + \mathbf{F}(0, t)\cdot u_t(0, t)\mathbf{i}]\, dt. \tag{3}$$

Thus

$$W_e = \int_0^{t_1} [E(L)\sigma(L)u_x(L, t)u_t(L, t) - E(0)\sigma(0)u_x(0, t)u_t(0, t)]\, dt. \tag{4}$$

* Recall that work is done at a rate equal to the scalar product of the force and velocity vectors. We are extending this definition of point mechanics with the goal of obtaining some meaningful generalizations of classical results.

If τ denotes the internal force per unit length, the internal work, W_i, will be given by

$$W_i = \int_0^{t_1} \int_0^L \tau u_t \, dx \, dt.$$

To evaluate τ, consider an arbitrary portion of the bar defined by

$$\xi \leq x \leq \xi + l.$$

The force exerted on this portion by the remaining parts of the bar is given by

$$T(\xi + l, t)\sigma(\xi + l) - T(\xi, t)\sigma(\xi) = \int_\xi^{\xi + l} [T(x, t)\sigma(x)]_x \, dx.$$

Thus the internal force, τ, per unit length at ξ is given by

$$\begin{aligned} \tau(\xi, t) &= \lim_{l \to 0} (l)^{-1} \int_\xi^{\xi + l} [T(x, t)\sigma(x)]_x \, dx \\ &= [T(\xi, t)\sigma(\xi)]_\xi . \end{aligned} \tag{5}$$

Therefore,

$$W_i = \int_0^{t_1} \int_0^L (T\sigma)_x u_t \, dx \, dt = \int_0^{t_1} \int_0^L (E\sigma u_x)_x u_t \, dx \, dt. \tag{6}$$

We shall define the **change in potential energy** in the time interval $0 \leq t \leq t_1$ as the difference between the work of the external forces and the internal forces in that time interval. In order to find an analytic expression for the change in potential energy, let us return to the formula for W_i. Integrating by parts, we find that

$$\begin{aligned} W_i &= \int_0^{t_1} E\sigma u_x u_t \Big|_0^L dt - \int_0^L E\sigma \int_0^{t_1} u_x u_{xt} \, dt \, dx \\ &= \int_0^{t_1} E\sigma u_x u_t \Big|_0^L dt - \int_0^L \tfrac{1}{2} E\sigma u_x^2 \Big|_0^{t_1} dx \\ &= W_e - \tfrac{1}{2} \int_0^L E(x)\sigma(x)u_x^2(x, t) \Big|_0^{t_1} dx. \end{aligned}$$

Thus we see that the change in potential energy in the time interval $0 \leq t \leq t_1$ is given by

$$\tfrac{1}{2} \int_0^L E(x)\sigma(x)u_x^2(x, t_1) \, dx - \tfrac{1}{2} \int_0^L E(x)\sigma(x)u_x^2(x, 0) \, dx.$$

Consequently, the potential energy P at time t is given by

$$P(t) = \tfrac{1}{2} \int_0^L E(x)\sigma(x)u_x^2(x, t) \, dx + C. \tag{7}$$

We shall take $C = 0$ so that the arbitrary zero of potential energy is identified with the strain-free state ($u_x \equiv 0$).

Let us return now to the equation of motion

$$(E\sigma u_x)_x = \rho\sigma u_{tt}.$$

If we multiply by u_t and integrate over both x and t, we then have

$$\int_0^{t_1}\int_0^L (E\sigma u_x)_x u_t \, dx \, dt = \int_0^L \int_0^{t_1} \rho\sigma u_{tt} u_t \, dt \, dx.$$

Integrating, we obtain

$$\int_0^{t_1} E\sigma u_x u_t \Big|_0^L dt - \int_0^{t_1}\int_0^L E\sigma u_x u_{tx} \, dx \, dt = \tfrac{1}{2}\int_0^L \rho\sigma u_t^2 \Big|_0^{t_1} dx.$$

The second integral on the left-hand side can be simplified as

$$\int_0^L E\sigma \int_0^{t_1} u_x u_{xt} \, dt \, dx = \tfrac{1}{2}\int_0^L E(x)\sigma(x)u_x^2(x, t)\Big|_0^{t_1} dx.$$

Thus we may conclude that

$$\int_0^{t_1} E\sigma u_x u_t \Big|_0^L dt = \tfrac{1}{2}\int_0^L \rho\sigma u_t^2 \Big|_0^{t_1} dx + \tfrac{1}{2}\int_0^L E\sigma u_x^2 \Big|_0^{t_1} dx. \tag{8}$$

We note that (8) states that the *work done by the external forces equals the change in kinetic energy plus the change in potential energy*. This result is referred to as the **work–energy principle**.

A VIBRATION PROBLEM

We now turn our attention to a special problem that can serve as a prototype for a whole class of questions that appear frequently in this text. Let us consider a bar of length L, fixed at the end $x = 0$, and attached to a linear spring at the other end. We assume that there are no external forces. Thus the displacement satisfies the following differential equation and boundary conditions:

$$\begin{gathered} (E\sigma u_x)_x = \rho\sigma u_{tt}, \qquad 0 < x < L, \\ u(0, t) = 0, \\ E(L)\sigma(L)u_x(L, t) + k^2 u(L, t) = 0. \end{gathered} \tag{9}$$

We ask whether it is possible to obtain *standing wave solutions with harmonic time dependence*; i.e., are there solutions to (9) of the form

$$u(x, t) = w(x) \exp(i\omega t)?$$

Substituting this assumed solution into the boundary value problem (9) and dropping the exponential factor that is common to every term, we find

$$\boxed{\begin{aligned} (E\sigma w')' = -\omega^2\rho\sigma w, \qquad 0 < x < L,\\ w(0) = 0,\\ E(L)\sigma(L)w'(L) + k^2 w(L) = 0, \end{aligned}} \tag{10}$$

where $'$ denotes differentiation with respect to x.

Equations (10) define a **Sturm–Liouville problem**, which, in general, has only the trivial solution $w = 0$, corresponding to the uninteresting case of no motion. There are, however, certain values of the frequency ω for which there are nontrivial solutions. Such values of ω are **eigenvalues** and the corresponding solutions are **eigenfunctions**. A number of general results for Sturm–Liouvillepr oblems have been given in Section 5.2. Our discussion here is largely independent. We concentrate on a particular method for practical estimation of the eigenvalues.

THE RAYLEIGH QUOTIENT

Let us examine some of the properties enjoyed by the eigenvalue and the corresponding eigenfunction $w(x)$. We begin by multiplying the differential equation in the system (10) by $w(x)$ and integrating from 0 to L, with the result

$$\int_0^L (E\sigma w')'w\,dx + \omega^2 \int_0^L \rho\sigma w^2\,dx = 0.$$

Integrating the first integral by parts, we obtain

$$E\sigma w'w\Big|_0^L - \int_0^L E\sigma w'^2\,dx + \omega^2 \int_0^L \rho\sigma w^2\,dx = 0.$$

Now

$$E(0)\sigma(0)w'(0)w(0) = 0$$

from the first of the boundary conditions in (10) and

$$E(L)\sigma(L)w'(L)w(L) = -k^2w^2(L)$$

from the second of the boundary conditions. Thus

$$-k^2w^2(L) - \int_0^L E\sigma w'^2\,dx + \omega^2 \int_0^L \rho\sigma w^2\,dx = 0,$$

or

$$\omega^2 = R(w), \tag{11}$$

where

$$R(w) = \frac{\int_0^L E\sigma w'^2\,dx + k^2 w^2(L)}{\int_0^L \rho\sigma w^2\,dx}$$

is known as the **Rayleigh quotient**. Equation (11) will be seen to be a very useful relation connecting the eigenfunction and the eigenvalue.

The Rayleigh quotient $R(w)$ has an interesting physical interpretation. To see this, multiply both the numerator and the denominator of R by $\frac{1}{2}$. Then upon substitution of $u = w \exp(i\omega t)$, one can see at once that R is equal to the potential energy of internal strain (7) plus a potential energy equal to one-half the force* $k^2 w(L)$ on the stretched spring times the distance $w(L)$ stretched, all divided by the kinetic energy (2). In short, *the Rayleigh quotient is the ratio of total potential to kinetic energy for a given mode of vibration.*

PROPERTIES OF EIGENVALUES AND EIGENFUNCTIONS

Let us return to the eigenvalue problem (10) and assume that we have a second eigenvalue μ, corresponding to the eigenfunction $y(x)$. We also assume that ω^2 and μ^2 are distinct from one another. Thus μ and $y(x)$ satisfy

$$\begin{aligned} (E\sigma y')' + \mu^2 \rho\sigma y &= 0, \\ y(0) &= 0, \\ E(L)\sigma(L)y'(L) &= -k^2 y(L). \end{aligned} \tag{12}$$

By an argument virtually the same as that used to prove (5.2.3), one can show that eigenfunctions corresponding to distinct eigenvalues are orthogonal, in the sense that

$$\int_0^L \rho\sigma w y\,dx = 0. \tag{13}$$

Also by reasoning which differs only in detail from that used in Chapter 5, one can show that the eigenvalues ω^2 of (10) are necessarily real and that the eigenfunctions can always be taken to be real. But is it possible for ω^2 to be negative, in which event ω would be purely imaginary? Should this prove to be the case, the temporal behavior would no longer be oscillatory but would rather exhibit exponential growth or decay. [If $\omega = i\sigma$, then $\exp(i\omega t) = \exp(-\sigma t)$ grows (decays) if σ is negative (positive).] This question can be settled by noting that the expression (11), which equates ω^2 to the Rayleigh quotient, demonstrates that since the eigenfunction corresponding to the eigenvalue ω^2 is real, then ω^2 is positive.

Finally, we note that the eigenfunctions corresponding to a single eigenvalue are unique to within a multiplicative constant (see Exercise 1).

* For a spring, the potential energy can be regarded as the average force required to stretch it a given amount, times the amount stretched. See Exercise 5.

AN EXACT SOLUTION WHEN PROPERTIES ARE CONSTANT

Next, we shall consider the very simple example that arises when one considers a bar with constant properties which is built in at $x = 0$ and free at $x = L$. Here E, σ, and ρ are constant and $k^2 = 0$. Boundary value problem (10) becomes

$$w'' + \left(\frac{\omega}{c}\right)^2 w = 0,$$

$$w(0) = w'(L) = 0,$$

where, as usual, $c^2 = E/\rho$. The general solution of the differential equation is

$$w(x) = M \sin\left(\frac{\omega}{c} x\right) + N \cos\left(\frac{\omega}{c} x\right).$$

The first boundary condition implies that $N = 0$, and the second boundary condition then requires that

$$\frac{\omega}{c} \cos\left(\frac{\omega}{c} L\right) = 0.$$

Noting that $\omega = 0$ yields a trivial solution, we see that

$$\frac{\omega}{c} L = \frac{(2n + 1)\pi}{2},$$

where $n = 0, 1, 2, \ldots$. Thus there are a denumerable number of eigenvalues ω_n and a denumerable number of corresponding eigenfunctions $w_n(x)$ given by

$$\omega_n = \frac{(2n + 1)\pi c}{2L},$$

$$w_n(x) = \sin\left[\frac{(2n + 1)\pi x}{2L}\right],$$

$n = 0, 1, 2, \ldots$.

CHARACTERIZATION OF THE LOWEST EIGENVALUE AS THE MINIMUM OF THE RAYLEIGH QUOTIENT

The preceding example illustrates the fact that the exact calculation of the eigenvalues and eigenfunctions depends on finding explicit solutions to the governing differential equation, a task that is not always easily accomplished. A number of techniques have been developed for finding eigenvalues approximately, and we shall close this chapter by considering one of these methods briefly.

In Chapter 12 of II we shall consider a number of minimum properties that are related to eigenvalue problem (10). One result, which we shall state without proof, is the following. Let U be the class of functions $u(x)$ that are twice differentiable and vanish at $x = 0$. Designate the lowest eigenvalue of (10) by ω_1^2. Let

$$\mu = \min_{u \in U} R(u) \tag{14}$$

and let $w(x)$ denote the corresponding minimizing function. Then $\mu = \omega_1^2$ and $w(x)$ is the corresponding eigenfunction.

The minimum principle has the following physical interpretation. Functions in class U are called **virtual displacements** (perhaps "conceivable displacements" would be a better phrase). They are smooth and satisfy the boundary condition at $x = 0$. The corresponding Rayleigh quotient gives the ratio of virtual potential energy to virtual kinetic energy. The lowest eigenvalue is the minimum of this ratio.

It is interesting to note that the functions which belong to the class U are not required to satisfy the second of the boundary conditions in (10). But it can be shown that the minimizing function will automatically satisfy this boundary condition. Such a boundary condition is termed **natural** and appears in a wide variety of problems associated with minimum principles. Further discussion, which includes a proof of the characterization of ω^2, as the minimum of the Rayleigh quotient (if it has one), will be found in Section 12.2 of II.

The minimum principle stated above, as well as the similar theorems for the higher eigenvalues, gives rise to a very effective technique, the *Rayleigh–Ritz method*, for obtaining approximate values for the eigenvalues. Here we shall only give an indication of the power of the technique by finding an estimate for the lowest eigenvalue in a particular problem. The idea is simple. If the lowest eigenvalue ω_1^2 is the smallest value of $R(u)$ among all $u \in U$, then an upper bound for ω_1^2 can be computed by evaluating R for any function $\bar{u}$ whatever which belongs to U. Presumably, the cleverer we are in guessing a $\bar{u}$ that approximates the correct eigenfunction ω, the closer our upper bound will be to the exact eigenvalue.

ESTIMATE OF THE LOWEST EIGENVALUE FOR A WEDGE

We now consider the special example of a bar that is of unit thickness but wedge-shaped in plan form, as shown in Figure 12.21. The bar is free at the pointed end and clamped at its base. The area σ is given by

$$\sigma = \frac{bx}{L}.$$

The eigenvalue problem for this configuration is

$$\left(\frac{Ebxw'}{L}\right)' = -\frac{\omega^2 \rho bxw}{L} \qquad \text{for } 0 < x < L,$$

$$w(L) = 0, \qquad\qquad xw' \to 0 \text{ as } x \to 0.$$

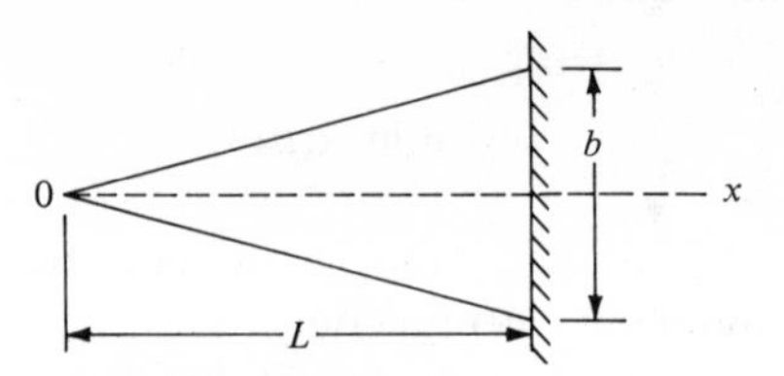

FIGURE 12.21. *A wedge that is built in at its broad end and is free at its pointed end.*

The second boundary condition is by no means a trivial one, for it states that the solution w cannot be as singular as $\log x$ in the neighborhood of the origin. Since E, b, ρ, and L are constants, the boundary value problem reduces to

$$(xw')' + \frac{\omega^2 \rho}{E} xw = 0 \qquad \text{for } 0 < x < L,$$

$$w(L) = 0, \qquad xw' \to 0 \text{ as } x \to 0.$$

The solution to the differential equation satisfying the second boundary condition is

$$w = J_0\left[\omega\left(\frac{\rho}{E}\right)^{1/2} x\right],$$

where J_0 is the Bessel function of zero order. The first boundary condition requires that

$$J_0\left[\omega\left(\frac{\rho}{E}\right)^{1/2} L\right] = 0.$$

There are a denumerable infinity of solutions ω to the above equation, one for each zero of the Bessel function J_0. Taking the smallest zero from a table in one of the various collections, we find that

$$\omega_1 = \left(\frac{E}{\rho}\right)^{1/2} 2.4048\, L^{-1}.$$

Let us pretend that we do not know the above solution and try to estimate ω_1 by means of the minimum principle. The Rayleigh quotient corresponding to this eigenvalue problem is found to be

$$R(u) = \frac{\int_0^L Exu'^2\, dx}{\int_0^L \rho x u^2\, dx},$$

with the required boundary condition that $u(L) = 0$. If we take as our trial function

$$\bar{u} = L - x$$

and denote the corresponding value of the Rayleigh quotient by Λ^2, a simple calculation shows that

$$\Lambda^2 = \frac{6E}{\rho L^2},$$

or

$$\Lambda = \left(\frac{E}{\rho}\right)^{1/2} 2.45\, L^{-1}.$$

We thus see that our very simple approximation provides an upper bound which is in error by less than 2.8 per cent.

It should be quickly pointed out that the example just cited is somewhat too successful. It will be found, in practice, that such simple trial functions rarely lead to such close estimates. Nevertheless, it is hoped that this case will convince the reader that the minimum characterization is an extremely valuable one and that comparatively elementary calculations can lead to quite decent estimates of the eigenvalues. Experience shows that if the trial function reflects a physically reasonable displacement, the estimated eigenvalue will be much closer than one resulting from an arbitrarily chosen function. There is an art in using the Rayleigh–Ritz method efficiently, and prior reflection frequently produces rewards in accuracy and ease of computation.

EXERCISES

1. Establish the fact that the eigenfunctions of (10) corresponding to a given eigenvalue are unique to within a multiplicative constant by the following argument. Suppose that there are two linearly independent eigenfunctions corresponding to the same eigenvalue. Show that there is a linear combination, $u(x)$, with the property $u(0) = u'(0) = 0$. Apply the uniqueness theorem for second order equations (see, e.g., Boyce and DiPrima, 1969, p. 99).
2. The eigenvalue problem for a bar fixed at $x = 0$ and carrying a mass M at $x = L$ is given by

$$(E\sigma v')' + \omega^2 \rho\sigma v = 0, \qquad 0 < x < L,$$
$$v(0) = 0,$$
$$E(L)\sigma(L)v'(L) = \omega^2 M v(L).$$

What is the Rayleigh quotient for this problem? What is the orthogonality relation that eigenfunctions corresponding to distinct eigenvalues satisfy?

3. Consider the eigenvalue problem discussed in the text for a bar of constant properties. The eigenvalue problem takes the form

$$v'' + \lambda^2 v = 0, \qquad 0 < x < \tfrac{1}{2}\pi,$$
$$v(0) = 0,$$
$$v'(\tfrac{1}{2}\pi) + \alpha^2 v(\tfrac{1}{2}\pi) = 0.$$

(a) Find the transcendental equation satisfied by λ. What is the lowest nontrivial eigenvalue if $\alpha^2 = \tfrac{3}{2}$?
(b) Find an approximate eigenvalue by taking

$$\bar{v} = x.$$

(c) Find an improved eigenvalue by taking

$$\bar{v} = x + ax^3,$$

where a is determined so that $R[\bar{v}]$ is as small as possible. What are the percentage errors when compared with the previous exact calculation? (The calculations here are somewhat lengthy. Excellent results can frequently be obtained by the Rayleigh–Ritz method if a number of parameters are employed, but then use of a computer is essential.)

4. The transverse vibrations of a simply supported beam of length L are governed by the eigenvalue problem

$$[E(x)I(x)w'']'' = \omega^2 \rho\sigma(x)w,$$
$$w(0) = w''(0) = W(L) = w''(L) = 0.$$

(The boundary conditions correspond to simply supported ends.)
(a) Show that the corresponding Rayleigh quotient is

$$R(w) = \frac{\int_0^L E(x)I(x)w''(x)^2\,dx}{\int_0^L \rho\sigma(x)w(x)^2\,dx}.$$

Begin, as in the text, by multiplying the governing differential equation by w.
(b) What is the orthogonality relation that eigenfunctions corresponding to distinct eigenvalues satisfy?

5. The displacement $x(t)$ of a mass m restrained by a spring constant k^2 satisfies $d^2x/dt^2 = -k^2x$. [Here $x = 0$ when the spring is unstretched.] Multiply the equation by dx/dt and integrate. Interpret in terms of a kinetic energy $\tfrac{1}{2}m(dx/dt)^2$ and a potential energy $\tfrac{1}{2}k^2x^2$. Compare these definitions of energy with the continuum versions of kinetic and potential energy defined in the text. [Note that the potential energy can be interpreted as stored work, where this work is the product of the distance the spring is extended times the average of the initial force (zero) and the final force in the extension process.]

6. At the end of Section 12.2 we obtained a complex solution to a wave reflection problem. The real part of this solution was

$$u = \cos(x - c_1 t) + A \cos(x + c_1 t), \qquad x < 0;$$
$$u = B \cos[\beta(x - c_2 t)], \qquad x > 0.$$

[The constants A and B are given in (2.14).] Provide a full discussion of the distribution of energy.

(a) In particular, find the total energy per period for $x < 0$ and for $x > 0$.

(b) Your results in (a) should be independent of time; show from general considerations that this must be so.

(c) Making allowances for the fact that the transmitted wave does not have the same period as the incident and the reflected waves, show that the energy in the incoming wave balances the sum of the energies in the transmitted and reflected waves.

Chapter 13
The Continuous Medium

HISTORIC moments usually culminate one long series of events and begin another. One such moment came in 1687 with the publication of Newton's *Principia*. Dramatically positioned at the head of Book 1 were the three laws.

I. Every body continues in its state of rest, or of uniform motion straight ahead, unless it be compelled to change that state by forces impressed upon it.

II. The change of motion is proportional to the motive force impressed, and it takes place along the right line in which that force is impressed.

III. To an action there is always a contrary and equal reaction, or, the mutual actions of two bodies upon each other are always equal and directed to contrary parts.

By taut and clear deductions from these laws Newton summed up a century's work on the motion of one or two bodies in a vacuum. "What Newton writes is correct, clear, and short. In earlier works the brilliant diamonds of discovery lie concealed in an opaque matrix of wordy special cases, laborious details, metaphysics, confusion, and error; while Newton follows a vein of pure gold."*

Unlike his work on the one and two body problems, Newton's work on the three body problem was not centered on the formulation and solution of the governing differential equations. Indeed "he shows no sign of an attempt to set up equations of motion. That he obtained some correct inequalities and groped his way to the major approximate results under this handicap is one more tribute to his peerless grasp of the physical essence of mechanics and to the might of his brain. It does not show, however, that his formulation of the general laws of mechanics was adequate. History proves the contrary."† Further progress on the three body problem required formulation of the general equations of celestial mechanics and these equations were not published until 1749 (by Euler).

Newton's definitive and lasting work concerned bodies that could be idealized as rigid bodies or as point masses. In Book II Newton struggled with motion in a resisting medium, but most of his work has since been discarded. There is no disgrace in this. It took at least a century to achieve even a general understanding of the problems Newton grappled with, and many important questions remain to this day.

* C. Truesdell, "A Program Toward Rediscovering the Rational Mechanics of the Age of Reason," *Arch. Hist. Exact Sci. 1*: 1, 5 (1960).

† Ibid., p. 7.

Motion in a resisting medium is a major problem in the subject that is now termed continuum mechanics. A turning point in the history of continuum mechanics occurred in 1776, *nearly a century after the publication of Principia*, with the publication of Euler's laws of mechanics. These state that for every body or system of bodies, and for every part of every body, whether the bodies are conceived of as points or as filling space, whether deformable or rigid:

1. The total force acting upon the body equals the rate of change of the total momentum.
2. The total torque acting upon the body equals the rate of change of the total moment of momentum, where both the torque and the moment are taken with respect to the same fixed point.*

Euler's laws form the core of the restrictions governing the motion of continuous media, but they are not the only ones. It is the purpose of this chapter and the next to present a mathematical formulation of all such restrictions that need to be considered, whatever the constitution of the medium, if one ignores electric, magnetic, chemical, and certain other effects. The few historical remarks we have made indicate that formulation of these restrictions, or laws, required more than a century of endeavor by some of the most brilliant men of their age. Further evidence for this assertion is provided in the historical discussion of Appendix 14.2.

As this book is being written, the equations of continuum mechanics are an active subject for research. Older equations are being clarified and unified by the invention of deeper and further-reaching concepts, while new equations are being formulated to encompass ever-widening classes of important scientific phenomena. Thus, it is our hope that from the discussion to follow the reader will not only gain a good understanding of "laws of nature" that have great historic importance and great present importance, but that he will also gain an understanding of the process of formulating the equations that embody such laws which may aid him in a future attempt to formulate new equations.

We begin the present chapter with a discussion of the continuum model. We then introduce the material and spatial descriptions of the motion of a continuous medium and some related mathematical concepts and formulas. With this background, the equations that embody the basic laws can be efficiently presented. This is done in Chapter 14.

13.1 The Continuum Model

In earlier discussions we have used models of nature in which a substance or quality is regarded as continuously distributed in space. For example, temperature and mass have been regarded as smoothly varying functions

* Ibid., p. 31.

of position. In this section we shall subject the concept of a continuum to closer scrutiny, particularly as it is used in mechanics.

Particle mechanics deals with the equilibrium and motion of systems of point masses. In continuum mechanics one deals with a **continuous medium** wherein smoothly varying properties like density and velocity are assigned to each point in space that is instantaneously occupied by the body under consideration.

Most of us probably view materials like water or steel as being composed of assemblages of molecules. This corpuscular (*corpus* = body) viewpoint and the continuum viewpoint are both illuminated by considering how dependent variables in the second can be regarded as a suitable average of corresponding variables in the first. To fix ideas, let us start by considering how the density of a fluid is related to its molecular structure.

MOLECULAR AVERAGES

At time t consider a cube C, centered about the point (x_0, y_0, z_0), whose sides have length h.

$$C: |x - x_0| \leq \tfrac{1}{2}h, \qquad |y - y_0| \leq \tfrac{1}{2}h, \qquad |z - z_0| \leq \tfrac{1}{2}h.$$

Within C there is a certain amount of mass M_h and hence a mean density $\rho_h \equiv M_h/h^3$. In order to define a density $\rho(x_0, y_0, z_0, t)$ at the point (x_0, y_0, z_0) and at the time t, we examine what happens as h becomes smaller and smaller. We obtain the graph of ρ_h as a function of h that is depicted in Figure 13.1. This graph does not record the results of actual experiments; it is impossible to make a sequence of measurements at a fixed time t. Rather,

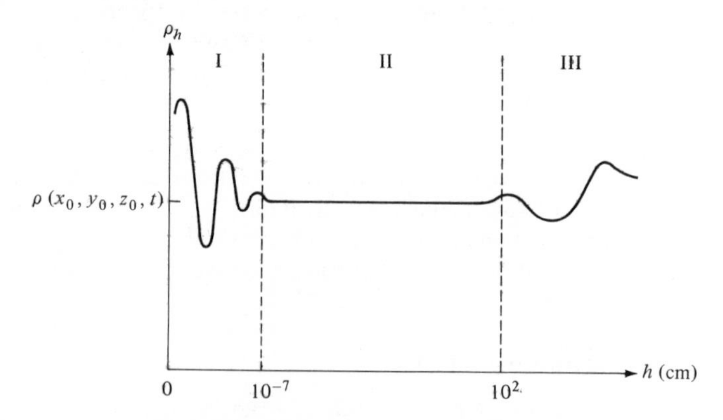

FIGURE 13.1. *Graph of mean density in a box of side h, versus h. The box is centered on a fixed point* (x_0, y_0, z_0), *and in a "thought experiment" all measurements are made at a certain time t. The scale on the abscissa is logarithmic.*

the figure presents the conclusions of a "thought experiment" based on the following general considerations.

The density varies in region III because of the nonuniform nature of the fluid. For example, suppose that (x_0, y_0, z_0) were a point in the Gulf Stream. If C were very large, it would contain some relatively warm Gulf Stream water and some relatively cool ordinary ocean water, and would have a higher density than a smaller cube containing only Gulf Stream water. But one would anticipate only very small density fluctuations as h decreased from 10 to 1 cm. The transition between regions III and II might be expected to occur when h is of the order of magnitude of hundreds of meters.*

In region II the graph of ρ_h is horizontal because one would certainly not expect fluctuations in ρ_h as h decreased, say, from about 10 cm to about 10^{-5} cm.

If h were so small that there were only a few molecules in the cube C, then a small decrease in h might well cause a large change in the mean density ρ_h. This change would occur when a previously counted molecule was excluded by a slight shrinkage of the box. To gain an idea of when this effect is significant, note that 1 cm^3 of water contains about 3×10^{22} molecules. Consequently, when h is approximately 10^{-7} cm one expects about 30 molecules in C. For values of h less than about 10^{-7} cm, one therefore expects the rapid variations of region I.

Thus, as h decreases from large values, fluctuations in ρ_h will only become negligible at values of h small compared to L, the length scale of density nonuniformities. (This scale is a typical distance over which significant changes in density take place. Compare Section 6.3.) Rapid fluctuations again appear when h is of the order of a length scale ℓ describing the typical distance between molecules. Region II will exist providing that $\ell \ll L$. The quantity $\rho(x_0, y_0, z_0, t)$ *should be taken to be the value of* ρ_h *in region II.* From the present molecular average point of view, it thus seems possible to consider mass to be continuously distributed if the scale of distance between molecules is small compared to the scale of density nonuniformities.

Similar considerations lead to the identification of the continuum velocity $\mathbf{v}(x_0, y_0, z_0, t)$ with the average velocity of the molecules that at time t are located in a small cube centered about (x_0, y_0, z_0). The cube should contain many molecules, but it should be small compared to the scale of velocity nonuniformities. (Incidentally, "the" scale of nonuniformity actually depends on what types of changes are of interest in a particular problem. For example, the scale of velocity variation in the study of global air circulation would be far different than the scale that is relevant to an analysis of insect flight.)

* We have in mind the grossest density variations. Certain small variations would be masked by coarse averaging. If these are of interest, the transition between regions II and III would occur at a much smaller value of h.

MASS DISTRIBUTION FUNCTIONS

A more fundamental approach to averaging out molecular detail involves the **mass distribution function** $\psi(x, y, z, u, v, w, t)$. This function is defined by the requirement that the integral

$$\int_{w_1}^{w_2} \int_{v_1}^{v_2} \int_{u_1}^{u_2} \int_{z_1}^{z_2} \int_{y_1}^{y_2} \int_{x_1}^{x_2} \psi(x, y, z, u, v, w, t,)\, dx\, dy\, dz\, du\, dv\, dw \tag{1}$$

gives the mass of material at time t that satisfies the following restrictions on its location and its velocity components:

$$\begin{aligned} x_1 \le x \le x_2, \qquad y_1 \le y \le y_2, \qquad z_1 \le z \le z_2, \\ u_1 \le u \le u_2, \qquad v_1 \le v \le v_2, \qquad w_1 \le w \le w_2. \end{aligned} \tag{2}$$

To simplify our brief discussion of the mass distribution function, let us restrict our consideration to the case of a single spatial component x and a corresponding velocity component u. In this case an informal version of the above definition is that for very small Δx and Δu, $\psi(x, u, t)\, \Delta x\, \Delta u$ gives the mass located between x and $x + \Delta x$ which has a velocity component between u and $u + \Delta u$. Note that the total mass located between $x = a$ and $x = b$ is given by

$$\int_{-\infty}^{\infty} \int_a^b \psi(x, u, t)\, dx\, du.$$

Consider a system composed of a single particle of mass m_1 and velocity u_1, located at x_1. Since there is no material with positions other than x_1 and velocities other than u_1, the distribution function must satisfy

$$\psi(x, u, t) = 0 \qquad \text{for} \quad x \neq x_1, \quad u \neq u_1.$$

In addition, since the total mass of the system is m_1, we have

$$\iint_{R_1} \psi(x, u, t)\, dx\, du = m_1,$$

where R_1 is any region of the xu plane that contains (x_1, u_1). A "function," like the one just considered, which vanishes everywhere except at a point yet has a nonzero and finite integral, is called a **delta function**.

Now consider a system composed of N particles having masses m_i, velocities u_i, and locations x_i, respectively; $i = 1, 2, \ldots, N$. The corresponding distribution function must vanish except at the points (x_i, u_i), where it can be regarded as having infinitely tall, slim peaks such that

$$\iint_{R_i} \psi(x, u, t)\, dx\, du = m_i.$$

Here R_i is a region that contains (x_i, u_i) but not other points (x_k, u_k), $k \neq i$.

In many situations N is very large, but there is a distance increment Δx and a velocity increment Δu such that the following two requirements are satisfied: (i) Each rectangle of dimension Δx by Δu in xu space contains many peaks of ψ; and (ii) the increments Δx and Δu are small compared to interesting length and velocity changes. Since details are not of interest, it is appropriate to define an average mass distribution function $\Psi(x, u, t)$. One way to do this is to require that the mass corresponding to Ψ is correct in an appropriate Δx by Δu region:

$$\Psi(x, u, t)\, \Delta x\, \Delta u = \int_{u-(1/2)\Delta u}^{u+(1/2)\Delta u} \int_{x-(1/2)\Delta x}^{x+(1/2)\Delta x} \psi(\xi, \eta, t)\, d\xi\, d\eta. \tag{3}$$

So defined, the function Ψ no longer has infinite peaks, but it does have small discontinuities when the boundary of the Δx by Δu rectangle about (x, u) crosses one of the (x_i, u_i). Since the integral of a piecewise continuous function is continuous, however, another averaging would remove these discontinuities. The first derivative of the new averaged function would still be discontinuous, but this defect can be removed by another averaging, etc. The details are not important here. What should be realized is that for large scale phenomena one can achieve considerable simplification by focusing attention on a smooth averaged distribution function. In Chapter 1 we discussed the use of such an average, Ψ, in stellar dynamics. Also see Exercises 14.1.12 and 14.1.13.

Further simplification results if it is also true that the distance that a particle travels before it is sensibly deflected by another is small compared to the scale of interesting phenomena. Then, for example, one can write equations for $\rho(x, y, z, t)$ and $\mathbf{v}(x, y, z, t)$, rather than $\Psi(x, y, z, u, v, w, t)$. This matter will be discussed further in Section 14.5.

THE CONTINUUM AS AN INDEPENDENT MODEL

Another approach to the continuum, although superficially naïve, is no less valid than the molecular-average approach just discussed. This approach starts by questioning the assumptions underlying the justification of a continuum *model* by reference to the "real" molecular structure of nature. Indeed, are molecules "real"? Are those assemblages of spheres and rods which the chemists use "real"? Do not the physicists tell us that molecules are "really" complicated structures of atoms, which are themselves tiny solar systems of electrons in orbit about a nucleus? When pressed a little, physicists in fact caution us that the tiny solar system is a model for what is more accurately described by certain clouds of probability densities which give the statistical behavior of fundamental particles. When pressed still further, they aver that there may be something more fundamental than fundamental particles but urge that the questioner return in a few years when the present stimulating confusion may have given way to a clearer understanding of the foundations of physics.

From this point of view the continuum is that one of a number of models used in science which should obviously be the first employed in attempting to explain what "appear to be" macroscopic phenomena. This model is neither inferior nor superior to other models of nature. Like all of them, it is to be judged by experimental examination of the deductions made from it. In particular, there is no necessity to refer to the molecular model in developing the continuum model. The latter can stand on its own.

Nevertheless, one might assert that the molecular model is superior to the continuum model in the same way that quantum mechanics or relativistic mechanics is superior to classical mechanics, because the latter can be derived as an average or a special case of the former. This is partially true. By taking appropriate averages of the equations used in the kinetic theory of perfect gases, for example, one can derive the corresponding continuum equations. "In theory" one could also do this for all materials, but in practice this is not at present possible; the molecular details of the liquid and solid states are imperfectly understood and the required mathematical techniques are not fully developed. Indeed, a great part of the utility of the continuum model lies in its *ability to disregard molecular details.*

A more subtle point is this. The measurements needed for quantum mechanics and molecular mechanics are made with instruments. To interpret the needle deflections or oscilloscope traces, one must use the laws of continuum mechanics. From this point of view the roles of "superior" and "inferior" in the previous paragraph are reversed. It appears that the various models of nature are probably better conceived as lying on a circle, equally near and far from "reality," than as being linearly ordered by their degree of reality.

Most scientists on most occasions do not concern themselves with the thorny philosophical questions that emerge from a searching examination of what lies at the foundation of their endeavors. Our brief look at the foundations of continuum mechanics has revealed that they are no less secure than the foundations of other branches of science. The reader is probably ready to cover over what defects there are in the foundations and, hoping for the best, to commence building.

EXERCISE

1. Demonstrate that the value for $\rho(x_0, y_0, z_0, t)$ would be unchanged if spheres, rather than cubes, were used in the computation of ρ_h. Generalize.

13.2 Kinematics of Deformable Media

We turn to the study of the motion of a continuum without regard to the forces that produce the motion. This is called **kinematics** as opposed to **dynamics**, the study of the influence of forces on motion.

POINTS AND PARTICLES

As a continuum moves, it is natural to try to specify the position of each point as a function of the time t. To identify a given point, we specify its position $\mathbf{A}$ at a certain initial time t_0. If $\mathbf{x}$ is the position vector of this point at t, we describe the motion of the medium by the equations

$$\mathbf{x} = \mathbf{x}(\mathbf{A}, t), \qquad \mathbf{x}(\mathbf{A}, t_0) \equiv \mathbf{A}. \tag{1a, b}$$

In words, $\mathbf{x}$ is the position at time t of a point that was initially (at $t = t_0$) located at $\mathbf{A}$. The word **particle** is used to denote the point that starts at a fixed initial position $\mathbf{A}$ and moves according to (1). For fixed $\mathbf{A}$ the curve

$$\mathbf{x} = \mathbf{x}(\mathbf{A}, t), \qquad t \geq t_0,$$

is called a **particle path**.

We emphasize that in the continuum model a particle is not a molecule. An example gives an idea of the magnitudes involved. Recall from the conclusion of Section 4.1 that in time t a substance diffuses roughly $(Dt)^{1/2}$ units, where D is the diffusivity. The observed value of D for carbon dioxide diffusing in air is $0.14\ \text{cm}^2/\text{sec}$ at $0°\text{C}$. Thus molecules of carbon dioxide that were initially confined to a very small volume in 1 min would diffuse so as to be located mainly within a sphere of radius about 3 cm. Diffusion into a sphere of radius 1 cm would take place in just 7 sec.

Let $\mathbf{A}(\mathbf{x}, t)$ be the function of $\mathbf{x}$ and t obtained by solving (1) for $\mathbf{A}$. The result

$$\mathbf{A} = \mathbf{A}(\mathbf{x}, t)$$

states that $\mathbf{A}$ is the initial position of the particle now (at time t) located at $\mathbf{x}$. Since $\mathbf{A}(\mathbf{x}, t)$ and $\mathbf{x}(\mathbf{A}, t)$ are a pair of inverse functions,

$$\mathbf{A}[\mathbf{x}(\mathbf{A}, t), t] \equiv \mathbf{A}, \qquad \mathbf{x}[\mathbf{A}(\mathbf{x}, t), t] \equiv \mathbf{x}. \tag{2a, b}$$

We shall postulate **impenetrability of matter**: $\mathbf{A}(\mathbf{x}, t)$ and $\mathbf{x}(\mathbf{A}, t)$ will be assumed to be single-valued invertible functions so that, at time t, for each $\mathbf{A}$ there exists a unique $\mathbf{x}$ and vice versa. We shall also postulate **smoothness of motion**: $\mathbf{A}(\mathbf{x}, t)$ and $\mathbf{x}(\mathbf{A}, t)$ will be assumed to have as many continuous partial derivatives as required for the various operations we shall perform on these functions.

In some contexts it is necessary to relax these assumptions on certain curves and surfaces to allow for the splitting or coalescence of material. Moreover, consideration of discontinuous shock waves or the formation and collapse of bubbles require that the invertibility assumption be dropped for certain domains.

As one example of the many dependent variables that will enter our discussion, consider the density. Let $\delta(\mathbf{A}, t)$ denote the mass per unit volume at time t of the particle initially located at $\mathbf{A}$. Let $\rho(\mathbf{x}, t)$ denote the mass per unit volume at time t, at the fixed point $\mathbf{x}$. *We assert that at time t the value of a dependent variable at the* **point x** *equals the value of the same dependent variable for the* **particle** *located at* **x** *at time t. This assertion could be called* **point–particle interchangeability**. It is really part of the definition of "particle."

In the case of density, point–particle interchangeability gives

$$\rho(\mathbf{x}, t) = \delta[\mathbf{A}(\mathbf{x}, t), t]. \tag{3}$$

To see this, observe that, by the definition of δ, the expression $\delta[\mathbf{A}(\mathbf{x}, t), t]$ describes the density now associated with a particle that was initially located at the initial location of the particle now at $\mathbf{x}$. ("Now" means "at time t.") This is the same as saying that $\delta[\mathbf{A}(\mathbf{x}, t), t]$ is the density associated with a particle now at $\mathbf{x}$.

MATERIAL AND SPATIAL DESCRIPTIONS

Substituting $\mathbf{x} = \mathbf{x}(\mathbf{A}, t)$ into (3) and using (2a), we find

$$\rho[\mathbf{x}(\mathbf{A}, t), t] = \delta(\mathbf{A}, t). \tag{4}$$

Equations (3) and (4) illustrate the transformation between the **spatial description** and **material description**, where these terms* are defined as follows.

Spatial description: $\mathbf{x}, t$ independent variables.
Material description: $\mathbf{A}, t$ independent variables.

To obtain a spatial description of air motion, one would measure (as time progresses) velocity, pressure, temperature, etc., at various fixed stations on the ground and in towers. To obtain a material description, one would make such measurements in various freely traveling balloons.

Let us consider the relation between material and spatial descriptions of velocity. In analogy with elementary point mechanics, we define the velocity of a particle by the rate of change of that particle's position with time:

$$\mathbf{V}(\mathbf{A}, t) \equiv \frac{\partial \mathbf{x}(\mathbf{A}, t)}{\partial t}. \tag{5}$$

To preserve point–particle interchangeability, we must define $\mathbf{v}(\mathbf{x}, t)$, the velocity in spatial coordinates, by

$$\mathbf{v}(\mathbf{x}, t) = \mathbf{V}[\mathbf{A}(\mathbf{x}, t), t].\dagger \tag{6a}$$

* Instead of "spatial" and "material" description the terms *Eulerian* and *Lagrangian* description are often used. Both descriptions were originated by Euler, but Lagrange's contributions to the development of particle mechanics were sufficient to justify his name being used.

† Equation (6a) must hold for the same reasons as those used to confirm Equation (3). Because the argument is involved, we repeat it in somewhat different terms. (The required verbalization deepens understanding of material and spatial variables.) Note first that $\mathbf{V}(\mathbf{A}, t)$ is the velocity at time t of the particle initially located at $\mathbf{A}$. Also, in the equation $\mathbf{A} = \mathbf{A}(\mathbf{x}, t)$, the expression $\mathbf{A}(\mathbf{x}, t)$ denotes the initial position of the particle that now (at time t) is located at $\mathbf{x}$. Therefore, $\mathbf{V}[\mathbf{A}(\mathbf{x}, t), t]$ is the velocity at time t of the particle whose initial position is that of the particle now at $\mathbf{x}$. In other words, $\mathbf{V}[\mathbf{A}(\mathbf{x}, t), t]$ is the velocity at time t of the *particle* that is located at $\mathbf{x}$ at time t. By point–particle interchangeability, this must be the velocity at time t at the fixed spatial *point* $\mathbf{x}$.

Equations (6a) and (2a) imply that

$$\mathbf{v}[\mathbf{x}(\mathbf{A}, t), t] = \mathbf{V}(\mathbf{A}, t). \tag{6b}$$

When all dependent variables, like $\mathbf{v}(\mathbf{x}, t)$ and $\rho(\mathbf{x}, t)$, are independent of t, the flow is called **steady**.* Note that the term "steady" implies time independence in the spatial description. Thus in steady flow of water through a pipe of varying cross section, the velocity measured by a *fixed* probe will always be the same. But the velocity of a given fluid particle will vary as the particle speeds up through constrictions and slows down at wide sections of the pipe.

Throughout this chapter (and most of this book) we shall determine the components of vectors by referring to a fixed right-handed cartesian coordinate system. For theoretical purposes numerical subscripts are best, but for particular problems it is often convenient to adopt a simpler notation. The two notations that we shall use are summarized in Table. 13.1.

TABLE 13.1. Two notations.

	Theoretical discussions	Particular problems
Unit vectors along coordinate axes	$\mathbf{e}^{(1)}, \mathbf{e}^{(2)}, \mathbf{e}^{(3)}$	$\mathbf{i}, \mathbf{j}, \mathbf{k}$
Position vector components		
Material	A_1, A_2, A_3	A, B, C
Spatial	x_1, x_2, x_3	x, y, z
Velocity vector components		
Material	V_1, V_2, V_3	U, V, W
Spatial	v_1, v_2, v_3	u, v, w

Theorem 1. If the particle paths are known, the spatial velocity field $\mathbf{v}(\mathbf{x}, t)$ can be determined, and vice versa.

Proof. Given $\mathbf{x}(\mathbf{A}, t)$, we can find $\mathbf{V}(\mathbf{A}, t)$ from (5). If we express $\mathbf{A}$ in terms of $\mathbf{x}$ and t and substitute into $\mathbf{V}(\mathbf{A}, t)$, then we obtain $\mathbf{v}(\mathbf{x}, t)$, by (6a).

On the other hand, suppose that $\mathbf{v}(\mathbf{x}, t)$ is known. Combining (5) and (6b), we obtain

$$\frac{\partial \mathbf{x}(\mathbf{A}, t)}{\partial t} = \mathbf{v}[\mathbf{x}(\mathbf{A}, t), t]. \tag{7}$$

* The notion of a steady flow is used mainly in fluid mechanics. Displacement rather than velocity is the central dependent variable in solid mechanics, and the displacement generally depends on time even when the velocity is steady. Compare Exercise 12.1.8.

If we no longer explicitly indicate the dependence of $\mathbf{x}$ on its initial position $\mathbf{A}$ and on the time t, we can rewrite (7) more simply as

$$\frac{\partial \mathbf{x}}{\partial t} = \mathbf{v}(\mathbf{x}, t). \tag{8}$$

Since $\mathbf{v}(\mathbf{x}, t)$ is assumed known, (8) can be regarded as a first order nonlinear differential equation for $\mathbf{x}$ (or as a system of three equations for the three components of $\mathbf{x}$). The required initial conditions are simply

$$\text{at } t = t_0, \qquad \mathbf{x} = \mathbf{A}. \tag{9}$$

Provided these equations have a solution (as we shall assume), we can find $\mathbf{x}$. □

In dealing with differential equations, one usually does not explicitly indicate the dependence of the unknown function on the initial conditions. For this reason, in many texts (8) is written

$$\frac{d\mathbf{x}}{dt} = \mathbf{v}(\mathbf{x}, t). \tag{10}$$

Here $\mathbf{x}$ is regarded as a function of t only, with "fixed" initial point $\mathbf{A}$. The partial derivative notation of (8), which says "keep $\mathbf{A}$ constant," is therefore not deemed necessary.

STREAMLINES AND PARTICLE PATHS

At time t a **streamline** is defined as a curve whose tangent vector at each point is parallel to the instantaneous velocity vector at that point. Let s be the arc length of the streamline $\mathbf{x} = \mathbf{x}(s, t)$, which emanates from a point $\mathbf{x}_0$ at time t. Choose the sign of s so that $d\mathbf{x}/ds$ points in the same direction as the velocity vector, not in the opposite direction. Let $s = 0$ at $\mathbf{x}_0$. Then $\mathbf{x}(s, t)$ is determined by the following equation and initial condition:*

$$\frac{d\mathbf{x}(s)}{ds} = \frac{\mathbf{v}[\mathbf{x}(s), t]}{|\mathbf{v}[x(s), t]|}, \quad \mathbf{x}(0) = \mathbf{x}_0 \qquad (|\mathbf{v}| \neq 0). \tag{11}$$

(We have used the fact that $d\mathbf{x}/ds$ is a unit tangent vector.) If written in subscript notation, the differential equations of (11) become

$$\frac{dx_i}{ds} = \frac{v_i}{|\mathbf{v}|}; \qquad i = 1, 2, 3; \quad |\mathbf{v}|^2 \equiv \sum_{i=1}^{3} v_i^2.$$

If $v_1 \neq 0$,

$$\frac{v_2}{v_1} = \frac{dx_2/ds}{dx_1/ds} = \frac{dx_2}{dx_1}, \qquad \frac{v_3}{v_1} = \frac{dx_3}{dx_1}.$$

* As above, we shall often use $t = 0$, rather than $t = t_0$, for the initial time.

Similarly, if $v_2 \neq 0$ or $v_3 \neq 0$, one can write

$$\frac{v_1}{v_2} = \frac{dx_1}{dx_2}, \quad \frac{v_3}{v_2} = \frac{dx_3}{dx_2} \qquad \text{or} \qquad \frac{v_1}{v_3} = \frac{dx_1}{dx_3}, \quad \frac{v_2}{v_3} = \frac{dx_2}{dx_3}.$$

Such equations are often compactly summarized by

$$\frac{dx_1}{v_1} = \frac{dx_2}{v_2} = \frac{dx_3}{v_3}. \tag{12}$$

It can be shown that *in steady flow, streamlines and particle paths coincide* (Exercise 5).

Example 1. Suppose the material description of a certain two-dimensional motion is given by

$$x_1 = A_1 e^t, \qquad x_2 = A_2 e^{-t}, \qquad t \geq 0. \tag{13}$$

Find the spatial description of the motion and show that the motion is steady. Reverse the process, and find the material description from the spatial. Find particle paths and streamlines.

Solution

$$\frac{\partial \mathbf{x}}{\partial t} = A_1 e^t \mathbf{i} - A_2 e^{-t} \mathbf{j} = \mathbf{V}(\mathbf{A}, t). \tag{14}$$

Inverting (13) is trivial:

$$A_1 = x_1 e^{-t}, \qquad A_2 = x_2 e^t, \, t \geq 0. \tag{15}$$

Substitution of (15) into (14) yields the spatial description

$$\mathbf{V}[\mathbf{A}(\mathbf{x}, t), t] = \mathbf{v}(\mathbf{x}, t) = (x_1 e^{-t}) e^t \mathbf{i} - (x_2 e^t) e^{-t} \mathbf{j} = x_1 \mathbf{i} - x_2 \mathbf{j}. \tag{16}$$

Since $\mathbf{v}$ is independent of t, the motion is a steady one.

Working backward, given the spatial description

$$\mathbf{v} = x_1 \mathbf{i} - x_2 \mathbf{j}, \tag{17}$$

we find that

$$\frac{d\mathbf{x}}{dt} = \mathbf{v} \qquad \text{or} \qquad \frac{dx_1}{dt} = x_1, \quad \frac{dx_2}{dt} = -x_2. \tag{18}$$

Solving (18) subject to the initial condition $\mathbf{x}(0) = \mathbf{A}$, we obtain

$$x_1 = A_1 e^t, \qquad x_2 = A_2 e^{-t}. \tag{19}$$

We thereby recover (13). As t increases from 0 to ∞, one can thus determine from (19) the path of the particle that was located at $\mathbf{x} = \mathbf{A}$ when $t = 0$. In this case the parameter t can easily be eliminated giving for the particle path a portion of the hyperbola

$$x_1 x_2 = A_1 A_2. \tag{20}$$

Suppose that $A_1 > 0$, $A_2 < 0$. Then, for $t > 0$, $x_1 = A_1 e^t$ implies that $x_1 > A_1$, while $x_2 = A_2 e^{-t}$ implies that $A_2 < x_2 < 0$; therefore the particle path is the "upper" portion of that part of $x_1 x_2 = A_1 A_2$ which lies in the fourth quadrant. See Figure 13.2. (This example illustrates the care that must always be taken in passing from a parametric representation of a curve to a nonparametric representation.)

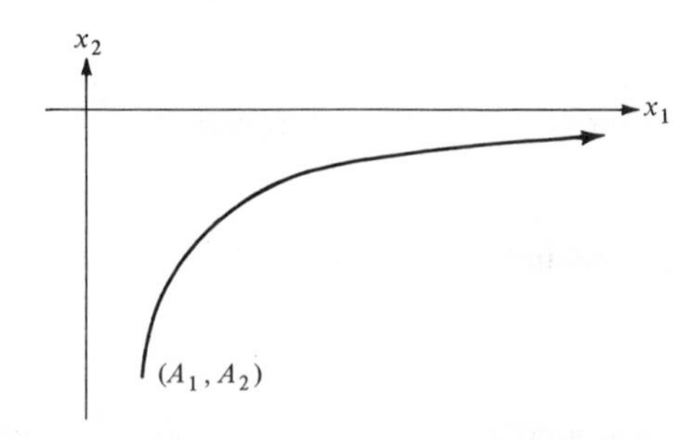

FIGURE 13.2. *The particle path determined in Example* 1. *As t increases, x_1 increases, so the arrow indicates the direction of motion.*

The *streamlines* are found by using (12). Thus

$$\frac{dx_1}{x_1} = -\frac{dx_2}{x_2}.$$

Hence

$$\ln|x_1| = -\ln|x_2| + \text{constant} \qquad \text{or} \qquad x_1 x_2 = \text{constant}. \tag{21}$$

Comparison of (21) and (20) shows that the coincidence of particle paths and streamlines for steady flows is verified in this particular case.

The passage from particle paths to spatial velocity field requires as a key step the inversion

$$\mathbf{x} = \mathbf{x}(\mathbf{A}, t) \to \mathbf{A} = \mathbf{A}(\mathbf{x}, t).$$

In the very simple example just discussed, this was easily accomplished; but can it be accomplished in general? Another difficult mathematical question arises in attempting to compute particle paths from spatial velocity fields. Under what conditions is there a solution to the differential equation (8) subject to the initial conditions (9)? We cannot delve into these matters here, but we refer the reader to Chapter 2, where a start was made on presenting some of the mathematics involved.

There are concrete problems that are easier to solve if the material description is employed. Examples can be found in the nonlinear theory of one-dimensional compressible flow phenomena. (See Courant and Friedrichs,

1948.) Such problems are the exception, however, since there is usually not sufficient compensation for the extra effort required to ascertain the fate of each particle. But although the spatial description is generally more convenient than the material* when it comes to investigating specific phenomena, the material description offers the more natural framework for the formulation of the basic physical laws. It is therefore necessary to transform certain expressions from material to spatial coordinates. Consideration of such transformations begins in the next section.

A SIMPLE KINEMATIC BOUNDARY CONDITION

Materials are often subject to exterior constraints of a kinematic nature. A simple but important example occurs when the material is in contact with an impermeable boundary (wall). Since no material can penetrate such a boundary, the velocity component normal to the wall must be zero. (No penetration is caused by relative tangential velocity.) If $\mathbf{n}(\mathbf{x}, t)$ is a unit normal to the wall, the boundary condition for $\mathbf{x}$ on the wall is

$$\mathbf{v}(\mathbf{x}, t) \cdot \mathbf{n}(\mathbf{x}, t) = \text{prescribed normal velocity of wall.} \tag{22}$$

When the wall is stationary, of course, $\mathbf{v} \cdot \mathbf{n} = 0$. Other kinematic boundary conditions arise (e.g., at the interface between two media), but at this point it is sufficient to indicate the existence of this type of constraint.

EXERCISES

‡1. Suppose that the spatial coordinates x_i of a particle at time t are given by

$$x_1 = A_1 + t, \qquad x_2 = A_2 e^t, \qquad x_3 = A_3 + tA_1.$$

Find the velocity vector $\mathbf{v}(\mathbf{x}, t)$; i.e., transform from the material to the spatial description.

2. In a two-dimensional steady flow the velocity components are given by

$$u = y, \qquad v = x.$$

(a) Show that the streamline passing through (x_0, y_0) is a hyperbola.
‡(b) Find the particle path that commences at (x_0, y_0). Note the coincidence of the particle path and the streamline.

‡3. In a two-dimensional *unsteady* flow the velocity components are given by

$$u = t, \quad v = x + 1 \qquad (t \text{ is the time}).$$

(a) Find the streamline passing through (x_0, y_0).
(b) Find the particle path that commences at (x_0, y_0).

* For certain classes of models, notably those of linear elasticity, the difference between material and spatial variables is of the order of terms already neglected. Consequently, the distinction between these variables is essentially negligible.

‡**4.** In a two-dimensional flow, the x and y velocity components are given by

$$u = e^{-y} \cos t, \qquad v = 1.$$

(a) Find the streamlines at time t.
(b) Find the particle paths, expressing the constants of integration in terms of the initial coordinates A and B.

5. Give a physical argument showing that particle paths and streamlines coincide in steady flow.

6. Interpret (6b) in the manner of the text's interpretation of (6a).

13.3 The Material Derivative

To provide physical motivation for the mathematical content of this section, consider the fact that fishing boats in Norwegian fjords sometimes act sluggishly because they generate internal waves on the underwater interface between fresh fjord water and heavier sea water beneath. The time in which significant diffusion of salt takes place is very large compared to the period of the waves generated. As a consequence, one can assume that each packet of fluid maintains its density as it moves. Formally, this **incompressibility assumption** is

$$\frac{\partial \delta(\mathbf{A}, t)}{\partial t} = 0, \tag{1}$$

where $\delta(\mathbf{A}, t)$ is the density at time t of a water particle that was initially at $\mathbf{A}$. The reader should contrast (1) with the assumption of uniform density

$$\delta(\mathbf{A}, t) \equiv \text{constant}, \tag{2}$$

which is *not* appropriate for this problem.

In Section 15.2 we discuss waves in layers of incompressible fluid whose density varies with height. Here we wish only to examine how equations like (1) can be transformed from material to spatial variables. We shall obtain a formula that is fundamental not only for the problem of internal density waves but for the entire formulation of continuum mechanics.

Before proceeding, the reader should review the chain rule (Appendix 13.1). Use of the careful notation discussed in this Appendix is essential for secure understanding of what follows.

Recall that in spatial variables, we denote the density by $\rho(\mathbf{x}, t)$ so that

$$\rho(\mathbf{x}, t) = \delta[\mathbf{A}(\mathbf{x}, t), t], \qquad \delta(\mathbf{A}, t) = \rho[\mathbf{x}(\mathbf{A}, t), t], \tag{3}$$

as in (2.3) and (2.4). In material variables the density is described by $\delta(\mathbf{A}, t)$. When we compute a partial derivative of δ with respect to t, we regard the initial position $\mathbf{A}$ as fixed. On the other hand, when we compute such a partial derivative of $\rho(\mathbf{x}, t)$, the density described in spatial variables, we regard the spatial position $\mathbf{x}$ as fixed. Since $\mathbf{A}$ and $\mathbf{x}$ are related, we can relate the two derivatives by the chain rule [compare (5), Appendix 13.1]:

$$\frac{\partial \delta(\mathbf{A}, t)}{\partial t} = \frac{\partial \rho}{\partial x_1}\bigg|_{\mathbf{x}=\mathbf{x}(\mathbf{A}, t)} \frac{\partial x_1}{\partial t} + \frac{\partial \rho}{\partial x_2}\bigg|_{\mathbf{x}=\mathbf{x}(\mathbf{A}, t)} \frac{\partial x_2}{\partial t} + \frac{\partial \rho}{\partial x_3}\bigg|_{\mathbf{x}=\mathbf{x}(\mathbf{A}, t)} \frac{\partial x_3}{\partial t} + \frac{\partial \rho}{\partial t}\bigg|_{\mathbf{x}=\mathbf{x}(\mathbf{A}, t)}. \tag{4}$$

As in (2.5), $\partial x_i/\partial t$ evaluated at $(\mathbf{A}, t)$ is simply $V_i(\mathbf{A}, t)$, the ith velocity component expressed in material variables; $i = 1, 2, 3$. Our goal is an expression for the left-hand side of (4) in spatial coordinates. To obtain this, we must make the substitution $\mathbf{A} = \mathbf{A}(\mathbf{x}, t)$, which yields

$$\frac{\partial \delta(\mathbf{A}, t)}{\partial t}\bigg|_{\mathbf{A}=\mathbf{A}(\mathbf{x}, t)} = \sum_{i=1}^{3} v_i(\mathbf{x}, t) \frac{\partial \rho(\mathbf{x}, t)}{\partial x_i} + \frac{\partial \rho(\mathbf{x}, t)}{\partial t}, \tag{5}$$

where we have used the relation

$$V_i(\mathbf{A}, t)\,|_{\mathbf{A}=\mathbf{A}(\mathbf{x}, t)} = v_i(\mathbf{x}, t). \tag{6}$$

The notation D/Dt is frequently employed for the **material** or **substantial derivative** which appears on the left-hand side of (5). With this we have

$$\frac{\partial \delta(\mathbf{A}, t)}{\partial t}\bigg|_{\mathbf{A}=\mathbf{A}(\mathbf{x}, t)} \equiv \frac{D\rho(\mathbf{x}, t)}{Dt}. \tag{7}$$

$$\frac{D\rho(\mathbf{x}, t)}{Dt} = \frac{\partial \rho(\mathbf{x}, t)}{\partial t} + \mathbf{v}(\mathbf{x}, t) \cdot \nabla \rho(\mathbf{x}, t). \tag{8}$$

It is important that the reader fully understand the meaning of (7) and (8). Remember that the function $\partial\delta(\mathbf{A}, t)/\partial t$ gives the rate of change with time of the density of the particle whose initial position was $\mathbf{A}$. If the substitution $\mathbf{A} = \mathbf{A}(\mathbf{x}, t)$ is made in the afore-mentioned function of $\mathbf{A}$ and t, then a function of $\mathbf{x}$ and t results. This gives the rate of change with respect to time of the density of the particle now located at $\mathbf{x}$. With (8) the right side of (7) is a calculable function of $\mathbf{x}$ and t. Thus (7) and (8) allow computation of the rate of change of *particle* density when all variables are referred to *spatial* coordinates. We see that this computation requires knowledge of the spatial description of density $\rho(\mathbf{x}, t)$ and of the spatial description of velocity $\mathbf{v}(\mathbf{x}, t)$. We remark that in spatial variables, the condition for incompressibility is $D\rho/Dt = 0$. A motion satisfying this condition is sometimes called *isochoric*.

We have used the density throughout our discussion only for concreteness. For any quantity, the material derivative tells the time rate of change of that quantity for a fixed particle. Thus, to give an important example, if $\mathbf{v}(\mathbf{x}, t)$

denotes the velocity of the particle at position $\mathbf{x}$ and time t, then $D\mathbf{v}(\mathbf{x}, t)/Dt$ denotes the acceleration *of the particle* located at position $\mathbf{x}$ at time t. That is,

$$\left.\frac{\partial \mathbf{V}(\mathbf{A}, t)}{\partial t}\right|_{\mathbf{A}=\mathbf{A}(\mathbf{x}, t)} = \frac{D\mathbf{v}(\mathbf{x}, t)}{Dt}; \qquad \frac{D\mathbf{v}}{Dt} = \frac{\partial \mathbf{v}}{\partial t} + (\mathbf{v} \cdot \nabla)\mathbf{v}. \tag{9}$$

The ith component of $D\mathbf{v}/Dt$ is given by the formula

$$\frac{Dv_i}{Dt} = \frac{\partial v_i}{\partial t} + \sum_{j=1}^{3} v_j \frac{\partial v_i}{\partial x_j}.$$

We have derived the important formulas (7) and (8) by a careful application of rather standard mathematics. Having accomplished this, we shall reexamine the material derivative in a way that is less careful but which perhaps sheds more light on its physical meaning.

The time rate of change of the density of a particle is the result of the following two effects:

(a) At the spatial position $\mathbf{x}$ the density changes with time.

(b) The particle at $\mathbf{x}$ at time t arrived there from another part of the material, so there may be a density change even in a steady motion for which

$$\frac{\partial \rho(\mathbf{x}, t)}{\partial t} \equiv 0.$$

Effect (a) is clearly given by the first term of (9). We now show that effect (b) is given by the second term of this equation.

In the very short time interval $(t, t + \Delta t)$, the particle located at $\mathbf{x}$ at time t moves a distance approximately given by $|\mathbf{v}(\mathbf{x}, t)|\ \Delta t$. This motion is (approximately) in the direction of the unit vector $\mathbf{v}(\mathbf{x}, t)/|\mathbf{v}(\mathbf{x}, t)|$. The change of ρ *per unit length* in this direction (at a fixed time) is $(\mathbf{v}/|\mathbf{v}|) \cdot \nabla\rho$.* The change of ρ due to motion through the distance $|\mathbf{v}|\ \Delta t$ is the product of the distance traveled and the change per unit distance, i.e.,

$$(|\mathbf{v}|\ \Delta t)\cdot\left(\frac{\mathbf{v}}{|\mathbf{v}|} \cdot \nabla\rho\right) = (\mathbf{v} \cdot \nabla\rho)\ \Delta t.$$

This is the change in time Δt. To find the change per unit time, we must divide by Δt, which gives the desired result.

EXERCISES

‡1. Suppose that $\rho(\mathbf{x}, t) = x_1 + x_2 \sin t$; $x_1 = A_1 + t$, $x_2 = A_2 + t^2$, $x_3 = 0$. Verify (8) in this case by explicitly computing both sides of the equations. [Find $\delta(\mathbf{A}, t)$; compute $\partial\delta(\mathbf{A}, t)/\partial t$; etc.]

2. (a) Show that the flow given by

$$x = A + \varepsilon \sin \omega t, \quad y = B \qquad (\varepsilon \text{ and } \omega \text{ constants})$$

is not a steady one. Describe it.

(b) If the density is given by $\rho(x, t) = \rho_0/(1 + x^2)$, calculate $D\rho/Dt$.

* Recall that the directional derivative of a function is obtained by taking the scalar product of the gradient of the function with a unit vector in the appropriate direction.

3. Consider the two-dimensional motion whose material description is, for some constant ω,

$$x_1 = A_1 \cos \omega t - A_2 \sin \omega t, \qquad x_2 = A_1 \sin \omega t + A_2 \cos \omega t.$$

(a) Show that in spatial coordinates, the velocity $\mathbf{v}(\mathbf{x}, t)$ is given by

$$\mathbf{v} = -\omega x_2 \mathbf{i} + \omega x_1 \mathbf{j}$$

so that the flow is steady.

(b) Show that the particle paths are circles about the origin.

(c) *Calculate* the streamlines at time t and thus show that they coincide with the particle paths.

(d) Find $\mathbf{V}(\mathbf{A}, t)$ and verify (9).

4. In the Eulerian description, a flow is given by the steady velocity field

$$v_1 = 2x_1 + 3x_2, \qquad v_2 = x_1 - x_2, \qquad v_3 = 0.$$

(a) Find the corresponding Lagrangian description.

(b) Plot the path of particles that pass through the point $x_1 = 2$, $x_2 = 0$, sometime during the motion. Note that all such particles have the same path, since the motion is steady.

(c) Find the acceleration of the particle that is instantaneously located at (3, 4). Comment on the fact that the answer is nonzero, although the flow is steady.

5. Let $x_1 = A_1 \exp t$, $x_2 = A_2 \exp(-t)$ (as in Example 1 of Section 13.2). Show that $\rho(\mathbf{x}, t) \equiv x_1 x_2$ is the density of an incompressible motion in two ways:

(a) By computing $D\rho(\mathbf{x}, t)/Dt$.

(b) By computing $\partial\delta(\mathbf{A}, t)/\partial t$.

13.4 The Jacobian and Its Material Derivative

Having introduced via $\mathbf{x} = \mathbf{x}(\mathbf{A}, t)$ the material description of a moving continuum, we can *make precise the notion that a specified portion of material occupies the region $R(t)$ at time t*. We specify the portion of material by giving its location at some instant of time, say $t = 0$. Denote by $R(0)$ the region occupied at $t = 0$ by the designated portion of material. Let the subsequent motion of any point $\mathbf{x}$ be described by $\mathbf{x} = \mathbf{x}(\mathbf{A}, t)$. Then the region $R(t)$ occupied by the designated material at time t is in mathematical language the image of $R(0)$ at time t under the mapping $\mathbf{x} = \mathbf{x}(\mathbf{A}, t)$. More informally, a point $\mathbf{x}$ is in $R(t)$ if, at time t, $\mathbf{x} = \mathbf{x}(\mathbf{A}, t)$ for a point $\mathbf{A}$ in $R(0)$.

Let $\mathscr{V}(t)$ denote the volume of $R(t)$. Then

$$\mathscr{V}(0) = \iiint_{R(0)} dA_1\, dA_2\, dA_3 \qquad \text{and} \qquad \mathscr{V}(t) = \iiint_{R(t)} dx_1\, dx_2\, dx_3. \quad (1\text{a, b})$$

Although any letters could have been used for the three dummy variables of integration in (1), it is suggestive to use the "initial" variables A_1, A_2, and A_3 in (1a) and to use in (1b) the variables x_1, x_2, and x_3 associated with the time t. In (1b) we can change the variables of integration to A_1, A_2, and A_3, where $\mathbf{A} = \mathbf{A}(\mathbf{x}, t)$. According to the Jacobian rule

$$\mathscr{V}(t) = \iiint_{R(0)} J(A_1, A_2, A_3, t)\, dA_1\, dA_2\, dA_3, \tag{2}$$

where the Jacobian is defined by the following determinant:

$$J(A_1, A_2, A_3, t) \equiv \frac{\partial(x_1, x_2, x_3)}{\partial(A_1, A_2, A_3)} = \begin{vmatrix} \dfrac{\partial x_1}{\partial A_1} & \dfrac{\partial x_1}{\partial A_2} & \dfrac{\partial x_1}{\partial A_3} \\ \dfrac{dx_2}{\partial A_1} & \dfrac{\partial x_2}{\partial A_2} & \dfrac{\partial x_2}{\partial A_3} \\ \dfrac{\partial x_3}{\partial A_1} & \dfrac{\partial x_3}{\partial A_2} & \dfrac{\partial x_3}{\partial A_3} \end{vmatrix}.$$

No absolute value sign is needed on J, since the components of $\mathbf{x}$ and $\mathbf{A}$ are both determined in right-handed coordinate systems.

Using the integral mean value theorem (Appendix 13.2), we see that (2) implies that

$$\mathscr{V}(t) = J(\bar{A}_1, \bar{A}_2, \bar{A}_3, t) \iiint_{R(0)} dA_1\, dA_2\, dA_3 = J(\bar{A}_1, \bar{A}_2, \bar{A}_3, t)\mathscr{V}(0) \tag{3}$$

for some point $(\bar{A}_1, \bar{A}_2, \bar{A}_3)$ in $R(0)$.

We now wish to consider (3) as R continues to "shrink" about a fixed point (A_1, A_2, A_3). To make this idea precise, let us apply (3) to each of a sequence of similar regions $R_n(0)$ whose volume $\mathscr{V}_n(0)$ approaches zero as $n \to \infty$. (Roughly speaking, "similar" means "having the same shape," but the idea of similarity, one that we shall use several times, is made precise in Appendix 13.3.) Every region R_n is required to contain the fixed point (A_1, A_2, A_3). According to (3),

$$\mathscr{V}_n(t) = J(\bar{A}_1^{(n)}, \bar{A}_2^{(n)}, \bar{A}_3^{(n)}) \cdot \mathscr{V}_n(0); \quad \bar{A}^{(n)} \in R_n; \qquad n = 1, 2, 3, \ldots.$$

But, since the R_n are similar and have ever-decreasing volume,

$$\lim_{n \to \infty} (\bar{A}_1^{(n)}, \bar{A}_2^{(n)}, \bar{A}_3^{(n)}) = (A_1, A_2, A_3),$$

as is proved in Appendix 13.3. Thus, if J is continuous at (A_1, A_2, A_3),

$$\lim_{\mathscr{V}_n(0) \to 0} \frac{\mathscr{V}_n(t)}{\mathscr{V}_n(0)} = J(A_1, A_2, A_3, t). \tag{4}$$

More geometrically but less precisely, one can say that the Jacobian at point $\mathbf{A}$ and time t represents the **dilatation** of an infinitesimal volume initially at $\mathbf{A}$, where the dilatation is the ratio of volume occupied by the infinitesimal material region at time t to its initial volume. The importance of the Jacobian is confirmed by our ability to describe it, as in (4), in a manner that does not depend on the use of a particular coordinate system.

In later work we shall frequently need an expression for $\partial J(\mathbf{A}, t)/\partial t$, the rate of change of the dilatation.* We shall now find this expression in a straightforward manner for the two-dimensional case. The three-dimensional case will be left as an exercise. A compact and elegant derivation is afforded by the machinery of cartesian tensors (Example 3 in Section 1.2 of II).

We shall employ the following formula for the derivative of a determinant:

$$\frac{d}{dt}\begin{vmatrix} A(t) & B(t) \\ C(t) & D(t) \end{vmatrix} = \begin{vmatrix} \dfrac{dA}{dt} & \dfrac{dB}{dt} \\ C & D \end{vmatrix} + \begin{vmatrix} A & B \\ \dfrac{dC}{dt} & \dfrac{dD}{dt} \end{vmatrix}, \tag{5}$$

which the reader can easily verify. Using (5), we have the following expression for the partial derivative of the Jacobian with respect to time, keeping the initial coordinates A_1 and A_2 fixed:

$$\frac{\partial J(A_1, A_2, t)}{\partial t} = \frac{\partial}{\partial t}\begin{vmatrix} \dfrac{\partial x_1}{\partial A_1} & \dfrac{\partial x_1}{\partial A_2} \\ \dfrac{\partial x_2}{\partial A_1} & \dfrac{\partial x_2}{\partial A_2} \end{vmatrix} = \begin{vmatrix} \dfrac{\partial V_1}{\partial A_1} & \dfrac{\partial V_1}{\partial A_2} \\ \dfrac{\partial x_2}{\partial A_1} & \dfrac{\partial x_2}{\partial A_2} \end{vmatrix} + \begin{vmatrix} \dfrac{\partial x_1}{\partial A_1} & \dfrac{\partial x_1}{\partial A_2} \\ \dfrac{\partial V_2}{\partial A_1} & \dfrac{\partial V_2}{\partial A_2} \end{vmatrix}. \tag{6}$$

We have employed the result $\partial \mathbf{x}/\partial t = \mathbf{V}$ of (2.5). We have also assumed x_1 and x_2 to be twice continuously differentiable so that orders of differentiation can be interchanged. For example,

$$\frac{\partial}{\partial t}\frac{\partial x_1}{\partial A_1} = \frac{\partial}{\partial A_1}\frac{\partial x_1}{\partial t} = \frac{\partial V_1}{\partial A_1}.$$

By the chain rule, the first of the two determinants on the right-hand side of (6) can be written

$$\begin{vmatrix} \dfrac{\partial v_1}{\partial x_1}\dfrac{\partial x_1}{\partial A_1} + \dfrac{\partial v_1}{\partial x_2}\dfrac{\partial x_2}{\partial A_1} & \dfrac{\partial v_1}{\partial x_1}\dfrac{\partial x_1}{\partial A_2} + \dfrac{\partial v_1}{\partial x_2}\dfrac{\partial x_2}{\partial A_2} \\ \dfrac{\partial x_2}{\partial A_1} & \dfrac{\partial x_2}{\partial A_2} \end{vmatrix}. \tag{7}$$

* Some authors define the dilatation as the limit of $[\mathscr{V}_n(t) - \mathscr{V}_n(0)]/\mathscr{V}_n(0)$. Whether this definition or (4) is used, $\partial J/\partial t$ still gives the rate of change of the dilatation.

In (7) we have indicated by a vertical line that $\mathbf{x}$ must be expressed in terms of $\mathbf{A}$ and t in $\partial v_1/\partial x_1$ and $\partial v_1/\partial x_2$.

To obtain the final simplification in a generalizable way, we notice that (7) can be written as the sum of two determinants:

$$\begin{vmatrix} \dfrac{\partial v_1}{\partial x_1}\bigg|\dfrac{\partial x_1}{\partial A_1} & \dfrac{\partial v_1}{\partial x_1}\bigg|\dfrac{\partial x_1}{\partial A_2} \\ \dfrac{\partial x_2}{\partial A_1} & \dfrac{\partial x_2}{\partial A_2} \end{vmatrix} + \begin{vmatrix} \dfrac{\partial v_1}{\partial x_2}\bigg|\dfrac{\partial x_2}{\partial A_1} & \dfrac{\partial v_1}{\partial x_2}\bigg|\dfrac{\partial x_2}{\partial A_2} \\ \dfrac{\partial x_2}{\partial A_1} & \dfrac{\partial x_2}{\partial A_2} \end{vmatrix}. \tag{8}$$

The second determinant in (8) is zero, since it equals $\partial v_1/\partial x_2$ times a determinant with two identical rows. The first determinant on the right side of (6) thus reduces to $(\partial v_1/\partial x_1)J$. Similarly, the second determinant on the right side of (6) reduces to $(\partial v_2/\partial x_2)J$, giving

$$\frac{\partial J(A_1, A_2, t)}{\partial t} = \left(\frac{\partial v_1}{\partial x_1} + \frac{\partial v_2}{\partial x_2}\right)\bigg|_{\mathbf{x}=\mathbf{x}(\mathbf{A},\, t)} J(A_1, A_2, t). \tag{9a}$$

Analogous calculations (Exercise 3) yield the three-dimensional result

$$\frac{\partial J(A_1, A_2, A_3, t)}{\partial t} = \left(\frac{\partial v_1}{\partial x_1} + \frac{\partial v_2}{\partial x_2} + \frac{\partial v_3}{\partial x_3}\right)\bigg|_{\mathbf{x}=\mathbf{x}(\mathbf{A},\, t)} J(A_1, A_2, A_3, t). \tag{9b}$$

If we make the substitution $\mathbf{A} = \mathbf{A}(\mathbf{x}, t)$ [i.e., if we write (9) in spatial variables] and use the definition of $\nabla \cdot \mathbf{v}$, we have the **Euler expansion formula**,

$$\frac{DJ[\mathbf{A}(\mathbf{x}, t), t]}{Dt} = (\nabla \cdot \mathbf{v})J[\mathbf{A}(\mathbf{x}, t), t]. \tag{10}$$

Equations (9) and (10) look simpler when written as follows.

$$\text{In material variables:} \quad \frac{\partial J}{\partial t} = [(\nabla \cdot \mathbf{v})|_{\mathbf{x}=\mathbf{x}(\mathbf{A},\, t)}]J. \tag{11}$$

$$\text{In spatial variables:} \quad \boxed{\frac{DJ}{Dt} = (\nabla \cdot \mathbf{v})J}\,. \tag{12}$$

We emphasize that the derivatives required by the divergence $\nabla \cdot$ are with respect to spatial variables.

The reader may choose to remember the Euler expansion formula as (11) or as (12) or as: "The derivative of the Jacobian is the divergence of the velocity times the Jacobian." When actually using the Euler formula, however, he risks trouble unless he keeps the independent variables straight, as in (9) and (10).

EXERCISES

1. (a) Give an example of a sequence of regions (R_n) with the following properties:

(i) A and B are distinct *fixed* points (independent of n) interior to each region R_n, $n = 1, 2, \ldots$.

(ii) The volume of R_n approaches zero as $n \to \infty$.

2. Use induction to generalize (5) to a result for an $n \times n$ determinant.

3. Generalize the proof given in the text so that (10) holds for the three-dimensional case $J = J(A_1, A_2, A_3, t)$.

‡4. Calculate the Jacobian of the two-dimensional motion and verify (12) in the following cases:

(a) $x_1 = A_1 \exp t, \qquad x_2 = A_2 \exp(-t), \qquad x_3 = A_3.$

(b) $x_1 = A_1^{1/2} \exp t, \qquad x_2 = A_2 \exp(-t), \qquad x_3 = A_3.$

5. By direct calculation show that $\partial(x_1, x_2)/\partial(A_1, A_2) \equiv 1$ for the motion of Exercise 3.3. What is the significance of this result?

Appendix 13.1 On the Chain Rule of Partial Differentiation

By means of an example we illustrate the careful notation that is essential if confusion is to be avoided in all but simple applications of the chain rule.

Suppose that a certain quantity is given as a function of the Cartesian coordinates x and y. For definiteness let us consider the temperature of a certain body B. In the present discussion we shall use the notation $T = T(x, y)$.

Suppose that polar coordinates (r, θ) are introduced so that

$$x = r \cos \theta, \qquad y = r \sin \theta.$$

Denoting the temperature of B as described in polar coordinates by $\tau(r, \theta)$, we have

$$\tau(r, \theta) = T[x(r, \theta), y(r, \theta)], \quad T(x, y) = \tau[r(x, y), \theta(x, y)]. \tag{1}$$

For example, if

$$T(x, y) = x^2 + \sin y, \tag{2}$$

then on introducing polar coordinates, we find

$$\tau(r, \theta) = r^2 \cos^2 \theta + \sin(r \sin \theta). \tag{3}$$

Note that

$$T(r, \theta) = r^2 + \sin \theta \not\equiv \tau(r, \theta). \tag{4}$$

If we wished to find $\partial\tau/\partial\theta$, the chain rule could be used as follows:

$$\frac{\partial \tau}{\partial \theta} = \left.\frac{\partial T}{\partial x}\right|_{\substack{x = r\cos\theta \\ y = r\sin\theta}} \cdot \frac{\partial x}{\partial \theta} + \left.\frac{\partial T}{\partial y}\right|_{\substack{x = r\cos\theta \\ y = r\sin\theta}} \cdot \frac{\partial y}{\partial \theta}. \tag{5}$$

The notation is unambiguous. For example, since T is a function of x and y, y is held constant in computing $\partial T/\partial x$. The equations associated with the vertical line indicate that after computing $\partial T/\partial x$, one should substitute $r\cos\theta$ for x and $r\sin\theta$ for y. Thus (5) becomes

$$\frac{\partial \tau}{\partial \theta} = 2x\Big|_{\substack{x=r\cos\theta\\ y=r\sin\theta}}(-r\sin\theta) + \cos y\Big|_{\substack{x=r\cos\theta\\ y=r\sin\theta}}(r\cos\theta)$$

$$= -2r^2\sin\theta\cos\theta + \cos(r\sin\theta)\cdot r\cos\theta.$$

The same result can, of course, be obtained directly from (3).

A proof of the chain rule is given in calculus books, but sometimes care is not taken to distinguish between expressions such as $T(r, \theta)$ and $\tau(r, \theta)$. In these books an implicit change in the usual functional notation is made in discussing certain topics such as the chain rule. When r and θ are used as arguments of T, then T refers to the function in (3) that we call τ. When x and y are used as arguments, the original function (2) is meant. Thus the symbol T is ambiguous. In our work such ambiguity would lead to confusion.

EXERCISES

1. Derivation of the following equation is requested in Exercise 2 of Appendix 14.1:

$$\frac{\partial \Theta}{\partial V} + \Theta\frac{\partial P}{\partial E} - P\frac{\partial \Theta}{\partial E} = 0 \qquad (V, E \text{ independent}). \tag{6}$$

Here V is the volume of a system, Θ is its temperature, E its internal energy, and P its pressure. Any two of these variables can be regarded as dependent and the other two as independent. Show that for the five other choices of independent variables the above relation implies that

$$\Theta - P\left(\frac{\partial \Theta}{\partial P}\right) + \frac{\partial(\Theta, E)}{\partial(V, P)} = 0 \qquad (V, P \text{ independent}), \tag{7}$$

$$\frac{\partial E}{\partial V} - \Theta\frac{\partial P}{\partial \Theta} + P = 0 \qquad (V, \Theta \text{ independent}), \tag{8}$$

$$\frac{\partial E}{\partial P} + \Theta\frac{\partial V}{\partial \Theta} + P\frac{\partial V}{\partial P} = 0 \qquad (\Theta, P \text{ independent}), \tag{9}$$

$$\Theta\frac{\partial(P, V)}{\partial(\Theta, E)} - P\frac{\partial V}{\partial E} - 1 = 0 \qquad (\Theta, E \text{ independent}), \tag{10}$$

$$\frac{\partial \Theta}{\partial P} - \Theta\frac{\partial V}{\partial E} - P\frac{\partial(V, \Theta)}{\partial(E, P)} = 0 \qquad (P, E \text{ independent}). \tag{11}$$

A good way to proceed is as follows. First consider a general variable change and specialize later to all the particular cases. To rewrite (6) using new independent variables x and y, define $\overline{\Theta}(x, y)$ by

$$\Theta[E(x, y), V(x, y)] = \overline{\Theta}(x, y).$$

Recall the chain rule results

$$\overline{\Theta}_x = \Theta_E E_x + \Theta_V V_x, \qquad \overline{\Theta}_y = \Theta_E E_y + \Theta_V V_y.$$

Express Θ_E as the following quotient of Jacobians:

$$\Theta_E = \frac{\partial(\Theta, V)/(\partial(x, y)}{\partial(E, V)/\partial(x, y)}.$$

Find similar expressions for Θ_V and P_E. Proceed by successively identifying (x, y) with the various independent variable pairs of (7)–(11).

NOTE. Equations (7)–(11) have been written in the "usual" notation in which, to give an example, no distinction is made between the internal energy regarded as a function of P and V, or as a function of P and Θ.

‡2. Using the notation of (1), by chain rule express

$$\left[\frac{\partial^2 T}{\partial x^2} + \frac{\partial^2 T}{\partial y^2}\right]_{x = r\cos\theta,\, y = r\sin\theta}$$

in terms of $\tau(r, \theta)$ and its derivatives.

3. Exercise 2 requested the transformation of the Laplacian into polar coordinates. Make the transformation into spherical coordinates by expressing

$$\left[\frac{\partial^2 T}{\partial x^2} + \frac{\partial^2 T}{\partial y^2} + \frac{\partial^2 T}{\partial z^2}\right]_{x = \rho\sin\phi\cos\theta,\, y = \rho\sin\phi\sin\theta,\, z = \rho\cos\phi}$$

in terms of $\tau(\rho, \phi, \theta)$ and its derivatives, using one of the following possibilities.

‡(a) Do the problem in full generality (a rather long calculation).
(b) Assume that $\partial/\partial\phi \equiv 0$.
(c) Assume that $\partial/\partial\phi \equiv \partial/\partial\theta \equiv 0$.

Appendix 13.2 The Integral Mean Value Theorem

Theorem. Let R be a connected closed bounded region of three-dimensional space with volume V, $0 < V < \infty$. Let f be continuous on R and denote by m and M the maximum and minimum values of $f(\mathbf{x})$ for $\mathbf{x}$ in R. Then there exists a point $\tilde{\mathbf{x}}$ in R such that

$$\iiint_R f(\mathbf{x})\, d\tau = f(\tilde{\mathbf{x}}) \cdot V. \tag{1}$$

Proof. From the definition of the triple integral as a limit of sums, it follows that

$$\iiint_R m\,d\tau \le \iiint_R f(\mathbf{x})\,d\tau \le \iiint_R M\,d\tau,$$

so

$$mV \le \iiint_R f(\mathbf{x})\,d\tau \le MV \qquad \text{or} \qquad m \le V^{-1}\iiint_R f(\mathbf{x})\,d\tau \le M. \quad \text{(2a, b)}$$

Since a continuous function defined over a closed bounded region is bounded and takes on its upper and lower bounds, there exist points $\mathbf{x}^{(m)}$ and $\mathbf{x}^{(M)}$ in R that satisfy

$$f(\mathbf{x}^{(m)}) = m, \qquad f(\mathbf{x}^{(M)}) = M.$$

The hypothesis of connectedness assures the existence of a space curve $\mathbf{x} = \mathbf{x}(t)$ that joins $\mathbf{x}^{(m)}$ and $\mathbf{x}^{(M)}$ and lies within R:

$$\mathbf{x}(t^{(m)}) = \mathbf{x}^{(m)}, \quad \mathbf{x}(t^{(M)}) = \mathbf{x}^{(M)}; \qquad \mathbf{x}(t) \text{ in } R \text{ for } t^{(m)} \le t \le t^{(M)}.$$

Since $f[\mathbf{x}(t)]$ is a continuous function of the parameter t, it must take on every value between m and M, by the intermediate value theorem. In particular, from (2b), there exists a parameter value $\tilde{t}$ between $t^{(m)}$ and $t^{(M)}$ such that

$$f[\mathbf{x}(\tilde{t})] = V^{-1}\iiint_R f(\mathbf{x})\,d\tau. \tag{3}$$

The conclusion (1) follows immediately from (3) and the definition $\tilde{\mathbf{x}} = \mathbf{x}(\tilde{t})$. □

REMARKS. The integral mean value theorem holds for n-dimensional connected closed bounded regions R in m-dimensional space; $n \le m$, $m = 1, 2, \ldots$. Only slight changes in the wording of the proof are necessary. If $\mathbf{F}(\mathbf{x})$ is a vector function of $\mathbf{x}$, then (1) obviously holds for each component F_i of $\mathbf{F}$. There will generally be a different point $\tilde{\mathbf{x}}$ for each component.

Appendix 13.3 Similar Regions

In this appendix we shall sketch the derivation of certain useful properties of "star-shaped" regions. We then generalize to unions of such regions.

Definition 1. A closed region of space R is called **star-shaped** if there exists an interior point of R (called O) that can be connected to each point of R by a straight line lying within R. Figure 13.3 illustrates a planar star-shaped region and a possible point O.

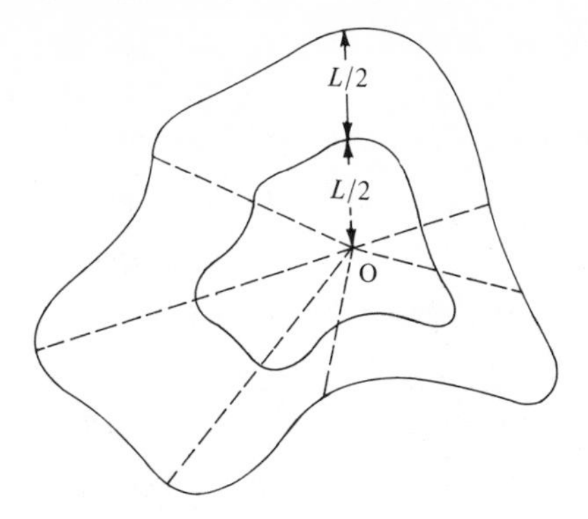

FIGURE 13.3. *Planar star-shaped region. The inner region is similar to the outer.*

If we choose O as the origin, then R can be described using spherical coordinates (ρ, ϕ, θ) by the relations

$$0 \leq \rho \leq f(\phi, \theta), \qquad 0 \leq \phi \leq \pi, \quad 0 \leq \theta \leq 2\pi.$$

(The coordinate ρ merely provides the distance from point O to a given point in R, along the line joining the two points.) We shall assume that f is continuous and piecewise continuously differentiable with $f(\phi, 0) = f(\phi, 2\pi)$.

Definition 2. Two star-shaped regions $R^{(1)}$ and $R^{(2)}$ are called **similar** if, after possible rotation and translation, they can be described by the equations

$$0 \leq \rho \leq f^{(i)}(\phi, \theta), \qquad i = 1, 2, \tag{1}$$

where $f^{(2)} = Lf^{(1)}$ for a positive constant L (Figure 13.3).

We now show that the volumes $V[R^{(i)}]$ and the surface areas $S[R^{(i)}]$ of the similar star-shaped regions $R^{(i)}$ are in the ratios L^3 and L^2, respectively.

Theorem 1. $V[R^{(2)}] = L^3 V[R^{(1)}]$.

Proof. Using the formula for volume in spherical coordinates, we have

$$\begin{aligned} V[R^{(2)}] &= \int_0^{2\pi} \int_0^{\pi} \int_0^{f^{(2)}(\phi, \theta)} \rho^2 \sin\phi \, d\rho \, d\phi \, d\theta \\ &= \tfrac{1}{3} \int_0^{2\pi} \int_0^{\pi} [f^{(2)}(\phi, \theta)]^3 \sin\phi \, d\phi \, d\theta. \end{aligned} \tag{2}$$

The result follows at once from the relation $f^{(2)} = Lf^{(1)}$. ∎

Theorem 2. $S[R^{(2)}] = L^2 S[R^{(1)}]$.

Proof. Using (1), if a point $\mathbf{x} = (x_1, x_2, x_3)$ is on the boundary of $R^{(i)}$, it can be described by the relation $\mathbf{x} = \mathbf{x}^{(i)}$, where

$$\begin{aligned} x_1^{(i)} &= f^{(i)}(\phi, \theta) \sin\phi \cos\theta, \\ x_2^{(i)} &= f^{(i)}(\phi, \theta) \sin\phi \sin\theta, \\ x_3^{(i)} &= f^{(i)}(\phi, \theta) \cos\phi. \end{aligned} \tag{3}$$

We regard (3) as a parametric representation of the bounding surfaces and use the appropriate formula for surface area to write*

$$S[R^{(i)}] = \int_0^{2\pi} \int_0^{\pi} \left| \frac{\partial \mathbf{x}^{(i)}}{\partial \phi} \wedge \frac{\partial \mathbf{x}^{(i)}}{\partial \theta} \right| d\phi \, d\theta. \tag{4}$$

The result follows from the relation $\mathbf{x}^{(2)} = L\mathbf{x}^{(1)}$. ☐

Another result that will prove useful involves a sequence of mutually similar star-shaped regions $R^{(n)}$.

$$R^{(n)}: \rho \le f^{(n)}(\phi, \theta), \quad f^{(n)} = L_n f^{(1)}; \qquad n = 1, 2, 3, \ldots. \tag{5a, b}$$

Let D_n be the maximum distance between two points in $R^{(n)}$.

Theorem 3

$$\text{As } n \to \infty, \qquad V[R^{(n)}] \to 0 \quad \text{if and only if } D_n \to 0. \tag{6}$$

Proof. Using Theorem 1, $V[R^{(n)}] = L_n^3 V[R^{(1)}]$, so $V[R^{(n)}] \to 0$ if and only if $L_n \to 0$. By (5b), $L_n \to 0$ if and only if $f^{(n)} \to 0$. This implies that

$$V[R^{(n)}] \to 0 \qquad \text{if and only if } \rho_{\max}^{(n)} \to 0, \tag{7}$$

where

$$\rho_{\max}^{(n)} = \max_{\substack{0 \le \phi \le \pi \\ 0 \le \theta \le 2\pi}} f^{(n)}(\phi, \theta).$$

[The existence of $\rho_{\max}^{(n)}$ is guaranteed, since $f^{(n)}(\phi, \theta)$ is continuous in the closed bounded region $0 \le \phi \le \pi$, $0 \le \theta \le 2\pi$.] It is not difficult to show (Exercise 2) that

$$\rho_{\max}^{(n)} \le D_n \qquad \text{and} \qquad D_n \le 2\rho_{\max}^{(n)}. \tag{8a, b}$$

The result follows from (7) and (8). ☐

Theorem 4. The three previous theorems remain true for regions that are the union of a finite number of star-shaped regions.

Proof. Left to the reader.

* The letter x is employed so frequently that, to avoid confusion, we use a wedge to denote the cross product.

EXERCISES

1. (a) Use polar coordinates to define *planar* star-shaped regions.
(b) Show that "similar triangles" are similar in the sense used here.
(c) State and prove versions of the above theorems which are appropriate for planar star-shaped regions.

‡2. (a) Prove (8b).
(b) Prove (8a).

3. Show that for the family of mutually similar regions $R^{(n)}$,

$$V[R^{(n)}] = \lambda_R L_n^3, \qquad S[R^{(n)}] = \mu_R L_n^2,$$

where λ_R and μ_R are independent of n. (This result is used in Section 14.2.)

4. Formulate and prove for regions that are the unions of star-shaped regions:
(a) Theorem 1.
(b) Theorem 2 (note that corresponding regions must be "similarly located," in a sense that should be explained).
(c) Theorem 3.

CHAPTER 14
Field Equations of Continuum Mechanics

IN THIS chapter we shall provide careful derivations of the **field equations** that restrict the behavior of any continuous medium. These equations express conservation of mass, and balance of linear momentum, of angular momentum, and of energy. In addition there are Gibbs relations, which couple heat flow with entropy and temperature, and couple work done with pressure and volume. Finally, there is an entropy inequality.

Only "ordinary" mechanical effects will be considered—electrical, magnetic, and chemical effects will be ignored.

In the concluding section of the chapter we give a short discussion of a topic to be taken up much more fully in the remainder of the book, the different constitutive equations that distinguish different media. It is important to bear in mind that such equations (indeed, all physically meaningful equations) must not differ essentially if a new coordinate system is employed. This matter is discussed briefly, although it is only taken up fully in II. We conclude with a few additional remarks on the validity of the continuum model.

14.1 Conservation of Mass

Derivation of the equations governing continuum mechanics is not an easy matter. Hard study is necessary before real understanding is gained. Although it may initially be confusing to contemplate more than one approach to a topic, nevertheless we shall present four ways to derive the equation of mass conservation. Consideration of the advantages and deficiencies of all four approaches will lead the reader to a deeper understanding of the issues involved in formulating equations that describe natural processes. We aim not to give an excuse for using a "well-known" equation but rather to attempt a derivation that will satisfy a critical person. We hope thereby to aid those readers who, sometime in the future, might attempt to derive equations governing phenomena not previously formulated in mathematical terms.

The first method begins with a statement of mass conservation for an arbitrary material region. The central technical content of the derivation is a computation of the time derivative of an integral over a moving *material* region. Ultimately, a volume integral over an arbitrary region is shown to vanish. The desired differential equation follows by the Dubois–Reymond lemma.

The second method, an application of the divergence theorem, is based on a statement of mass conservation for an arbitrary *spatial* region. Both the third and fourth methods are based on invoking mass conservation for a fixed rectangular box. In the third approach the box is infinitesimal; in the fourth, suggestive but dangerous manipulation of infinitesimals is avoided by means of the integral mean value theorem.

INTEGRAL METHOD: ARBITRARY MATERIAL REGION

Section 13.4 gave a precise characterization of the idea that a specified portion of material occupies the region $R(t)$ at time t. *That the mass of a portion of material does not vary as time increases is one way to state what is meant by conservation of mass.* Thus we hypothesize that

$$\frac{d}{dt}\iiint_{R(t)} \rho(x_1, x_2, x_3, t)\, dx_1\, dx_2\, dx_3 = 0, \qquad R(t) \text{ arbitrary}, \tag{1}$$

where $\rho(\mathbf{x}, t)$ denotes the density at point $\mathbf{x}$, time t.

Since (1) holds over an arbitrary material region $R(t)$, we shall be able to transform it into the more tractable form of a differential equation by means of the **Dubois–Reymond lemma**, which follows.

Suppose that

$$\iiint_{R} f(\mathbf{x})\, dx_1\, dx_2\, dx_3 = 0 \tag{2}$$

for every region R contained in a domain D. If $f(\mathbf{x})$ is continuous for $\mathbf{x}$ in D, then $f(\mathbf{x}) \equiv 0$ for $\mathbf{x}$ in D.

The simple proof is given in Section 4.1.

To take the derivative inside the integral in (1), as required for (2), we shall transform to initial coordinates $\mathbf{A}$ via the material description $\mathbf{x} = \mathbf{x}(\mathbf{A}, t)$. The region of integration will thereby no longer depend on time. Thus, denoting the integral in (1) by $I(t)$ (the integral is a function only of time), we have

$$\frac{dI}{dt} \equiv \frac{d}{dt}\iiint_{R(t)} \rho(\mathbf{x}, t)\, dx_1\, dx_2\, dx_3 = \frac{d}{dt}\iiint_{R(0)} \rho[\mathbf{x}(\mathbf{A}, t), t] J(\mathbf{A}, t)\, dA_1\, dA_2\, dA_3. \tag{3}$$

Here $R(0)$ is the region occupied by the material when $t = 0$. We have changed the element of volume by using the Jacobian rule.

We can write (3) more simply as

$$\frac{dI}{dt} = \frac{d}{dt}\iiint_{R(0)} \delta(\mathbf{A}, t) J(\mathbf{A}, t)\, d\tau$$

by employing the material density $\delta(\mathbf{A}, t) \equiv \rho[x(\mathbf{A}, t), t]$ of (13.3.3). We now take the time differential inside the integral and use partial derivative notation to indicate that $\mathbf{A}$ is held constant during the differentiations with respect to t:

$$\frac{dI}{dt} = \iiint_{R(0)} \left[\frac{\partial \delta(\mathbf{A}, t)}{\partial t} J(\mathbf{A}, t) + \delta(\mathbf{A}, t) \frac{\partial J(\mathbf{A}, t)}{\partial t}\right] dA_1\, dA_2\, dA_3. \tag{4}$$

Using the Euler expansion formula (13.4.11), we obtain

$$\frac{dI}{dt} = \iiint_{R(0)} \left\{\frac{\partial \delta(\mathbf{A}, t)}{\partial t} + \delta(\mathbf{A}, t)(\nabla \cdot \mathbf{v})\Big|_{\mathbf{x}=\mathbf{x}(\mathbf{A}, t)}\right\} J(\mathbf{A}, t)\, dA_1\, dA_2\, dA_3.$$

Making the variable change $\mathbf{A} = \mathbf{A}(\mathbf{x}, t)$ and using (13.3.7), (13.3.3), and the relation between $\mathbf{x}$ and $\mathbf{A}$ given in (13.2.2b), we then obtain

$$\frac{dI}{dt} = \iiint_{R(t)} \left\{\frac{D\rho}{Dt}(\mathbf{x}, t) + \rho(\mathbf{x}, t)[\nabla \cdot \mathbf{v}(\mathbf{x}, t)]\right\} dx_1\, dx_2\, dx_3. \tag{5}$$

But $dI/dt = 0$, by (1). From (5) we thereby conclude that

$$\boxed{\frac{D\rho}{Dt} + \rho \nabla \cdot \mathbf{v} = 0,} \tag{6}$$

where we have assumed that the integrand in (5) is continuous and therefore have used the Dubois–Reymond lemma. Equation (6), often called the **continuity equation**, is the desired differential equation form of **mass conservation**. Since $D\rho/Dt = \partial\rho/\partial t + \mathbf{v} \cdot \nabla\rho$, an alternative form of (6) is

$$\boxed{\frac{\partial \rho}{\partial t} + \nabla \cdot (\rho \mathbf{v}) = 0.} \tag{7}$$

We have seen how the time derivative of an integral over the moving material region $R(t)$ can be calculated by the device of shifting from $R(t)$ to $R(0)$ and back. It is appropriate here to present two useful formulas that essentially record the results of using this device in commonly encountered contexts. Proofs are left to the reader (Exercise 1).

For a sufficiently smooth function F and material region $R(t)$, assuming conservation of mass, it is not difficult to show that

$$\frac{d}{dt} \iiint_{R(t)} F\rho\, d\tau = \iiint_{R(t)} \frac{DF}{Dt} \rho\, d\tau. \tag{8}$$

From a formal point of view the above result can be regarded as permitting one to take d/dt inside the integral provided that one recognizes the "moving" variables of integration by changing this operator to the material derivative D/Dt. If this were permitted, we would have

$$\frac{d}{dt} \iiint_{R(t)} F\rho\, d\tau = \iiint_{R(t)} \frac{DF}{Dt} \rho\, d\tau + \iiint_{R(t)} F \frac{D}{Dt}(\rho\, d\tau). \tag{9}$$

Comparison of (8) with (9) shows that the validity of the well-established former equation requires $D(\rho\, d\tau)/Dt = 0$. But this is reasonable, as $D(\rho\, d\tau)/Dt$ can be interpreted as giving the change with time of the mass of a moving material element.

A result closely related to (8) is the following. For any sufficiently smooth function G and material region $R(t)$,

$$\frac{d}{dt}\iiint_{R(t)} G\, d\tau = \iiint_{R(t)} \frac{\partial G}{\partial t}\, d\tau + \oint\!\!\oint_{\partial R(t)} G\mathbf{v}\cdot\mathbf{n}\, d\sigma.^* \tag{10}$$

Here, as usual, $\mathbf{n}$ is the unit exterior normal to R, $d\tau$ is a volume element, $d\sigma$ is a surface element, and ∂R denotes the boundary of R. Both (8) and (10) are valid if F and G are replaced by vector fields $\mathbf{F}$ and $\mathbf{G}$ [Exercise 1(d)].

Equation (10) is the **Reynolds transport theorem**. Its interpretation is illuminating. Let $\bar{R}_t$ be a region *fixed in space* that coincides with $R(t)$ at time t. Certainly,

$$\frac{d}{dt}\iiint_{\bar{R}_t} G\, d\tau = \iiint_{\bar{R}_t} \frac{\partial G}{\partial t}\, d\tau \equiv \iiint_{R(t)} \frac{\partial G}{\partial t}\, d\tau. \tag{11}$$

(A partial derivative notation must be used under the integral sign, to show that the integration variables are held constant during the t differentiation.) But in computing the left side of (10), the contribution of (11) must be supplemented by

$$\oint\!\!\oint_{\partial R(t)} G\mathbf{v}\cdot\mathbf{n}\, d\sigma \equiv \oint\!\!\oint_{\partial \bar{R}_t} G\mathbf{v}\cdot\mathbf{n}\, d\sigma. \tag{12}$$

We assert that the surface integral (12) can be regarded as giving the rate at which a substance of concentration G units per unit volume is being carried by the material as it flows through the boundary of $\bar{R}_t$. To see this, observe that $\mathbf{v}\cdot\mathbf{n}$ provides the component of the velocity $\mathbf{v}$ that points outward from the boundary. Suppose, for the moment, that $\mathbf{v}$ is independent of time. In U units of time, a column of material $U\mathbf{v}\cdot\mathbf{n}$ units long and $d\sigma$ in cross-sectional area passes through surface element of area $d\sigma$. (Tangential flow does not cause material to pass through the element.) As a consequence, material passes through the element at the rate $\mathbf{v}\cdot\mathbf{n}\, d\sigma$ volume units per unit time. Even if $\mathbf{v}$ depends on time, $\mathbf{v}\cdot\mathbf{n}$ gives the instantaneous **volume flux density** through the surface element. (Flux density means flow rate per unit area.) If the material contains a substance of concentration G units per unit volume, then the flux of this substance through the surface element is $G\mathbf{v}\cdot\mathbf{n}\, d\sigma$. The total flux of G is given by (12), a result that we shall often use.

Putting the pieces of our discussion of (10) together, we see that *the derivative with respect to time of* $\iiint_{R(t)} G\, d\tau$ *is equal to the derivative with*

* As in (10), we frequently employ the traditional notation to indicate that an integral is taken over a *closed* surface. But we have not attempted to be completely consistent in this practice.

respect to time of the integral of G over the stationary region $\bar{R}_t$ that instantaneously coincides with the material region $R(t)$, plus the flux of G through the boundary of $\bar{R}_t$.

INTEGRAL METHOD: ARBITRARY SPATIAL REGION

We start with the statement that conservation of mass requires that *for a region fixed in space the rate of increase of the mass contained in this region must equal the net mass flux into the region.* Only flow of mass across the boundary can cause a change in the mass of the material contained in a fixed region, since mass is neither created nor destroyed. In mathematical terms, then, the italicized statement is

$$\frac{d}{dt}\iiint_R \rho(\mathbf{x}, t)\, d\tau = -\oiint_{\partial R} \mathbf{n}(\mathbf{x}, t)\cdot\mathbf{v}(\mathbf{x}, t)\rho(\mathbf{x}, t)\, d\sigma. \tag{13}$$

Here R is an arbitrary region, fixed in space; ∂R is its boundary; and $\mathbf{n}$ is a unit exterior normal to ∂R. Using the divergence theorem, we can write (13) as

$$\iiint_R \left\{\frac{\partial\rho(\mathbf{x}, t)}{\partial t} + \nabla\cdot[\mathbf{v}(\mathbf{x}, t)\rho(\mathbf{x}, t)]\right\} d\tau = 0. \tag{14}$$

(Partial derivative notation is used in the first term, because the variable of integration, $\mathbf{x}$, must be kept fixed during the time differentiation.) Since R is arbitrary, if the integrand is continuous, then (7) follows by the Dubois–Reymond lemma.

SMALL BOX METHOD

The ideas we wish to illustrate are more clearly seen if we confine ourselves to two-dimensional situations, with velocity components u and v in the directions of increasing x and increasing y, respectively. We shall assume that there is no velocity component in the z direction, and that u, v, and the density ρ are independent of z.

We again apply the idea that the rate at which mass in a region increases must equal the mass influx across its boundary. In contrast to the previous method, here we consider a special region. This is a "small box" of unit thickness in the z direction but of small dimensions Δx and Δy in the x and y directions (Figure 14.1). The box is an imaginary one; fluid freely passes through faces F_1, F_2, F_3, and F_4 shown in Fig. 14.1. Because of the assumption of two-dimensionality, no fluid passes through the faces that are parallel to the xy plane.

We shall reason informally. The reader is asked to abandon skepticism for the moment. The mass flux across face F_1 *into* the box is $(\rho u)_0\,\Delta y$, where the subscript indicates that ρu is to be evaluated at (x_0, y_0). (The z coordinate can be ignored, since there is no variation in the z direction.) Since Δy is

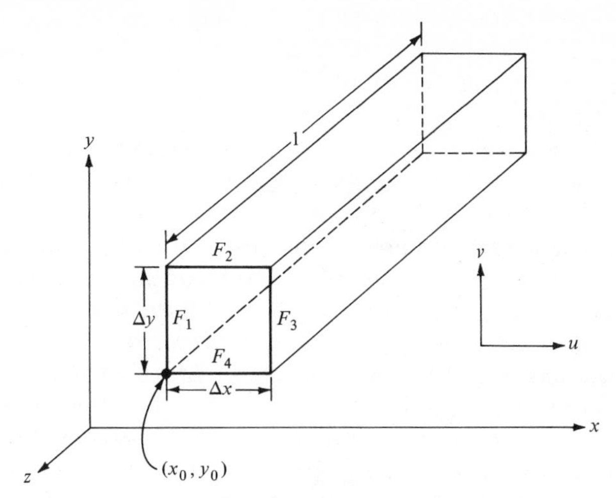

FIGURE 14.1. *Imaginary box used for the derivation of the continuity equation in cartesian coordinates.*

very small, we can probably ignore the fact that ρu is not strictly constant over F_1.

The mass flux *into* the box across F_3 is

$$-\left\{(\rho u)_0 + \left[\frac{\partial(\rho u)}{\partial x}\right]_0 \Delta x\right\} \Delta y. \tag{15}$$

The second term in the square brackets is a correction to account for the fact that ρu is to be evaluated on F_3, not F_1. Since $\partial(\rho u)/\partial x$ gives the change in ρu per unit length, we should get the required correction by multiplying by Δx, the actual distance between F_1 and F_3. As Δx is small, it is probably permissible to evaluate $\partial(\rho u)/\partial x$ at $x = x_0$. The negative sign in (15) comes from the fact that at F_3 a positive u velocity component means net outward mass flux.

Adding the contributions from F_1 and F_3, we find a net mass influx of

$$-\left[\frac{\partial(\rho u)}{\partial x}\right]_0 \Delta x\, \Delta y. \tag{16}$$

Similarly, or by symmetry, the net mass influx across F_2 and F_4 is

$$-\left[\frac{\partial(\rho v)}{\partial y}\right]_0 \Delta x\, \Delta y. \tag{17}$$

The sum of (16) and (17) must equal the net rate of mass increase as expressed by means of a change in density, i.e.,

$$\left\{-\left[\frac{\partial(\rho u)}{\partial x}\right]_0 - \left[\frac{\partial(\rho v)}{\partial y}\right]_0\right\} \Delta x\, \Delta y = \left(\frac{\partial \rho}{\partial t}\right)_0 \Delta x\, \Delta y. \tag{18}$$

Since (x_0, y_0) was arbitrary, we have at any point

$$\frac{\partial \rho}{\partial t} + \frac{\partial}{\partial x}(\rho u) + \frac{\partial}{\partial y}(\rho v) = 0. \tag{19}$$

Equation (19) is the two-dimensional version of the continuity equation (7). Extension of the argument just given to three dimensions should present no difficulty.

This type of derivation was used by the early masters of continuum mechanics, yet a critical person cannot be satisfied with it. These early masters were certainly geniuses, and a genius might be defined as a person who gets the right answer for inadequate reasons, or the wrong reasons. Not having a genius's almost unerring instinct for truth, and the self-confidence that goes with it, the ordinary person is filled with doubt. Why can we ignore the variation of ρu across F_1 when it is essential to take into account the variation of ρu as one passes from F_1 to F_3? How can we be sure that we are not neglecting something important when we evaluate $\partial\rho/\partial t$ at (x_0, y_0)? What confidence can we have in our ability to discern negligible terms after learning that in polar coordinates one cannot neglect Δr in the expression $(r_0 + \Delta r)\,\Delta\theta$ for the length of one side of the appropriate box? (See Exercise 10.) Our final derivation retains the spirit of the previous method but overcomes our doubts by a more rigorous procedure.

LARGE BOX METHOD

We still consider two-dimensional motion through a box such as that in Figure 14.1, but we no longer require Δx and Δy to be "very small." In fact, to avoid unwarranted simplifications, it is best to think of Δx and Δy as large.

Consider again the mass flux inward through F_1. At time t this is

$$\int_{y_0}^{y_0+\Delta y} \rho(x_0, y, t)u(x_0, y, t)\,dy, \tag{20}$$

where we have used the fact that $x = x_0$ on F_1. Similarly, the mass flux inward through F_3 is

$$-\int_{y_0}^{y_0+\Delta y} \rho(x_0 + \Delta x, y, t)u(x_0 + \Delta x, y, t)\,dy.$$

The net inward mass flux contribution from F_1 and F_3 is, using the integral mean value theorem,

$$\int_{y_0}^{y_0+\Delta y} M(y)\,dy = M(y_1)\int_{y_0}^{y_0+\Delta y} dy = \Delta y M(y_1). \tag{21}$$

Here

$$M(y) \equiv \rho(x_0, y, t)u(x_0, y, t) - \rho(x_0 + \Delta x, y, t)u(x_0 + \Delta x, y, t), \tag{22}$$

and y_1 is a constant that satisfies $y_0 \le y_1 \le y_0 + \Delta y$. In employing the notation $M(y)$ above, we have not explicitly indicated the dependence on x_0 and t in order to emphasize the y dependence relevant to application of the integral mean value theorem.

In like manner, the mass flux inward through faces F_4 and F_2 is

$$\int_{x_0}^{x_0+\Delta x} [\rho(x, y_0, t)v(x, y_0, t) - \rho(x, y_0 + \Delta y, t)v(x, y_0 + \Delta y, t)]\, dx$$
$$= \Delta x[\rho(x_1, y_0, t)v(x_1, y_0, t) - \rho(x_1, y_0 + \Delta y, t)v(x_1, y_0 + \Delta y, t)]. \quad (23)$$

On the other hand, the rate of change of mass contained in the "big box" is

$$\frac{d}{dt}\int_{y_0}^{y_0+\Delta y}\int_{x_0}^{x_0+\Delta x} \rho(x, y, t)\, dx\, dy$$
$$= \int_{y_0}^{y_0+\Delta y}\int_{x_0}^{x_0+\Delta x} \frac{\partial\rho}{\partial t}(x, y, t)\, dx\, dy = \frac{\partial\rho}{\partial t}(x_2, y_2, t)\, \Delta x\, \Delta y. \quad (24)$$

The second equality in (24) follows from the integral mean value theorem for areas, so x_2 and y_2 are constants that satisfy

$$x_0 \le x_2 \le x_0 + \Delta x, \qquad y_0 \le y_2 \le y_0 + \Delta y. \quad (25)$$

Contemplation of the factor $\Delta x\, \Delta y$ in (24) makes one wish that there were a similar factor in (22) and (23). This can be arranged by applying the derivative mean value theorem, which states that if f' exists in an interval containing $[y_0, y_0 + \Delta y]$, then

$$f(y_0 + \Delta y) - f(y_0) = f'(\xi)\, \Delta y, \quad (26)$$

for some ξ satisfying

$$y_0 < \xi < y_0 + \Delta y.$$

To apply this to (23), we identify $f(y_0)$ with $\rho(x_1, y_0, t)v(x_1, y_0, t)$. The right hand side of (23) can then be written

$$-\Delta x\, \Delta y\, \frac{\partial[\rho(x_1, y_3, t)v(x_1, y_3, t)]}{\partial y}, \quad (27)$$

where $y_0 < y_3 < y_0 + \Delta y$. Similarly, from (22) the contribution from F_1 and F_3 can be written

$$-\Delta x\, \Delta y\, \frac{\partial[\rho(x_3, y_1, t)u(x_3, y_1, t)]}{\partial x}, \quad (28)$$

where $x_0 < x_3 < x_0 + \Delta x$. Conservation of mass requires that the sum of the expressions in (27) and (28) equals the right-hand side of (24). Canceling $\Delta x\, \Delta y$, we find that for any Δx and Δy, however large or small,

$$-\left.\frac{\partial(\rho u)}{\partial x}\right|_{\substack{x=x_3\\ y=y_1}} - \left.\frac{\partial(\rho v)}{\partial y}\right|_{\substack{x=x_1\\ y=y_3}} = \left.\frac{\partial\rho}{\partial t}\right|_{\substack{x=x_2\\ y=y_2}} \quad (29)$$

for some x_i and y_i satisfying

$$x_0 \le x_i \le x_0 + \Delta x, \qquad y_0 \le y_i \le y_0 + \Delta y, \qquad i = 1, 2, 3.$$

The desired equation (19) now follows for continuous u, v, and ρ by taking equations (29) in the limit as $\Delta x \to 0$, $\Delta y \to 0$.

In performing our derivations, we require differing degrees of smoothness in order that we can be guaranteed that various theorems hold. It is inappropriate to ask whether the velocity components, density, etc., "really" have the required number of derivatives. What we actually would like to know is whether a model in which, say, thrice-differentiable velocity components are assumed can accurately describe the phenomena in which we are interested. Only experience in solving problems can give the answer, so for the present, we merely need to keep in mind that our derivations require "sufficient" smoothness. Looking ahead, we can remark that experience indicates that no smoothness difficulties arise for most solutions, but that in certain very important problems discontinuous solutions definitely must be allowed and appropriate modifications made in our derivations.

EXERCISES

1. (a) Derive (8) from first principles.
(b) Derive (10) from first principles.
(c) Show that (8) follows from (10) and vice versa (assuming mass conservation).
(d) Show that (8) and (10) remain true if F and G are replaced by vector fields $\mathbf{F}$ and $\mathbf{G}$.

2. Carry through the following alternative approach to the manipulations of the large box method. Show that (21) can be written as

$$-\int_{y_0}^{y_0+\Delta y}\int_{x_0}^{x_0+\Delta x} \frac{\partial[\rho(x, y, t)u(x, y, t)]}{\partial x}\, dx\, dy.$$

Combine with a similar expression for the first line of (23), and with the left-hand side of (24), to obtain a double integral that vanishes over the arbitrary rectangle of Figure 14.1. Deduce (19).

3. Here is another approach to conservation of mass that is analytically neat but somewhat unsatisfying physically.
(a) Show that in one dimension mass conservation requires that

$$(\rho u)(x_0, y_0, t) - (\rho u)(x_0 + \xi, y_0, t) = \int_{x_0}^{x_0+\xi} \rho_t\, dx.$$

Differentiate with respect to ξ. Obtain the final equation by setting $\xi = 0$ or by noting that $x_0 + \xi$ and y_0 are arbitrary.
(b) Repeat for the two-dimensional case.

4. Derive the three-dimensional version of (19) using the following methods.
(a) The small box method.
(b) The large box method.

5. According to a certain cosmological theory, matter is continually being created in the universe. Assuming that the material in the universe can be taken as a continuum, write a modified mass conservation equation.

6. (a) Interpret

$$\iiint_{R(0)} \delta(\mathbf{A}, 0)\, dA_1\, dA_2\, dA_3 = \iiint_{R(t)} \rho(\mathbf{x}, t)\, dx_1\, dx_2\, dx_3 \qquad (30)$$

as a form of the mass conservation requirement. Deduce the **mass conservation equation in material coordinates.**

$$\rho[\mathbf{x}(\mathbf{A}, t), t]J(\mathbf{A}, t) = \delta(\mathbf{A}, 0) \qquad \text{or} \qquad \delta(\mathbf{A}, t)J(\mathbf{A}, t) = \delta(\mathbf{A}, 0). \qquad (31)$$

(b) Show that for an incompressible motion $J(\mathbf{A}, t) = 1$.
(c) Show that $J(\mathbf{A}, 0) \equiv 1$.

7. Derive the spatial continuity equation from the material, and vice versa, by direct change of variables. Start with

$$\frac{\partial}{\partial t}[\delta(\mathbf{A}, t)J(\mathbf{A}, t)] = 0.$$

Cartesian tensors, the subject of Chapter 2 of II, provide an efficient framework for formulating physical laws that must be invariant in some sense, regardless of which cartesian coordinate system is used. Rules for changing from one cartesian system to another are formulated in this subject. General tensors permit manipulation of general coordinate changes. The next four exercises give a hint of what is involved.

8. Show by direct calculation that the two-dimensional mass conservation equation (19) has the same form in variables appropriate to new axes rotated counterclockwise through an angle θ with respect to the original axes. Follow these steps.
(a) Denote the original unit vectors by $\mathbf{i}$ and $\mathbf{j}$ and the new unit vectors by $\mathbf{i}'$ and $\mathbf{j}'$. Show that

$$\mathbf{i} \cdot \mathbf{i}' = \cos\theta, \qquad \mathbf{i} \cdot \mathbf{j}' = \sin\theta, \qquad \mathbf{j} \cdot \mathbf{i}' = \sin\theta, \qquad \mathbf{j} \cdot \mathbf{j}' = \cos\theta. \qquad (32)$$

(b) Let

$$\mathbf{v}(x, y) = u(x, y)\mathbf{i} + v(x, y)\mathbf{j} = u'(x, y)\mathbf{i}' + v'(x, y)\mathbf{j}'. \qquad (33)$$

Thus $u(x, y)$ and $u'(x, y)$ are the velocity components at (x, y) in the directions of $\mathbf{i}$ and $\mathbf{i}'$, respectively. Show that

$$u(x, y) = u'(x, y) \cos \theta - v'(x, y) \sin \theta, \tag{34a}$$

$$v(x, y) = u'(x, y) \sin \theta + v'(x, y) \cos \theta, \tag{34b}$$

$$u'(x, y) = u(x, y) \cos \theta + v(x, y) \sin \theta, \tag{34c}$$

$$v'(x, y) = -u(x, y) \sin \theta + v(x, y) \cos \theta. \tag{34d}$$

(c) Let (x', y') be the coordinates in the primed system of a point whose coordinates in the unprimed system are (x, y). Show that

$$x = x' \cos \theta - y' \sin \theta, \qquad y = x' \sin \theta + y' \cos \theta, \tag{35a, b}$$

$$x' = x \cos \theta + y \sin \theta, \qquad y' = -x \sin \theta + y \cos \theta. \tag{35c, d}$$

(d) Use a bar to denote the use of primed coordinates so that, for example,

$$u(x, y) = \bar{u}[x'(x, y), y'(x, y)],$$

$$\bar{u}(x', y') = u[x(x', y'), y(x', y')].$$

Employ the chain rule to establish that (19) implies

$$\frac{\partial \bar{\rho}}{\partial t} + \frac{\partial (\bar{\rho} \bar{u})}{\partial x'} + \frac{\partial (\bar{\rho} \bar{v})}{\partial y'} = 0, \tag{36}$$

where $\bar{\rho}(x', y')$ gives the density in primed coordinates.

9. Consider a two-dimensional flow ($w = \partial/\partial z = 0$) of a compressible fluid of density $\rho(x, y)$. By passing from the known cartesian equation, using changes of variables and the chain rule, show that in polar coordinates

$$\frac{\partial R}{\partial t} + \frac{1}{r} \frac{\partial}{\partial r} (rUR) + \frac{1}{r} \frac{\partial}{\partial \theta} (VR) = 0. \tag{37}$$

Here

$$x = r \cos \theta, \qquad y = r \sin \theta, \qquad R(r, \theta) = \rho(r \cos \theta, r \sin \theta),$$

and $U(r, \theta)$ and $V(r, \theta)$ are the velocities in the directions of r increasing and θ increasing, respectively. If you need a hint, refer to Appendix 15.2.

10. Derive (37) by direct application of the box argument to the element of area pictured in Figure 14.2. (After the derivation has been completed, note that the difference in length between the outer and inner curved boundaries can be ignored in taking the area of the box but not in computing the flux across these boundaries.)

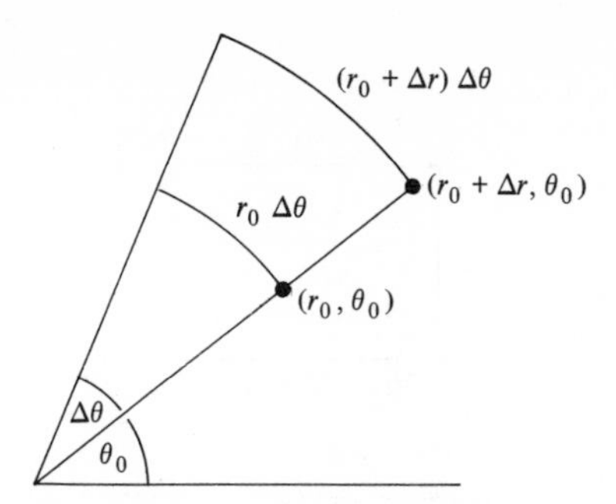

FIGURE 14.2. *Front view of an imaginary box used for the derivation of the continuity equation in polar coordinates. (The box should be regarded as extending a unit distance into the paper.)*

11. By considering a moving point to have coordinates $[x(t), y(t)]$ or alternatively $[r(t), \theta(t)]$, derive (34a) and (34b) from $x = r\cos\theta$, $y = r\sin\theta$.

12. The object of this exercise is to explore how the concepts of continuum and conservation can be applied to stellar dynamics. These concepts are applied to the average mass distribution function Ψ. (See the discussion in Section 13.1.) What we wish to do is to derive the fundamental conservation equation (1.2.4). No knowledge of Section 1.2 is required to do the exercise.

For simplicity we shall restrict consideration to a function that depends on spatial coordinate x, the corresponding velocity component u, and the time t. Thus $\Psi(x, u, t)$ is a smooth function with the property that to sufficient accuracy

$$\int_{u_1}^{u_2}\int_{x_1}^{x_2} \Psi(x, u, t)\, dx\, du$$

gives the mass of material which is located between x_1 and x_2, and which has a velocity between u_1 and u_2.

(a) Consider the "big box" in (x, u) space pictured in Figure 14.3. Explain fully what is meant by "the mass of the fluid occupying the box." That is, what can one say about the position and velocity of such fluid? This is an easy question.

(b) Write a double integral giving the mass of fluid in the box at time t.

(c) Show that the time rate of change of (b) is $\Psi_t(x^*, u^*, t)\,\Delta x\,\Delta u$, where

$$x_0 \le x^* \le x_0 + \Delta x, \qquad u_0 \le u^* \le u_0 + \Delta u.$$

Remember that u is an independent variable.

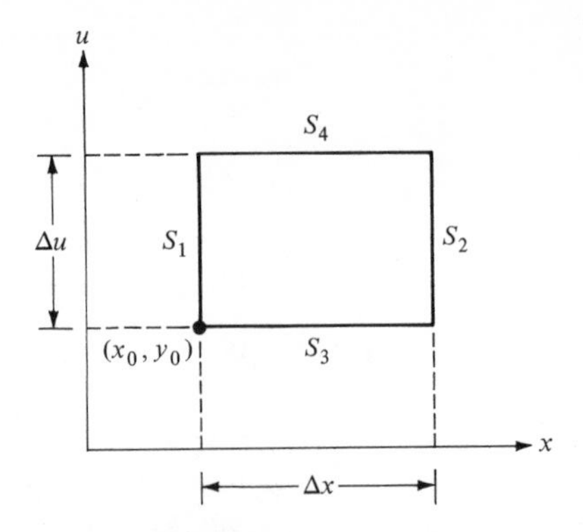

FIGURE 14.3. *Imaginary box used to derive an equation for the distribution function* $\Psi(x, u, t)$.

(d) What is meant by "the net amount of fluid mass passing inward through the side labeled S_1 in the time interval $(t, t + \Delta t)$"?

(e) Explain why (d) is given by

$$\int_{u_0}^{u_0+\Delta u} \left[\int_{x_0 - u\,\Delta t}^{x_0} \Psi(x, u, t)\, dx \right] du.$$

(f) From (e) show that the rate of mass flow inward through side S_1 at time t is

$$\int_{u_0}^{u_0+\Delta u} u\Psi(x_0, u, t)\, du.$$

(g) Derive a similar expression valid for side S_2.

(h) Combine (f) and (g) to form an expression proportional to $\Delta x\, \Delta u$.

(i) Same as (d), only substitute S_3 for S_1.

(j) Explain why (i) is given by

$$\int_{x_0}^{x_0+\Delta x} \left[\int_{u_0 - a(x,t)\,\Delta t}^{u_0} \Psi(x, u, t)\, du \right] dx,$$

where $a(x, t)$ is the acceleration in the direction of x increasing. As only gravitational forces need be considered, the acceleration (force per unit mass) is independent of u.

(k) Use the methods of (f)–(h) to find an expression for the mass flux inward through S_3 and S_4.

(l) Since collisions between stars can be ignored, the final answer is obtained by equating (c) with (h) + (k) and taking the appropriate limit.

13. The object of this problem is to derive the same final differential equation (1.2.4) that was obtained in Exercise 12, but in a different manner. Here we shall imitate the first method that was used above for deriving the ordinary mass conservation equation.

We shall use the notation

$$x = x(X, U, t), \qquad u = u(X, U, t), \tag{38a, b}$$

to describe motion in xu space. We shall assume that (38a) and (38b) can be inverted to yield

$$X = X(x, u, t), \qquad U = U(x, u, t). \tag{39a, b}$$

(a) What is the physical meaning of (38a) and (39b)?

(b) We make the definition

$$A(X, U, t) = \frac{\partial u(X, U, t)}{\partial t}.$$

What is the physical meaning of A? How should we define $a(x, u, t)$, the "spatial" version of A?

(c) Let

$$\overline{\Psi}(X, U, t) = \Psi[x(X, U, t), u(X, U, t), t]. \tag{40}$$

Define the generalized material derivative $\mathscr{D}/\mathscr{D}t$ by

$$\frac{\mathscr{D}\Psi(x, u, t)}{\mathscr{D}t} = \frac{\partial \overline{\Psi}(X, U, t)}{\partial t}\bigg|_{\substack{X = X(x, u, t) \\ U = U(x, u, t)}}. \tag{41}$$

Show (using careful and explicit notation) that

$$\frac{\mathscr{D}\Psi}{\mathscr{D}t} = \frac{\partial \Psi}{\partial t} + U\frac{\partial \Psi}{\partial x} + a\frac{\partial \Psi}{\partial u}. \tag{42}$$

(d) Define

$$\mathscr{J}(X, U, t) \equiv \begin{vmatrix} \dfrac{\partial x}{\partial X} & \dfrac{\partial x}{\partial U} \\[2ex] \dfrac{\partial u}{\partial X} & \dfrac{\partial u}{\partial U} \end{vmatrix}. \tag{43}$$

Using careful and explicit notation, derive

$$\frac{\partial \mathscr{J}}{\partial t} = \mathscr{J} \cdot \frac{\partial a}{\partial u}\bigg|_{\substack{x = x(X, U, t) \\ u = u(X, U, t)}}. \tag{44}$$

(e) In the absence of "collisions," a final differential equation can now be obtained by using the procedure that we followed in deriving the usual equation for mass conservation. That is, we define a suitably generalized "material region" $R(t)$ and assume that

$$\frac{d}{dt}\iint_{R(t)} \Psi(x, u, t)\, dx\, du = 0. \tag{45}$$

The final differential equation is obtained by manipulating (45). Perform this manipulation.

14.2 Balance of Linear Momentum

We shall now apply a version of Newton's second law to the motion of a continuum.

AN INTEGRAL FORM OF NEWTON'S SECOND LAW

In particle mechanics a particle P that has mass m and moves with velocity $\mathbf{v}$ is said to possess (linear) momentum $m\mathbf{v}$. According to Newton's second law, the rate of change of the linear momentum of P is equal to the net force acting on P. To generalize this law to continuum mechanics, we define the linear momentum at time t possessed by material of density $\rho(\mathbf{x}, t)$ that occupies a region $R(t)$ as

$$\iiint_{R(t)} \rho(\mathbf{x}, t)\mathbf{v}(\mathbf{x}, t)\, d\tau, \tag{1}$$

where $\mathbf{v}$ is the velocity vector. Again, the time rate of change of the integral in (1) is hypothesized to be equal to the net forces acting on the material in R.

What kind of forces are expected? The force of gravity is the most familiar example of **body forces**, which in particle mechanics are proportional to the mass of a point particle. In continuum mechanics it is natural to pass from points to regions by means of an integral. We thus consider body forces on the material in R that are of the form

$$\iiint_{R(t)} \rho(\mathbf{x}, t)\mathbf{f}(\mathbf{x}, t)\, d\tau, \tag{2}$$

where $\mathbf{f}(\mathbf{x}, t)$ is the body force vector per unit mass. We also consider **surface forces**. These have the form

$$\iint_{\partial R(t)} \mathbf{t}(\mathbf{x}, t, \mathbf{n})\, d\sigma, \tag{3}$$

where $\mathbf{x}$ is the position vector of the surface element, t is the time, and $\mathbf{n}$ is the unit exterior normal to ∂R at $\mathbf{x}$. The dependent variable $\mathbf{t}$ is called the **stress vector**.

In writing (3), we have implicitly assumed that force per unit area (*stress*) approaches a limit as area approaches zero. To see this, let S_k $(k = 1, 2, 3, \ldots)$ denote a sequence of smooth similar surfaces of decreasing area, each a portion of ∂R and each containing the point $\mathbf{x}^{(0)}$. Let $\mathbf{n}^{(0)}$ be the unit exterior normal to ∂R at $\mathbf{x}^{(0)}$. Then if A_k is the area of S_k, the ith component of the force per unit area on S_k is

$$\frac{1}{A_k}\iint_{S_k} t_i(\mathbf{x}, t, \mathbf{n})\, d\sigma = t_i(\tilde{\mathbf{x}}^{(k)}, t, \tilde{\mathbf{n}}^{(k)}); \qquad i = 1, 2, 3,$$

where we have used the integral mean value theorem. Here $\tilde{\mathbf{n}}^{(k)}$ is the unit exterior normal at $\tilde{\mathbf{x}}^{(k)}$, a point of S_k. If t is a continuous function of its variables,

$$\lim_{\substack{A_k \to 0 \\ \mathbf{x}^{(0)} \text{ in } S_k}} \frac{1}{A_k}\iint_{S_k} t_i(\mathbf{x}, t, \mathbf{n})\, d\sigma = t_i(\mathbf{x}^{(0)}, t, \mathbf{n}^{(0)}), \tag{4}$$

since $\tilde{\mathbf{x}}$ must approach $\mathbf{x}^{(0)}$ as $A_k \to 0$. [Note that $\mathbf{n}$ is determined at every point $\mathbf{x}$ *on a given surface* so that in this situation $\mathbf{t}(\mathbf{x}, t, \mathbf{n})$ is actually a function only of $\mathbf{x}$ and t.]

Why ought one to assume the existence of a stress? Historically, this assumption emerged as a generalization of what were originally special assumptions concerning special materials. In stretching a metal bar, for example, it has been assumed since Hooke that the relative elongation is proportional to the applied force per unit area. The randomly colliding elastic spheres model of a perfect gas leads to the introduction of a normal force per unit area (pressure), as the reader doubtless recalls from his introductory physics course. Some historical remarks can be found in Appendix 14.2, but an adequate appreciation of the long and interesting history of continuum mechanics requires further reading.

The definitions of $\mathbf{f}$ and $\mathbf{t}$ are given fully by the assumptions that forces of the form (2) and (3) act on R. But it may be useful to provide more verbal descriptions of these important quantities.

Body force. At time t, a force per unit mass $\mathbf{f}(\mathbf{x}, t)$ is assumed to act at each point $\mathbf{x}$ of R.

Surface stress. At time t, consider a point $\mathbf{x}$ on the boundary ∂R of region R. Let $\mathbf{n}$ be the unit exterior normal* to R at $\mathbf{x}$. Then $\mathbf{t}(\mathbf{x}, t, \mathbf{n})$ is the force per unit area exerted at the point $\mathbf{x}$ of ∂R *by* the material exterior to R *on* the material interior to R.

* Whenever we employ the unit exterior normal $\mathbf{n}$, it virtually goes without saying that we assume that $\mathbf{n}$ exists. That is, we assume thet the boundary ∂R of R divides space into two regions, a bounded connected set of points called the *interior* of R, and the complement of this region called the *exterior*. We generally assume that $\mathbf{n}$ varies smoothly as $\mathbf{x}$ moves about the surface.

Combining our speculations thus far, we put forth the hypothesis that the rate of change of linear momentum in a material region $R(t)$ equals the contribution of the body forces plus the contribution of the surface forces:

$$\frac{d}{dt}\iiint_{R(t)} \rho\mathbf{v}\, d\tau = \iiint_{R(t)} \rho\mathbf{f}\, d\tau + \iint_{\partial R(t)} \mathbf{t}\, d\sigma. \tag{5}$$

This fundamental equation, due to Cauchy, can be viewed in the following way. The effect of the rest of the universe on the linear momentum of material occupying $R(t)$ can be replaced, it is hypothesized, by the effect of a long range *body force* per unit mass $\mathbf{f}$, acting throughout $R(t)$, plus the effect of stresses $\mathbf{t}$, acting on the boundary of $R(t)$ and representing the additional effect (*contact force*) of material adjacent to $R(t)$. In other words, as far as linear momentum is concerned, our assumption is that we can replace the effect of the universe outside $R(t)$ by appropriate body forces $\mathbf{f}$ and stresses $\mathbf{t}$.

Alternatively, (5) simply states that linear momentum increases because of a momentum source per unit volume $\rho\mathbf{f}$ and a momentum influx per unit area $\mathbf{t}$. Such a lofty point of view has a certain appeal, but a descent is probably necessary when one wishes to identify momentum sources and fluxes with the pushes and pulls of everyday experience.

To attain an appropriate degree of respect for (5), the reader might (a) mull it over until he is convinced that it is obviously correct, and then (b) consult Section 7.1 in II which shows that (5) is *incorrect* under certain circumstances (when surface tension must be taken into account).

LOCAL STRESS EQUILIBRIUM

By the vector version of (1.8) we can take the time derivative in (5) inside the integral:

$$\frac{d}{dt}\iiint_{R(t)} \rho\mathbf{v}\, d\tau = \iiint_{R(t)} \rho\left(\frac{D\mathbf{v}}{Dt}\right) d\tau. \tag{6}$$

With this, (5) can be rewritten as

$$\iiint_{R(t)} \rho\left(\frac{D\mathbf{v}}{Dt}\right) d\tau - \iiint_{R(t)} \rho\mathbf{f}\, d\tau = \iint_{\partial R(t)} \mathbf{t}\, d\sigma. \tag{7}$$

Important consequences can be derived from (7) even without a clear idea of the forces involved. To do this, consider a family of similar regions R_L characterized by the length L. (An example is provided by the family of rectangular parallelepipeds with sides of lengths aL, bL, and cL.) As mentioned in Exercise 3 of Appendix 13.3, such regions have the same shape and their volume $\mathscr{V}_L$ and area $\mathscr{S}_L$ satisfy

$$\mathscr{V}_L = \lambda_R L^3, \qquad \mathscr{S}_L = \mu_R L^2,$$

for some constants λ_R and μ_R, which depend only on the shape of the regions. Hence for any continuous function G, the integral mean value theorem implies that

$$\iiint_{R_L} G(\mathbf{x})\,d\tau = \lambda_R L^3 G(\tilde{\mathbf{x}}), \qquad \tilde{\mathbf{x}} \text{ in } R_L. \tag{8}$$

Let us apply (8) to the right side of (7), component by component. Since all volume integrals are proportional to L^3, by dividing by L^2 and then letting $L \to 0$, we find that

$$\lim_{L\to 0} \frac{1}{L^2} \iint_{\partial R_L(t)} \mathbf{t}\,d\sigma = \mathbf{0}. \tag{9}$$

Equation (9) is an important one. Insight into its meaning can be acquired by recognizing that if the material in the region R_L were subject to surface stresses and to no other forces, and if this material were in equilibrium, then necessarily the total force on the material would have to vanish. That is,

$$\iint_{\partial R_L} \mathbf{t}\,\partial\sigma = \mathbf{0}.$$

Actually, each region R_L is moving and is subject to body forces. But as $L \to 0$, the effect of the surface stress term becomes more and more dominant compared to the volume integrals

$$\iiint_{R_L} \rho \frac{D\mathbf{v}}{Dt}\,d\tau \qquad \text{and} \qquad \iiint_{R_L} \rho\mathbf{f}\,d\tau,$$

which represent the effects of motion (inertia) and body forces. In the limit one obtains (9), which states, roughly speaking, that the surface force on an infinitesimal region, divided by its area $\mu_R L^2$, vanishes—just *as if* the region were in equilibrium. Equation (9) is thus known as the **principle of local stress equilibrium.**

ACTION AND REACTION

We shall derive consequences of the principle of local stress equilibrium by applying it to various regions. We first consider a rectangular parallelipiped F_L, containing a fixed point $\mathbf{x}$ on one corner and having dimensions εL by L by L, $\varepsilon \ll 1$ (Figure 14.4). Let $\mathbf{n}$, $-\mathbf{n}$, $\mathbf{n}_T$, $-\mathbf{n}_T$, $\mathbf{n}_R$, and $-\mathbf{n}_R$ be the unit exterior normals to the front, rear, top, bottom, right, and left faces of this "flake," respectively. (See Figure 14.4.) By the integral mean value theorem, the ith component of the surface integral in (9) can be written as

$$L^2[t_i(\mathbf{x}_F, t, \mathbf{n}) + t_i(\mathbf{x}_R, t, -\mathbf{n}) + \varepsilon Q_i]; \qquad i = 1, 2, 3, \tag{10}$$

where

$$Q_i = t_i(\mathbf{x}_T, t, \mathbf{n}_T) + t_i(\mathbf{x}_B, t, -\mathbf{n}_T) + t_i(\mathbf{x}_{Ri}, t, \mathbf{n}_R) + t_i(\mathbf{x}_L, t, -\mathbf{n}_R).$$

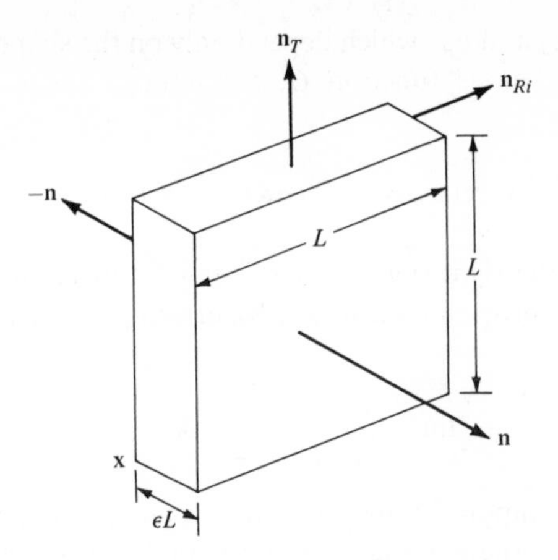

FIGURE 14.4. *The "flake" F_L. Application of the principle of local stress equilibrium to this family of regions gives the "action–reaction" law* (12).

Here $\mathbf{x}_F$, $\mathbf{x}_R$, $\mathbf{x}_T$, $\mathbf{x}_B$, $\mathbf{x}_{Ri}$, and $\mathbf{x}_L$ are points in the front, rear, top, bottom, right, and left faces of the flake, respectively. Equation (10) comes from multiplying t_i at some point on each face by the area of that face. The area is L^2 for the front and rear faces, εL^2 for the remaining faces.

Combining (9) and (10), we obtain

$$t_i(\mathbf{x}, t, \mathbf{n}) + t_i(\mathbf{x}, t, -\mathbf{n}) + \varepsilon Q_i = 0; \qquad i = 1, 2, 3.$$

The first argument in each t_i is $\mathbf{x}$, since $\mathbf{x}_F$, $\mathbf{x}_B$, etc., all approach $\mathbf{x}$ as $L \to 0$, and since *we assume that t is continuous* in a domain containing $\mathbf{x}$. Using the continuity and hence the boundedness of Q_i, we observe that εQ_i can be made arbitrarily small so that

$$t_i(\mathbf{x}, t, \mathbf{n}) + t_i(\mathbf{x}, t, -\mathbf{n}) = 0; \qquad i = 1, 2, 3; \tag{11}$$

or, suppressing the dependence on $\mathbf{x}$ and t,

$$\mathbf{t}(-\mathbf{n}) = -\mathbf{t}(\mathbf{n}). \tag{12}$$

We have thus *deduced* the counterpart of Newton's third (action–reaction) law for continua.

Consider Figure 14.5. At time t the stress at $\mathbf{x}$ representing the action of adjacent material exterior to R_1 on R_1 is $\mathbf{t}(\mathbf{x}, t, \mathbf{n})$. We have shown that this stress is equal in magnitude and opposite in direction to $\mathbf{t}(\mathbf{x}, t, -\mathbf{n})$, which represents the effect at $\mathbf{x}$ on R_2 of adjacent material exterior to R_2.

THE STRESS TENSOR

We now show that $\mathbf{t}(\mathbf{n})$ can be expressed as a linear combination of $\mathbf{t}(\mathbf{e}^{(1)})$, $\mathbf{t}(\mathbf{e}^{(2)})$, and $\mathbf{t}(\mathbf{e}^{(3)})$. To do this, we apply (9) to the tetrahedron of Figure 14.6. The mutually perpendicular faces of area S_i are normal to the ortho-

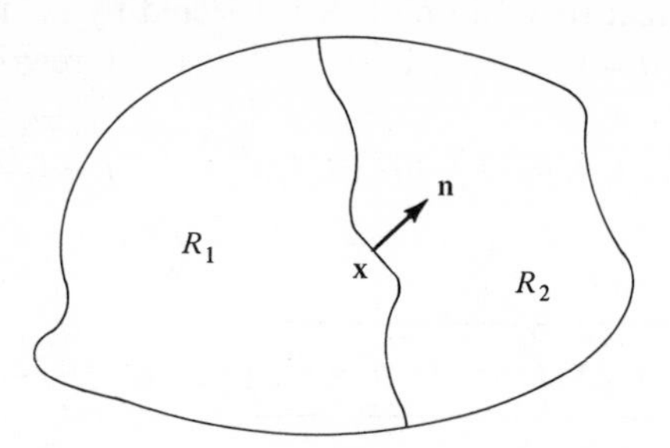

FIGURE 14.5. *At* $\mathbf{x}$ *the stress on* R_1*, due to the material of* R_2*, is equal and opposite to the stress on* R_2*, due to the material of* R_1*.*

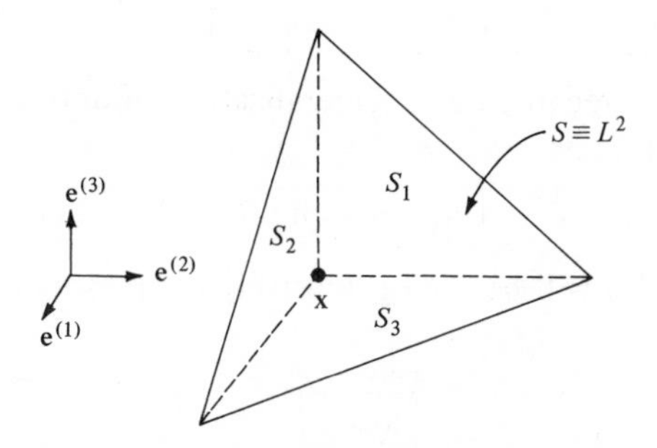

FIGURE 14.6. *A tetrahedron whose slanting face has area* $S = L^2$*. Application of the principle of local stress equilibrium as* $L \to 0$ *gives the relation* (17) *between the stress vector and the stress tensor.*

normal basis vectors $\mathbf{e}^{(i)}$. These faces meet at $\mathbf{x}$. The "slanting" face has area S. We define L by $L \equiv S^{1/2}$.

The jth component of the force acting on the face of area S_i is, by the integral mean value theorem,

$$S_i t_j(\mathbf{x}^{(i)}, t, -\mathbf{e}^{(i)}); \qquad i, j = 1, 2, 3,$$

where $\mathbf{x}^{(i)}$ is a point in this face. Let us denote by $\mathbf{n}$ the unit exterior normal to the slanting face of area S. Then, for some point $\mathbf{x}^{(0)}$ in the slanting face, $St_j(\mathbf{x}^{(0)}, t, \mathbf{n})$ is the jth component of the force acting on this face. Since S_i is the projection of S on the ith coordinate plane,

$$S_i = [\mathbf{n} \cdot \mathbf{e}^{(i)}]S = n_i S, \tag{13}$$

where n_i is the ith component of $\mathbf{n}$. Applying (9), with $L^2 \equiv S$, we obtain

$$\sum_{i=1}^{3} n_i t_j(\mathbf{x}, t, -\mathbf{e}^{(i)}) + t_j(\mathbf{x}, t, \mathbf{n}) = 0. \tag{14}$$

We have used the fact that all points $\mathbf{x}^{(q)}$ selected by the mean value theorem must approach $\mathbf{x}$; $q = 0, 1, 2, 3$. Using (12), we can rewrite (14) as

$$t_j(\mathbf{x}, t, \mathbf{n}) = \sum_{i=1}^{3} n_i t_j(\mathbf{x}, t, \mathbf{e}^{(i)}), \qquad j = 1, 2, 3. \tag{15}$$

We define T_{ij} by

$$\boxed{T_{ij}(\mathbf{x}, t) \equiv t_j(\mathbf{x}, t, \mathbf{e}^{(i)}),} \qquad i, j = 1, 2, 3. \tag{16}$$

(See Figure 14.7.) Now we rewrite (15) finally as

$$\boxed{t_j(\mathbf{x}, t, \mathbf{n}) = \sum_{i=1}^{3} n_i T_{ij}(\mathbf{x}, t),} \qquad j = 1, 2, 3. \tag{17}$$

Using what can be regarded as a generalization of dot-product notation, we sometimes write (17) as

$$\mathbf{t}(\mathbf{x}, t, \mathbf{n}) = \mathbf{n} \cdot \boldsymbol{T}(\mathbf{x}, t).$$

(This *direct notation* will be used extensively for tensors in II.)

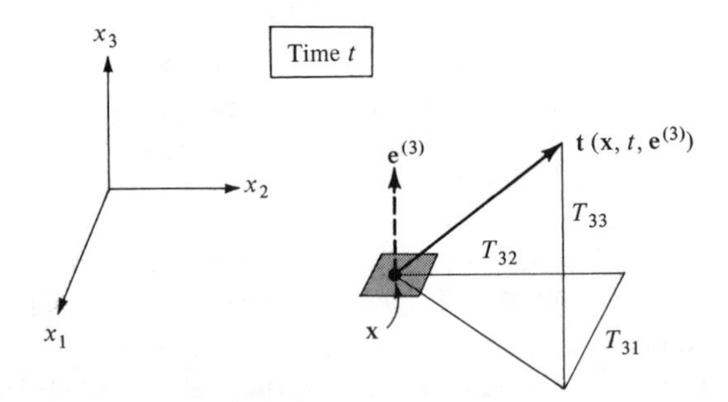

FIGURE 14.7. *Some components of the stress tensor. The shaded area represents a surface element at point* $\mathbf{x}$ *with unit normal* $\mathbf{e}^{(3)}$. *The effect at time t of material above the element on material below it is given by the stress vector* $\mathbf{t}(\mathbf{x}, t, \mathbf{e}^{(3)})$. *The three components of this vector are* T_{31}, T_{32}, *and* T_{33}.

The *very important relation* (17) shows that the dependence of the stress vector $\mathbf{t}$ on the exterior unit normal vector $\mathbf{n}$ is necessarily of a special linear form. This dependence is a consequence of the principle of local stress equilibrium. The nine quantities T_{ij}; $i, j = 1, 2, 3$ are called **components of the stress tensor** $\boldsymbol{T}$. The significance of the term *tensor* will be discussed in detail in Chapter 2 of II. It will, however, be worth remembering that the *first* component of T_{ij} refers to the unit vector that is normal to the *face* on which the *stress*, denoted by the *second* component of T_{ij}, is acting. [Refer to definition (16).] A good mnemonic is

First
Face

Second
Stress .

For example, T_{13} is the third component of the stress vector acting on the face (area element) whose exterior normal points in the direction of increasing x_1.

NEWTON'S SECOND LAW IN DIFFERENTIAL EQUATION FORM

As a first consequence of (17) we observe that the jth component of the surface term in the momentum conservation equation (7) can be written

$$\iint_{\partial R(t)} t_j\, d\sigma = \iint_{\partial R(t)} \sum_{i=1}^{3} n_i T_{ij}\, d\sigma = \iiint_{R(t)} \sum_{i=1}^{3} \frac{\partial}{\partial x_i}(T_{ij})\, d\tau, \tag{18}$$

where in the last equation we have used the divergence theorem. With this (7) can be written in component form as

$$\iiint_{R(t)} \left[\rho \frac{Dv_j}{Dt} - \rho f_j - \sum_{i=1}^{3} \frac{\partial}{\partial x_i}(T_{ij})\right] d\tau = 0, \qquad j = 1, 2, 3.$$

Since $R(t)$ is arbitrary, if we assume that the integrand is continuous we can write

$$\boxed{\rho \frac{Dv_j}{Dt} = \rho f_j + \sum_{i=1}^{3} \frac{\partial}{\partial x_i}(T_{ij}),} \qquad j = 1, 2, 3. \tag{19}$$

In direct notation, (19) is written

$$\rho \frac{D\mathbf{v}}{Dt} = \rho \mathbf{f} + \nabla \cdot T. \tag{20}$$

Relations (19) or (20) are **Cauchy's differential equations** *expressing for any continuum the balance* of linear momentum.* The only hypotheses used in obtaining these equations were sufficient smoothness of the functions involved and mass conservation [employed to obtain (6)].

EXERCISES

1. Let S denote the common material boundary of the regions R_1 and R_2 depicted in Figure 14.5. Write the integral equations of linear momentum balance for R_1, for R_2, and for $R_1 + R_2$. Infer that

$$\iint_S [\mathbf{t}(\mathbf{x}, t, -\mathbf{n}) + \mathbf{t}(\mathbf{x}, t, \mathbf{n})]\, d\sigma = \mathbf{0},$$

* If momentum were conserved, a particle's momentum would change only as a result of momentum flux across its boundary. The body force term in (19) provides an additional source of momentum. Hence it has become customary to speak of "balance" rather than "conservation."

and hence obtain an alternative derivation of (11). [Compare the derivation of Equation (12.1.45).]

‡**2.** In a certain two-dimensional problem, the stress tensor has four nonzero components: T_{11}, T_{12}, T_{21}, and T_{22}. In terms of these four quantities, find the force exerted by the material in the direction of the vector $\mathbf{i}-\mathbf{j}$ on the surface element of area dA whose exterior normal has the direction of the vector $2\mathbf{i}+\mathbf{j}$.

3. (a) Show that if no body forces are acting, then

$$T_{11}=T_{22}=T_{33}=T_{12}=T_{21}=0, \qquad T_{13}=T_{31}=x_2(x_1-1),$$

and

$$T_{32}=T_{23}=\tfrac{1}{2}(x_1^2+2x_1-x_2^2)$$

can be the components of the stress tensor for a material at rest.

(b) Consider an infinite equilateral triangular cylinder whose cross section is depicted in Figure 14.8. Show that the net force on each of the bounding planes is zero, if the stress tensor is that of part (a).

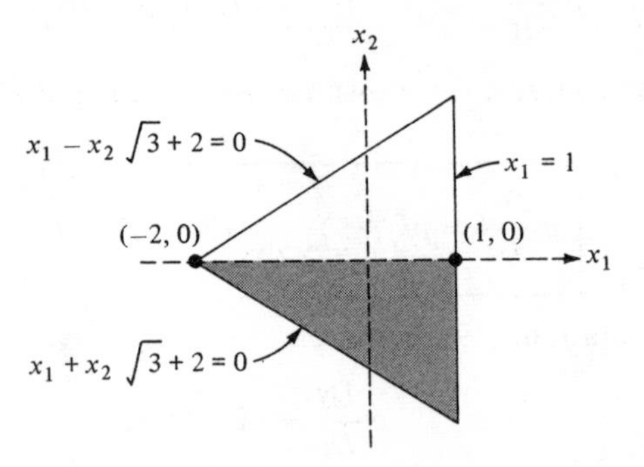

FIGURE 14.8. *Cross section of a triangular cylinder.*

‡(c) Find the net force (per unit length in the x_3 direction) exerted by the lower shaded region on the upper unshaded region.

4. (a) Using conservation of mass, show that the ith component of $\rho D\mathbf{v}/Dt$ is

$$\sum_{j=1}^{3}\frac{\partial(v_i\rho v_j)}{\partial x_j}+\frac{\partial(\rho v_i)}{\partial t}.$$

Deduce from (19) the *linear momentum transfer equation*

$$\frac{d}{dt}\iiint_R \rho\mathbf{v}\,d\tau=\iiint_R \rho\mathbf{f}\,d\tau+\oiint_{\partial R}[\mathbf{t}-\rho\mathbf{v}(\mathbf{v}\cdot\mathbf{n})]\,d\sigma. \tag{21}$$

Here R is a region *fixed in space*.

(b) The ith component of $\rho\mathbf{v}(\mathbf{v}\cdot\mathbf{n})$ is $\sum_{j=1}^{3} \rho \mathbf{v}_i \mathbf{v}_j \mathbf{n}_j$. The nine quantities $\rho v_i v_j$ $(i, j = 1, 2, 3)$ are called the components of the *momentum flux tensor*. Why is this an appropriate name? (Concentrate on the words "momentum flux." The significance of "tensor" will be discussed later. Also see Exercise 4.5.)

5. Suppose that fluid is in *steady motion* past a bounded obstacle. Let S be an imaginary surface enclosing the obstacle (Figure 14.9). Suppose that $\mathbf{f} \equiv 0$. Let $\mathbf{K}$ be the force acting on the obstacle. Use the linear momentum transfer equation to deduce $\mathbf{K} = \oiint_S [\mathbf{t} - \rho\mathbf{v}(\mathbf{v}\cdot\mathbf{n})]\, d\sigma$. (The force on an obstacle immersed in a steady flow can thus be deduced from a knowledge of what is happening far from the obstacle. See Exercise 6.)

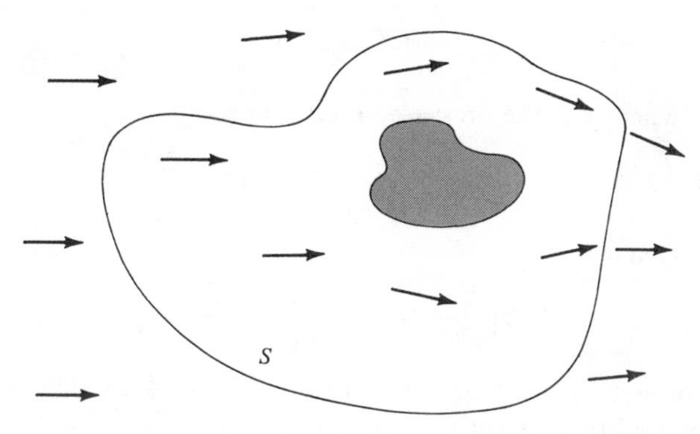

FIGURE 14.9. *Knowledge of the stress and velocity on the control surface S allows computation of the force on the obstacle.*

‡6. Consider a uniform steady flow past an obstacle. Suppose that the fluid is inviscid so that (Section 15.1) $\mathbf{t} = -p\mathbf{n}$, where p is the pressure. Suppose further that

$$\mathbf{v} = \mathbf{v}_\infty + O(|\mathbf{x}|^{-3}), \qquad p = p_\infty + O(|\mathbf{x}|^{-3}), \tag{22}$$

where the constants $\mathbf{v}_\infty$ and p_∞ represent the constant velocity and pressure at $|\mathbf{x}| = \infty$. [As stated in Appendix 3.1, $\mathbf{g}(\mathbf{x}) = O(|\mathbf{x}|^{-n})$ (with "$|\mathbf{x}| \to \infty$" understood) means that $\lim_{|\mathbf{x}|\to\infty} |\mathbf{x}|^n |\mathbf{g}(\mathbf{x})| = \text{constant}$.] Use the momentum transfer equation to prove that the force on the obstacle is zero. This result is essential in the proof of the D'Alembert paradox. (See Appendix 15.1.)

7. (a) For various purposes it is worthwhile writing the Cauchy equations (19) in material variables. We obtain

$$\delta \frac{\partial^2 x_j}{\partial t^2} = \delta F_j + \left[\sum_{i=1}^{3} \frac{\partial}{\partial x_i} (\mathcal{T}_{ij})\right]_{\mathbf{x}=\mathbf{x}(\mathbf{A},t)}. \tag{23}$$

The body force and stress tensor terms on the right side of (23) are defined by

$$F_j(\mathbf{A}, t) = f_j[\mathbf{x}(\mathbf{A}, t), t], \qquad \mathcal{T}_{ij}(\mathbf{A}, t) = T_{ij}[\mathbf{x}(\mathbf{A}, t), t].$$

What equations justify the transformation of the left side of (19) to the left side of (23)?

(b) A weakness in (23) is that it contains a differentiation with respect to the dependent variables x_i, while the independent variables are A_i. Make the necessary adjustment using the chain rule. To do this, use the chain rule to deduce expressions for the quantities $\partial A_j/\partial x_i$ from the result $\partial A_i/\partial A_j = \delta_{ij}$.

(c) For inviscid fluids

$$\sum_{i=1}^{3} \frac{\partial}{\partial x_i}(\mathcal{T}_{ij}) \qquad \text{becomes} \qquad -\frac{\partial}{\partial x_j} p,$$

where p is the pressure, and (23) becomes

$$\rho \frac{\partial^2 x_j}{\partial t^2} = \rho F_j - \frac{\partial p}{\partial x_j}.$$

Deduce

$$\sum_{j=1}^{3} \rho\left(\frac{\partial^2 x_j}{\partial t^2} - F_j\right)\frac{\partial x_j}{\partial A_k} = -\frac{\partial p}{\partial A_k}, \qquad k = 1, 2, 3. \tag{24}$$

8. Suppose that a primed coordinate system is rotated by an angle θ compared to an unprimed system so that

$$\begin{aligned} x_1 &= x_1' \cos\theta - x_2' \sin\theta \\ x_2 &= x_1' \sin\theta + x_2' \cos\theta \end{aligned} \qquad \text{and} \qquad \begin{aligned} x_1' &= x_1 \cos\theta + x_2 \sin\theta, \\ x_2' &= -x_1 \sin\theta + x_2 \cos\theta. \end{aligned}$$

(a) To obtain a relation between components of the two-dimensional stress tensor in the primed and unprimed coordinates, verify the following reasoning.

$T'_{22} \equiv$ stress component in $2'$ direction on face with normal in $2'$ direction.

The ith component of stress of face with exterior normal in $2'$ direction is given (in unprimed system) by

$$t_i = -\sin\theta\, T_{1i} + \cos\theta\, T_{2i}.$$

Therefore

$$T'_{22} = T_{11}\sin^2\theta - (T_{12} + T_{21})\sin\theta\cos\theta + T_{22}\cos^2\theta.$$

(b) Similarly, show that

$$T'_{12} = -T_{11}\sin\theta\cos\theta + T_{12}\cos^2\theta - T_{21}\sin^2\theta + T_{22}\sin\theta\cos\theta,$$

and find expressions for T'_{21} and T'_{11}.

14.3 Balance of Angular Momentum

In this section we shall establish an equation for the balance of angular momentum. To begin, let us review some material studied in elementary particle mechanics.

TORQUE AND ANGULAR MOMENTUM

The **torque M** (or **moment of force**) about the origin O due to a force **f** acting at point **x** is defined by the equation

$$\mathbf{M} \equiv \mathbf{x} \wedge \mathbf{f}, \tag{1}$$

where $\wedge$ denotes **vector product**. If

$$\mathbf{f} \equiv f_1\mathbf{i} + f_2\mathbf{j} + f_3\mathbf{k}, \qquad \mathbf{x} = x_1\mathbf{i} + x_2\mathbf{j} + x_3\mathbf{k},$$
$$\mathbf{M} = M_1\mathbf{i} + M_2\mathbf{j} + M_3\mathbf{k}, \tag{2}$$

then, in particular,

$$M_3 = x_1 f_2 - x_2 f_1. \tag{3}$$

The third component of **M** (the moment about the x_3 axis) can thus be computed by multiplying the force components f_1 and f_2 by their respective "lever arms" x_2 and x_1, and adopting the convention that counterclockwise and clockwise contributions are positive and negative, respectively. (See Figure 14.10.*) Analogous results hold for M_1 and M_2.

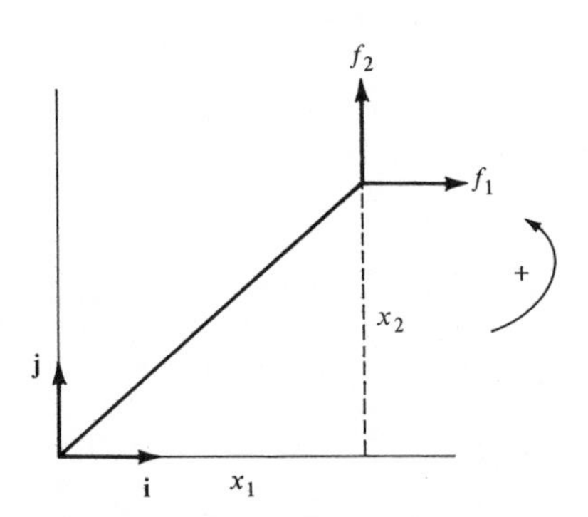

FIGURE 14.10. *The lever arms of the force components f_1 and f_2 are x_2 and x_1. Counterclockwise contributions to torque about the origin are regarded as positive.*

* The sign can also be remembered by the right-hand rule. Point your thumb along the x_3 axis (out of the paper). Then the fingers of your right hand will point in the direction of a positive contribution to the moment about this axis.

If a mass point at $\mathbf{x}$ possesses momentum $\mathbf{P}$, its **moment of momentum** or **angular momentum** about O is defined to be $\mathbf{x} \wedge \mathbf{P}$. Suppose a system of mass points is subject to mutual *central* forces (like gravity) which are directed along the lines joining the interacting particles. It is then a *theorem* of point mechanics that the rate of change of each particle's moment of momentum is equal to the sum of the torques acting on the particle.

In continuum mechanics we deal not only with central acting body forces like gravity, but also with contact forces that differ greatly from material to material. Now we must make the *assumption* that in any "nonmoving" or inertial coordinate system, the rate of change of moment of momentum is equal to the torques exerted by body forces and surface stresses. To formulate this assumption more precisely, we must generalize the concepts of moment of momentum and torque used in point mechanics, but this is straightforward. We thus postulate

$$\frac{d}{dt}\iiint_{R(t)} (\mathbf{x} \wedge \rho\mathbf{v})\, d\tau = \iiint_{R(t)} (\mathbf{x} \wedge \rho\mathbf{f})\, d\tau + \iint_{\partial R(t)} (\mathbf{x} \wedge \mathbf{t})\, d\sigma. \tag{4}$$

POLAR FLUIDS

Assumption (4) is *not* obviously true. In fact, it is sometimes not true at all. For *polar fluids*, which typically contain long molecules, it is necessary to introduce an internal angular momentum per unit mass $\mathbf{l}$. Thus the total angular momentum contains a contribution from internal angular momentum in addition to the usual contribution from the moment of linear momentum. We assumed in (4) that the moment of momentum was changed only by the same body forces and surface stresses which change linear momentum. Now we allow the presence of a body torque per unit mass $\mathbf{g}$ and a couple stress per unit area $\mathbf{c}$. Instead of (4), then, for polar fluids one assumes that

$$\frac{d}{dt}\iiint_{R(t)} (\mathbf{x} \wedge \mathbf{v} + \mathbf{l})\rho\, d\tau = \iiint_{R(t)} (\mathbf{x} \wedge \mathbf{f} + \mathbf{g})\rho\ + d\tau \iint_{\partial R(t)} (\mathbf{x} \wedge \mathbf{t} + \mathbf{c})\, d\sigma. \tag{5}$$

Manipulation of (5) is not difficult once cartesian tensors can be used (see Exercise 2.3.12 of II). It was introduced here to counteract complacent acceptance of (4). We shall therefore not consider (5) further in this section. References that show the possible usefulness of (5), to analyze experiments involving liquid crystals, are articles by H. Gasparoux and J. Proust [*J. Phys. 32*, 953 (1971)] and by H. Tseng, D. Silver, and B. Finlayson [*Phys. Fluids 15*, 1213 (1972)].

SYMMETRY OF THE STRESS TENSOR

In our studies of mass conservation and linear momentum balance, we started with an integral statement of the appropriate law and derived a differential equation as a consequence. From the angular momentum balance relation (4), an even more striking consequence can be derived, namely,

$$T_{ij}(\mathbf{x}, t) = T_{ji}(\mathbf{x}, t). \tag{6a}$$

In words, "the stress tensor is symmetric when the moment of momentum equals the sum of torques due to body forces and surface stresses." In direct notation the symmetry is indicated by the equation

$$\boldsymbol{T} = \boldsymbol{T}^{\mathrm{Tr}}, \tag{6b}$$

where Tr indicates transpose.

Once the machinery of cartesian tensors has been mastered, proof of (6) can be relegated to an exercise (Exercise 2.3.11 of II). Some will therefore prefer to take the symmetry of $\boldsymbol{T}$ for granted temporarily.

An alternative proof of (6) proceeds in a way that is analogous to the development in the previous section. A principle of local *moment* equilibrium is established. Application of the principle to a cube yields the desired result. The argument is not particularly short, but it confers the advantage of generating familiarity with the notions of "stress tensor" and "moment." This argument is given in the remainder of this section.

THE PRINCIPLE OF LOCAL MOMENT EQUILIBRIUM

In analogy with our treatment of linear momentum, we first apply the fundamental integral assumption (4) to a family of similar regions R_L characterized by the length L. For the body force term in (4) one can obtain the estimate

$$\left| \iiint_{R_L(t)} (\mathbf{x} \wedge \rho \mathbf{f})\, d\tau \right| \leq |\mathbf{x} \wedge \rho \mathbf{f}|_{\max} \cdot \text{volume of } R_L(t), \tag{7}$$

where $|\mathbf{P}(\mathbf{x})|_{\max}$ denotes the largest value of $|\mathbf{P}|$ for $\mathbf{x}$ in R_L. [See Exercise 3(a).]

As we have mentioned, the volume of R_L is $\lambda_R L^3$, where λ_R depends only on the shape of the regions R_L, not on their size. Let ψ denote the angle between $\mathbf{x}$ and $\mathbf{f}$. Since $|\sin \psi| \leq 1$,

$$|\mathbf{x} \wedge \rho \mathbf{f}|_{\max} = [|\mathbf{x}|\ |\rho \mathbf{f}|\ |\sin \psi|]_{\max} \leq |\mathbf{x}|_{\max} |\rho \mathbf{f}|_{\max}.$$

Thus

$$\left| \iiint_{R_L(t)} (\mathbf{x} \wedge \rho \mathbf{f})\, d\tau \right| \leq \lambda_R L^3 |\rho \mathbf{f}|_{\max} |\mathbf{x}|_{\max}.$$

By Exercise 3(b), we can transform the inertia term in (4) as follows:

$$\frac{d}{dt} \iiint_{R(t)} (\mathbf{x} \wedge \rho \mathbf{v})\, d\tau = \iiint_{R(t)} \left(\mathbf{x} \wedge \rho \frac{D\mathbf{v}}{Dt} \right) d\tau. \tag{8}$$

Hence the inertia term in (4) can be treated just like the body force term, and it is also bounded by an expression proportional to $L^3 |\mathbf{x}|_{\max}$. We thus obtain the **principle of local moment equilibrium**

$$\lim_{L \to 0} \frac{1}{L^2 |\mathbf{x}|_{\max}} \iint_{\partial R_L(t)} (\mathbf{x} \wedge \mathbf{t})\, d\sigma = \mathbf{0}. \tag{9}$$

LOCAL MOMENT EQUILIBRIUM FOR A CUBE

Let us now apply (9) to a cube C of side L. We choose our cartesian coordinate system so that the origin is at one corner of the cube and the axes point along its edges (Figure 14.11). The faces of the cube have exterior normals $\pm\mathbf{i}$, $\pm\mathbf{j}$, $\pm\mathbf{k}$ so that the components of the stress vectors acting on these

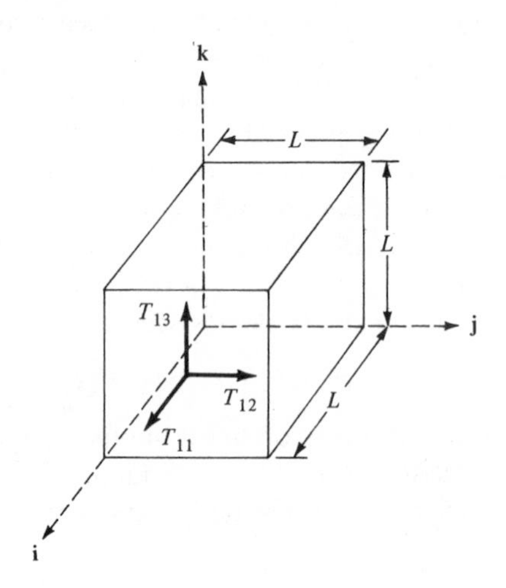

FIGURE 14.11. *A cube for application of the principle of local moment equilibrium. The three components of the force on the front face are equal in magnitude to the indicated components of the stress tensor.*

forces are appropriately signed components of the stress tensor T_{ij}. For C, $|\mathbf{x}|_{\max} = \sqrt{3}L$, so the principle of local moment equilibrium (9) becomes

$$\lim_{L \to 0} \frac{1}{L^3} \iint_{\partial C} (\mathbf{x} \wedge \mathbf{t})\, d\sigma = \mathbf{0}. \tag{10}$$

We shall discuss the moment about the x_3 axis in detail, but we shall leave the analogous treatment of the moments about the other axes to the reader (Exercise 5). Since we consider the situation at time t throughout, we shall not explicitly indicate the time dependence of the various quantities until we obtain our final result.

On the face of C in the plane $x = L$, the stress vector $\mathbf{t}(L, y, z)$ has the components $T_{1i}(L, y, z)$, $i = 1, 2, 3$. Because of (3) the contributions to the moment about $\mathbf{k}$ of these three components are

$$-\int_0^L \int_0^L T_{11}(L, y, z) y\, dy\, dz, \qquad L \int_0^L \int_0^L T_{12}(L, y, z)\, dy\, dz, \qquad 0, \tag{11}$$

respectively. Of course, T_{13} makes no contribution to the moment about the

k axis. The contributions due to stresses acting on the face in the plane $x = 0$ are

$$\int_0^L \int_0^L T_{11}(0, y, z)y \, dy \, dz, \qquad 0, \qquad 0. \tag{12}$$

The T_{12} contribution to (12) is zero because its lever arm has zero length. The sign of the T_{11} contribution requires careful consideration. Since $\mathbf{t}(\mathbf{i}) = -\mathbf{t}(-\mathbf{i})$, $\mathbf{t}(0, y, z)$ has the components $-T_{1i}(0, y, z)$, $i = 1, 2, 3$. The first component, $-T_{11}$, points in the direction of **i**, so the product of $-T_{11}$ and its lever arm y must be preceded by a minus sign.

Adding (11) and (12) and then applying the integral mean value theorem, we find a net contribution of

$$-[T_{11}(L, y_1, z_1) - T_{11}(0, y_1, z_1)]y_1L^2 + T_{12}(L, y_2, z_2)L^3, \tag{13}$$

where

$$0 \le y_i, z_i \le L, \qquad i = 1, 2. \tag{14}$$

Using the mean value theorem for derivatives, we find that

$$T_{11}(L, y_1, z_1) - T_{11}(0, y_1, z_1) = \frac{\partial T_{11}}{\partial x}(x_1, y_1, z_1)L; \qquad 0 \le x_1 \le L.$$

Hence the absolute value of the first term in (13) is less than

$$\left| \frac{\partial T_{11}}{\partial x}(x_1, y_1, z_1) \right|_{\max} L^4.$$

When the limit of (10) is applied, this $O(L^4)$ term will vanish. (We assume that $\partial T_{11}/\partial x$ and similar quantities are bounded.)

Let us now consider the contributions to the moment about **k** caused by the stresses acting on the faces in $y = 0$ and $y = L$. Arguments of the kind we have just presented show that the T_{2j} contributions are

$$-T_{21}(x_3, L, z_3)L^3, \quad O(L^4), \quad 0; \qquad 0 \le x_3, z_3 \le L. \tag{15}$$

In like manner the contributions from the faces in $z = 0$ and $z = L$ are all $O(L^4)$. Applying (10) to the cube C, we find that

$$T_{12}(0, 0, 0) = T_{21}(0, 0, 0). \tag{16}$$

Similarly (Exercise 5),

$$T_{31}(0, 0, 0) = T_{13}(0, 0, 0) \qquad \text{and} \qquad T_{23}(0, 0, 0) = T_{32}(0, 0, 0). \tag{17}$$

Since the origin was taken at the arbitrary point that we initially selected for consideration, (16) and (17) hold for any point. Thus (6) has been demonstrated.

EXERCISES

1. Let $\mathbf{f} = f_1\mathbf{i} + f_2\mathbf{j}$ be a force that acts at $\mathbf{x} = x_1\mathbf{i} + x_2\mathbf{j}$. Show that the formula $M_3 = |\mathbf{f}|d$, where $|\mathbf{f}| = (f_1^2 + f_2^2)^{1/2}$, is an alternative to (3). Here d is the length of the perpendicular from the origin to the "line of action" of $\mathbf{f}$, as pictured in Figure 14.12.

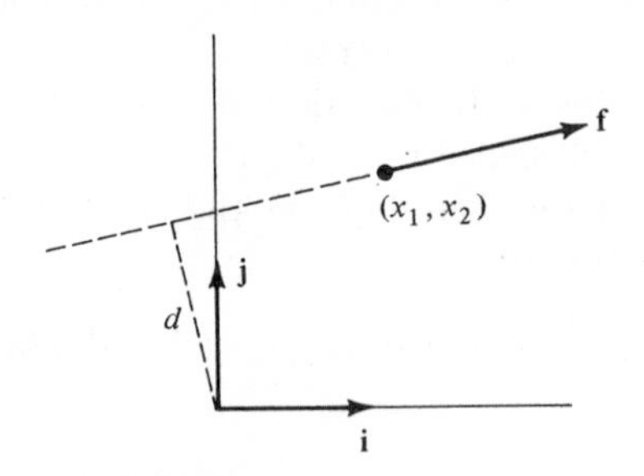

FIGURE 14.12. *For calculation of the moment about the origin, the "lever arm" of* **f** *has length d.*

2. Show that (4) remains true if the coordinates are translated so that $\mathbf{x}$ becomes $\mathbf{x} + \mathbf{k}$, where $\mathbf{k}$ is a constant vector.

3. ‡(a) Prove (7).
(b) Derive (8).

4. (a) Verify the results stated in the text concerning the contributions to the moment about $\mathbf{k}$ caused by the faces of the cube C in the planes $y = 0$, $y = L$, $z = 0$, $z = L$.
(b) Using the data of Exercise 2.3 (Figure 14.8), find the torque about the origin which is exerted on the material below $x_3 = 0$ by material above $x_3 = 0$.

5. Show that
(a) $T_{31} = T_{13}$.
(b) $T_{23} = T_{32}$.

14.4 Energy and Entropy

Completion of the set of field equations that are satisfied for any continuous medium requires introduction of thermodynamic concepts. To do this with any degree of thoroughness requires considerable effort, and the reader may be wearying of general considerations. For those who feel refreshed by frequent contact with particular problems, we begin this section with a discussion of the special case of ideal gases.* Indeed, except for certain important

* This material could be read after Section 14.2.

general results that are the subject of exercises, the material on perfect fluids is the only thermodynamical material that is used subsequently (in discussions of sound waves, Sections 15.3 and 16.3).

Energy and entropy considerations must be included for completeness, and such considerations play a major role in many areas of applied mathematics. Those who wish to have a strong foundation in continuum mechanics must at some time return to the main part of this section, beginning with the portion labeled "Equilibrium Thermodynamics." There they will find discussions of the general differential equation of energy balance, Gibb's relations, the Clausius–Duhem law of entropy increase, etc.

IDEAL GASES

It is an experimental fact that at low pressures the product of the pressure P and volume V_n of n moles* of a gas is equal to the product of n, the temperature Θ, and a constant $\mathscr{R}$ whose value is the same for all gases. Moreover, at low pressures the internal energy E is almost completely determined by the temperature. With support from a simple kinetic theory model to which the reader probably has been exposed, one is thereby led to define an **ideal gas**, a "model" material with the equations of state

$$PV_n = n\mathscr{R}\Theta, \qquad E = E(\Theta) \qquad (\mathscr{R} = 8.3143 \text{ J/mole deg}). \qquad (1a, b)$$

Common gases can be regarded as ideal at pressures below about 2 atm if errors of a few per cent are tolerable.

In particular problems (1a) takes a more convenient form if it is divided by M_n, the mass of n moles of gas. We may therefore write

$$P = \frac{M_n}{V_n} R\Theta, \qquad (2a)$$

where $R = n\mathscr{R}/M_n$ is a constant for any particular gas. We also note that it can be shown that (1b) can be replaced by the relation

$$E = C_V \Theta + \text{constant}, \qquad (2b)$$

where C_V is the specific heat at constant volume [Exercise 1(a)].

As is usual in elementary treatments of thermodynamics, our discussion thus far has concerned a container filled with gas under uniform conditions. We are aiming at a theory for a gas wherein conditions vary, but we shall proceed in the simplest way possible by assuming that (2a) and (2b) still hold locally. Thus we write

$$p = \rho R\theta, \qquad e = c_v \theta + \text{constant}, \qquad (3a, b)$$

* Recall that a mole of a gas has a mass in grams that is numerically equal to the molecular weight of the gas. It contains a number of molecules equal to Avogadro's number: 6.0225×10^{23}.

where $\rho = \rho(\mathbf{x}, t)$ is the local mass per unit volume, $\theta = \theta(\mathbf{x}, t)$ is the local temperature, $p = p(\mathbf{x}, t)$ is the local pressure, $e = e(\mathbf{x}, t)$ is the local internal energy per unit mass, and c_v is the local specific heat.*

Elementary physics texts give arguments which show that the assumption

$$\mathbf{t}(\mathbf{x}, t, \mathbf{n}) = -p(\mathbf{x}, t)\mathbf{n} \tag{4}$$

for the stress vector $\mathbf{t}$ is a reasonable one for an ideal gas. We shall accept this assumption here. It is discussed at some length in Section 15.1.

With (4), the equation of momentum balance (2.20) plus the equation of mass conservation (1.6) provide four equations for five unknowns: three components of velocity, the density, and the pressure. The introduction of the equation of state (3a) introduces one more relation, but also one more variable, the temperature. According to (3b), the temperature is related to the internal energy of the system. Hence we need to introduce a new physical law: the principle of energy balance, or the first law of thermodynamics.†

Consider a *unit mass of gas* of volume V. The first law states that

$$dQ = dE + P\,dV, \tag{5}$$

where dQ is the heat introduced to the unit mass and E is its internal energy.

Before we attempt to transform (5) into a partial differential equation, let us note an important relationship. If we consider the ratio dQ/Θ, we have (using $dE = C_V\,d\Theta$ and $P = R\Theta/V$)

$$\frac{dQ}{\Theta} = C_V \frac{d\Theta}{\Theta} + R\frac{dV}{V}. \tag{6}$$

Assume that C_V is constant. Then the right-hand side of (6) equals $d(C_V \log \Theta + R \log V)$. Thus dQ/Θ is a *total differential* in terms of the state variables Θ and V. (Q itself is not a state variable.) Equation (6) can be regarded as a *very* special case of the second law of thermodynamics, which states that the quantity dQ/Θ is a total differential of the state variables for all reversible (infinitesimally slow) exchanges of heat. Thus, if we follow a reversible process and restore a medium to its original state, we have

$$\oint \frac{dQ}{\Theta} = 0, \tag{7}$$

where the integral is taken, for example, along a closed curve in the (V, Θ)-state plane. For reversible processes we may consequently define a new function S of any state Σ, the *entropy*, by

$$S - S_0 = \int_0^\Sigma \frac{dQ}{\Theta}, \tag{8}$$

* Often the specific heats can be considered constants, so there is no difference between C_V and c_v.

† In many cases the density of a fluid particle, or even of the whole fluid, can be regarded as constant. (See Section 15.2 for an example.) Then there is no need to introduce thermodynamic concepts.

where the integral is taken along any path in a state plane, starting from a reference state 0 and ending at Σ. In particular, for an ideal gas, (8) yields

$$S = C_V \log \Theta + R \log V + \text{constant} \tag{9}$$

if the specific heat at constant volume is independent of temperature. From (9) it follows that the equation of state $P = R\Theta/V$ may be rewritten in the form

$$P = Ke^{S/C_V} V^{-\gamma}. \tag{10}$$

Here

$$\gamma = \frac{C_P}{C_V} = \frac{R + C_V}{C_V} \tag{11}$$

is the ratio of specific heats (at constant pressure and constant volume, respectively), and K is a constant characteristic of the gas. (See Exercise 1.)

We now turn to the reformulation of (5) as a partial differential equation for the flow field. We regard (5) as referring to a given fluid particle, and we invoke the substantial derivative. Since V, the volume of a unit mass, is the reciprocal of the density, we are led to write

$$\frac{De}{Dt} + p\frac{D}{Dt}\left(\frac{1}{\rho}\right) = \frac{Dq}{Dt}. \tag{12}$$

Here we have used local variables. In particular, e is the internal energy per unit mass. Also Dq/Dt is the rate at which heat is introduced to the gas per unit mass. From our previous experience with heat conduction, we expect that (in the absence of radiation effects)

$$\frac{Dq}{Dt} = \frac{1}{\rho}\nabla \cdot (k\nabla\theta). \tag{13}$$

If we introduce the entropy per unit mass s, and if we assume that local processes are reversible, we may also reformulate (8) as

$$\frac{Ds}{Dt} = \theta^{-1}\frac{Dq}{Dt}. \tag{14}$$

Moreover, (10) becomes

$$p = Ke^{s/C_V}\rho^{\gamma}. \tag{15}$$

Since e is a function of the state variables, a combination of (9) and (10),

$$\frac{De}{Dt} + p\frac{D(1/\rho)}{Dt} = \frac{1}{\rho}\nabla \cdot (k\nabla\theta), \tag{16}$$

together with the equations of continuity and motion, form a system of *five* differential equations for the three components of velocity and two of the state variables. The three equations (3a), (3b), and (15) relate the five state variables p, ρ, θ, e, and s.

Hopefully, the above discussion will make it easier for the reader to understand the situation for a general medium. The energy equation and, consequently, other thermodynamic relations must be introduced to complete the formulation of the basic equations. To show that we can put the full set of equations to practical use, we point out (as mentioned above) that they can be used to study acoustic waves, to give just one example. If we consider small disturbances in a homogeneous gas originally at density ρ_0 so that

$$\rho = \rho_0(1 + \sigma)$$

and

$$\sigma \ll 1,$$

then one can show that σ satisfies, in the linear approximation, the partial differential equation of wave motion

$$\frac{\partial^2 \sigma}{\partial t^2} = c_0^2 \frac{\partial^2 \sigma}{\partial x^2}.$$

Here c_0 is the wave speed, a constant in the medium. The detailed discussions, which the reader may wish to study at this point, are given in Section 15.3.

Having completed our treatment of an important special case, we now turn to a general discussion of thermodynamical considerations in mechanics. Some of the points made above will be repeated, but in a fuller and more general context.

EQUILIBRIUM THERMODYNAMICS

Elementary treatments of thermodynamics normally consider only materials whose properties are uniform in space and time. The term **thermostatics** has been introduced to emphasize that only infinitesimally slow changes can be directly handled by the elementary theory. We shall retain the older terminology **equilibrium thermodynamics**.

We wish to study materials whose properties change from moment to moment and from point to point, so we cannot restrict ourselves to equilibrium concepts. On the other hand, these simpler concepts must provide the foundation of our discussion. Although we have generally assumed a knowledge of basic physical principles, it is our experience that thermodynamic concepts such as entropy are far less familiar than corresponding mechanical concepts such as force, work and kinetic energy. Consequently, equilibrium thermodynamics is briefly reviewed in Appendix 14.1.*

Some salient ideas of equilibrium thermodynamics are these: (i) Energy is conserved in the sense that an increase in a system's internal energy E is equal to the heat added to the system minus the work done by the system.

* Energy and entropy considerations must be included for completeness, but further use of the results we obtain is restricted to our discussion of compression waves and to various exercises.

(ii) For a reversible process, excluding electrical and chemical effects (as we shall throughout), an element of work dW done by the system on a unit area can be computed by forming the product of the pressure P and the element of distance ds moved by the substance on which the pressure acts. (Summing such contributions, one obtains the formula $dW = P\,dV$, where dV is a change in volume.) (iii) If heat Q is reversibly added to a system, the corresponding entropy change $S_2 - S_1$ is given by the integral $\int \Theta^{-1}\,dQ$, where Θ is the absolute temperature. If the change in entropy is irreversible, then $S_2 - S_1$ exceeds $\int \Theta^{-1}\,dQ$. (iv) The entropy of an isolated system does not decrease, and only in reversible processes does it remain the same.

EFFECTS OF INHOMOGENEITY AND MOTION

Situations in which properties change from point to point are handled in a familiar manner. One defines point functions (intensive variables) whose integral provides the corresponding overall (extensive) quantities. Thus, if R is a region with boundary ∂R, at some point $\mathbf{x}$ in R or on ∂R, and at a time t, $e(\mathbf{x}, t)$ = internal energy density per unit mass, at point $\mathbf{x}$ and time t, and $h(\mathbf{x}, t, \mathbf{n})$ = heat efflux defined as follows. At time t, consider a point $\mathbf{x}$ on the boundary ∂R of region R. Let $\mathbf{n}$ be the unit exterior normal to R at $\mathbf{x}$. Then $h(\mathbf{x}, t, \mathbf{n})$ is the rate of heat flow per unit area through ∂R at $\mathbf{x}$, *from* the interior *to* the exterior.

The net internal energy in R at time t is $\iiint_R e(\mathbf{x}, t)\,d\tau$; the net heat efflux is given by the surface integral $\oiint_{\partial R} h(\mathbf{x}, t, \mathbf{n})\,d\sigma$.

In contrast to equilibrium thermodynamics we must consider the effects of motion. One effect is that the internal energy must be supplemented by an energy of motion. This is the familiar kinetic energy with a density per unit mass of $\frac{1}{2}\mathbf{v} \cdot \mathbf{v}$, where $\mathbf{v}$ is the velocity vector. Motion must also be considered when net work is computed. Extending (ii) above, we assume that the rate of doing work at a point is the product of the force at that point times the local velocity. There will thus be contributions to the net work from an integral over R of the body force $\rho\mathbf{f}$ times the velocity, and an integral over the boundary ∂R of the stress $\mathbf{t}$ times the velocity.

We cannot limit ourselves to the homogeneous states of equilibrium thermodynamics, where, for example, a single temperature characterizes the entire domain under consideration. But we shall assume that the various dependent variables have a sufficiently weak dependence on both space and time that we can regard our domain as composed of a number of volume elements, each in equilibrium but changing reversibly due to slight property differences between adjacent elements. If this **local equilibrium** assumption is justified, we can postulate with some confidence that the various classical equilibrium relations hold for a material particle. Presumably, our results will be valid if elements that are small compared to length scales of interest will reach equilibrium in times that are short compared to time scales of interest. Deeper study is certainly needed to build confidence in the equations, but the reader can be assured that they have been successfully used in very many instances. On the other hand, the search for a further-ranging thermodynamics continues to the present day, spurred particularly by the "far-from equilibrium" behavior that occurs in biology.

ENERGY BALANCE

Taking into account nonhomogeneity and motion, then, we postulate the following extension of the energy balance* condition (i), for any material region $R(t)$:

$$\frac{d}{dt}\iiint_{R(t)}\left(\frac{1}{2}\mathbf{v}\cdot\mathbf{v}+e\right)\rho\,d\tau=\iiint_{R(t)}(\mathbf{f}\cdot\mathbf{v})\rho\,d\tau+\oiint_{\partial R(t)}(-h+\mathbf{t}\cdot\mathbf{v})\,d\sigma. \tag{17}$$

The heat efflux h is preceded by a negative sign because *inward* flow of heat increases the energy in R.

It is instructive to contemplate the analogy between (17) and the angular momentum balance equation for polar fluids (3.5):

$$\frac{d}{dt}\iiint_{R(t)}(\mathbf{x}\wedge\mathbf{v}+\mathbf{l})\rho\,d\tau$$

$$=\iiint_{R(t)}(\mathbf{x}\wedge\mathbf{f}+\mathbf{g})\rho\,d\tau+\oiint_{\partial R(t)}(\mathbf{x}\wedge\mathbf{t}+\mathbf{c})\,d\sigma.$$

Both equations have the form of a general law of balance, namely,

rate of change of quantity
= rate of volume creation + rate of flux through surface.

But further, each term† is composed of a macroscopic part and a microscopic part. For example, the energy is divided into kinetic and internal. Kinetic energy can be thought of as being due to the gross macroscopic motion of the material, while internal energy can be regarded as being due to microscopic internal degrees of freedom. The distinction between macroscopic and microscopic seems to blur on more intense scrutiny, however, for is it not true that all continuum quantities are macroscopic manifestations of microscopic behavior? A deeper view of the distinction we seek seems to involve the notion of observer invariance. For example, internal energy looks the same to two observers moving with constant velocity relative to one another; kinetic energy does not. We shall not pursue these matters further, except to note the remarkable fact that *all* the fundamental balance laws of mechanics can be deduced from the energy balance postulate and suitable requirements of observer invariance. (See Exercise 2.3.13 of II.)

* We assume a nonpolar fluid; otherwise there would have to be additional terms to represent, e.g., the rate of working of the couple stresses $\mathbf{c}$.

† In (17) $\mathbf{f}\cdot\mathbf{v}$ could be supplemented by a heat generation contribution r. Such contributions, however, will not considered here.

As in previous manipulations of balance laws we wish to derive a differential equation from the integral equation (17). To accomplish this, we must first parallel the procedure used to obtain (2.18) and write

$$\oiint_{\partial R} \mathbf{t} \cdot \mathbf{v}\, d\sigma \equiv \oiint_{\partial R} \sum_{i=1}^{3} t_i v_i\, d\sigma = \iiint_R \sum_{j=1}^{3} \frac{\partial}{\partial x_j}\left(\sum_{i=1}^{3} T_{ji} v_i\right) d\tau$$

$$\equiv \iiint_R \nabla \cdot (\mathbf{T} \cdot \mathbf{v})\, d\tau. \tag{18}$$

In the usual way [as in (1.8)] we can bring the time derivative on the left side of (17) inside the volume integral. We can then consider (17) for a sequence of similar regions R_L and deduce that

$$\lim_{L \to 0} \frac{1}{L^2} \oiint_{\partial R_L} h\, d\sigma = 0. \tag{19}$$

Applying (19) to a flake and a tetrahedron, we can show that there exists a **heat flux density vector** $\mathbf{q}(\mathbf{x}, t)$ such that

$$h(\mathbf{x}, t, \mathbf{n}) = \mathbf{n} \cdot \mathbf{q}(\mathbf{x}, t). \tag{20}$$

Using (18) and (20) in (17) and assuming continuity of the integrand, we can finally obtain the **differential equation of energy balance**:

$$\boxed{\rho \frac{D}{Dt}\left(\frac{1}{2}\mathbf{v} \cdot \mathbf{v} + e\right) = \rho(\mathbf{f} \cdot \mathbf{v}) - \nabla \cdot (\mathbf{q} - T \cdot \mathbf{v}).} \tag{21}$$

The detailed derivations of (19), (20), and (21) are so similar to the derivations of (2.9), (2.17), and (2.19) that the reader should have no trouble in supplying them, as he is requested to do in Exercise 2(a).

ENTROPY, TEMPERATURE, AND PRESSURE

Roughly speaking, (21) is a generalization of the equilibrium version of the first law

$$\frac{dE}{dt} = \frac{dQ}{dt} - \frac{dW}{dt} \qquad \text{for reversible changes.}$$

We must also generalize the relations that connect heat input to entropy through temperature, and work to volume change through pressure.

In the reversible case, these **Gibbs relations** are

$$\frac{dQ}{dt} = \Theta \frac{dS}{dt}, \quad \frac{dW}{dt} = P \frac{dV}{dt} \qquad \text{for reversible changes.}$$

Thus the first law can be written

$$\frac{dE}{dt} = \Theta \frac{dS}{dt} - P \frac{dV}{dt}. \tag{22}$$

Temperature Θ and pressure P were regarded as known concepts in the informal discussion from which (22) was generated as (9b) of Appendix 14.1. (Other notation: E = internal energy, S = entropy, V = volume.) An alternate approach is to arrive at (22) by starting with the definitions

$$\Theta = \left(\frac{\partial E}{\partial S}\right)_V, \qquad P = -\left(\frac{\partial E}{\partial V}\right)_S, \tag{23}$$

where the subscripts indicate the variable that is held constant in the partial differentiation. (It is not without intuitive appeal, for example, that the partial derivative of internal energy with respect to volume V is the negative of the pressure.) Now two state functions determine any third state function. In particular, if $E = E(S, V)$, then

$$\frac{dE}{dt} = \left(\frac{\partial E}{\partial S}\right)_V \frac{dS}{dt} + \left(\frac{\partial E}{\partial V}\right)_S \frac{dV}{dt}, \tag{24}$$

and (22) follows with the aid of (23).

To generalize, let us now use capital letters to denote quantities that are regarded as functions of material variables $\mathbf{A}$ and t. Functions of thermodynamic variables will be written with an overbar. For example, $E(\mathbf{A}, t)$ gives the internal energy at time t of the particle which was initially at $\mathbf{A}$. On the other hand, $\bar{E}(S, V)$ provides the internal energy of a particle whose entropy and specific volume are S and V, respectively. We assume that if two state variables are known *for a particle*, then any third state variable is uniquely determined. In particular,

$$E(\mathbf{A}, t) = \bar{E}[S(\mathbf{A}, t), V(\mathbf{A}, t)].$$

Paralleling (24), we have by chain rule that*

$$\frac{\partial E}{\partial t} = \frac{\partial \bar{E}}{\partial S}\frac{\partial S}{\partial t} + \frac{\partial \bar{E}}{\partial V}\frac{\partial V}{\partial t}. \tag{25}$$

We can use the following counterpart of (23) to define temperature Θ and pressure P:

$$\Theta \equiv \frac{\partial \bar{E}}{\partial S}, \qquad P = -\frac{\partial \bar{E}}{\partial V}. \tag{26}$$

With this, (25) becomes

$$\frac{\partial E}{\partial t} = \Theta \frac{\partial S}{\partial t} - P \frac{\partial V}{\partial t}. \tag{27}$$

* Since the dependence on S, V, $\mathbf{A}$, and t has been exhibited, there is no need to make explicit the variables that are held constant in the partial differentiations.

Let us return to spatial coordinates, making a corresponding shift to small letters for the various variables. Entropy per unit mass, for example, satisfies

$$s(\mathbf{x}, t) = S[\mathbf{A}(\mathbf{x}, t), t].$$

A partial derivative with respect to time keeping $\mathbf{A}$ constant becomes a material derivative. In spatial coordinates, then, (27) is

$$\boxed{\frac{De}{Dt} = \theta \frac{Ds}{Dt} - p \frac{D(\rho^{-1})}{Dt}.} \tag{28}$$

In (28), ρ denotes density as usual. Mass per unit volume turns out to be a more natural variable in continuum mechanics than its reciprocal, the volume per unit mass.

The energy balance requirement (21) and the Gibbs relation (28) complete the list of *field equations* that express fundamental physical laws that are valid for any continuum for which chemical, electrical, and relativistic effects are negligible. There is another fundamental law, however, an *inequality* which generalizes the thermostatic idea that entropy increase equals or exceeds heat addition divided by absolute temperature. This is,

$$\frac{d}{dt} \iiint_{R(t)} \rho s \, d\tau \geq - \oiint_{\partial R(t)} \theta^{-1} \mathbf{q} \cdot \mathbf{n} \, d\sigma, \tag{29}$$

where as usual $\mathbf{n}$ is a unit exterior normal to the material region $R(t)$. In differential equation form, assuming appropriate continuity, we have [Exercise 2(c)]

$$\boxed{\rho \frac{Ds}{Dt} \geq -\nabla \cdot (\theta^{-1} \mathbf{q}).} \tag{30}$$

Equations (29) and (30) are two versions of what is sometimes referred to as the **Clausius–Duhem inequality**.

INTERNAL ENERGY AND DEFORMATION RATE

Some manipulations of the basic energy and thermodynamic equations provide insight into their implications and also provide alternative and often more convenient forms. These manipulations give rise to expressions that contain quantities D_{ij} defined by the following combination of derivatives of velocity components:

$$D_{ij} = \frac{1}{2} \left(\frac{\partial v_i}{\partial x_j} + \frac{\partial v_j}{\partial x_i} \right). \tag{31}$$

The quantities D_{ij} are called the **components of the rate of deformation tensor** $\boldsymbol{D}$. As their name indicates, these components give information on how and at what rate the material is being deformed. Our later study of cartesian tensors provides a background against which the meaning of the D_{ij} is more clearly seen. This meaning is discussed in detail in Section 3.1. of II. The exercises for that section request proofs of several results which elucidate the relation between rate of deformation and energy charges. One of these results will be noted in the next paragraph.

A relation involving kinetic energy is implied by the momentum balance requirement (2.19)

$$\rho \frac{Dv_j}{Dt} = \rho f_j + \sum_{i=1}^{3} \frac{\partial}{\partial x_i}(T_{ij}).$$

By taking the scalar product of this equation with the velocity vector $\mathbf{v}$ and integrating over a material volume R, one finds that

$$\frac{d}{dt}\iiint_R \frac{1}{2}\mathbf{v}\cdot\mathbf{v}\rho\, d\tau = \iiint_R \mathbf{f}\cdot\mathbf{v}\rho\, d\tau + \oiint_{\partial R} \mathbf{t}\cdot\mathbf{v}\, d\sigma - \iiint_R \boldsymbol{T}:\boldsymbol{D}\, d\tau,$$

$$\boldsymbol{T}:\boldsymbol{D} \equiv \sum_{i,j=1}^{3} T_{ij} D_{ij}. \tag{32}$$

(A proof is requested in Exercise 3.1.7 of II). Thus the kinetic energy in a given piece of material is increased by the rate of working of body forces and surface stresses and decreased by a term involving the interaction of stress and deformation. This result can be converted into the following differential equation by a process used several times:

$$\rho \frac{D}{Dt}\left(\frac{1}{2}\mathbf{v}\cdot\mathbf{v}\right) = \rho\mathbf{f}\cdot\mathbf{v} + \nabla\cdot(\boldsymbol{T}\cdot\mathbf{v}) - \boldsymbol{T}:\boldsymbol{D}. \tag{33}$$

Upon subtracting (33) from the energy balance law (21), one obtains

$$\rho \frac{De}{Dt} = -\nabla\cdot\mathbf{q} + \boldsymbol{T}:\boldsymbol{D}. \tag{34}$$

This equation shows how internal energy results from heat flux and from the stress-deformation term $\boldsymbol{T}:\boldsymbol{D}$ that appeared in (33). This term is thus seen to represent the conversion of kinetic to internal energy.

ENERGY AND ENTROPY IN FLUIDS

Thermodynamic relations are particularly important in fluid mechanics, so it will be worthwhile to abandon complete generality and derive a few more relations concerning energy and entropy in fluids. This will require anticipation of the characterization of fluids to be given later, but this should not cause difficulty.

We split the components of the stress tensor into a part proportional to the Kronecker delta and a "leftover" part with components V_{ij}:

$$T_{ij} = -p\delta_{ij} + V_{ij}. \tag{35}$$

For compressible fluids, p can be identified with the pressure. When viscous effects are ignored, $V_{ij} = 0$. Substitution of (35) into (34) yields

$$\rho \frac{De}{Dt} = -p\nabla \cdot \mathbf{v} - \nabla \cdot q + \Phi, \qquad \Phi \equiv V{:}\mathbf{D}. \tag{36}$$

From this, one can deduce [Exercise 3(b)]

$$\rho \frac{Ds}{Dt} = -\theta^{-1}\nabla \cdot \mathbf{q} + \theta^{-1}\Phi, \tag{37}$$

which shows how the entropy of a material particle changes. Both terms on the right side presumably represent entropy change due to heat flow divided by temperature. In the second Φ must therefore represent heat generation by deformation. Indeed, the **dissipation function** Φ represents the rate per unit volume at which mechanical energy is dissipated into heat.

Comparison of (37) with the second law of thermodynamics in the differential equation form (30) leads to the requirement

$$\theta^{-1}\Phi - \theta^{-2}\mathbf{q} \cdot \nabla\theta \geq 0. \tag{38}$$

Sufficient conditions for satisfaction of this requirement, and therefore sufficient conditions for the validity of the second law, are

$$\mathbf{q} \cdot \nabla\theta \geq 0, \qquad \Phi \geq 0. \tag{39}$$

The first condition states that heat does not flow against a temperature gradient, the second that deformation never converts heat into mechanical energy.

EXERCISES

1. (a) In general, the **specific heats** C_V and C_P are defined as the rate of change of heat with temperature at constant volume and pressure, respectively. That is,

$$C_V = \left(\frac{\partial Q}{\partial \Theta}\right)_V, \qquad C_P = \left(\frac{\partial Q}{\partial \Theta}\right)_P,$$

where we use subscripted parentheses to remind us of the variable held constant in the partial differentiation. Use the first law to justify $C_V = (\partial E/\partial\Theta)_V$. Deduce that an ideal gas satisfies (2b).

(b) Show that consideration of changing temperature at constant pressure gives the result $dQ = C_P\, d\Theta$; hence $C_P = C_V + R$. [Use (a).]

2. (a) Supply detailed derivations of equations (19), (20), and (21).

(b) Verify (33) and (34).

(c) Derive (30) by extending the Dubois–Reymond lemma.

3. (a) Verify (36).
(b) Use the continuity equation and (28) to derive (37).
(c) Verify (38).

‡**4.** If the velocity $\mathbf{v}$ of a motion satisfies $\mathbf{v} = \nabla\phi$ for a scalar function ϕ, then the motion is said to be **irrotational**. Suppose that a material has constant density and occupies a simply connected region R. Let K be the kinetic energy of the irrotational motion of such a material and K^* the kinetic energy of any other motion having the same normal velocity at the boundary. Prove that $K \le K^*$.

5. Note that in their differential equation versions the law of mass conservation, and the laws of momentum and energy balance, all have the form

$$\rho \frac{DF}{Dt} = \rho Q - \nabla \cdot \mathbf{J}. \tag{40a}$$

(a) What is the general interpretation of equations with this form? In particular, why is the Q term missing in the mass *conservation* equation but not in the other two *balance* equations?
(b) Why can $\mathbf{v}$ be interpreted as a mass flux vector, and $-\boldsymbol{T} \cdot \mathbf{v}$ as an energy flux vector? (The flux of a scalar is a vector. The flux of the momentum vector must be a "higher" quantity in some sense, and analogy with the other equations shows this flux to be given by the negative stress tensor $-\boldsymbol{T}$.)
(c) Explain why

$$\frac{d}{dt}\iiint_{R(t)} \rho F \, d\tau = \iiint_{R(t)} \rho Q \, d\tau + \iint_{\partial R(t)} j(\mathbf{n}) \, d\sigma \tag{40b}$$

is the integral form of a general balance law; in particular, what is $j(\mathbf{n}) \equiv j(\mathbf{x}, t, \mathbf{n})$? Show that $j(-\mathbf{n}) = -j(\mathbf{n})$, and hence that $j(\mathbf{x}, t, \mathbf{n}) = \mathbf{n} \cdot \mathbf{J}(\mathbf{x}, t)$ for some vector $\mathbf{J}$. Deduce that a continuity assumption allows one to derive (40a) from (40b). [Exercise 8 shows what happens in the presence of discontinuities.]

The field equations were initially formulated as integral equations involving arbitrary material regions. When the integrands are continuous, differential equations can be deduced. When the integrands are discontinuous (as may be the case), one can derive relations between the various discontinuities. This is done in the next four exercises.

6. Let $Q(t)$ be the set of points $\mathbf{x}$ that were in a given region $Q(t_0)$ at time t_0 and whose movements are governed by a certain function $\mathbf{y}$. That is,

$$Q(t) = \{\mathbf{x} \,|\, \mathbf{x} = \mathbf{y}(\boldsymbol{Y}, t) \text{ for every } \mathbf{Y} \text{ in } Q(t_0), \text{ where } \mathbf{y}(\mathbf{Y}, t_0) = \mathbf{Y}\}.$$

Show that if F and ρ are smooth functions of space and time, and if f is defined by

$$f(t) = \iiint_{Q(t)} F\rho \, d\tau,$$

then

$$\frac{df}{dt} = \iiint_{Q(t)} \frac{\partial}{\partial t}(F\rho) \, d\tau + \iint_{\partial Q(t)} F\rho w_n \, d\sigma, \tag{41}$$

where $\mathbf{w} \equiv \partial \mathbf{y}/\partial t$, $w_n \equiv \mathbf{w} \cdot \mathbf{n}$, and $\mathbf{n}$ is the unit exterior normal to ∂Q.

This result is similar to the Reynolds transport theorem (1.10). The present problem deals with the change of a quantity that is integrated over a region moving according to a context-free rule given by the abstract function $\mathbf{y}$, whereas (1.10) deals with the case $\mathbf{y} \equiv \mathbf{x}$, $\mathbf{Y} \equiv \mathbf{A}$, so that the region moves according to the transformation $\mathbf{x} = \mathbf{x}(\mathbf{A}, t)$. Recall that when this particular notation is used we speak of Q as a *material region* and think of the motion of a continuum with velocity $\mathbf{v} \equiv \partial \mathbf{x}/\partial t$.

The interpretation of (41) is as follows. The instantaneous rate of change at time t of a quantity obtained by integrating $F\rho$ over a moving region $Q(t)$ is caused by two factors:

(a) The change of $F\rho$ with time at all points that are included in the region Q at time t.

(b) The net distortion of Q due to the motion of its boundaries with velocity $\partial \mathbf{y}/\partial t \equiv \mathbf{w}$. Locally tangential motion of the boundary adds no new points to the region of integration; hence the appearance of the normal component w_n in (41).

7. Let R be a *material region* that is divided into subregions R^- and R^+ by an arbitrarily moving surface Σ. Dependent variables, including density and velocity, may be discontinuous across Σ. Let $S^-(S^+)$ be the portion of ∂R, the boundary of R, which is also a boundary of $R^-(R^+)$. Thus

$$\partial R = S^- + S^+, \qquad \partial R^- = S^- + \Sigma, \qquad \partial R^+ = S^+ + \Sigma.$$

Let $\mathbf{n}$ be the exterior normal to R and let $\mathbf{n}_\Sigma$ be the normal to Σ which points into R^+. Let $\mathbf{n}^-(\mathbf{n}^+)$ be the exterior normal to $R^-(R^+)$. (See Figure 14.13.)

If $\mathbf{a}$ is a point on Σ, we define for any scalar or vector quantity E

$$E^+(\mathbf{a}) \equiv \lim_{\substack{\mathbf{x}\to\mathbf{a} \\ \mathbf{x} \text{ in } R^+}} E(\mathbf{x}), \qquad E^-(\mathbf{a}) \equiv \lim_{\substack{\mathbf{x}\to\mathbf{a} \\ \mathbf{x} \text{ in } R^-}} E(\mathbf{x}), \qquad [\![E(\mathbf{a})]\!] \equiv E^+(\mathbf{a}) - E^-(\mathbf{a}).$$

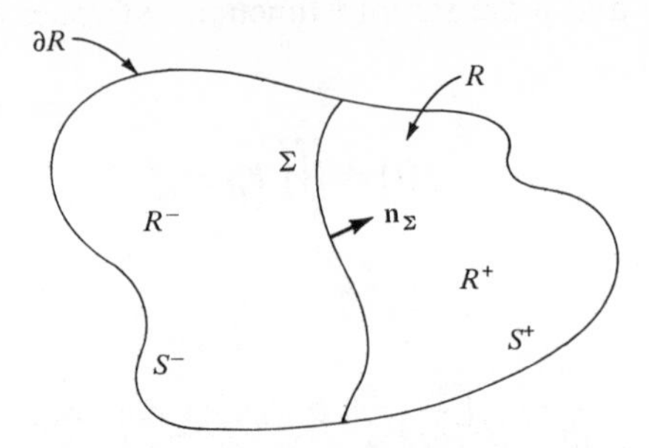

FIGURE 14.13. *Diagram for the calculation of relations between discontinuities across the surface* Σ.

(a) Apply the result of Exercise 6 to R^+ and R^-, then add. (Be careful of the normals.) Derive

$$\frac{d}{dt}\iiint_{R(t)} F\rho\, d\tau = \iiint_{R(t)} \frac{\partial}{\partial t}(F\rho)d\tau + \oiint_{\partial R} F\rho v_n\, d\sigma - \iint_{\Sigma} [\![\rho F]\!] w_n\, d\sigma. \quad (42)$$

Here $v_n = \mathbf{v}\cdot\mathbf{n}$ where $\mathbf{v}$ is the velocity of a particle.

(b) Let $v_n^{\pm} = \mathbf{v}\cdot\mathbf{n}^{\pm}$. For $\mathbf{a}$ on Σ, introduce $s_n^{\pm}(\mathbf{a}) \equiv w_n(\mathbf{a}) - v_n^{\pm}(\mathbf{a})$, the relative normal speeds of the discontinuity surface. Derive

$$\frac{d}{dt}\iiint_{R(t)} F\rho\, d\tau \equiv \iiint_{R(t)} \frac{\partial}{\partial t}(F\rho)\, d\tau + \oiint_{\partial R^+} F^+\rho^+ v_n^+\, d\sigma + \iint_{\partial R^-} F^-\rho^- v_n^-\, d\sigma$$

$$- \iint_{\Sigma} [\![s_n \rho F]\!]\, d\sigma. \quad (43)$$

(c) Assume that there are no discontinuities except across Σ. Derive finally, *now identifying* ρ *with the density*, that

$$\frac{d}{dt}\iiint_{R(t)} F\rho\, d\tau = \iiint_{R(t)} \frac{DF}{Dt}\rho\, d\tau - \iint_{\Sigma} [\![\rho F s_n]\!]\, d\sigma. \quad (44)$$

8. (a) Define $[\![j]\!]$ by

$$[\![j(\mathbf{x}, t, \mathbf{n})]\!] \equiv j^+(\mathbf{x}, t, \mathbf{n}_\Sigma) - j^-(\mathbf{x}, t, \mathbf{n}_\Sigma), \quad (45)$$

where we have used notation from Exercise 7. Deduce that

$$[\![\rho F s_n]\!] + [\![f]\!] = 0 \quad (46)$$

at each point of a surface of discontinuity.

(b) Deduce the following "jump conditions."

Conservation of mass: $[\![\rho s_n]\!] = 0.$ (47a)

Balance of linear momentum: $[\![\rho\mathbf{v}s_n + \mathbf{t}]\!] = \mathbf{0}.$ (47b)

Balance of energy: $[\![\rho(e + \frac{1}{2}\mathbf{v}\cdot\mathbf{v})s_n]\!] + [\![\mathbf{t}\cdot\mathbf{v} - h]\!] = 0.$ (47c)

(c) Show that for a nonpolar material, balance of moment of momentum adds no new restrictions.

9. (a) Suppose that the discontinuity surface is actually a material surface. Show that the jump condition derived from the mass conservation equation in the previous exercise is no longer a restriction and that an arbitrary discontinuity in density is allowable.

(b) Suppose that a material surface divides two *inviscid* fluids, where $\mathbf{t} = -p\mathbf{n}$. (Here p is the pressure; see Section 15.1.) Show that the momentum balance jump condition leads to the requirement that the pressure be continuous across the discontinuity surface.

For more applications of the jump conditions, see the exercises for Section 15.3.

14.5 On Constitutive Equations, Covariance, and the Continuum Model

A wide view of continuum mechanics is provided in this section. We begin by recapitulating the field equations so far derived. We see that there are many fewer equations than unknowns, and we give a few simple examples of the constitutive equations that are needed to fill out the picture by providing the nature of a particular material's behavior.

When formulating constitutive equations, perhaps the only general restriction is the requirement that the equations not depend essentially on the choice of a number of possible, equivalent, coordinate systems. This is an instance of the principle of covariance. In a brief discussion of this principle, we give as a primary example the "derivation" of the Lorentz transformation from the postulate that the velocity of light be the same in systems that display uniform relative motion.

It is appropriate to conclude the chapter by reviewing the appropriateness of the continuum model. We shall see that this model is only applicable if the typical distance between molecular collisions is small compared to length scales of interest.

RECAPITULATION OF FIELD EQUATIONS

Let us recapitulate the equations we have obtained for a nonpolar single-phase continuous medium, which does not exhibit electric, magnetic, or chemical effects.

Conservation of mass:

$$\frac{D\rho}{Dt} + \rho\nabla \cdot \mathbf{v} = 0. \tag{1.6}$$

Balance of linear momentum:

$$\rho \frac{D\mathbf{v}}{Dt} = \rho\mathbf{f} + \nabla \cdot \boldsymbol{T}. \tag{2.20}$$

Balance of moment of momentum:

$$\boldsymbol{T} = \boldsymbol{T}^{\mathrm{Tr}}. \tag{3.6}$$

Balance of energy:

$$\rho \frac{D}{Dt}\left[\left(\frac{1}{2}\mathbf{v} \cdot \mathbf{v} + e\right)\right] = \rho(\mathbf{f} \cdot \mathbf{v}) - \nabla \cdot (\mathbf{q} - \boldsymbol{T} \cdot \mathbf{v}). \tag{4.21}$$

Gibbs relations:

$$\frac{De}{Dt} = \theta \frac{Ds}{Dt} - p \frac{Dv}{Dt}; \tag{4.28}$$

$$v \equiv \rho^{-1}; \qquad \theta = \frac{\partial e(s, v)}{\partial s}; \qquad p = \frac{-\partial e(s, v)}{\partial v}.$$

Second law of thermodynamics:

$$\rho \frac{Ds}{Dt} \geq -\nabla \cdot (\theta^{-1}\mathbf{q}). \tag{4.30}$$

Kinematic boundary condition:

$$\mathbf{v} \cdot \mathbf{n} \quad \text{prescribed on boundary.} \tag{13.2.22}$$

The unknowns are density ρ, velocity vector $\mathbf{v}$, pressure p, specific volume v, heat flux vector $\mathbf{q}$, temperature θ, specific internal energy e, specific entropy s, and stress tensor $\boldsymbol{T}$. Regarded as given is the specific body force $\mathbf{f}$.

There are 21 unknown scalar functions, counting three components for the vectors $\mathbf{v}$ and $\mathbf{q}$ and nine components for the stress tensor $\boldsymbol{T}$. There are the equivalent of 12 scalar equations, one inequality, and one boundary condition. We must reconcile the disparity between the number of equations and the number of unknowns with our feeling that if the initial state of a medium is known, then normally the laws of mechanics should uniquely determine how the medium moves and changes.

Predictions of existence and uniqueness of solutions from tallys of the number of equations and unknowns is dangerous, even for systems of linear algebraic equations. Our system is vastly more complicated—but an explanation of the excess number of unknowns does not lie in this direction.

INTRODUCTION TO CONSTITUTIVE EQUATIONS

It is as obvious as any statement about the material world ever is that additional equations are necessary to describe fully the restrictions on the motion of a given material. The deformation of a railroad bridge as a train passes over it and the meandering of the Gulf Stream are both limited by the *common* necessity to conserve mass, balance momentum and energy, etc. But there must be further *different* constraints which describe the fact that throughout the deformation steel remains steel and water remains water. **Constitutive equations** is the term given to the mathematical descriptions of the various kinds of solidity of solids, fluidity of fluids, and the exotic quality of exotic materials.

Here are some examples of constitutive equations. In a **rigid body**, the distance between any two particles remains constant. In an **incompressible medium** the density of each particle is constant, so $D\rho/Dt = 0$. For a **perfect gas** (with constant specific heats), the pressure p and the density are related by $p = K\rho^{\gamma}$, where K and γ are constants for a given gas. For ordinary materials the heat flux vector $\mathbf{q}$ is related to the temperature θ by the **Newton–Fourier law of cooling** $\mathbf{q} = -\kappa\nabla\theta$, where the scalar thermal conductivity κ depends on θ, and to a small extent (usually) on the pressure p. In Chapter 15 we consider an **inviscid fluid** for which the stress vector $\mathbf{t}$, the normal vector $\mathbf{n}$, and the pressure p are related by $\mathbf{t} = -p\mathbf{n}$.

Some constitutive equations, such as those for rigid bodies and incompressible materials, are reasonably obvious idealizations. Others, such as Newton's law of cooling, are generally regarded as obtained by experiment.*

But to obtain the constitutive equations such as those for ordinary elastic solids and viscous fluids, one starts with assumptions that hardly seem obvious even in retrospect, and then simplifies by means of some reasonably sophisticated mathematical manipulations. These manipulations are most easily performed with the apparatus of tensors. That this apparatus is not necessary is demonstrated by the fact that it was not used by the men who originally formulated the constitutive equations for elastic solids and viscous fluids. Tensors do seem a necessary conceptual tool, however, in modern attempts to find constitutive equations for more exotic materials. We therefore postpone our major discussions of constitutive equations until II, after tensors have been introduced.

THE PRINCIPLE OF COVARIANCE

The study of constitutive equations perhaps brings into the sharpest focus the need to stress the principle of covariance. Indeed, many of the modern attempts to find constitutive equations for more exotic materials are based on this principle.

The principle of covariance is most conveniently formulated in terms of tensor calculus. A detailed discussion of covariance and its relation to cartesian tensors is provided in II. At this point, let us explain the principle only briefly.

* This is often an oversimplified view. For example, see J. A. Ruffner's article, "Reinterpretation of the Genesis of Newton's Law of Cooling." *Arch. Hist. Exact Sci.* 2, 138–52 (1964). (Note that, as above, the name Fourier is often also associated with this cooling law.)

Basically, the principle of covariance (or invariance) states that all physical laws must take on the same form when stated in "equivalent" frames of reference of the observers. Thus two observers who choose two different sets of coordinate axes, stationary relative to each other, should certainly find the same physical laws stated in the same form. Examples are provided in Exercises 1.8 through 1.11. To give another example, consider the equation of heat conduction in two spatial dimensions:

$$\frac{\partial T}{\partial t} = \kappa\left(\frac{\partial^2 T}{\partial x^2} + \frac{\partial^2 T}{\partial y^2}\right). \tag{1}$$

If a second observer employs another Cartesian coordinate system, he will find the equation

$$\frac{\partial T'}{\partial t} = \kappa\left(\frac{\partial^2 T'}{\partial x'^2} + \frac{\partial^2 T'}{\partial y'^2}\right). \tag{2}$$

However, there is also a *mathematical* connection $\mathbf{x} = \mathbf{x}(\mathbf{x}')$ and $\mathbf{x}' = \mathbf{x}'(\mathbf{x})$ between the coordinate systems (x, y) and (x', y'). Since $\kappa^{-1}\,\partial T/\partial t$ is independent of the coordinate system chosen, it must follow that

$$\frac{\partial^2 T}{\partial x^2} + \frac{\partial^2 T}{\partial y^2} = \left[\frac{\partial^2 T'}{\partial x'^2} + \frac{\partial^2 T'}{\partial y'^2}\right]_{\mathbf{x}'=\mathbf{x}'(\mathbf{x})}, \tag{3}$$

as can indeed be verified (Exercise 1).

When vectorial quantities are involved, the situation becomes much more complicated. The formalism of vector analysis avoids this issue. However, this is only superficial. *One must make sure that a certain quantity, such as a velocity, is a vector.* Not every quantity with a direction and a magnitude *is* a vector. For example, an infinitesimal rotation or an angular velocity is a vector, but a finite rotation is not. (It is a *tensor*, as we shall see later.)

Besides tensor analysis, the principle of covariance can underly other theories. Thus the variational formulation of mechanics provides an illustration of an extensive branch of applied mathematics that is developed in a manner independent of a particular choice of coordinates. This leads to the development of the subject in terms of generalized and canonical coordinates. The classical developments of the theories of Lagrange, Hamilton, Jacobi, and others may be found in treatises of mechanics.

As an example of the usefulness of the principle of covariance, consider the Michelson–Morley experiment, which shows that light propagates at a constant speed whether the frame of reference is moving with constant velocity or not. We shall now demonstrate that the Lorentz transformation is the only linear transformation which leaves the propagation speed of waves invariant in two (relativistically equivalent) frames of reference that are in uniform relative motion.

The wave equation

$$\frac{\partial^2 u}{\partial t^2} - c^2 \frac{\partial^2 u}{\partial x^2} = 0 \tag{4}$$

describes light propagation in the frame of reference of the observer (x, t). Now, if another observer, moving at a speed v relative to the first, attempts to describe the same phenomenon with his own frame of reference (x', t'), he finds that the light waves are also propagating with the same speed c (not $c \pm v$). We wish to show that this can be true if the two frames (x', t') and (x, t) are related by the Lorentz transformation

$$x' = \frac{x - vt}{\sqrt{1 - v^2/c^2}}, \tag{5a}$$

$$t' = \frac{t - vx/c^2}{\sqrt{1 - v^2/c^2}}. \tag{5b}$$

We observe that in the new frame of reference, the equation of wave propagation for $u'(x', t') = u[x(x', t'), t(x', t')]$ takes on the standard form

$$\frac{\partial^2 u'}{\partial t'^2} - c^2 \frac{\partial^2 u'}{\partial x'^2} = 0. \tag{6}$$

One can readily verify that the linear transformation (5) does transform (4) into (6) (Exercise 2). However, we wish to "discover" (5).

Let us recall, as we just mentioned above in connection with the heat equation, that the Laplace operator remains invariant under a rotation of axes. Thus, if we define

$$y = ict, \qquad y' = ict', \tag{7}$$

(4) and (6) would transform into each other if the variables are related by a rotation between the (x, y) system and the (x', y') system; i.e., if for some angle θ,

$$x' = x \cos \theta + y \sin \theta, \tag{8a}$$

$$y' = -x \sin \theta + y \cos \theta. \tag{8b}$$

If the equations of (8) are to represent real relations between (x, t) and (x', t'), then θ must be complex. Indeed, if $\theta = i\phi$, where ϕ is real, we derive from (8) the real relationships

$$x' = x \cosh \phi - ct \sinh \phi, \tag{9a}$$

$$ct' = -x \sinh \phi + ct \cosh \phi. \tag{9b}$$

To obtain a physical interpretation for ϕ, we note that to an observer using the (x, t) system, a point $x' =$ constant fixed in the (x', t') system satisfies

$$\frac{dx}{dt} = c \tanh \phi.$$

Thus

$$\tanh \phi = \frac{v}{c},$$

where v is the velocity of the frame (x', t') as measured in the frame (x, t). Conversely, for x fixed, we can verify from (9) that

$$\frac{dx'}{dt'} = -c \tanh \phi = -v,$$

which confirms the symmetry of the two systems. Equations (9) can now be easily rewritten into the form (5) (Exercise 3).

We are all well informed of the revolutionary changes of our fundamental concepts caused by the special theory of relativity, whose essential mathematical basis is the Lorentz transformation. Here is a powerful demonstration that physical issues must be examined and appreciated beyond pure mathematical formalism. At the same time, it is a demonstration that the seemingly innocuous principle of covariance can lead to important consequences.

VALIDITY OF CLASSICAL CONTINUUM MECHANICS

Now that the foundations of continuum mechanics have been laid, it is appropriate to reconsider the relationship between the continuum and the particulate models of a material. Complete understanding of this relationship is still an unattained goal; here we confine ourselves to some qualitative remarks.

As we pointed out in Section 13.1, continuum variables such as density and velocity can be defined, a priori or by an averaging process, providing that a typical interparticle distance is small compared to a typical length scale of the phenomenon under investigation. Other continuum variables can be defined in the same fashion. For example, the components of the stress tensor at a point can be related to average values of momentum flux. But for a continuum description to be of use, not only must continuum variables be definable, but it must be possible to relate these variables by equations that determine them completely.

In classical continuum theories the values of the various variables at a given point of space and time are determined by a set of partial differential equations. These equations relate the values of the unknowns and their derivatives at each of the various *points* of the spatial and temporal domain of interest. Implicit in these equations, then, is the assumption that the average variables are **locally determined**. For if the change in average quantities at a point were influenced by faroff configurations, then no set of equations at that point could completely determine these quantities.

Local determinism requires that particles do not "remember" their past for long. To be more precise, the "forgetting time" must be short compared to time scales of interest if ordinary continuum theories are to be employed. An example illustrating a situation when this is not the case is provided by the very high frequency waves which make up ultrasonic phenomena.*

The process of effectively "forgetting" the details of past configurations is accomplished by the interactions between particles. One can imagine a group of people who are blindfolded and set out into a crowded room. After a few bumps into others, each blindfolded person will become disoriented. Similarly, after a few interparticle interactions (which are usually called **collisions**), the fate of a given set of particles will scarcely be predictable from their initial configuration.†

The term **mean free path** is used to denote a typical distance between collisions. Another way of saying that "forgetting time" must be short compared to time scales of interest is that the mean free path must be short compared to distance scales of interest.

To summarize our all-too-brief discussion, *for classical continuum theories to be applicable both the typical interparticle distance and the mean free path must be small compared to length scales of interest.* When the first of these conditions holds, but not the second, a modified continuum approach can be used—as in the treatment of stellar dynamics in Section 1.2.

EXERCISES

1. Use (8) to verify (3).
2. Verify that if a change of variables is described by (5), then (4) transforms into (6).
3. Show that (5) is an equivalent form of (9).

Appendix 14.1 Thermodynamics of Spatially Homogeneous Substances

We discuss here the thermodynamics of continuous substances that are spatially homogeneous.§ By a **spatially homogeneous** substance we mean one that can be regarded as characterized by a single value of such variables as temperature, pressure, etc. (Thus we can speak of *the* temperature of the

* Considerable effort is being devoted to formulating generalizations of classical continuum theories that will provide improved predictions in situations such as those found in ultrasonics. A good illustration is provided by the papers of Ph. Selwyn and I. Oppenheim, *Physica 54*, 161–94, 195–222 (1971).

† "In principle" the future configuration of a set of particles may be completely determined by initial conditions. After a few collisions, however, inevitable slight errors in specification of the initial conditions will have brought about such a large cumulative error that further calculation would be meaningless.

§ Our presentation has been considerably influenced by H. C. Van Ness's excellent elementary explanation of thermodynamic concepts in the slim paperback *Understanding Thermodynamics* (New York: McGraw-Hill, 1969).

substance and *the* pressure.) In the present context, homogeneity will also be regarded as implying lack of a preferred direction of molecular motion—thus only motionless substances will be discussed. A substance that remains in the same homogeneous state is said to be at **equilibrium.**

An example of the type of system to be considered is provided in Figure 14.14. There a motionless quantity of gas is contained between thermally insulating walls and an insulated piston. With the gas can be associated various quantities, such as its volume V, the pressure P it exerts, its temperature Θ (we reserve the letter T for stress), and its internal energy E. It is an experimental fact that substances having the same values of any pair of these **state variables** will always have the same value of the remaining variables. By contrast, there are measurable quantities, such as the heat Q, which are not determined by the values of a pair of state variables. These quantities depend on the history of the manipulations to which the material was subjected.

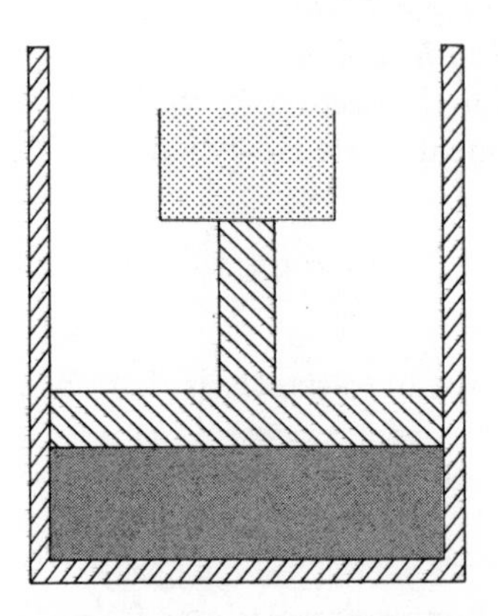

FIGURE 14.14. *Gas contained by a piston that is weighted down by a container filled with sand. The system is regarded as perfectly insulated so that no heat can flow into or out of the gas.*

EXPERIMENTS WITH A PISTON

Consider some thought experiments on the configuration of Figure 14.14. There the weight pushing down on the piston is partly composed of fine sand located in the container pictured in the figure. Suppose that a single speck of sand is removed. (We do not wish to complicate our discussion by considerations of work against gravity, so we imagine that the sand is transferred, without changing its height, to a shelf.) On removal of the speck, the piston would move upward a very small amount (assuming a frictionless piston). There would only be a tiny amount of nonuniformity introduced into the system, and that for a very short time. If the remaining grains of sand are removed, one at a time, the piston would gradually become elevated. This elevation is a result of the work W done by the gas on the piston and the sand. Now $W = \int dW$, where, by definition, $dW = F\,ds$. Here F is the force and ds is the element of distance moved against that force. Because each tiny step of the process barely disturbs the uniformity of the system, with little error we can continue to speak of *the* pressure P in the system at each volume V. It is this pressure that is responsible for the force on the piston. If the area of the piston is A, the magnitude of the force is PA. Suppose that the initial gas volume was V_1 and the final volume V_2. Now $F\,ds = PA\,ds$ and $A\,ds = dV$, the element of volume. Therefore,

$$\begin{array}{l}\text{work done by the gas on the piston in slowly} \\ \text{expanding from volume } V_1 \text{ to volume } V_2\end{array} = \int_{V_1}^{V_2} P(V)\,dV. \tag{1}$$

If the specks of sand were placed back on the piston, one by one, we assert that the piston would be found to return to a position that would be indistinguishable from its original position. Although the temperature of the gas would be found to drop as it expanded, the temperature would also return to its original value when the sand specks were replaced.

Suppose, in contrast, that all the sand were suddenly removed. The piston would shoot upward, fall back, rebound upward a little, oscillate a bit more perhaps, and finally settle to a new level. During the course of this process the pressure and temperature would *not* remain uniform throughout the gas. (Remember, heat flow is prevented by insulation.) In particular, since one could not speak of *the* pressure P at a given volume V, one could not use (1) to evaluate the work done. One would observe in addition the qualitative result that the piston would not rise as far as it did in the very slow process considered in the previous paragraph. Also, the temperature would not decrease as much. If the sand were suddenly dumped back onto the piston, the apparatus would not return to the position it was in before the experiment started. Neither would the temperature return to its initial value. It would, in fact, be higher.

We shall return later to a discussion of the **irreversible** processes, such as that which occurs when all the sand is suddenly removed. For the present it is sufficient to note that the course of such processes cannot be described by a formalism restricted to homogeneous states. On the other hand, the final result obtained by carrying out an irreversible process and then waiting a time of sufficient duration *is* a homogeneous state and can be described. Furthermore, we have seen that a very slow process virtually does preserve homogeneity at all times, and that such processes are virtually reversible. We shall be able to describe the course of limiting infinitely slow **reversible** processes. "Limiting" here is used essentially in its mathematical sense, so there should be no cavils that an infinitely slow process will never achieve a change. The slower an actual process is, the more nearly it will be approximated by the limiting reversible case.

For simplicity we considered an insulated cylinder and hence an **adiabatic** process in which no heat was added to the system or removed from it. But one can also consider essentially reversible processes in which heat is slowly added by means of tiny temperature differences between the inside of the container and the outside.

We shall now formulate the first of the two main postulates of equilibrium thermodynamics. Like all physical laws, these postulates cannot be "proved" by experiment. Rather, they *conjecture* that a certain amount of experiment and experience has revealed results of a universal character. The validity of the conjectures is established by comparing deductions from them with experiment. Consequently, it is not profitable to engage in exacting scrutiny of the heuristic reasoning that leads to the conjectures. Some feeling for the heurism must be acquired, however, if only because "laws" that are posited without supporting discussion have little psychological appeal.

Two versions will be given for each of the thermodynamic "laws." The first will be a comparison of two homogeneous states, the second a statement about the behavior of slowly varying systems.

FIRST LAW OF EQUILIBRIUM THERMODYNAMICS

First Form. Let two homogeneous states of a system have internal energies E_1 and E_2, respectively. Then if $\Delta E \equiv E_2 - E_1$,

$$\Delta E = Q - W,$$

where Q is the heat added *to* the system and W is the work done *by* the system in passing from state 1 to state 2. (Note that nothing is said about the process by which state 2 is attained from state 1. This process may be irreversible, so that nonhomogeneities are present in the course of the process.)

Second Form. For reversible processes, a shorthand* for the computation required to compute changes in internal energy from the amount of heat transferred and work done is

$$dE = dQ - dW. \tag{2}$$

Equation (2) reminds one of the line integral that is required for calculation of the change in E. In the present discussion, let us regard P and V as independent state variables. Suppose that a system passes from a state $\sum_1$ with internal energy E_1, pressure P_1, and volume V_1 to a state $\sum_2$ with internal energy E_2, pressure P_2, and volume V_2. Suppose that $\sum_2$ is reached from

* The "shorthand" is analogous to the formula $ds^2 = dx^2 + dy^2$ for the arc length of a plane curve. If the curve is given by $x = x(t)$, $y = y(t)$, $t_1 \leq t \leq t_2$, then of course the "long-hand" formula is $s = \int ds = \int_{t_1}^{t_2} [(dx/dt)^2 + (dy/dt)^2]^{1/2}\, dt$.

$\sum_1$ along a state curve $\sum$ that specifies the P–V values traversed. (See Figure 14.15.) The first law (2) tells us that

$$E_2 - E_1 = \int_\Sigma \frac{\partial Q}{\partial P}\,dP + \frac{\partial Q}{\partial V}\,dV - \int_\Sigma \frac{\partial W}{\partial P}\,dP + \frac{\partial W}{\partial V}\,dV. \tag{3}$$

As usual, a line integral can be converted to an ordinary single integral if a parametrization of the curve $\sum$ is available. Suppose that such a parametrization is given by

$$P = \bar{P}(t), \quad V = \bar{V}(t), \qquad \text{where } t \text{ is time,} \quad t_1 \le t \le t_2. \tag{4}$$

Then

$$E_2 - E_1 = \int_{t_1}^{t_2} \left(\frac{\partial Q}{\partial P}\frac{d\bar{P}}{dt} + \frac{\partial Q}{\partial V}\frac{d\bar{V}}{dt}\right) dt - \int_{t_1}^{t_2} \left(\frac{\partial W}{\partial P}\frac{d\bar{P}}{dt} + \frac{\partial W}{\partial V}\frac{d\bar{V}}{dt}\right) dt. \tag{5}$$

[In (5), $\bar{P}(t)$ and $\bar{V}(t)$ are substituted for P and V, the arguments in the derivatives of $Q(P, V)$ and $W(P, V)$.]

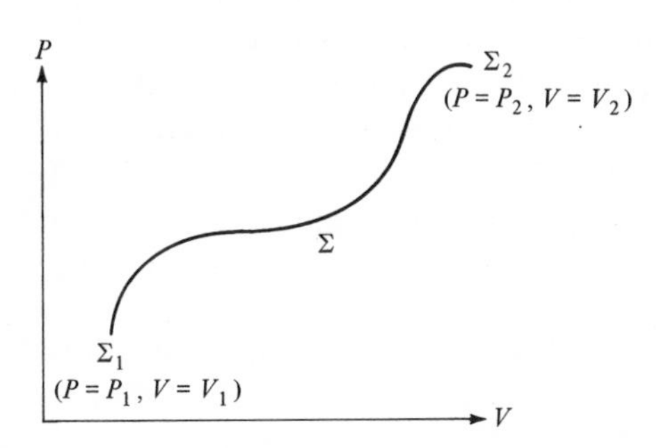

FIGURE 14.15. *The P–V state plane. The passage of the system from state Σ_1 to state Σ_2 is fully described by specification of the state curve Σ.*

Since, for example,

$$\frac{d}{dt}\,Q[\bar{P}(t), \bar{V}(t)] = \frac{\partial Q}{\partial P}\frac{d\bar{P}}{dt} + \frac{\partial Q}{\partial V}\frac{d\bar{V}}{dt},$$

in (5) we can replace t_2 by a variable upper limit t, differentiate with respect to t, and obtain the first law as the differential equation

$$\frac{dE}{dt} = \frac{dQ}{dt} - \frac{dW}{dt}. \tag{6}$$

This equation is valid only in the limit of very slow changes.

In the piston system of Figure 14.14 we have seen that $dW = P\,dV$. Generalization of our earlier argument establishes the same expression for the work of expansion in such systems as balloons of gas. More general systems can accomplish chemical and electrical work, but we shall not introduce such generality here.

A very important point is that neither the heat Q nor the work W is a state variable. As can be verified by experiment, Q and W in a reversible process depend on the *process*, not just on the initial and final states. In particular, it is profitless to try to ascertain "the heat content" of a piece of material; for, suppose that to establish a reference point the material were regarded as having zero heat at some moment. A later value of the material's "heat content" would presumably be equal to the measurable quantity "net heat added." But if the exact details of the entire history were ever lost, then the value of its heat content could not be checked—since the present state of a system gives no clue to the way heat was added to it in the past. The same argument shows that it is profitless to try to assign "work content" to a piece of material.

Since W and Q are not state variables, the line integrals $\int_\Sigma dW$ and $\int_\Sigma dQ$ would give different values for different state curves $\sum$. [But the first law guarantees that the difference of these two integrals, $\int_\Sigma (dW - dQ) = \int_\Sigma dE$, depends only on the initial and final points of $\sum$, for E is a state variable.] In other words, neither dW nor dQ is an exact differential. Note, however, that (in a reversible process) multiplication of dW by the integrating factor P^{-1} provides an exact differential, for $P^{-1}\,dW = dV$.* Empirical establishment of the fact that $P^{-1}\,dW$ is the differential of a state function might be accomplished by forcing some substance to pass in various ways from state $\sum_1$ to state $\sum_2$ by doing work very slowly. Suppose that the pressure was measured frequently so that plots of P^{-1} versus W could be constructed, as in Figure 14.16. Then the area under these curves would all be the same, for

$$\int_\Sigma P^{-1}\,dW = \int_\Sigma dV = V_2 - V_1, \tag{7}$$

and the difference between volumes does not depend on the process by which the system passes from its initial state to its final state.

ENTROPY

An important observation is not yet reflected in our theoretical construction, namely that there is a "natural" direction to processes. For example, if a cold metal bar is brought into a room, heat always flows into the metal until the temperature of the bar and the room are equalized. No energy has been lost, it has just been spread uniformly around. But uniformly distributed energy is *not available* to flow and do useful work. Indeed work has to be done (say by means of a refrigerator) to restore the original nonuniform state. The Nobel physicist Richard Feynman sums up the observations in a

* Since $dW = P\,dV$ for a reversible process, only a change in V causes an alteration in W. Thus $\partial W/\partial P = 0$ in (3) and (5).

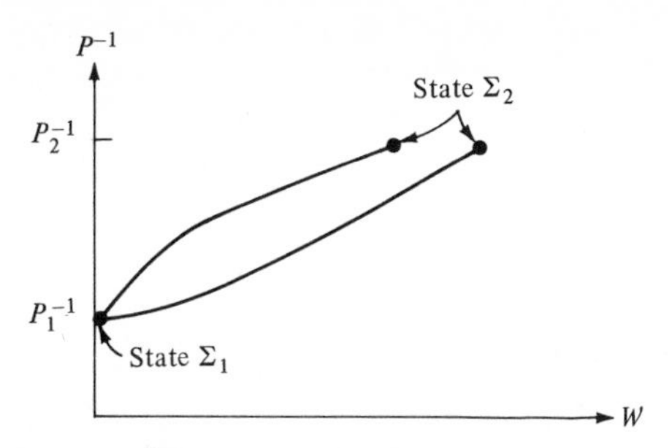

FIGURE 14.16. *Results from two hypothetical reversible experiments plotted as curves of* $1/P$ *versus* W. *Although two different values of* W *are found at the same state* Σ_2, *the area under the two curves is the same (up to their respective termination points). But if* $\int_\Sigma P^{-1}\,dW$ *depends only on the initial and final points of* Σ, *then* $P^{-1}\,dW$ *is an exact differential of a state function (the volume* V*).*

picturesque way* by comparing energy with water that initially is located all over a man who has just emerged from a bath. Two handy but tiny towels are dry. The man pats himself first with one towel and then the other, but after awhile no further transfer of water can take place (in this idealized model). The man and the towels are all equally damp. All the water (energy) is still there, but no further natural "flow" will take place.

None of the state functions that we have discussed changes in a "one-way" fashion. With hindsight, at least, it is natural to speculate that there is a missing state function, and that this function will indeed change in a directional manner. We have pointed the way toward discovery of this function in an earlier paragraph. Multiplication of dW by the integrating factor P^{-1} converted it into an exact differential, the differential of the state function volume. In addition to dW, the first law contains another nonexact differential, dQ. And indeed, experiments *for reversible processes* show that an exact differential results if dQ is multiplied by an integrating factor, in this case Θ^{-1} (Figure 14.17). The state function whose differential is given by $\Theta^{-1}\,dQ$ is called the **entropy** S. Thus

$$dS = \Theta^{-1}\,dQ \qquad \text{(reversible process).} \tag{8}$$

NOTE. For (8) to hold, temperature must be measured with respect to a zero value at the lowest possible temperature. Thus in all calculations involving thermodynamics an absolute scale (such as the Kelvin scale) is used in ascribing a value to Θ. We emphasize that *only for reversible processes does* (8) *provide a shorthand for a method to calculate entropy changes by a line integral.*

* In *The Character of Physical Law* (Cambridge, Mass.: MIT Press, 1967).

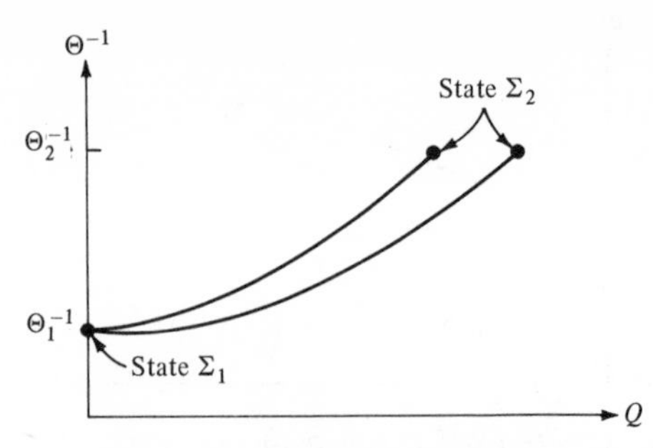

FIGURE 14.17. *Results from two hypothetical reversible experiments plotted as curves of Θ^{-1} versus Q. The area under the two curves is the same. Thus $\Theta^{-1}\,dQ$ is an exact differential of a state function (the entropy S).*

With (8) and the expression $dW = P\,dV$, which holds for reversible expansion and compression processes, the first law can be written

$$dE = \Theta\,dS - P\,dV, \tag{9a}$$

or for slow changes,

$$\frac{dE}{dt} = \Theta\frac{dS}{dt} - P\frac{dV}{dt}. \tag{9b}$$

Is there a directionality to the change of S? To obtain some insight, consider two experiments with the insulated piston of Figure 14.14. In both cases we remove the sand from the piston so that the gas expands from its initial state $\sum_1$ to its final state $\sum_2$. Since the piston is insulated, the expansion is adiabatic.

CASE 1. *Slow reversible adiabatic process.* Here

$$S_2 - S_1 = \int\frac{dQ}{\Theta}.$$

The system is insulated; so, certainly, no heat is added. Thus $S_2 = S_1$: the entropy does not change. (Consequently, the process is called **isentropic**.)

CASE 2. *Irreversible adiabatic process.* Consider the P–Θ diagram of Figure 14.18. For reference, the reversible expansion from $\sum_1$ to some state $\sum_3$, caused by slow removal of the sand, is represented by a solid curve. The course of the irreversible process cannot be wholly characterized on the diagram, because at a given time no single values of P and Θ characterize the system. It is helpful to consider average values of P and Θ, represented by the dotted curve, but only the end point $\sum_2$ is of importance. The P coordinate of this point is the same as the corresponding coordinate in the reversible case, because the same amount of sand has been removed. But

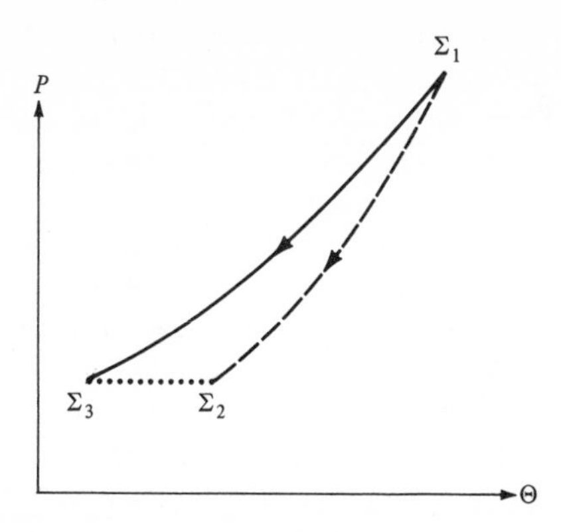

FIGURE 14.18. *The curve from Σ_1 to Σ_3 represents a reversible expansion of a piston. The dashed curve represents the average course of an irreversible expansion that ultimately results in a gas state Σ_2 with the same pressure as Σ_3. Properties of Σ_2 can be deduced by imagining the effect of the reversible cooling necessary to bring the system from Σ_2 to Σ_3 (dotted curve).*

experiment shows that there is a "wasteful" extra heating in the fast irreversible expansion, hence the higher temperature coordinate at point $\sum_2$.

To determine information about the state of the system at $\sum_2$, we adopt the frequently used stratagem of imagining a reversible process that would bring the system to a state with known properties. The state of the system could be brought from $\sum_2$ to $\sum_3$ by a slow reversible cooling. Since heat would have to be removed from the system, dQ is negative in the formula $dS = dQ/\Theta$ and the cooling would result in a decrease in entropy relative to the uniform value of the entropy which is characteristic of the states along the reversible curve between $\sum_1$ and $\sum_3$. This reasoning shows that the *entropy increases in the irreversible adiabatic expansion* from $\sum_1$ to $\sum_2$.

The same reasoning applies to adiabatic compression. If it is reversible it is isentropic, but *irreversible adiabatic compression increases entropy* (Exercise 1). Similar results are obtained in other experiments and thought experiments with systems, like the piston, which are insulated from their surroundings so that there is no heat transfer across their boundaries. Indeed, presumably one must exclude the transfer of anything at all across the boundaries. In isolated systems, then, one is led to the hypothesis that the entropy never decreases, and only stays the same in reversible processes.

A related hypothesis is that the entropy in the universe never decreases, for the universe is "isolated" in the sense that nothing can enter it from "elsewhere." But we must beware of generalizing too far from our mundane experiences. In most heat transfer experiments, it is hard to trace any influence of surroundings farther away than the building, or perhaps the city in which the experiment is being conducted. It is perhaps not surprising, therefore, that there are unanswered questions involved in truly cosmological thermodynamics.* To evade such questions, let us hypothesize that *the entropy of a system plus its surroundings never decreases*, and leave somewhat vague the definition of "surroundings."

A useful property of entropy is implied by the statement in italics that concludes the last paragraph. This property applies to systems that change irreversibly but slowly enough so that they can still be characterized by a single temperature Θ. (An example is provided by a system composed of two constituents which are allowed to mix, or by a system containing a slowly moving but "nonfrictionless" piston.) We know that if the change were reversible, then the entropy change would be given by $\int \Theta^{-1}\, dQ$. We assert that in the irreversible case the entropy change exceeds this quantity.

To back up our assertion, we begin with the observation that the heat flow into the system should have an effect that is independent of the type of surrounding. We therefore imagine surroundings that have a uniform temperature $\bar{\Theta}$ which is equal at any time to the temperature of the system.† No irreversible aspects will be allowed in our imagined surroundings, so the slow temperature change of the surroundings will be reversible.

Let S_1 and S_2 denote the values of entropy in the initial and final states of the system, and let $\bar{S}_1$ and $\bar{S}_2$ denote the corresponding entropies of the surroundings. Using the fact that the heat flowing *into* the system is flowing reversibly *out of* the surroundings, we have

$$\bar{S}_2 - \bar{S}_1 = \int \bar{\theta}^{-1}\, dQ.$$

But by the italicized statement above

$$S_2 + \bar{S}_2 - (S_1 + \bar{S}_1) \geq 0.$$

Consequently, since $\Theta = \bar{\Theta}$,

$$S_2 - S_1 \geq \int \Theta^{-1}\, dQ.$$

Our discussion can be summarized by the postulation of another law of nature.

SECOND LAW OF EQUILIBRIUM THERMODYNAMICS

First Version. Suppose that an *isolated* system changes from a homogeneous state with entropy S_1 to a homogeneous state with entropy S_2. Then $S_2 \geq S_1$, with equality holding only for a reversible change.

* See, for example, B. Gal-or, "The Crisis About the Origin of Irreversibility and Time Anisotropy," *Science 176*, 11–17 (1972).

† No heat flows in the absence of a temperature difference. Thus it would be more accurate to speak of the limit as the heat conductivity of the surroundings increases without bound, so that a vanishingly small temperature difference between system and surroundings is required to generate dQ.

Second Version. Suppose that a *nonisolated* system changes *slowly* from a homogeneous state with entropy S_1 to a homogeneous state with entropy S_2. (The change must be slow enough so that at any moment the system is characterized by a single temperature Θ.) Then

$$S_2 - S_1 \geq \int \Theta^{-1}\, dQ,$$

with equality holding only for a reversible change.

As for the connection between entropy and the availability of energy for work, we mention a revealing deduction by Lord Kelvin. Kelvin considered processes that operate among various heat reservoirs at temperatures Θ_0, $\Theta_1, \Theta_2, \ldots,$ where $\Theta_0 \leq \Theta_1 \leq \Theta_2 \cdots$. He showed that such a process causes an amount of energy to become unavailable for further work that is equal in magnitude to Θ_0 times the change in entropy brought about by the process.

EXERCISES

1. The text discusses entropy changes in reversible and irreversible expansion experiments that use the insulated piston apparatus depicted in Figure 14.14. Provide a similar discussion for the corresponding compression experiments.
2. If $S = S(E, V)$, deduce the following equation from the facts (i) that $dS = \Theta^{-1}P\, dV + \Theta^{-1}\, dE$ is an exact differential, and (ii) that the order of second partial derivatives is immaterial (if these derivatives are continuous):

$$\frac{\partial \Theta}{\partial V} + \Theta \frac{\partial P}{\partial E} - P \frac{\partial \Theta}{\partial E} = 0 \qquad (V, E \text{ independent}).$$

(Similar relations are valid when other variables are regarded as independent. Derivation of some of these provides good practice in the technique of partial differentiation. See Exercise 1 of Appendix 13.1.)

Appendix 14.2 Some Historical Remarks

In the next paragraphs we shall sketch very briefly some points in the early historical development of continuum mechanics so that the reader may begin to see that it was developed only after years of effort by brilliant men. In doing so, we lean heavily on a characteristically penetrating and eloquent article by Clifford Truesdell,* to which we refer the reader for much more detail and also for assessment of the role of experiment in the development of mechanics.

* "A Program Toward Rediscovering the Rational Mechanics of the Age of Reason," *Arch. Hist. Exact Sci.* *1*(1), 3–36 (1960).

That Newton's laws are only the vital first step in the development of continuum mechanics is apparent from the fact that Newton himself could not solve a number of the major problems of the subject. Indeed, many of these problems are not solved today.

For example, Newton tried to find the force on a body immersed in flowing air by considering the motion of an elastic solid moving through a lattice of equally spaced elastic particles. The resulting theory gave a resistance proportional to the density of the air, the cross-section area of the body, and the square of the speed. It accounts poorly for the force on the front of the body and not at all for the relatively low pressures at the rear that are responsible for the major part of the resistance. Late in the nineteenth century, Newton's theory was used to "prove" that airplane flight was impossible because of the large resistance that would be encountered. Its influence lingered into the twentieth century in wind pressure calculations required by building codes.* This is not surprising, since it was work at the beginning of the twentieth century by Prandtl that opened the way for the good understanding of the resistance of streamlined bodies that we have today. Even this understanding is incomplete, while bluff body resistance remains largely an unsolved problem. (See Chapter 3 in II.)

Newton did *not* try to ascertain the resistance of a body by attempting to find an approximate solution to certain partial differential equations that followed from his laws. Instead he tried to solve the problem by means of various particular assumptions and special insights. This is typical. Only after years of experience with particular problems did it become clear which assumptions were valid and which invalid. More study was necessary before it was possible to say which of the valid assumptions were fundamental and which could be deduced from fundamental assumptions. Most difficult of all was the distillation of "ways of looking at things" into clearly defined principles.

An instance of a "failure" to recognize what later turned out to be a vital general principle is provided by some work on the vibrating string by the English mathematician Brook Taylor. In 1713 Taylor applied Newton's law to a differential element of the string, but he did not see that equations of motion can be derived by applying Newton's second law to each infinitesimal part of a body. "He did not recognize the result as a differential equation of motion, and his (further) work rests on confusing and partially erroneous assumptions" (Truesdell).

The first explicit derivation of an equation of motion by application of Newton's laws to an infinitesimal element appeared in John Bernoulli's *Hydraulics* (1739). In this treatment of the one-dimensional motion of incompressible inviscid fluid, Bernoulli introduced as a dependent variable

* P. F. Nemenyi, "The Main Concepts and Ideas of Fluid Dynamics in Their Historical Development," *Arch. Hist. Exact Sci.*, 2(1), 52–86 (1960).

what he called *internal force* and what we would now call pressure. The pressure exerted by a wall or by a column of liquid in hydrostatics had been used before. Nevertheless, although he explained it badly and hardly exploited it, John Bernoulli made a major conceptual advance by introducing the *unknown* internal force.

We come now to the titan Leonard Euler (1707–1783), who "calculated without apparent effort, as men breathe, or as eagles sustain themselves in the wind" (Arago). Euler's works, comprising numerous volumes, codified great quantities of earlier material and made profound contributions to every kind of mathematics known in his time. "His notation is almost modern—or perhaps we had better say that our notation is almost Euler's" (Struik). Even total blindness during the last 17 years of his life seemed only to heighten the quality of Euler's work.*

We can only touch upon a few of Euler's most important discoveries. Two vital contributions appeared during the 1750s. First, a sharpened appreciation of the difference between kinematics and dynamics led Euler to an understanding of mass conservation. Second, he applied John Bernoulli's barely exploited concept of internal force for one-dimensional flows to flow in pumps and turbines, and then generalized to our modern notion of the internal normal pressure in three-dimensional flow of inviscid fluid.

For years Euler tried to reconcile the theory of flexible lines with that of elastic bands. The former uses equilibrium of forces, the latter equilibrium of moments. In 1771 he saw that a shear as well as a tension is necessary to account for the force exerted on an isolated element of an elastic line. This was the first use of a general stress vector.

It appears that Euler was led to the postulation of the conservation of moment of momentum by the work just mentioned and by work in 1744 on many linked bars in a plane. Thus decades of effort culminated in 1776 with the publication of Euler's laws of mechanics.

Although Euler introduced an unknown stress vector in his study of elasticity, the vital general concept of stress vector was not introduced for 56 more years. This was accomplished by no less a person than Cauchy, who also did fundamental work in the theory of complex functions, formulated the calculus as we now learn it, and made a number of important early investigations of finite groups.

Even the few paragraphs we have devoted to some historical points suffice to show that the development of continuum mechanics required years of work by men of the highest ability. One more point remains to be made. It is that although the principles of continuum mechanics can now be reduced to a few lines, thanks to a distillation by men of genius, nevertheless, secure understanding requires much study.

* But even geniuses are not flawless. For example, Newton and Euler made an incorrect physical assumption when they tried to calculate the speed of sound. See Section 15.3.

By contrast, calculational methods are an example of ingenious discoveries that are readily appreciated. For instance, consider Euler's discovery that linear ordinary differential equations with constant coefficients have exponential solutions. In a few hours an able college sophomore can learn all there is to know about this matter. It is usually otherwise with a new physical concept. To a student, at first the new way of looking at things often seems simple and natural. Experience indicates, however, that if an individual is to gain a thorough understanding of a hard-won concept he must personally experience some of the painful intellectual struggle that led to it. A lecture or a text can do no more than point the way.

The phrase "intuitively obvious" therefore does not belong in a discussion of the basic principles of continuum mechanics, or in a discussion of the basic principles of any branch of modern science. A more appropriate watchword is that found on one of the humorous cards that are sometimes displayed in offices and laboratories: "Anyone who remains calm amidst all this confusion simply does not understand what is going on."

Chapter 15
Inviscid Fluid Flow

As was discussed in Section 14.5, the field equations describe properties, such as conservation of mass, which are common to all continua, whereas constitutive equations describe the qualities that distinguish one particular kind of continuum from another. We begin this chapter with a discussion of one of the simplest constitutive equations, the one for an inviscid or perfect fluid. (As we shall see, this equation holds for *all* fluids when there is no motion.) We then turn to the formulation and solution of some specific problems and thus begin to understand the power and defects of the inviscid fluid model.

The first specific class of problems treated deals with the fate of small disturbances to a quiescent horizontal fluid layer whose density varies with height. The important concept of stability is introduced, the technique of linearization is used, and the resulting eigenvalue problems are examined for qualitative behavior or are solved in special cases. We see that the inviscid model provides reasonable predictions for situations of physical interest.

We next consider compression waves in gases. Our discussion touches on small-amplitude sound waves and large-amplitude shock waves. We encounter solutions to the linearized wave equation which propagate without changing shape, and we examine deforming simple wave solutions to the appropriate system of nonlinear partial differential equations. Solutions of the latter type develop such rapid variation that virtually discontinuous shock waves result. All these predicted flows are well verified experimentally.

Finally, we employ separation of variables to find a two-dimensional flow of a uniform stream past a circular cylinder. This flow does not seem in accord with experience, for a calculation indicates that the net drag force on the cylinder is zero. Further investigation shows that the same result holds for the flow of a uniform stream past quite general bodies. For this problem, then, the inviscid model does not seem satisfactory. A discussion of how viscous forces cause drag is a central motif of Chapter 3 in II.

15.1 Stress in Motionless and Inviscid Fluids

The central object of the present section is to characterize the stress vector in a motionless fluid and to give evidence that the result obtained could also be useful for a class of fluids in motion. A molecular model will provide the basis for our initial inference.

MOLECULAR POINT OF VIEW

Consider an assemblage of many tiny mass points moving rapidly about with no preferred direction and colliding elastically. This is a reasonable and familiar "perfect gas" molecular model of a gas *at rest*. (Molecules have a preferred direction in a moving gas.) From this model, as in elementary physics texts, we deduce that the molecular bombardment of a surface element is responsible for a purely normal stress on that element. The tangential force components add to zero because there is no preferred direction. For the same reason, the magnitude of the **pressure** (normal force per unit area) on an element should be independent of the element's orientation. Thus we assume that the stress vector $\mathbf{t}$ satisfies

$$\mathbf{t}(\mathbf{x}, t, \mathbf{n}) = -p(\mathbf{x}, t)\mathbf{n}, \tag{1}$$

where p is the pressure.

The elastic mass point molecular model of the previous paragraph is obviously inappropriate for liquids. A summary of progress in the study of more appropriate molecular models can be found, for example, in Seeger and Temple (1965). But without delving into the kinetic theory of liquids, we can guess that some sort of jostling will replace the elastic collisions of the perfect gas. There will still be no preferred direction so that (1) can be expected to hold in a motionless liquid.

Although solids are not the object of our present investigation, we should perhaps mention the very different behavior enforced by their molecular structure. Molecules that are normally constrained to more or less preferred locations can generally adjust to any time-independent force by a fixed displacement, with a concommitant alteration of the various molecular interactions. Consequently, just because the solid is motionless, there is no reason to expect a simple stress distribution within it.

It is also appropriate here to recall that many materials are not easily classified as "solids," "liquids," or "gases." An example is the children's toy Silly Putty, which bounces if a ball of it is thrown on the floor but slowly flows into the shape of its container if it is left undisturbed. We shall not concern ourselves here with such borderline cases. We shall use the words "solid, liquid, gas" according to their meaning in ordinary discourse. (Thus a "fluid" is either a liquid or a gas.) Our constitutive relations will apply to typical representatives of each class of material: the solid steel, the liquid water, and the gas air.

CONTINUUM POINT OF VIEW

Let us reconsider stress in a motionless fluid, this time using common experience of fluid as a continuum. In doing so, we shall find it useful to distinguish between **shear stresses** that act on an element in a direction perpendicular to its normal $\mathbf{n}$ and **normal stresses** that act along $\mathbf{n}$.

It is noteworthy that shear stresses are particularly apparent in liquids like cold molasses, which we call "very viscous." As an illustration, think of how difficult it is to slice through cold molasses with a knife. The edge of the knife blade is so thin that this resistance must be due to shear stresses acting on the flat part of the blade (Figure 15.1). The faster you try to slice, the harder it is.

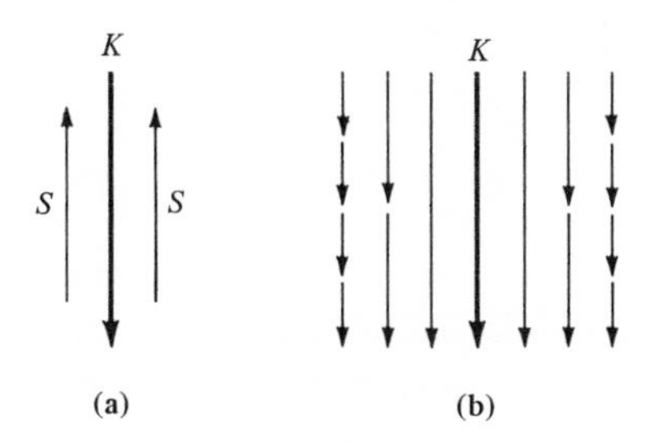

FIGURE 15.1. *Knife K slicing downward through cold molasses.* (a) *S denotes the resisting shear stress exerted on the knife by the fluid.* (b) *Arrows indicate typical fluid velocities.*

If you turn the knife sideways and try to "spread" the molasses, the faster you want to make one layer of molasses flow over another, the harder you have to push (Figure 15.2). These observations could be summed up in the qualitative graph of Figure 15.3. Note that small relative speeds are associated with small shear stresses. One expects that the shear stresses will vanish for a motionless fluid, wherein relative velocities of adjacent material are always zero.

We are led to assume that *for a motionless fluid, stresses are purely normal.* That is, $\mathbf{t}$ must be in the direction of $\mathbf{n}$, so

$$\mathbf{t}(\mathbf{x}, t, \mathbf{n}) = -p(\mathbf{x}, t, \mathbf{n})\mathbf{n}, \tag{2}$$

for a scalar function p. The present continuum argument thus leads to an assumption that is the same as (1) except that at the moment we cannot rule out

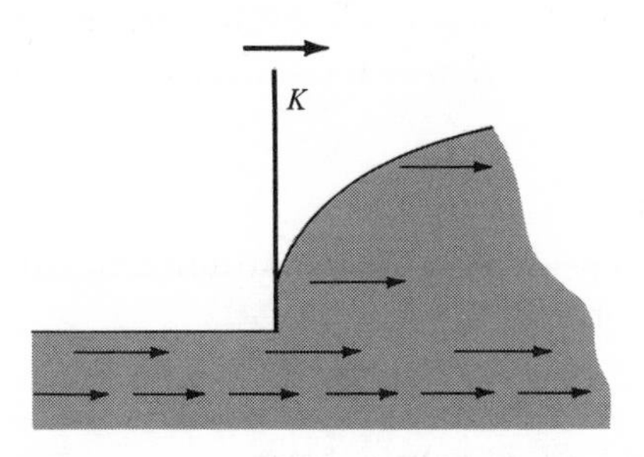

FIGURE 15.2. *Knife K moving to the right and spreading molasses. Arrows in the fluid indicate typical velocity vectors.*

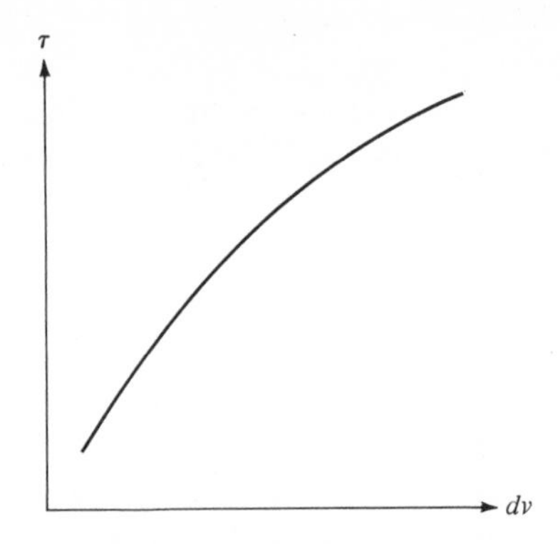

FIGURE 15.3. *Qualitative plot, deduced from ordinary experience, of the shear stress* τ *versus dv, the relative velocity of adjacent layers of fluid.*

a dependence of p on $\mathbf{n}$. But from the relation between the stress vector and the stress tensor

$$t_j(\mathbf{x}, t, \mathbf{n}) = \sum_{i=1}^{3} n_i T_{ij}(\mathbf{x}, t)$$

we find using (2) that

$$-p(\mathbf{x}, t, \mathbf{n})n_j = \sum_{i=1}^{3} n_i T_{ij}(\mathbf{x}, t),$$

which implies that

$$T_{ij}(\mathbf{x}, t) = -p(\mathbf{x}, t, \mathbf{n})\delta_{ij}, \tag{3}$$

where δ_{ij} is the Kronecker delta. Since the left side of (3) is independent of $\mathbf{n}$, it must be that $p(\mathbf{x}, t, \mathbf{n})$ is actually independent of $\mathbf{n}$. Thus we find from a continuum viewpoint that the pressure p on an element of area located at point $\mathbf{x}$ must be independent of the normal vector $\mathbf{n}$—just as we inferred above from the lack of a preferred direction of molecular motion.

NOTE. Tiny gas bubbles in carbonated beverages are spherical. This confirms our view that the magnitude of the pressure in a motionless fluid is independent of the orientation of the surface element over which the stress acts.

HYDROSTATICS

Since p does not depend on $\mathbf{n}$, the constitutive equation for a motionless fluid takes the final form

$$T_{ij}(\mathbf{x}, t) = -p(\mathbf{x}, t)\delta_{ij}. \tag{4}$$

For a motionless fluid we, of course, have

$$\mathbf{v} = \mathbf{0}. \tag{5}$$

With (4) and (5), Equation (14.2.19) for linear momentum conservation becomes simply

$$\rho \mathbf{f} - \nabla p = \mathbf{0}, \tag{6}$$

which is the **hydrostatic equation**. It is easily seen that unless there are heat effects, all the other field equations are automatically satisfied.

As a simple application of hydrostatics let us check that our results are compatible with the fact that one never sees a block of fluid resting motionless on a solid surface. Suppose that there were such a motionless block of fluid, of height H and width W, and with a coordinate origin located in the center of its base (Figure 15.4). We attempt to solve (6), where the body force is due to gravity,

$$\mathbf{f} = -g\mathbf{k}, \tag{7}$$

and the fluid is subject to a uniform atmospheric pressure p_a:

$$p = p_a \qquad \text{when } 0 \le z \le H, \quad x = \pm\tfrac{1}{2}W \text{ (on the sides);} \tag{8a}$$

$$p = p_a \qquad \text{when } z = H, \quad -\tfrac{1}{2}W \le x \le \tfrac{1}{2}W \text{ (on the top).} \tag{8b}$$

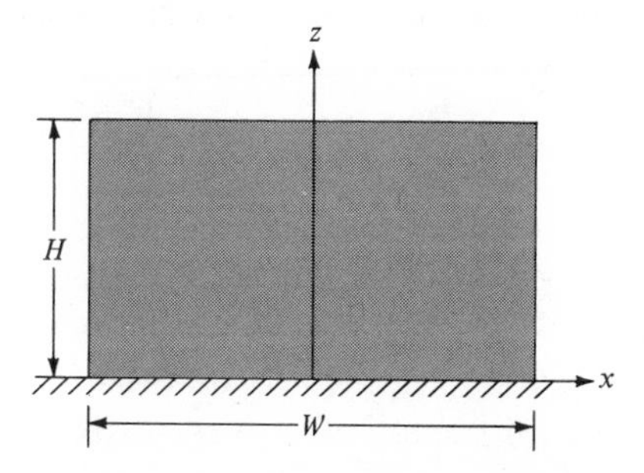

FIGURE 15.4. *Postulated stationary block of fluid resting on a solid.*

As is customary, when solving specific problems we use x, y, and z as spatial coordinates rather than the more cumbersome x_1, x_2, and x_3.

If we use (7), (6) becomes

$$-p_x = 0, \qquad -\rho g - p_z = 0. \tag{9}$$

Equations (9) and (8b) require that

$$p = -\rho g(z - H) + p_a,$$

but we are unable to satisfy (8a) with this pressure. As expected, our equations show that Figure 15.4 cannot depict a fluid at rest. On the other hand, the hydrostatic equations and boundary conditions would be satisfied if a wall were present at $x = \pm\tfrac{1}{2}W$, in which case (8a) should be deleted.

Suppose that a block of material is placed on a table and does not deform. Our argument would seem to allow us to rule out the hypothesis that the block is a fluid. But it could be that the block would undergo fluid-like deformation if observed for a longer period. We should say that "the block does not behave like a fluid when subjected to gravity-like forces over a period of hours." The more precise terminology reminds us that the abstraction "fluid" is useful for understanding a certain range of phenomena. It diminishes our surprise on learning that "solid rock" can usefully be regarded as fluid in the study of deformations that occur over geological time scales.*

INVISCID FLUIDS

Our discussion of slicing cold molasses associated large shear stresses with what are commonly called "fluids of high viscosity." On the other hand, it is easy to slice a knife through water or air. It therefore appears reasonable to assume, as a first approximation at any rate, that shear stresses are always zero in these important fluids. Thus we define a class of **inviscid fluids**, for which, *even when they are in motion, the stress tensor satisfies the constitutive equation* (4). The equation for conservation of linear momentum (14.2.19) now becomes the famous **Euler momentum equation for inviscid flow**:

$$\boxed{\rho \frac{D\mathbf{v}}{Dt} = -\nabla p + \rho\mathbf{f}.} \tag{10}$$

The remainder of this chapter contains some examples that show successes and failures of the inviscid† assumption. We shall use as governing equations the Euler momentum equation just derived, an equation for conservation of mass, and (in our study of compression waves) appropriate equations of

* In a note in *Physics Today* (Vol. 17, p. 62, 1964) M. Reiner ascribes the knowledge that rock flows to the Prophetess Deborah. "In her famous song after the victory over the Philistines, she sang, 'The mountains flowed before the Lord.' When, over 300 years ago, the Bible was translated into English, the translators . . . translated the passage as 'The mountains *melted* before the Lord'—and so it stands in the authorized version. But Deborah knew two things. First, that the mountains flow, as everthing flows. But second, that they flowed before the Lord, and not before man, for the simple reason that man in his short lifetime cannot see them flowing, while the time of observation of God is infinite. We may therefore well define as a nondimensional number the Deborah number

$$D = \frac{\text{time of relaxation}}{\text{time of observation}}.$$

The difference between solids and fluids is then defined by the magnitude of D."

From a scientific point of view, then, one should not ask whether a given piece of material *is* a solid or a fluid. If the Deborah number is large, the material *should be regarded* as a solid. Otherwise it should be regarded as a fluid.

† The term "perfect" used to be applied to fluids that satisfy (10). This term perhaps reflects an optimistic nineteenth-century view that only failures of its Maker could cause nature to deviate from behavior that is relatively accessible to mathematical analysis.

state. We shall require that no fluid penetrate the boundary. But in the absence of shear stresses, we have no reason to postulate any restriction on how our inviscid fluid can slip along the boundary.

EXERCISES

‡**1.** Show that the magnitude of the net pressure force on a body that is completely submerged in a liquid is equal to the weight of the liquid displaced by the body. Show that the direction of the force is upward (i.e., opposite to the direction of gravity).

2. Assuming that gravity acts in the direction opposite to that of the z axis, derive in detail *from first principles* the hydrostatic equation

$$\frac{dp}{dz} + \rho g = 0.$$

Do this by equating to zero the net vertical forces on a horizontal fluid slice.

The next two exercises deal with the useful concept of the **stream function.**

3. (a) Consider two-dimensional steady fluid flow. Show that the continuity equation implies that $\nabla \wedge (-\rho v \mathbf{i} + \rho u \mathbf{j}) = \mathbf{0}$. Thus there exists a function ψ such that $\rho u = \psi_y$, $\rho v = -\psi_x$. Why? Show that the same relations hold for unsteady flow when $\rho =$ constant.

(b) Justify the name "stream function" for ψ by showing that the curves $\psi =$ constant are streamlines. (Thus determination of ψ is tantamount to obtaining streamlines for the flow: also ρu and ρv can be found merely by differentiation.)

‡**4.** (a) Let C be a curve joining a pair of streamlines $\psi = \psi_0$ and $\psi = \psi_1$ of a two-dimensional flow. (See Figure 15.5.). Let the arc length s

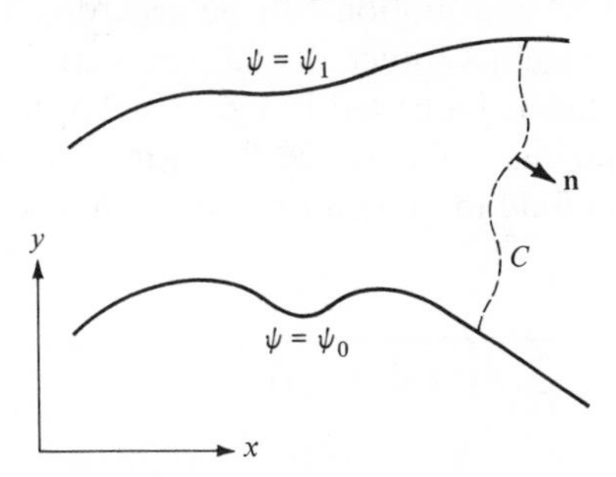

FIGURE 15.5. *Two streamlines $\psi = \psi_1$ and $\psi = \psi_0$. Computation of an integral along an arbitrary curve C (with unit normal* **n**) *provides a physical interpretation for the difference $\psi_1 - \psi_0$.*

of C have the values s_0 and s_1 when C intersects these two curves, $s_1 > s_0$. Show that

$$\int_{s_0}^{s_1} \rho \mathbf{v} \cdot \mathbf{n} \, ds = \psi_1 - \psi_0,$$

where $\mathbf{v}$ is the velocity vector and

$$\mathbf{n} = \frac{dy}{ds}\mathbf{i} - \frac{dx}{ds}\mathbf{j}.$$

Interpret physically.

(b) Consider two-dimensional motion of an inviscid fluid of constant density. Let a conservative body force $\mathbf{f}$ be given in terms of a potential Φ by $\mathbf{f} = \nabla\Phi$. By taking the curl of the momentum equation, show that

$$\frac{\partial(\nabla^2\psi)}{\partial t} + \frac{\partial(\nabla^2\psi, \psi)}{\partial(x, y)} = 0, \qquad \nabla^2 = \frac{\partial^2}{\partial x^2} + \frac{\partial^2}{\partial y^2}.$$

(The three components of a vector equation are thus replaced by a single scalar equation.)

‡5. A **barotropic fluid** is defined to be a fluid in which the pressure p depends only on the density: $p = p(\rho)$. (An example is provided by an ideal gas in isentropic flow.) In such a fluid show that $\nabla \wedge (\rho^{-1}\nabla p) = \mathbf{0}$. If $\nabla \wedge \mathbf{q} = \mathbf{0}$, then $\mathbf{q} = \nabla r$ for some r. Advanced calculus books give methods for determining r. Use such a method to determine the scalar function of which $\rho^{-1}\nabla p$ is the gradient in a barotropic fluid.

6. This exercise shows that a velocity potential continues to exist for a portion of inviscid barotropic fluid subject to a conservative body force if it exists initially for that portion of fluid. This justifies the assumed existence of the velocity potential in many problems, in particular those in which the motion can be regarded as commencing from a motionless state. (Another derivation of this important result, using tensor calculus, is requested in Exercise 2.3.10 in II.)

(a) The starting point is the following equation of motion for an inviscid fluid in *material variables* that was obtained in Exercise 14.2.7:

$$\sum_{j=1}^{3} \rho\left(\frac{\partial^2 x_j}{\partial t^2} - \mathscr{F}_j\right)\frac{\partial x_j}{\partial A_k} = -\frac{\partial p}{\partial A_k}; \qquad k = 1, 2, 3.$$

Suppose that external forces are conservative, so that the force vector is the gradient of some scalar potential Ω ($\mathscr{F}_j = -\partial\Omega/\partial A_j$). Suppose further that the fluid is barotropic [$p = p(\rho)$]. By integrating the above equations with respect to time between zero and t, show that they can be written

$$\sum_{j=1}^{3} \frac{\partial x_j}{\partial t} \frac{\partial x_j}{\partial A_k} - V_k^{(0)} = -\frac{\partial \chi}{\partial A_k}; \qquad k = 1, 2, 3,$$

where

$$\chi = \int_0^t \left[\int \frac{dp}{\rho} + \Omega - \tfrac{1}{2}(V_1^2 + V_2^2 + V_3^2) \right] dt, \qquad V_k \equiv \frac{\partial x_k}{\partial t}.$$

To do this, use integration by parts and show that

$$\left. \frac{\partial x_k}{\partial A_k} \right|_{t=0} = 1$$

directly from the definition of the partial derivative.

(b) Suppose a velocity potential exists at $t = 0$ for a finite portion of the fluid:

$$d\phi_0 = \sum_{i=1}^{3} V_i^{(0)} \, dX_i; \qquad V_i^{(0)} = V_i|_{t=0}.$$

Derive

$$\sum_{k=1}^{3} \frac{\partial x_k}{\partial t} dx_k - \sum_{i=1}^{3} V_k^{(0)} \, dA_k = -d\chi$$

and thereby show that a velocity potential exists for the same portion of fluid at all other times.

7. This exercise shows that under certain circumstances the momentum equation (10) for inviscid flows can be integrated, yielding the Bernoulli equation. As later exercises indicate, the Bernoulli equation provides a very useful relation between flow speed and pressure.

(a) Verify the identity

$$\mathbf{v} \cdot \nabla \mathbf{v} = \tfrac{1}{2} \nabla(|\mathbf{v}|^2) - \mathbf{v} \wedge (\nabla \wedge \mathbf{v})$$

by comparing the components of each side of the equation. (This result is Theorem 7 of Section 2.3 in II. There a more elegant proof is furnished by the methods of tensor calculus.)

(b) A fluid motion satisfying $\nabla \wedge \mathbf{v} = \mathbf{0}$ is termed **irrotational**. (A discussion at the beginning of Section 3.1 of II shows why the term is appropriate.) For irrotational fluids, $\mathbf{v} = \nabla\phi$ for some velocity potential ϕ. Also, if the force field $\mathbf{F}$ is conservative, then $\mathbf{F} = -\nabla\Phi$ for a force potential Φ.

For the irrotational motion of an inviscid barotropic fluid subject to a conservative body force, deduce the **unsteady Bernoulli equation**:

$$\frac{\partial \phi}{\partial t} + \frac{1}{2}|\mathbf{v}|^2 + \int \rho^{-1}\, dp + \Phi = \text{a function of time.}$$

NOTE. An important special case is steady flow of fluid with uniform density, in which case the Bernoulli equation clearly becomes

$$\frac{1}{2}|\mathbf{v}|^2 + \frac{p}{\rho} + \Phi = \text{constant.}$$

‡8. An inviscid fluid of uniform density is in steady irrotational flow along a horizontal pipe of variable cross section. Assume that the speed is uniform across any given cross section. If Δp is the pressure difference indicated by a differential pressure gauge connected at two locations having cross sections s_1 and s_2, find a formula for the mass m of liquid flowing per second. The answer is the square root of a certain quantity. Can this quantity be negative so that the mass flux is imaginary?

9. Consider some water in a tin can (right circular cylinder) rotating with constant angular velocity ω. The velocity field for a fluid rotating as a solid body is

$$u = -\omega y, \qquad v = \omega x, \qquad w = 0.$$

(This is in ordinary cartesian coordinates—*not* polar coordinates.) The above velocity field satisfies the boundary conditions on the tin

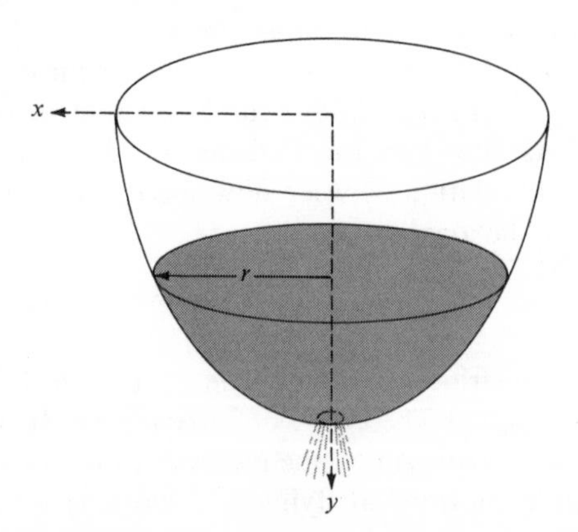

FIGURE 15.6. *Vessel in the form of a figure of revolution. At each value of y, the cross section is a circle of radius r and area $S = \pi r^2$. Liquid flows out through a small hole in the bottom. The vessel is to be designed so that its surface descends at a constant speed V.*

can. Show that the inviscid incompressible equations of motion, *with gravity*, are also satisfied. Find the form of the free surface, assuming that the pressure at the fluid surface equals the air pressure A (A a constant).

‡10. An axisymmetric hourglass is to be designed so that the surface of the incompressible inviscid liquid it contains descends at a constant rate V. (See Figure 15.6.) Show that $y \sim x^4$ and find the proportionality constant.

15.2 Stability of a Stratified Fluid

Slime mold amebae, after exhaustion of their food supply, suddenly begin to aggregate into more or less regularly spaced collection points. Fluid flowing smoothly through a pipe develops violent random eddies if the driving pressure gradient is increased too far. A metallic structure collapses if a slight additional load is placed on it. All these phenomena and many more can be regarded as exemplifying an **instability of equilibrium**. By a **state of equilibrium** we mean a steady state that is permitted by the governing equations of the problem. We say that a system is **perturbed** from equilibrium if it is put in a state that differs somewhat from the equilibrium state. The **perturbation** is what must be added to the equilibrium state to achieve the perturbed state. If an initial perturbation of equilibrium becomes larger in some sense as the system evolves, we say that the **equilibrium state is unstable to that perturbation**.

An instability problem of biological interest was briefly discussed in Section 1.3. It will help in motivation if that discussion has at least been skimmed. More important, before proceeding the reader should be familiar with the pendulum instability calculations presented (independently of surrounding material) in Section 11.1, for here we shall examine material that is related to, but is more complicated than, stability theory for the pendulum—namely, the study of the stability of bottom-heavy and top-heavy layers of fluid.

The simple pendulum is a mechanical system with but a single degree of freedom. Its state can be described in terms of a single variable, the angular deviation of the rod from a reference value. Thus the evolution of the pendulum in time is governed by a single ordinary differential equation. By contrast, a layer of fluid can be said to have an infinite number of degrees of freedom. In the present case, description of the state of the fluid at a given moment requires specification at every point in the layer of three velocity components, a pressure, and a density. The evolution of the system in time is governed by a set of partial differential equations. But in spite of these profound differences, one suspects (correctly) that there are many similarities in the behavior of a top-heavy fluid and an inverted pendulum, and in the behavior of a bottom-heavy fluid and a pendulum in "normal" position.

The motion of a pendulum is studied so carefully not because of its direct technical importance, but because it provides a relatively familiar and readily understandable prototype of many physical situations. Similarly, the inviscid fluid motion to be studied below, although of no small interest in itself, acquires considerable added importance when it is regarded as a prototype of the natural phenomena that are manifestations of instability in continuously distributed systems.

An extreme case of the sort of problem that we shall consider is a situation that involves two immiscible fluids, say oil and water. Each of the two fluids separately can be regarded as possessing a uniform density, but this is not true of the combination of fluids. During the course of motion, the density at a particular point in space might first be that of water, then later that of oil, again that of water, etc. In spite of this, the density associated with a given small packet of fluid particles remains constant as that packet moves about. Generalizing, we shall consider in this section a fluid that is incompressible but not of uniform density.

We shall see that if gravity acts vertically downward, then in principle it is possible for a horizontal layer of incompressible fluid to remain motionless for any distribution of density that depends solely on the vertical coordinate z. Hence we speak in the title of this section of a **stratified fluid**, a fluid "formed in layers" (of different densities). Assuming that the fluid is inviscid, we turn to an investigation of the stability of a stratified horizontal layer. We shall obtain some general results for classes of stratified layers and some particular results for special cases. The first goal of our analysis is to obtain a set of linear partial differential equations for the perturbations in velocity and density. Since the equilibrium solution depends only on the vertical coordinate z, we can then reduce the problem to a single ordinary differential equation of a function of z.

GOVERNING EQUATIONS AND THEIR EXACT EQUILIBRIUM SOLUTION

Consider, then, an incompressible inviscid fluid of varying density that is contained between two rigid horizontal planes located at $z^* = 0$ and $z^* = d$. (Here z^* is the vertical coordinate; x^* and y^* will be the horizontal coordinates.) The governing equations are the following:

Conservation of mass:

$$\frac{D\rho^*}{Dt^*} + \rho^* \nabla \cdot \mathbf{v}^* = 0. \tag{1a}$$

Balance of linear momentum for inviscid fluid:

$$\rho^* \frac{D\mathbf{v}^*}{Dt^*} = -\nabla p^* - \rho^* g \mathbf{k}. \tag{1b}$$

Incompressibility:

$$\frac{D\rho^*}{Dt^*} = 0. \tag{1c}$$

No penetration of horizontal planes:

$$\mathbf{v}^* \cdot \mathbf{k} = 0 \quad \text{for} \quad z^* = 0, d. \tag{1d}$$

All variables have their conventional meaning, except that an asterisk has been added to emphasize that variables are not dimensionless. As usual, $\mathbf{k}$ is the "vertical" coordinate vector and g is the acceleration due to gravity, so $-\rho^* g\mathbf{k}$ is the correct term to represent a constant gravitational force that acts vertically downward.

We introduce dimensionless variables in order to reduce the numbers of parameters that occur explicitly. (See Section 6.2 for a discussion of dimensional analysis.) At our disposal we have the length d and the gravitational acceleration g (dimension: length/time^2). In order to be able to obtain a quantity with the dimensions of mass, we employ a "typical" density R_0 (dimension: mass/length^3). As there is as yet no obvious way to assign a definite value to the constant R_0, we shall leave it unspecified for the present.

We select combinations of d, g, and R_0 so that variables without asterisks, defined as follows, are dimensionless†:

$$\mathbf{x} = \frac{\mathbf{x}^*}{d}, \quad t = \frac{t^*}{(d/g)^{1/2}}, \quad \rho = \frac{\rho^*}{R_0}, \quad \mathbf{v} = \frac{\mathbf{v}^*}{(dg)^{1/2}}, \quad p = \frac{p^*}{R_0\, dg}.$$

If we introduce these variables and also use (1c) to modify (1a), we obtain

$$\nabla \cdot \mathbf{v} = 0, \qquad \rho \frac{D\mathbf{v}}{Dt} = -\nabla p - \rho \mathbf{k}, \tag{2a, b}$$

$$\frac{D\rho}{Dt} = 0; \qquad w = 0 \quad \text{for} \quad z = 0, 1. \tag{2c, d}$$

Here w is the vertical component of velocity, and the operators ∇ and D/Dt are to be interpreted using the new (unstarred) variables. The equations of (2) are essentially the same as those of (1), except that in the nondimensional variables of (2) the distance between the bounding planes and the gravitational acceleration both have the value unity.

Does a motionless layer correspond to a possible solution of (2)? With $\mathbf{v} = 0$, (2a) and (2d) are identically satisfied, while the remaining equations become

$$p_x = 0, \quad p_y = 0, \quad p_z = -\rho, \quad \rho_t = 0, \tag{3a, b, c, d}$$

† Distances are compared to the width d of the layer. A factor of $2^{1/2}$ is omitted because it clutters the equations, but otherwise times are compared to the time $(2d/g)^{1/2}$ that it takes a freely falling particle to traverse the layer, starting from rest. Similarly, velocities are essentially referred to the terminal speed $(2dg)^{1/2}$ attained by a particle falling a distance d, starting from rest.

where the subscripts denote partial differentiation. Equations (3a) and (3b) require that p be independent of x and y, while (3d) requires that ρ be independent of t. With this information, (3c) becomes

$$p_z(z, t) = -\rho(x, y, z), \tag{4}$$

where we have explicitly indicated the remaining possible independent variables in ρ and p_z. If for any times t_1 and t_2 we have

$$p_z(z, t_1) \neq p_z(z, t_2),$$

then there is a contradiction with (4). Consequently, p_z cannot vary as t varies. Similarly, ρ cannot vary as x and y vary. Thus p_z and ρ can depend only on z.

Let

$$\rho \equiv R(z), \tag{5}$$

where $R(z)$ is an arbitrary function of z. Then from (4)

$$p(z) = -\int R(z)\, dz. \tag{6}$$

The constant of integration (which can depend on time) that is implicit in the indefinite integral of (6) can be used to make $p(z) = 0$ at some height $z = z_0$. This merely means that $p(z)$ is measured with respect to a zero value at the reference level z_0. Note that changing the reference level means changing the definition of p by adding a constant. This does not have an effect on the motion, since p occurs only in the form ∇p. Note also that there now appears a number of sensible ways to define the reference density R_0, e.g., $R_0 = R(\frac{1}{2})$ (the density at the center of the layer) or $R_0 = \int_0^1 R(z)\, dz$ (the average density).

We have shown that

$$\mathbf{v} \equiv \mathbf{0}, \qquad \rho = R(z), \qquad p = -\int R\, dz, \tag{7}$$

is an *exact solution* of the governing equations (2) for any (integrable) function $R(z)$. We shall now examine the stability of this solution to small perturbations in a manner analogous to our study of the stability of the equilibrium states of the pendulum.

LINEARIZED EQUATIONS FOR THE PERTURBATIONS

We first define perturbations $\mathbf{v}'$, ρ', and p' to the exact solution (7):

$$\mathbf{v} = \mathbf{0} + \mathbf{v}', \qquad \rho = R + \rho', \qquad p = -\int R\, dz + p'. \tag{8}$$

On substitution of (8) into the governing equations (2), we obtain a set of nonlinear partial differential equations for the perturbation quantities. For example, the incompressibility condition (2c) becomes

$$\begin{aligned} 0 &= (R + \rho')_t + u'(R + \rho')_x + v'(R + \rho')_y + w'(R + \rho')_z \\ &= \rho'_t + u'\rho'_x + v'\rho'_y + w'[R_z + \rho'_z]. \end{aligned} \tag{9}$$

As with the pendulum, we should get significant information even if we limit ourselves to perturbations so small that an expression containing the product of two perturbation terms can be neglected as insignificant compared with expressions containing as a factor only one perturbation term. It should be clearly understood that we restrict ourselves to small perturbations not because only these perturbations are physically significant, but because only with this restriction can we obtain a relatively simple linear problem to solve.

Application of the **linearization procedure*** (deletion of nonlinear terms) to (9) yields

$$\rho'_t + \frac{dR}{dz} w' = 0, \tag{10}$$

while application to the momentum balance condition (2b) gives the three scalar equations

$$Ru'_t = -p'_x, \qquad Rv'_t = -p'_y, \qquad Rw'_t = -p'_z - \rho'. \quad \text{(11a, b, c)}$$

The continuity equation (2a), without approximation, implies the linear equation

$$u'_x + v'_y + w'_z = 0, \tag{12}$$

and boundary condition (2d) requires

$$w' = 0 \qquad \text{for} \quad z = 0, 1. \tag{13}$$

Equations (10)–(13) comprise a linear problem for the perturbations. There are five partial differential equations for the five unknown functions u', v', w', p', and ρ'.

In the set of linear partial differential equations (10), (11), and (12), the independent variables x, y, and t appear only in the form of derivatives. To put it another way, as far as these variables are concerned, the partial differential equations have constant coefficients. Extending the procedure

* Those familiar with scaling (Section 6.3) can better appreciate the linearization. In (9), for example, terms like $u'\rho'_x$ and $v'\rho'_y$ can be neglected while a term like ρ'_t is retained if all dimensionless perturbation terms are small compared to unity. But the nondimensionalization process we used was based on scales comparable to those which describe free fall of a particle. Thus, for example, the dimensionless vertical velocity component v' will be small compared to unity if the actual velocity in the perturbed flow is small compared to the terminal velocity in free fall. This and similar arguments give an idea of what is meant by a "small perturbation."

If we wished to utilize the scaling technique in full, we should (for example) have chosen $(dg)^{1/2}$for ε a velocity scale, $0 < \varepsilon \ll 1$. This would explicitly indicate that we are considering a situation in which the maximum expected speed is small compared to the terminal speed of free fall. Similar introduction of a small parameter ε into the scales for the other dependent variables would lead to the appearance of ε before all nonlinear terms in the governing equations. The stage would then be set for a perturbation analysis, the first equations of which would form the linearized problem that we derived less formally in the main body of the text.

used for linear ordinary differential equations with constant coefficients, we can find solutions in which the dependence on *each* of the variables x, y, and t is exponential—again, on substitution each term will have an exponential factor that can ultimately be canceled.

The previous paragraph partially motivates our search for solutions of the form

$$\begin{bmatrix} u'(x, y, z, t) \\ v'(x, y, z, t) \\ w'(x, y, z, t) \\ \rho'(x, y, z, t) \\ p'(x, y, z, t) \end{bmatrix} = \begin{bmatrix} \hat{u}(z) \\ \hat{v}(z) \\ \hat{w}(z) \\ \hat{\rho}(z) \\ \hat{p}(z) \end{bmatrix} e^{i(k_1 x + k_2 y) + \sigma t}. \tag{14}$$

At this point it may seem that we are only considering a very small class of perturbations. Even if true, however, for these perturbations our equations reduce to ordinary differential equations, which makes their solution relatively easy.* A more telling point, which we shall discuss below, is that our special perturbations can be "added" to give a rather general perturbation.

We remark that as is always the case for linear problems, we can satisfy the requirement that velocity, pressure, and density be real by taking as our final solutions either the real or the imaginary part of the complex solution (14).

In (14), σ may be complex, but *we take k_1 and k_2 to be real.* If either k_1 or k_2 has a nonzero imaginary part, then the perturbations will grow without bound as $x \to \pm\infty$ or as $y \to \pm\infty$. For example, if

$$k_1 = k_1^{(r)} + ik_1^{(i)}, \qquad k_1^{(i)} \neq 0,$$

then

$$|\exp ik_1 x| = \exp(-k_1^{(i)}x) \to \infty \begin{cases} \text{as } x \to +\infty & \text{if } k_1^{(i)} < 0, \\ \text{as } x \to -\infty & \text{if } k_1^{(i)} > 0. \end{cases}$$

The restriction to real values of k_1 and k_2 limits the set of perturbations under discussion, but this set seems to contain enough elements to synthesize all perturbations of interest. It might be helpful to think of the contrasting case of disturbances generated at the entrance region of a long pipe. Here "interesting perturbations" are bounded in time but grow in space.†

Having established the form of perturbation to be considered we can proceed with the calculations. Upon substituting (14) into (10), (11), and (12),

* Investing a great deal of work to obtain a special solution might be unwise, but if only a relatively small amount of work is necessary, then consideration of a special case is probably worthwhile. This is a general rule of procedure in applied mathematics.

† In the last analysis, only complete specification of an initial boundary value problem can determine the class of perturbations that should be considered—but a more informal approach is appropriate in preliminary investigations such as the present one.

and canceling the common exponential factor, we obtain the linear ordinary differential equations

$$\sigma\hat{\rho} + \frac{dR}{dz}\hat{w} = 0, \tag{15}$$

$$R\sigma\hat{u} = -ik_1\hat{p}, \qquad R\sigma\hat{v} = -ik_2\hat{p}, \qquad R\sigma\hat{w} = -\frac{d\hat{p}}{dz} - \hat{\rho}, \quad \text{(16a, b, c)}$$

$$ik_1\hat{u} + ik_2\hat{v} + \frac{d\hat{w}}{dz} = 0, \tag{17}$$

and the boundary conditions

$$\hat{w}(0) = \hat{w}(1) = 0. \tag{18}$$

CHARACTERIZATION OF THE GROWTH RATE σ AS AN EIGENVALUE

Since the boundary conditions apply to $\hat{w}(z)$, we try to eliminate the other variables. This can be done as follows.* Solving (16a) and (16b) for $ik_1\hat{u}$ and $ik_2\hat{v}$ in terms of $\hat{p}$ and then substituting into (17), one obtains

$$\hat{p} = -R\sigma k^{-2}\frac{d\hat{w}}{dz} \tag{19}$$

where

$$k^2 \equiv k_1^2 + k_2^2. \tag{20}$$

One can then use (15) and (19) to write (16c) entirely in terms of $\hat{w}$. *To simplify our notation, we write* $W \equiv \hat{w}$. We finally obtain†

$$\frac{d}{dz}\left(R\frac{dW}{dz}\right) + k^2\left(\sigma^{-2}\frac{dR}{dz} - R\right)W = 0, \tag{21a}$$

$$W(0) = W(1) = 0. \tag{21b}$$

We note at once from (21) that k_1 and k_2 appear only in the combination $k^2 = k_1^2 + k_2^2$. Now the perturbations that we are considering are proportional to

$$\exp[i(k_1x + k_2y)] = \exp(i\mathbf{k}\cdot\mathbf{r}) \qquad \text{where } \mathbf{r} = x\mathbf{i} + y\mathbf{j}, \quad k = k_1\mathbf{i} + k_2\mathbf{j}.$$

* In a given problem, considerable trial and error might precede the discovery of an efficient elimination scheme.

† The equations of (21) form a Sturm–Liouville problem. We shall repeat some of the general discussion of this problem in Section 5.2. The repetition serves the dual purposes of illustrating general ideas in a particular context, and of making this section self-contained.

But [Exercise 1(a)]

$$\exp\left[i\mathbf{k}\cdot\left(\mathbf{r}+\frac{2\pi}{k}\frac{\mathbf{k}}{k}\right)\right]=\exp\left(i\mathbf{k}\cdot\mathbf{r}\right).$$

Thus the perturbations are constant along the lines $\mathbf{k}\cdot\mathbf{r}=$ constant (lines that are perpendicular to the **wave number vector** $\mathbf{k}$). They have a spatial period of $2\pi/k$, where $k=(k_1^2+k_2^2)^{1/2}=|\mathbf{k}|$. The fact that only k enters (21) means that the stability behavior depends only on the spatial period of the perturbations; it does not depend on their orientation. This is to be expected, because there is no preferred direction in the problem—but keep in mind that nonlinear interactions, not considered here, may act preferentially among modes with certain combinations of wave-number vectors.

NOTE. The terms **overall wave number** or **horizontal wave number** are sometimes used for k, the magnitude of $\mathbf{k}$.

There seems little possibility that we can find an explicit formula for the solution of (21) in terms of the function $R(z)$. We do know, however, that the linear second order ordinary differential equation (21) has a general solution of the form

$$W=C_1W^{(1)}(z;k^2,\sigma^2)+C_2W^{(2)}(z;k^2,\sigma^2), \tag{22}$$

where $W^{(1)}$ and $W^{(2)}$ are linearly independent solutions of (21). (We have explicitly indicated the dependence of the solution on the parameters k^2 and σ^2.)

Application of the boundary conditions (18) gives a pair of homogeneous linear algebraic equations for the constants C_1 and C_2:

$$\begin{aligned}C_1W^{(1)}(0;k^2,\sigma^2)+C_2W^{(2)}(0;k^2,\sigma^2)=0,\\C_1W^{(1)}(1;k^2,\sigma^2)+C_2W^{(2)}(1;k^2,\sigma^2)=0.\end{aligned} \tag{23}$$

These equations have the trivial solution $C_1=C_2=0$ corresponding to $W\equiv 0$. This must be the case, for $W\equiv 0$ implies that $u\equiv v\equiv p\equiv\rho\equiv 0$ and an identically zero perturbation to an exact solution is certainly permissible.

In order that there be nontrivial solutions to (23), we must have

$$\begin{vmatrix}W^{(1)}(0;k^2,\sigma^2) & W^{(2)}(0;k^2,\sigma^2)\\W^{(1)}(1;k^2,\sigma^2) & W^{(2)}(1;k^2,\sigma^2)\end{vmatrix}=0, \tag{24}$$

or, defining the function F,

$$F(k^2,\sigma^2)=0. \tag{25}$$

The (homogeneous) equation and boundary conditions (21) thus constitute an **eigenvalue problem** that possesses nontrivial solutions if and only if there is a relation between the parameters k^2 and σ^2. If k is regarded as given, (25) implicitly determines σ^2. Generally, there are many permitted values of σ that correspond to each k.

The consequences for stability are these. Given k_1 and k_2, the x and y wave numbers of the perturbation, we determine k by (20). For a given value of k, values of σ^2 can in principle be obtained from (25). Suppose that for some value of k there exists a corresponding σ with positive real part:

$$\sigma \equiv \sigma^{(r)} + i\sigma^{(i)}, \qquad \sigma^{(r)} > 0.$$

Since

$$|\exp \sigma t| = \exp (\sigma^{(r)} t), \tag{26}$$

the perturbation velocity, pressure, and density will increase exponentially with time. If we can find at least one such growing perturbation to the motionless layer, then we term the motionless state **unstable.** If all possible values of $\sigma^{(r)}$ are negative for a certain class of perturbations, we term the motionless state **stable to that class of perturbations.** If, for a certain class of perturbations, all values of $\sigma^{(r)}$ satisfy $\sigma^{(r)} \leq 0$ and at least one value satisfies $\sigma^{(r)} = 0$, then we term the motionless state **neutrally stable to that class of perturbations.** *

Let us return to the particular problem under study. This will enable us to illustrate the general discussion of the last four paragraphs and will prepare the ground for a somewhat more comprehensive discussion at the end of this section.

QUALITATIVE GENERAL DEDUCTIONS

Given the eigenvalue problem (21), there are two ways of proceeding. The first is to try to make some general deductions about the behavior of the eigenvalue σ, for classes of initial density distributions $R(z)$, without actually solving the equation. The second is to choose a particular function R, solve the problem in detail, and thereby to obtain a complete analysis of the behavior of σ for a particular problem. Let us start with the first approach.

Equation (21a) may have complex solutions. We are going to take as final answers only the real and imaginary parts of (14) so that there is no physical reason to require that W itself be real. With the possibility of complex solutions in mind, then, we multiply (21a) by its complex conjugate $\overline{W}$ and integrate from zero to 1:

$$\int_0^1 \overline{W} \frac{d}{dz}\left(R \frac{dW}{dz}\right) dz + k^2\sigma^{-2} \int_0^1 |W|^2 \frac{dR}{dz} dz - k^2 \int_0^1 R|W|^2 \, dz = 0. \tag{27}$$

* Unstable perturbations will eventually grow so large that linear theory becomes inconsistent. Even when disturbance growth is predicted, however, it presumably is always possible to select for examination initial perturbations which are so small that linearization is valid for a certain time interval. Thus an instability result derived from linear theory should be meaningful in its prediction of instability for the motionless state, in spite of the fact that the ultimate condition of the flow cannot be ascertained by linear theory in this case.

This is not an obvious thing to do, but, as we shall see, it yields some interesting information. The manipulations that we are in the process of relating are but the simplest example of a number of ways which have been developed over the years for obtaining qualitative information about the solutions to eigenvalue problems such as (21) without actually solving them explicitly.

If we integrate the first term in (27) by parts, we obtain

$$\overline{W}R\frac{dW}{dz}\bigg|_{z=0}^{z=1} - \int_0^1 \frac{d\overline{W}}{dz}\left(R\frac{dW}{dz}\right) dz.$$

The boundary conditions (21b) imply that $\overline{W}(0) = \overline{W}(1) = 0$, so we can deduce from (27) the equation

$$\sigma^{-2}\int_0^1 \frac{dR}{dz}\,|W|^2\,dz = \int_0^1 R|W|^2\,dz + k^{-2}\int_0^1 R\left|\frac{dW}{dz}\right|^2 dz. \tag{28}$$

Deduction 1. If $dR/dz > 0$ for $0 \le z \le 1$ (density continually increasing with height), then every value of σ^2 satisfies $\sigma^2 = P$ for some positive quantity P. Hence

$$\sigma = \pm\sqrt{P}. \tag{29}$$

This follows at once from (28), since k is real, once we make explicit the physical requirement that density be a positive quantity:

$$R(z) > 0, \qquad 0 \le z \le 1. \tag{30}$$

From (29) we see that eigenvalues of (21), *if they exist*, occur in pairs, one positive and one negative. In treatises on ordinary differential equations it is shown that an infinite number of eigenvalues do, in fact, exist provided that R satisfies certain conditions. It is sufficient, for example, that R satisfy (30) and also possess two continuous derivatives for $0 \le z \le 1$. Equilibrium states corresponding to such functions R are thus definitely unstable, owing to the presence of the positive root in (29). The reason for the negative root in (29) is entirely analogous to the corresponding reason in the case of the inverted pendulum. Whatever the initial density perturbation, the initial velocity perturbation can be chosen in exactly the right way so that the fluid finally ends up in its (unstable) equilibrium state wherein heavy fluid strata are at the top of the layer.

Deduction 2. If $dR/dz < 0$ for $0 \le z \le 1$, then we obtain $\sigma = \pm i\sqrt{P}$ for some positive quantity P. The motionless state is neutrally stable to the class of perturbations corresponding to eigenfunctions of (21). [Again, this follows at once from (28).] When the density continually decreases with height, velocity, density, and pressure oscillate in time, in analogy with the ordinary pendulum's periodic oscillation about equilibrium. Wave motion must thus be expected when a stably stratified fluid is disturbed. Such "internal waves" are frequent and important in various atmospheric and oceanographic phenomena.

We conclude our illustration of the qualitative deductions that can be made from (21) by stating the following noteworthy result (Ince, 1927, p. 237). If

$$\frac{dR}{dz} > 0 \quad \text{for } z_0 < z < z_1, \qquad \text{where } 0 \le z_0 < z_1 \le 1,$$

then there are an infinite number of positive eigenvalues σ^2. Thus a motionless layer of stratified inviscid fluid is unstable if it contains any sublayer whatever, no matter how thin, wherein the density increases with height.

DETAILED RESULTS FOR A PARTICULAR STRATIFICATION

We shall now consider (21) for a particular function R. If our interest in stratified fluids were directly motivated by a particular application, then the function R would be selected to represent some experimental measurements. Such a function would probably be rather complicated. Sophisticated analysis might yield a solution of the resulting version of (21), but it is quite possible that a numerical approach would be a more appropriate way to obtain the details that are often desired when one attempts to deal with a practical problem. Our interests here are more general. As a first step we therefore seek a particular function R which makes (21) as simple as possible. The reader might like to cover up the next paragraph and spend a few moments trying to select such a function.

As Lord Rayleigh was the first to observe, it turns out that if for some constant β

$$R(z) = \exp(\beta z),$$

then (21a) reduces to the following equation with constant coefficients:

$$\frac{d^2W}{dz^2} + \beta\frac{dW}{dz} + k^2(\beta\sigma^{-2} - 1)W = 0. \tag{31a}$$

We still have the same boundary conditions

$$W(0) = W(1) = 0. \tag{31b}$$

The constant coefficient linear equation (31a) has solutions of the form $\exp(rz)$, provided that the constant r is a root of the quadratic equation

$$r^2 + \beta r + k^2(\beta\sigma^{-2} - 1) = 0. \tag{32}$$

Thus

$$r = -\tfrac{1}{2}\beta \pm \tfrac{1}{2}\sqrt{Q}, \tag{33}$$

where

$$Q = \beta^2 - 4k^2(\beta\sigma^{-2} - 1). \tag{34}$$

It is convenient to consider separately the cases where the roots r_1 and r_2 of (32) are (1) real and unequal, (2) real and equal, and (3) complex conjugates.

CASE (1): $Q > 0$. The general solution to (31a) is

$$W = C_1 \exp(r_1 z) + C_2 (\exp r_2 z),$$

where

$$r_1 = -\tfrac{1}{2}\beta + \tfrac{1}{2}\sqrt{Q}, \qquad r_2 = -\tfrac{1}{2}\beta - \tfrac{1}{2}\sqrt{Q}.$$

In order that there be a nontrivial solution satisfying $W(0) = W(1) = 0$, direct calculation or application of (24) yields the requirement $\exp(r_1) = \exp(r_2)$, which is not satisfied for positive Q.

CASE (2): $Q = 0$. The general solution to (31) is

$$W = C_1 \exp(-\tfrac{1}{2}\beta z) + C_2 z \exp(-\tfrac{1}{2}\beta z).$$

The boundary conditions of (31b) imply that $C_1 = C_2 = 0$. As in case (1), there is no nontrivial solution.

CASE (3): $Q < 0$. We write $Q \equiv -\mu^2$ so that the general solution to (31a) is

$$W = \exp(-\tfrac{1}{2}\beta z)[C_1 \cos(\tfrac{1}{2}\mu z) + C_2 \sin(\tfrac{1}{2}\mu z)]. \tag{35}$$

The requirements $W(0) = W(1) = 0$ imply that

$$C_1 = 0, \qquad C_2 \sin(\tfrac{1}{2}\mu) = 0,$$

so for a nontrivial solution,

$$\tfrac{1}{2}\mu = n\pi, \qquad n = \pm 1, \pm 2, \pm 3, \ldots. \tag{36}$$

Since $Q = -\mu^2$, from (36) and (34) we deduce that if $\sigma^2 = \sigma_n^2$, where

$$\sigma_n^2 = 4k^2\beta(4k^2 + \beta^2 + 4n^2\pi^2)^{-1}, \tag{37}$$

then (31) has the nontrivial solutions (eigenfunctions)

$$W = \exp(-\tfrac{1}{2}\beta z) \sin(n\pi z), \qquad n = 1, 2, 3, \ldots. \tag{38}$$

Since $\sin(-n\pi z) = -\sin(n\pi z)$, no new eigenfunctions are obtained by allowing the possibilities $n = -1, -2, -3$ in (38).

NOTE. Equation (37) is an example of the general relation (25), which we asserted must connect σ^2 and k^2 if the eigenvalue problem is to have nontrivial solutions.

From (37), if $\beta < 0$ (density decreases exponentially with height), then σ is a purely imaginary number. The equilibrium state is neutrally stable with respect to the class of perturbations for which W is given by (38); and p, u, and v are found from (19) and (16). (Remember, $W \equiv \hat{w}$.)

In considering (37) for $\beta > 0$ (density increases exponentially with height) we note first that σ^2 is positive (instability) in accordance with deduction 1. Next, let us explore the effect of varying n. Remember that for fixed *horizontal* disturbance period $2\pi/k$ there is a disturbance mode of vertical period $2n$ for every positive integer n. Each such disturbance is multiplied by a different time factor $\exp(\sigma_n t)$, where σ_n is given by (37). The growth rate σ_n is largest when $n = 1$, and it approaches zero monotonically as $n \to \infty$.

The explanation for the variation of growth rate with n lies in the different streamline patterns for different values of n. Figure 15.7 shows these lines,

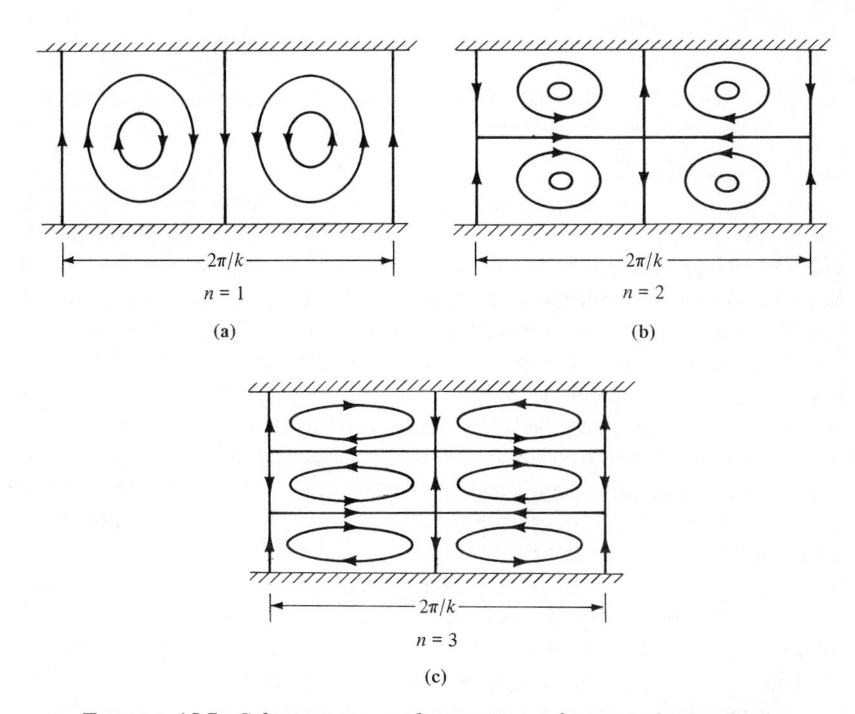

FIGURE 15.7. *Schematic streamline patterns for perturbation modes of various vertical periodicities. The perturbations are superimposed on a motionless layer of fluid whose density variation with height is given by* $\exp(\beta z)$, *where* β *is small and positive. The overturning or vortical motion is viewed along the axes of the vortices.*

along which the flow circulates with ever-increasing speed. (In this figure the density stratification exponent β has been taken to be small, to simplify the calculations required to obtain the streamlines. These calculations are requested in Exercise 2.) Evidentally, the motion accelerates faster when the heavy fluid falls in a single overturning vortex which fills the whole layer [Figure 15.7(a)] than it does when there is a double vortex [Figure 15.7(b)], etc.

For fixed n and k, σ^2 increases monotonically from zero to β as k increases from zero to infinity (Fig. 15.8). It is as expected that the maximum value of σ^2 increases as β increases. This means that, other things being equal, the

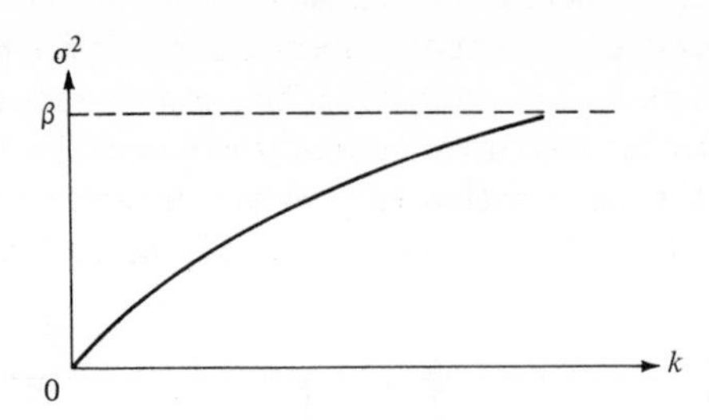

FIGURE 15.8. *Variation of the perturbation growth rate as a function of the horizontal perturbation wave number k.*

motion is faster the greater is the contrast in density between the heavy fluid above and the light fluid below. Less obvious is the reason why motion is faster for larger k's and hence thinner vortices. The explanation lies in the fact that the rate of increase in perturbation kinetic energy is equal to the rate of decrease of potential energy due to the falling of the relatively heavy top fluid (Exercise 3). Since gravity acts vertically, horizontal motions do not alter potential energy. Motion in narrow vortices is faster than motion in wide vortices because the latter contain a larger amount of horizontal motion, which is unproductive in the release of potential energy.

Similar observations flow from (37) when $\beta < 0$. These are left to the reader [Exercise 1(b)]. Even without further thought, however, the reader must agree that a surprisingly large amount of physical understanding has emerged from a careful examination of (37). One natural question remains: how much of this understanding applies only to the particular exponential function R for which (37) holds and how much applies to all functions R, or a large class of them? No convincing answer is possible without further investigation, but the general character of the explanations in the previous paragraph makes one fairly confident of their wide applicability.

Investigations of special problems are usually more productive than inexperienced persons anticipate. As in the mechanism for the release of potential energy, focusing one's attention on an unusual aspect of a particular problem frequently leads to insights having general applicability. More prosaic aspects of a particular problem also often turn out to be characteristic of large classes of problems. The fact that a class of problems shares a common aspect may only be suggested after several particular problems are solved and their solutions compared. Usually, a general argument can then be formulated, which proves that the suggestion is valid.

SUPERPOSITION OF NORMAL MODES

Our entire discussion thus far has been based on the behavior of certain particular solutions. For example, in the case $R(z) = \exp(\beta z)$ the only vertical variation considered was that given by one of the solutions $\exp(-\frac{1}{2}\beta z) \sin n\pi z$

of (31). But because the problem is linear, superposition of these solutions is permitted. How general a function can be represented by such a superposition? That is, for what class of functions f is it true that for appropriate constants d_n,

$$f(z) = \sum_{n=1}^{\infty} d_n \phi_n(z) \qquad \text{for} \quad 0 \le z \le 1, \tag{39}$$

where

$$\phi_n(z) = \exp\left(-\tfrac{1}{2}\beta z\right) \sin n\pi z? \tag{40}$$

If we multiply both sides of (39) by $\exp\left(\tfrac{1}{2}\beta z\right)$, the problem becomes one of selecting constants d_n to guarantee that

$$F(z) = \sum_{n=1}^{\infty} d_n \sin n\pi z, \qquad 0 \le z \le 1,$$

where

$$F(z) \equiv f(z) \exp\left(\tfrac{1}{2}\beta z\right).$$

This is a classical problem in Fourier series. We have discussed such problems at some length in Chapter 4 and shall not go further into them here.

From a larger point of view, as mentioned above we are considering one of the basic problems of the **Sturm–Liouville theory** discussed in Chapter 5. This theory deals with the properties of a class of eigenvalue problems of which (21) is a representative. Suppose, then, that we consider the expansion (39), where the $\phi_n(z)$ are eigenfunctions of the Sturm–Liouville problem (21). We assume only that $R(z)$ possesses two continuous derivatives for $0 \le z \le 1$. It can then be shown, as in Ince (1927) or Coddington and Levinson (1955), that if

$$d_n = \int_0^1 f(z)\phi_n(z)\, dz, \tag{41}$$

then the right-hand side of (39) converges uniformly to $f(z)$ for $0 \le z \le 1$, provided that $f(z)$ satisfies the same boundary conditions (31b) as the eigenfunctions and also possesses two continuous derivatives for $0 \le z \le 1$.

One can go further and attempt to deal with a general horizontal variation by superposing (adding together) solutions of the form (14) for various values of k_1 and k_2. An example of this procedure is given in the discussion of water waves found in Chapter 8 of II. For the present, suffice it to say that a very large class of perturbations can be handled by appropriate superposition of the eigenfunctions (14) which satisfy the homogeneous linear partial differential equations (10)–(12), the boundary condition (13), and the requirement that solutions remain bounded as $x^2 + y^2 \to \infty$. (The last requirement was used in discarding complex values of k_1 and k_2.)

Looking for eigenfunctions like those of (14) and hoping to be able to handle a general initial value problem by superposition is called the **normal mode approach** to stability. The eigenfunctions found by this approach are the **normal modes**. We observe that a rigorous normal mode treatment of the linearized problem requires not only a proof that a given class of initial conditions can be synthesized from the normal modes, but also that the corresponding infinite series satisfies the equations and boundary conditions.

REMARK. When the normal mode approach fails, as it sometimes does, the use of Laplace transforms usually saves the day. For an example, see G. F. Carrier and C. T. Chang, "On an Initial Value Problem Concerning Taylor Instability of Incompressible Fluids," *Q. Appl. Math. 16*, 436–39 (1959). [Some might find it helpful to view the situation in the following terms. Normal modes correspond to isolated eigenvalues (point spectrum). In many instances, a continuous spectrum is also present and the normal modes do not form a complete set. Integral superpositions are required to solve the initial value problem.]

NONLINEAR EFFECTS

We have treated a set of perturbations which are so small that nonlinear effects can be ignored. But is this set nonempty? That is, can one show that the behavior of perturbations of sufficiently small amplitude is given by the linearized theory? In many cases, recent work allows an affirmative answer to be given to this question. (We speak here and below of stability theory in general.)

If linear theory reveals even one perturbation that grows, we feel quite confident that the equilibrium state is unstable. Nonlinear terms might act to restrict the growth of a disturbance to very small amplitudes. Nevertheless, the resulting flow will differ from the equilibrium state, so the latter can properly be said to be unstable.

By contrast, if all disturbances die out according to linear theory, this is only evidence that *sufficiently small* disturbances ultimately will leave no effect upon the equilibrium state. This is a useful result, for it means that if one is sufficiently careful in isolating the equilibrium flow from disturbances then the flow should persist. But how careful does one have to be? Might it not be the case that the equilibrium flow will be destabilized by disturbances which are physically "moderate" but which are sufficiently large so that linear theory is not applicable? Such questions form the matter studied in *nonlinear stability theory* whose development "took off" in the 1960s. In a sentence one can report that some flows are extremely sensitive to large-amplitude perturbations, while for others increasing the amplitude of the perturbation does not hasten the onset of instability.

The computer is indispensable in studying large perturbations. As an example of the results achieved, consider a numerical study by B. J. Daly [(*Phys. Fluids 10*, 297 (1967)] of the stability of a heavy layer of fluid placed above a lighter layer. Daly reports that "at a density ratio of 10:1, the heavy fluid falls through the lighter fluid in the form of a sharp narrow spike whose amplitude accelerates with time. At density ratios of 2:1 or less, however, the tip of the spike is broadened . . . so that its late time velocity . . . is constant." A later calculation by the same author [*Phys. Fluids 12,* 1340 (1969)] examines the effect of surface tension upon the breaking up of the spike into separated drops.

Calculations of the initial development of the flow, when perturbations are small, agree with linear stability theory. Throughout, the calculations are in agreement with relevant experiments.

WORKED EXAMPLE: A MODEL OF VISCOUS FLOW INSTABILITY

A feature of linear stability theory has not yet been illustrated, namely, the prediction that at the onset of instability growing disturbances will have a particular finite wavelength. We have seen that for a stratified inviscid fluid, disturbances of all wavelengths are unstable when the fluid is top-heavy. Our study of the particular case $R(z) = \exp(\beta z)$ showed that growth speed decreased with wavelength, because thinner vortices are more efficient at using up the available potential energy. When viscosity is considered, however, very small wavelengths are inefficient because the resulting large velocity gradients dissipate a great deal of energy. A "compromise" leads to a "most dangerous disturbance" of intermediate wavelength.

Thus, when viscous fluid layers become sufficiently top-heavy, they begin a circulatory motion with a well-defined wavelength. Sufficient top-heaviness is required because the stabilizing nature of viscosity will otherwise prevent motion.

It is beyond our scope here to pursue the investigation of stability in top-heavy fluids. [For further information see S. Chandrasekhar's book *Hydrodynamic and Hydromagnetic Instability* (New York: Oxford U.P., 1961).] We do wish to point out, however, that there are many instances where instability sets in at a specific wavelength, when the destabilizing influence is sufficiently large. One such instance is illustrated in Exercise 10, a study of chemical reaction in the presence of diffusion. Another will now be outlined.

Simple viscous flows become unstable if they are sufficiently speeded up. This fact can be appreciated without studying viscous fluid motion, by an analysis of a model equation. This equation is contrived so that the analysis preserves the essential features of the full calculations but does not share their complication.

Example. Consider viscous flow (viscosity $= \nu$) between parallel plates a distance d apart. Let the bottom plate be stationary and the top plate move with speed U. *Pretend* that the dimensionless tangential speed $u(x, y, t)$ satisfies the following equation and boundary conditions:

$$R^{-1}[u_{xx} + u_{yy}] - u_{xxxx} - u_t = u_{xx} u_y, \qquad u(x, 0, t) = 0, \quad u(x, 1, t) = 1. \tag{42}$$

The true equations have the same general structure as (42). In particular, the dimensionless speed parameter or "Reynolds number" $R \equiv Ud/\nu$ appears in a very similar fashion.

It is easy to verify that (42) has the exact solution $u(x, y, t) = y$.† Consider small perturbations to this solution of the form $f(y) \cos kx \exp(\sigma t)$. Show that the (R, k) plane is divided into regions of stability and instability as shown in Figure 15.9. Find R_c and k_c. If the speed of the flow is slowly increased from a small value, instability will set in when $R = R_c$ at a wave number of k_c. Why?

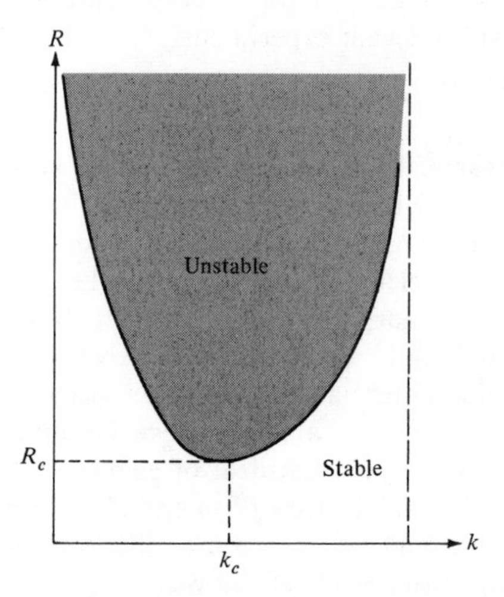

FIGURE 15.9. *The* **neutral curve** *in the plane of Reynolds number R versus wave number k divides the stability region from the instability region.*

Solution. If $u(x, y, t) = y + u'(x, y, t)$, the linearized equation for u' is

$$R^{-1}[u'_{xx} + u'_{yy}] - u'_{xxxx} - u'_t - u'_{xx} = 0.$$

If $u'(x, y, t) = f(y) \cos(kx) \exp(\sigma t)$, then f satisfies

$$f'' + \lambda f = 0, \quad f(0) = f(1) = 1, \qquad \text{where } \lambda = R\mu - R\sigma - R\mu^2 - \mu.$$

The only nontrivial solutions of the eigenvalue problem for f occur when $\lambda = n^2\pi^2$, $n = 1, 2, \ldots$. [The solutions are constant multiples of $\sin(n\pi z)$.] Thus

$$R\mu - R\sigma - R\mu^2 - \mu = n^2\pi^2$$

and $\sigma < 0$ (stability) if and only if

$$R < R(\mu, n) \qquad \text{where } R(\mu, n) = \frac{\mu + n^2\pi^2}{\mu - \mu^2}.$$

† The dimensional speed u^* and vertical coordinate y^* are related to u and y by $u^*/U = u$, $y^*/d = y$. Thus the solution $u = y$ corresponds to $u^* = Uy^*/d$. This is a flow whose speed increases linearly from zero at the motionless plate $y^* = 0$ to U at the moving plate $y^* = d$.

Instability will set in as R increases when the above inequality is reversed, at the smallest value R_c of $R(\mu, n)$. By inspection the smallest value occurs when $n = 1$. Minimization with respect to μ then gives

$$R_c = \frac{\mu_c + \pi^2}{\mu_c - \mu_c^2}, \qquad \text{where } \mu_c = k_c^2 = -\pi^2 + \pi\sqrt{\pi^2 + 1}.$$

For $R < R_c$, $\sigma < 0$ no matter what the wave number k of the perturbation. When R is just barely bigger than R_c, perturbations can grow, namely, those with wave numbers at or near k_c.

EXERCISES

1. (a) Remarks on the geometric significance of the wave number vector $\mathbf{k}$ were made under (21). Verify these remarks.

(b) Discuss in physical terms the variation of the oscillation frequencies that are given by (37), when $\beta < 0$.

2. This problem requests verification of the qualitative streamline picture of Figure 15.7. With no loss of generality the y axis can be aligned along the axis of the perturbation overturnings, so $v' = 0$, $k_2' = 0$. For simplicity restrict your consideration to the case of small β. It is virtually sufficient to determine the lines $x = \text{constant}$ and $z = \text{constant}$ along which the flow is vertical or horizontal.

3. From (11) deduce that if the flow pattern consists of the periodic repetition of flow in a domain D (e.g., a rectangle or a hexagon), then

$$\frac{\partial}{\partial t}\iiint_D \tfrac{1}{2}R(\mathbf{v}'\cdot\mathbf{v}')\,d\tau = -\iiint_D w'\rho'\,d\tau \qquad [\mathbf{v}' \equiv (u', v', w')].$$

(In the linear approximation the left side of the above equation is the rate of change of the perturbation kinetic energy; the right side is the rate at which work is done by the buoyancy force.)

4. An alternative to assuming (14) and taking the real part of the final answer is to assume that

$$u' = [\hat{u}_1 \cos(k_1 x + k_2 y) + \hat{u}_2 \sin(k_1 x + k_2 y)]e^{\sigma t},$$

with similar assumptions for the other dependent variables. Convince yourself that this is a bad alternative by attempting to carry through the required calculations. (If you complete the calculations, you should obtain the same result as that obtained using the complex notation, but with much more effort.)

‡5. In biology, the number of individuals p in a given population at time t is often approximately governed by the equation $dp/dt = kp$, where $k = k(t)$ is the reproduction rate. Consider the possibility

$$k = \alpha - \beta p \quad (\alpha, \beta \text{ positive constants}) \qquad \text{so } \frac{dp}{dt} = \alpha p - \beta p^2.$$

(The βp term in k represents the decline in population growth rate due to overcrowding and the concomitant decrease in the food supply, increase in level of toxins, etc.) Clearly, there are two exact *steady* solutions to the governing equation:

$$p = p_0 \qquad \text{where } p_0 = 0 \text{ or } p_0 = \frac{\alpha}{\beta}.$$

Investigate the stability of these solutions to small disturbances.

6. This problem concerns inviscid fluid of unit uniform density ($\rho \equiv 1$) contained between two infinite rigid parallel plates $y = 0$ and $y = 1$. Gravity will be ignored and only two-dimensional motions will be considered ($w = \partial/\partial z = 0$). The governing equations and boundary conditions

$$\frac{D\mathbf{v}}{Dt} = -\nabla p, \qquad \nabla \cdot \mathbf{v} = 0; \qquad \mathbf{v} = \mathbf{0} \quad \text{on} \quad y = 0, 1,$$

have the solution

$$u = U(y), \qquad v = 0, \qquad p = p_0,$$

where U is an arbitrary differentiable function and p_0 is a constant.

(a) Determine a set of *linear* partial differential equations that govern small perturbations to the above solution.

(b) Assume solutions proportional to $\exp[i\alpha(x - ct)]$ with α real. In particular, take $v = V(y) \exp[i\alpha(x - ct)]$. Eliminate pressure and horizontal velocity. Thereby show that the problem can be reduced to

$$(U - c)(V'' - \alpha^2 V) - U''V = 0; \qquad V(0) = V(1) = 0. \quad (43)$$

Why does the existence of nontrivial solutions to (43) imply a functional relation between c and α^2?

(c) Let $c = c_r + ic_i$ and assume that $c_i \neq 0$. Divide (43) by $U - c$. Multiply by $\bar{V}$, the complex conjugate of V, and integrate between zero and 1. Perform an integration by parts. Take the imaginary part of the resulting equation. Obtain

$$c_i \int_0^1 U''|V|^2|U - c|^{-2}\, dy = 0.$$

Deduce that U must have at least one inflection point if $c \neq_i 0$ and comment on the significance of the result.

7. As a model of fluid motion, J. M. Burgers considered the following system of equations for $u(t)$ and $U(t)$:

$$\frac{dU}{dt} = P - u^2 - \nu U, \qquad \frac{du}{dt} = Uu - \nu u.$$

Here P and ν are positive constants.

(a) Show that there is a solution $U = P/\nu$, $u = 0$, which is stable if and only if $P < \nu^2$.

(b) Show that when $P > \nu^2$, there is another pair of *stable* solutions with constant values for U and u. [It is reminiscent of real fluids that a relatively simple flow (with $u = 0$) becomes unstable at a critical value $P = \nu^2$ of the driving "pressure gradient" P and that a new type of flow emerges which is stable but more complicated.]

8. In this problem we consider the stability of an inviscid incompressible fluid layer of uniform density R_1, moving horizontally with uniform speed U_1 over a layer of uniform density R_2 and uniform horizontal speed U_2. Gravity acts vertically downward. We restrict ourselves throughout this problem to two-dimensional motions ($v = \partial/\partial y \equiv 0$.) It is reasonable to conjecture that our results will have something to do with the generation of water waves by wind.

(a) Find $p_1(z)$ and $p_2(z)$ so that if U_1, U_2, R_1, and R_2 are constants, then the following provides a solution to the inviscid incompressible flow equations (1a–c).

$$\text{For } 0 < z < \infty\text{: } u^* = U_1,\ w^* = 0,\ \rho^* = R_1,\ p^* = p_1(z).$$

$$\text{For } -\infty < z < 0\text{: } u^* = U_2,\ w^* = 0,\ \rho^* = R_2,\ p^* = p_2(z).$$

Also insure the continuity of pressure at the interface $z = 0$ between the two immiscible fluids [Exercise 14.4.8(b) with $s_n = 0$, $\mathbf{t} = -p\mathbf{n}$].

Account must be taken of the *kinematic boundary condition* $w^* = \zeta_t + u\zeta_x$ on $z = \zeta(x, t)$, where $z = \zeta(x, t)$ is the equation of the interface. (This condition is derived in Section 7.1 of II.) For the present simple solution, $z = 0$ is the equation of the interface and the kinematic boundary condition is clearly satisfied.

(b) In the upper layer, write

$$u^* = U_1 + u_1',\ w^* = w_1',\ \rho^* = R_1 + \rho_1',\ p^* = p_1(z) + p_1'. \tag{44}$$

Derive a linearized set of partial differential equations for the perturbations [the primed quantities in (44)]. Do the same for the lower layer perturbations. (Use the subscript 2 for these.)

(c) Write

$$w_i'(x, z, t) = \hat{w}_i(z) \exp[i\alpha(x - ct)] \qquad (\alpha \text{ real, } c \text{ complex}),$$

plus similar expressions for u_i', ρ_i', p_i'; $i = 1, 2$. Reduce each set of partial equations in (b) to an ordinary differential equation, one for $\hat{w}_1$ and one for $\hat{w}_2$.

(d) Let the perturbed interface between the two layers be written

$$z = \zeta(x, t) = \hat{\zeta} \exp[i\alpha(x - ct)], \qquad \hat{\zeta} \text{ a constant.}$$

From the kinematic boundary condition derive

$$\frac{\hat{w}_1}{U_1 - c} = \frac{\hat{w}_2}{U_2 - c} \qquad \text{at } z = \zeta.$$

(e) We are neglecting surface tension, so the pressure (*not* the perturbation pressure) must be continuous at $z = \zeta$. Deduce

$$R_1 \hat{w}_1 (U_1 - c)^{-1} + R_1 (U_1 - c) \frac{d\hat{w}_1}{dz}$$

$$= R_2 \hat{w}_2 (U_2 - c)^{-1} + R_2 (U_2 - c) \frac{d\hat{w}_2}{dz} \qquad \text{at } z = \zeta.$$

(f) It is consistent with our small perturbation assumption to apply the boundary conditions of (d) and (e) at $z = 0$. (Why?) It simplifies the algebra a little to consider the case $U_1 = -U_2 = V$. (Why does this involve no real loss of generality?) Use these two remarks to solve for $\hat{w}_1$ and $\hat{w}_2$ (impose sensible conditions as $|z| \to \infty$) and hence to deduce

$$c = c_m \pm (c_d + c_v)^{1/2}.$$

Here c_m is a weighted average velocity of the two layers

$$c_m \equiv \frac{R_1 U_1 + R_2 U_2}{R_1 + R_2},$$

and c_d and c_v result from the density and velocity discontinuities, respectively:

$$c_d \equiv \frac{g}{\alpha} \frac{R_1 - R_2}{R_1 + R_2}, \qquad c_v \equiv \frac{R_1 R_2 (U_1 - U_2)^2}{(R_1 + R_2)^2}.$$

(g) Show that there is always instability when heavy fluid is above lighter (*Taylor instability*).

(h) Compare the result for $V = 0$ with the text's results for exponential variation of density with height. As predicted here, an oil–water interface oscillates much more slowly than an air–water interface. This was noticed by Benjamin Franklin (Lamb, 1932, p. 371).

(i) Demonstrate the existence of instability when $R_1 = R_2$ (*Kelvin-Helmholtz instability*). Show that sufficiently short wavelengths are still unstable when $R_1 < R_2$ so that, contrary to experience, the slightest zephyr should cause the appearance of short waves in water. [When the approach of this exercise is modified by the inclusion of surface tension effects, theory predicts that a wind speed well in excess of 10 miles per hour is required before water waves are generated. This speed is much greater than the observed value so that a more subtle mechanism of wave generation must be sought. Much research has been devoted to this problem. See, e.g., A. Gupta, M. Landahl, and E. Mollo-Christensen, *J. Fluid Mech.* *33*, 673–91 (1968).]

‡9. In the case $R = \text{constant}$, the term involving the growth rate σ disappears from (21a). Reconsider this case. Since the assumption of exponential time dependence leads to difficulty, instead of (14) start with the assumption

$$u' = \hat{u}(z, t) \exp [i(k_1 x + k_2 y)]; \text{ etc.}$$

(a) Show that for some functions α, μ_0, v_0, and w_0,

$$\hat{u} = ik_1 k^{-2}\alpha'(z)t + u_0(z), \qquad \hat{v} = ik_2 k^{-2}\alpha'(z)t + v_0(z),$$
$$\hat{w} = \alpha(z)t + w_0(z), \qquad \hat{\rho} = R[k^{-2}\alpha''(z) - \alpha(z)], \qquad \hat{p} = -Rk^{-2}\alpha'(z).$$

(b) Interpret the results (i) when the problem is regarded as examining the stirring up of a layer of uniform density (so $\hat{\rho} \equiv 0$); (ii) when the problem is regarded as examining the fate of a layer whose initial density is nearly uniform (so $\hat{\rho}$ is some given function of z).

10. Publication of a paper* by the mathematician A. Turing (better known for his fundamental work in logic) has sparked interest in the possibility that chemical instabilities play important roles in biology. A contribution to this matter was made by I. Prigogine and R. Lefever,† who studied the (artificial) reaction scheme

$$A \xrightarrow{k_1} X, \qquad 2X + Y \xrightarrow{k_2} 3X, \qquad B + X \xrightarrow{k_3} Y + D, \qquad X \xrightarrow{k_4} E. \tag{45}$$

To obtain governing equations, diffusion terms are added to terms that are required by the law of mass action (Section 10.1). For simplicity, we consider only a single cartesian spatial coordinate r. The reader may wish to verify that one can obtain the following equations for X and Y.

$$\frac{\partial X}{\partial t} = k_1 A + k_2 X^2 Y - k_3 BX - k_4 X + D_X \frac{\partial^2 X}{\partial r^2}, \tag{46}$$

$$\frac{\partial Y}{\partial t} = -k_2 X^2 Y + k_3 BX + D_Y \frac{\partial^2 Y}{\partial r^2}. \tag{47}$$

Since the same number of molecules of X and Y are represented on the left of the expressions in (45) as on the right, (45) gives the over-all reaction scheme $A + B \to E + D$. We shall regard the initial constituents as having essentially fixed uniform concentrations A and B, so A and B

* *Proc. Roy. Soc.* (*London*) *B237*, 37 (1952).
† *J. Chem. Phys. 48*, 1695 (1968).

will be taken as constants in all that follows. The purpose of this problem is to show that there is a critical value of concentration at which an instability sets in.

(a) To make the calculations slightly easier, we shall assume that D_X and D_Y are equal and shall take $D_X = D_Y = D$. What is the physical meaning of this assumption? When might it be expected to be valid, or nearly so?

(b) Show that (46) and (47) have a unique *uniform* solution $X = X_0$, $Y = Y_0$, where the constants X_0 and Y_0 are given by

$$X_0 = \frac{k_1}{k_4} A, \qquad Y_0 = \frac{k_3 k_4}{k_1 k_2} \frac{B}{A}. \tag{48}$$

(c) Show that upon introduction of the change of variables

$$x = \frac{X}{X_0}, \qquad y = \frac{Y}{Y_0}, \qquad \tau = k_1 t, \qquad s = \frac{r}{\sqrt{D/k_1}}, \tag{49}$$

(46) and (47) become

$$\frac{\partial x}{\partial \tau} = \kappa + \beta x^2 y - \beta x - \kappa x + \frac{\partial^2 x}{\partial s^2}, \tag{50}$$

$$\frac{\partial y}{\partial \tau} = -\gamma x^2 y + \gamma x + \frac{\partial^2 y}{\partial s^2}, \tag{51}$$

with

$$\kappa = \frac{k_4}{k_1}, \qquad \beta = \frac{Bk_3}{k_1}, \qquad \gamma = \frac{k_1 k_2 A^2}{k_4^2}.$$

If you have studied the material on dimensional analysis and scaling, describe fully why this change of variables is a good idea. In doing so, mention what conclusions you can draw from (50) and (51) without further work, and state what simplifications could now be employed (if necessary) in a further study of these equations.

(d) Show that by writing $x = 1 + \bar{x}$, $y = 1 + \bar{y}$, with the further assumption

$$\bar{x} = \hat{x} \cos \mu s \, e^{\sigma\tau}, \qquad \bar{y} = \hat{y} \cos \mu s \, e^{\sigma\tau},$$

the stability of the uniform solution to small perturbations can be ascertained from a study of the equations

$$0 = \hat{x}(-\sigma - \tilde{A}) + \beta \hat{y}, \qquad 0 = -\gamma \hat{x} - (\tilde{B} + \sigma)\hat{y},$$

where

$$\tilde{A} = \kappa + \mu^2 - \beta, \qquad \tilde{B} = \gamma + \mu^2.$$

(e) Show that σ must satisfy an equation of the form $\sigma^2 + b\sigma + c = 0$. Demonstrate that disturbances die out if and only if $b > 0$ and $c > 0$. *In this part and the next we consider the possibility that instability arises due to a change of c from positive to negative.* Under these circumstances demonstrate that a necessary and sufficient condition for stability to a perturbation of wave number μ is $\tilde{A}\tilde{B} + \beta\gamma > 0$, or

$$\beta < \beta_c(m) \qquad \text{where } \beta_c(m) = (m + \kappa)(\gamma m^{-1} + 1), \qquad m \equiv \mu^2.$$

‡(f) Continuing part (e), find the value of β at which instability will first set in as β is slowly increased, and find the expected wavelength of this instability.

‡(g) Now consider the possibility of instability due to a change in the sign of b. Find the value of β at which this type of instability will first set in as β is slowly increased. Show that this value is less than the corresponding value of (f) and thereby deduce that instability will actually commence in the oscillatory manner described here in part (g). [If the assumption $D_X = D_Y$ is dropped, then there are parameter ranges such that instability begins in the fashion described in parts (e) and (f).]

15.3 Compression Waves in Gases

Compression waves in gases abound in nature. In everyday life we experience acoustic waves of human communication, noise and music. Explosions produce "shock waves," which are compression waves of high intensity. During the late stages of stellar evolution, a star explodes and sends out shock waves that eject much of its mass. In the solar atmosphere, acoustic waves are believed to be responsible for keeping the sun's corona region at a temperature of 1 or 2 million degrees, whereas the surface of the sun remains relatively cool at a temperature of about 6000°K.

In the present section we shall briefly describe the elementary theory of compression waves in ideal gases, beginning with the basic equations and the theory at small amplitudes.

A general discussion of compression waves leads to the development of the theory of partial differential equations of the "hyperbolic" type. We cannot go at all deeply into the general theory here, but we shall reproduce some of the early work of Riemann for one-dimensional compression waves of finite amplitude. We shall thereby afford one of the few glimpses at nonlinear effects which are offered in the present volume.* At present, fuller treatment of nonlinearity belongs in advanced courses. Moreover, new approaches to

* The perturbation methods of Part B are used on some nonlinear problems, but the nonlinearity does not play such a large role there.

nonlinearity are a central concern of contemporary research. Older books sometimes inculcated the attitude that the appearance of nonlinearity was a matter to be viewed with near despair, but we would like to leave the correct modern impression. Nonlinearity today offers a challenge to be viewed with enthusiasm. Perhaps existing numerical and analytical methods will work, at least after some adaptation; or perhaps a reputation can be made by inventing a new method.

INVISCID ISENTROPIC FLOW OF A PERFECT GAS

We shall employ the following basic equations of inviscid hydrodynamics: the equation of mass conservation (14.1.7),

$$\frac{\partial \rho}{\partial t} + \sum_{i=1}^{3} \frac{\partial}{\partial x_i}(\rho v_i) = 0; \tag{1}$$

the Euler equations of momentum balance (1.10), in the absence of body force,

$$\frac{\partial v_i}{\partial t} + \sum_{j=1}^{3} v_j \frac{\partial v_i}{\partial x_j} = -\frac{1}{\rho}\frac{\partial p}{\partial x_i}; \tag{2}$$

and the thermodynamic equation (14.4.14),

$$\theta\left(\frac{\partial s}{\partial t} + \sum_{j=1}^{3} v_j \frac{\partial s}{\partial x_j}\right) = 0. \tag{3}$$

Here ρ is density, the v_i are velocity components, p is pressure, θ is temperature, and s is entropy per unit mass.

In (3) we have omitted the term Dh/Dt, i.e., we have assumed that the heat flow into a given piece of gas is negligible during any time that this piece undergoes significant distortion. Thus the distortion process can be regarded as adiabatic. Moreover, we neglect the dissipation of mechanical energy into heat, an assumption that is consonant with the neglect of viscosity. (Compare Exercise 3.1.8 of II.) According to (3), the flow is isentropic.

The equations above are to be solved together with an equation of state. From Section 14.4 or from previous knowledge of elementary thermodynamics, we recall that we can use either

$$p = R\rho\theta \quad \text{or} \quad p = K_0\, e^{s/C_V} \rho^{\gamma}, \qquad K_0 \text{ a constant,} \tag{4a, b}$$

for either one of these equations can be derived if the other is regarded as known (and specific heats can be regarded as constant). Here C_V, the specific heat at constant volume, satisfies

$$C_V = (\gamma - 1)R, \tag{5}$$

where $\gamma = C_p/C_V$ is the ratio of specific heats and R is a constant for a given gas.

Equation (3) states that the entropy of a fluid particle is constant. We shall only consider cases (such as flow started from rest) where the entropy is uniform throughout the flow. In such cases $s \equiv$ constant and we can take as an equation of state (4b) in the form

$$p = K\rho^{\gamma}, \qquad K \text{ a constant.} \tag{6}$$

This is frequently called the **polytropic state equation**. It is sometimes replaced by the more general **barotropic** relation $p = p(\rho)$, since the theory is not greatly complicated thereby.

If we regard the pressure as a known function of density, then the phenomena are governed by (1) and (2), the equivalent of four scalar equations for the three velocity components and the density. The initial distribution of density and velocity must be prescribed, and the normal component of velocity must vanish on any fixed solid boundaries. On moving solid boundaries, the normal velocity of the gas must be the same as that of the boundary.

WAVES OF SMALL AMPLITUDES

We wish to consider small disturbances to a homogeneous motionless atmosphere. Let the undisturbed atmosphere have uniform density ρ_0. To consider a small disturbance, we write

$$\frac{\rho}{\rho_0} = 1 + \sigma. \tag{7}$$

The quantity σ, sometimes called the **condensation**, measures the departure of the dimensionless density ratio ρ/ρ_0 from its unperturbed value of unity.

We look for a solution where the condensation σ is small compared to unity and where the magnitudes of the velocity components are also small. The product of two small factors should thus be negligible, so our strategy will be to delete all nonlinear terms.

Substituting (7) into (1), we find that

$$\frac{\partial \sigma}{\partial t} + \sum_{i=1}^{3} \frac{\partial v_i}{\partial x_i} + \sum_{i=1}^{3} \frac{\partial}{\partial x_i}(\sigma v_i) = 0. \tag{8}$$

We neglect the third term compared with the second to obtain the linear equation

$$\frac{\partial \sigma}{\partial t} + \sum_{i=1}^{3} \frac{\partial v_i}{\partial x_i} = 0. \tag{9}$$

The linearization is valid if not only the condensation σ but also its spatial derivatives are small compared to unity. If we also linearize (2), we obtain [Exercise 1(a)]

$$\frac{\partial v_i}{\partial t} = -c_0^2 \frac{\partial \sigma}{\partial x_i}. \tag{10}$$

We have written

$$p(\rho) = p(\rho_0) + c_0^2(\rho - \rho_0) + \frac{p''(\rho_0)(\rho - \rho_0)^2}{2!} + \cdots, \tag{11}$$

where

$$c_0^2 \equiv p'(\rho_0). \tag{12}$$

What has been implied in the linearization that resulted in (10)? To see, let U and Σ be typical magnitudes of the velocity components and condensation. Let spatial derivatives multiply the magnitude of a quantity by about L^{-1} and temporal derivatives by about τ^{-1}. Then (9) and (10) imply that

$$\Sigma\tau^{-1} \approx UL^{-1}, \qquad U\tau^{-1} \approx c_0^2 L^{-1}\Sigma. \tag{13}$$

On elimination of Σ the above equations yield

$$L\tau^{-1} \approx c_0. \tag{14}$$

(This is a dimensionally correct result, for c_0 does have the dimensions of a speed.) Thus the convective acceleration $\mathbf{v}\cdot\nabla\mathbf{v}$ is, as assumed, small compared to the local acceleration $\partial\mathbf{v}/\partial t$ if

$$U^2L^{-1} \ll U\tau^{-1}, \qquad \text{i.e., if } U \ll c_0.$$

The ratio U/c_0 is called a **Mach number**. Our linearization appears to be valid if this ratio of fluid speed to sound speed is sufficiently small.

From (9) and (10) we derive the **wave equation** for the condensation σ:

$$\frac{\partial^2\sigma}{\partial t^2} = c_0^2 \sum_{i=1}^{3} \frac{\partial^2\sigma}{\partial x_i^2} \qquad \text{or} \qquad \frac{\partial^2\sigma}{\partial t^2} = c_0^2\,\nabla^2\sigma. \tag{15}$$

In general the velocity components v_i will not satisfy the wave equation but both $\partial v_i/\partial t$ and $\sum_{k=1}^{3} \partial v_k/\partial x_k$ do [Exercise 1(f)].

THE SPEED OF SOUND

If $s = s(x_1, t)$, at a given time the density is constant on the planes $x_1 =$ constant. The development of such **plane waves** is governed by

$$\frac{\partial^2\sigma}{\partial t^2} = c_0^2\frac{\partial^2\sigma}{\partial x_1^2}. \tag{16}$$

Solutions in unbounded space of this one-dimensional wave equation have been discussed in Section 12.2. There the equations governed small displacements in an elastic medium. The formal nature of the solutions is of course independent of their physical interpretation (and the discussion of the wave equation in Section 12.2 is independent of the rest of Chapter 12.)

Plane wave solutions of (16) propagate with unchanged shape at speed c_0. Thus c_0 is the **speed of sound** at the density ρ_0. This speed is not unequivocally specified by our analysis, for the assumed barotropic relation $p = p(\rho)$ can be obtained in more than one way (as we now discuss).

In his study of acoustic waves, Newton made the seemingly plausible assumption that the gas temperature does not change as it expands and contracts. Thus he regarded the equation of state as having the form $p = p(\rho, \theta)$ with $\theta =$ constant. For a perfect gas this would amount to using the equation (4a) with $\theta =$ constant, i.e., $p\rho^{-1} =$ constant. With this,

$$c^2 = \left(\frac{\partial p}{\partial \rho}\right)_\theta = \frac{p}{\rho}.$$

(It was actually Euler who derived the above result by appeal to the wave equation, thereby improving on Newton's less formal argument.)

Observation shows that the Newton–Euler isothermal sound speed is not correct. It was Laplace who pointed out that the expansion and compression in sound waves should be regarded as an adiabatic and hence isentropic process. [See the discussion of (3).] The state equation $p = K\rho^\gamma$ is thus the correct one for perfect gases, and

$$c^2 = \left(\frac{\partial p}{\partial \rho}\right)_s = \frac{\gamma p}{\rho}. \tag{17}$$

SPHERICAL WAVES

If the wave has spherical symmetry, the Laplace operator becomes

$$\nabla^2 = \frac{\partial^2}{\partial r^2} + 2r^{-1}\frac{\partial}{\partial r},$$

where $r = (x_1^2 + x_2^2 + x_3^2)^{1/2}$ is the distance of (x_1, x_2, x_3) from the origin. The wave equation takes the form

$$c_0^{-2}\frac{\partial^2 s}{\partial t^2} = \frac{\partial^2 s}{\partial r^2} + 2r^{-1}\frac{\partial s}{\partial r}. \tag{18}$$

To solve this equation, we introduce the transformation of variable

$$\sigma = \frac{\beta}{r}. \tag{19}$$

Then β satisfies the equation of plane waves

$$c_0^{-2}\frac{\partial^2 \beta}{\partial t^2} = \frac{\partial^2 \beta}{\partial r^2}, \tag{20}$$

whose solutions we have studied. Thus the general solution for spherical waves has the form

$$\sigma = r^{-1}f(r - ct) + r^{-1}g(r + ct). \tag{21}$$

The two terms represent outgoing and incoming waves, respectively.

The amplitude of the spherical waves decreases with distance according to the factor $1/r$. This could have been anticipated [and hence the change of variable (19) could have been motivated] by the following considerations. The energy in the wave is proportional to the square of its amplitude. No mechanisms for energy dissipation have been included in our model, so energy on a spherical shell will propagate outward on a surface whose area increases like r^2. Conservation of energy requires that the energy be proportional to r^{-2}, so the disturbance amplitude itself must be proportional to r^{-1}.

The solution (21) generally has a singularity at $r = 0$. This singularity usually does not appear in the solution to a physical problem however. For example, if we consider a pulsating sphere that sends out spherical waves, we exclude small values of r.

A solution of the form (21) that has special interest is found by taking

$$f(r - ct) = \sin k(r - ct), \qquad g(r + ct) = \sin k(r + ct), \tag{22}$$

where k is a constant. With (22), (21) becomes

$$s = 2r^{-1} \sin kr \cos ct. \tag{23}$$

This **standing wave** solution is finite throughout infinite space. Many other interesting solutions can be obtained by superimposing such standing spherical waves.

NONLINEAR WAVES IN ONE DIMENSION

In the linear theory of wave propagation, the wave shape is preserved. When the full nonlinear equations of hydrodynamics are used, as is appropriate when the amplitudes of the waves are no longer small, we do not expect the general nature of the propagation to be changed. We do expect, however, that the wave form may be distorted. We also expect that the waves propagating to the right and to the left may interact with each other. To demonstrate these effects, we consider the nonlinear problem following the method of Riemann.

We begin our investigation by looking for solutions that propagate in only one direction. In the linearized case, such waves are a function of a combination of x and t, either $x + ct$ or $x - ct$. Let us see if there are solutions in the nonlinear case that depend on *some* such combination. We seek **simple wave** solutions in which the density and velocity are functions of a single quantity, $\alpha(x, t)$.

For flows that depend on a single spatial coordinate x and have velocity $u(x, t)$, the continuity and momentum equations (1) and (2) can be written

$$\frac{\partial \rho}{\partial t} + u \frac{\partial \rho}{\partial x} + \rho \frac{\partial u}{\partial x} = 0, \tag{24}$$

$$\frac{\partial u}{\partial t} + u \frac{\partial u}{\partial x} + \frac{c^2}{\rho} \frac{\partial \rho}{\partial x} = 0, \tag{25}$$

where $c^2 = dp/d\rho$. Upon introducing

$$u = u(\alpha), \qquad \rho = \rho(\alpha), \tag{26}$$

we obtain

$$\rho'(\alpha)\left(\frac{\partial \alpha}{\partial t} + u\frac{\partial \alpha}{\partial x}\right) + \rho u'(\alpha)\frac{\partial \alpha}{\partial x} = 0, \tag{27a}$$

$$u'(\alpha)\left(\frac{\partial \alpha}{\partial t} + u\frac{\partial \alpha}{\partial x}\right) + \frac{c^2}{\rho}\rho'(\alpha)\frac{\partial \alpha}{\partial x} = 0. \tag{27b}$$

These equations are compatible if and only if

$$c^2\rho^{-2}\left(\frac{d\rho}{\partial \alpha}\right)^2 = \left(\frac{du}{\partial \alpha}\right)^2 \tag{28}$$

or

$$u = \pm \int c(\rho)\rho^{-1}\, d\rho. \tag{29}$$

The equation for α becomes

$$\frac{\partial \alpha}{\partial t} + (u \pm c)\frac{\partial \alpha}{\partial x} = 0. \tag{30}$$

Equation (30) is **quasi-linear**; i.e., it is linear in the derivatives. (It is not linear in the general sense, since the "coefficients" depend on the dependent variables u and c.) Such an equation can be treated by the "method of characteristics," which we shall not discuss in detail here. Instead we shall write down the solution of (30) and verify it by direct substitution.

We assert that a general solution of (30) is

$$\alpha = F(\mu), \qquad \mu \equiv x - (u \pm c)t, \tag{31}$$

which is obtained from (30) *as if* $u \pm c$ were a constant. To verify that (31) is indeed a solution, we note that

$$\frac{\partial \alpha}{\partial x} = F'(\mu)\left[1 - t(u' \pm c')\frac{\partial \alpha}{\partial x}\right],$$

$$\frac{\partial \alpha}{\partial t} = F'(\mu)\left[-(u \pm c) - t(u' \pm c')\frac{\partial \alpha}{\partial t}\right].$$

From this it follows that $\partial\alpha/\partial x$ and $\partial\alpha/\partial t$ are in the ratio $-(u \pm c)$, and (30) is satisfied.

The interpretation of (31) can be accomplished in a manner similar to that for linear waves. *Given* values of α, and hence given values of u and c, displace to the right at the speed $u \pm c$. That is, *the particular state corresponding to a given α propagates at a speed $u + c$ for one simple wave and $u - c$ for another.*

In other words, the wave propagation relative to the moving fluid element has a speed $\pm c$. To this extent, the result is similar to that in the linear theory.

However, in contrast to the results of linear theory, the propagation involves a *distortion* of the wave shape. This is a result of the fact that the propagation speeds $u \pm c$ vary with α and hence with the physical quantities u and c, the latter being uniquely related to the density of the gas.

A relatively clear picture of the distortion can be obtained in the polytropic case. It is convenient to write the state equation $p = K\rho^\gamma$ in the form

$$\frac{p}{p_0} = \left(\frac{\rho}{\rho_0}\right)^\gamma, \tag{32}$$

where p_0 and ρ_0 are the values of pressure and density at some reference state. Using this, the definition $c^2 = dp/d\rho$, and (29) in the more explicit form

$$u = \pm \int_{\rho_0}^{\rho} c(r) r^{-1}\, dr, \tag{33}$$

it is not difficult to show [Exercise 1(e)] that

$$u = \pm \frac{2}{\gamma - 1}(c - c_0) \qquad \text{so } c = \pm \frac{\gamma - 1}{2} u + c_0 . \tag{34a, b}$$

From this it follows that

$$u \pm c = \tfrac{1}{2}(\gamma + 1)u \pm c_0 , \tag{35}$$

so from (26), (31), and (35),

$$u = F\{x - [\tfrac{1}{2}(\gamma + 1)u \pm c_0]t\}, \tag{36}$$

where the positive (negative) sign is taken for disturbances that propagate rightward (leftward) with respect to the moving medium.

If the term proportional to $\frac{1}{2}(\gamma + 1)$ were not present in (36), then the graph of u would propagate either to the left or to the right with speed c_0. The presence of $\frac{1}{2}(\gamma + 1)u$ distorts the graph by adding to the velocity of

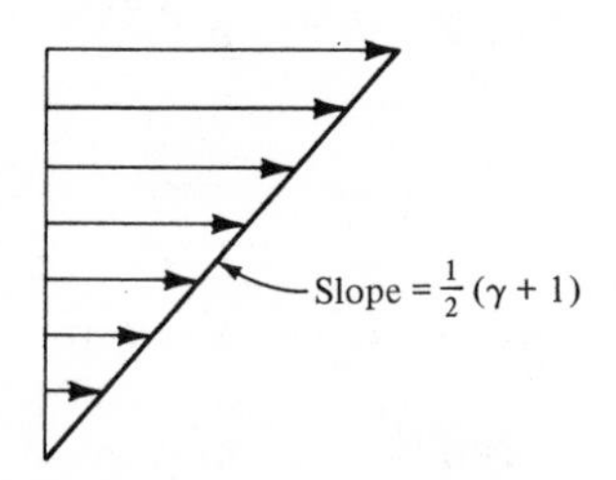

FIGURE 15.10. *Velocity profile of a shear flow. The arrows represent velocity vectors; speed increases linearly with height. The graph of* (36) *changes just as if it were a material line placed in such a flow.*

propagation a term proportional to the ordinate. The graph distorts just as if it were composed of dye particles placed in a stream whose speed increased linearly with height, as in Figure 15.10.

It can be shown that as long as the speed of sound increases in some fashion with density, the same qualitative features hold as in the polytropic case.

Although we have found two simple waves, it is not possible to solve the nonlinear initial value problem by superposition of these solutions. There is a well-developed theory for this problem, but we shall not pursue it further.

Example. (Details are requested of the reader in Exercise 4.) Consider a piston gradually started from rest and pushed into an infinitely long tube at a speed substantially lower than the speed of sound. Let the tube be filled with an ideal gas at density ρ_0.

We shall consider the early stages of the solution, during which the leading part of the disturbance should be of very small magnitude. Thus the leading edge of the disturbance should travel forward with the sound speed, so we expect

$$\text{for } x \geq c_0 t, \qquad \rho \equiv \rho_0 \quad \text{and} \quad u \equiv u_0.$$

Let the position ξ of the piston at time τ be described by $\xi = f(\tau)$. Then it can be shown that the appropriate simple wave solution can be parametrically described by

$$u = f'(\tau), \qquad x - f(\tau) = (u + c)(t - \tau); \quad f(\tau) \leq x \leq c_0 t. \tag{37}$$

Here c is related to u by (34b), taking the positive sign.

SHOCK WAVES

As the distortion of shape progresses, a point will be reached when the solution becomes three-valued (Figure 15.11). To have three values of density at a point is unphysical, and some spectacular phenomenon must occur that can no longer be described without refinement of the theory. Indeed, a *shock wave* is formed whose description by inviscid theory requires acceptance of discontinuous solutions.

The formation of a shock must involve some physical processes hitherto ignored. These are the diffusive processes of viscosity and heat conduction. However small the coefficients of viscosity and heat conduction are, just before the solution develops an infinite gradient [Figure 15.11(b)], derivatives are so large that diffusive processes are no longer negligible. If one assumes that the distortion process continues outside this diffusive region, it appears that a *discontinuity* would tend to form; this is the shock wave.

One can visualize the formation of the shock in another manner. Consider the piston problem mentioned in the example above. The disturbances have a front progressing at speed c_0. The gas behind the front, being compressed, has a higher local sound speed. Thus the later disturbances tend to catch up with the earlier ones and produce stronger compressions.

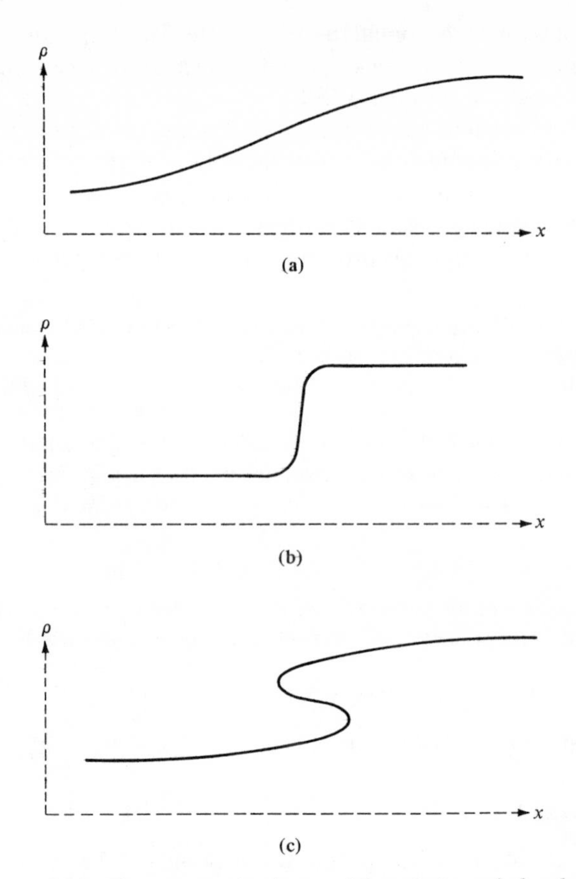

FIGURE 15.11. *Three stages in the spatial variation of the density ρ according to simple wave theory.*

Analysis of the detailed propagation of compression waves, including shocks in general, requires a numerical procedure. But we shall confine ourselves to a brief examination of conditions at the discontinuity.

To see how variables "jump" across shock discontinuities, consider an observer moving with the shock. This observer sees a shock with a fixed location. Let flow conditions *ahead* of the shock be designated with a subscript a and those *behind* by a subscript b. Thus the flow into the shock has a speed u_a, and the flow leaving the shock has a speed u_b. It is easy to see [Exercise 5(a)] that conservation of mass requires that

$$\rho_a u_a = \rho_b u_b \equiv m. \tag{38}$$

Other exercises show that the momentum relation requires that

$$p_a + \rho_a u_a^2 = p_b + \rho_b u_b^2, \tag{39}$$

while the energy equation requires that

$$p_a u_a + m(\tfrac{1}{2}u_a^2 + e_a) = p_b u_b + m(\tfrac{1}{2}u_b^2 + e_b), \tag{40}$$

where e is the internal energy per unit mass.

It is a remarkable fact that the detailed events within the thin shock region can be ignored, and that various compressible flows can be well described by treating shocks as discontinuities in the solutions. Values of the dependent variables on either side of the discontinuity are related by "jump conditions" like (38), (39), and (40).

The successful use of discontinuous solutions in compressible flow theory emphasizes the fact that theoreticians always deal with *models* of natural phenomena. One might even grant the assertion "nature *really* changes continuously" but this would not bar the use of discontinuous solutions, for (as in compressible flow) such solutions might provide excellent models when regions of rapid change are anticipated.

The use of discontinuous solutions in one-dimensional elasticity has been treated in Section 12.3. Exercises on deducing general jump conditions across discontinuities have been given in Section 14.4. Specialization of the general results to shocks is requested in Exercises 8 and 9. The most important result is this. As a consequence of the Clausius–Duhem integral inequality (14.4.29), *when an inviscid fluid passes through a shock its entropy cannot decrease** [Exercise 9(d)].

EXERCISES

1. Verify the following equations:
(a) (10).
(b) (20).
(c) (27).
(d) (28), (29), and (30).
(e) (34) and (35).
(f) Verify the statement after (15).

2. Formulate and solve a problem concerned with the radiation of sound waves from a pulsating sphere.

3. If you have studied the material on nondimensionalization and scaling, apply these ideas to the derivation of the wave equation for the condensation. Your answer should take the form of a brief essay.

4. (a) Derive (37).
(b) Discuss the solution given by (37) in the special case $f(\tau) = a\tau^2$, where a is a constant.

* According to the energy equation (3), a particle's entropy remains constant. But the essence of the treatment of shocks as discontinuity surfaces is to regard all dissipative processes [which were neglected in (3)] as present but only in narrow shock regions. In particular, (3) cannot be used to determine jumps across the shock.

‡(c) Comment on the statement: "This problem is without physical significance because there is no such thing as an infinite tube."

5. (a) Derive (38) by equating the rate at which mass flows into and out of a unit area of shock surface.

(b) Derive (39) by equating force to rate of increase of momentum.

(c) Derive (40) by equating the rate of working of pressure to a rate of energy change.

6. Suppose that a gas in an unbounded medium is initially at rest but has a perturbation in condensation with a nonuniform distribution $s = F(x)$. Verify that the solution is

$$s = \tfrac{1}{2}[F(x - ct) + F(x + ct)].$$

Sketch when $F(x)$ is given by a triangular "bump" at the origin.

7. Consider gas in an infinite tube with one end closed. If the boundary at the closed end moves back and forth according to the law $\xi = G(t)$, show that the induced motion in the gas is given according to linearized theory by

$$u_1 = G'\left(t - \frac{x}{c_0}\right).$$

Find the corresponding density distribution.

8. Consider an inviscid fluid with no heat flux ($h = r = 0$). This exercise and the next one apply the results of Exercise 14.4.8 to shocks, thereby obtaining a number of results concerning the jumps of various quantities.

Since fluid crosses a shock, in the notation of Exercise 14.4.7, $s_n^+ \neq 0$, $s_n^- \neq 0$. Let $\mathbf{v}_t$ denote the vector obtained by projecting the velocity vector $\mathbf{v}$ onto the shock surface. Manipulate the formulas of Exercise 14.4.8(b) to obtain

$$[\![\rho s_n]\!] = 0, \quad [\![\rho s_n^2 + p]\!] = 0, \quad [\![\mathbf{v}_t]\!] = \mathbf{0}, \quad [\![\tfrac{1}{2}s_n^2 + I]\!] = 0, \quad \text{(41a, b, c, d)}$$

where $I \equiv e + p\tau$ is the specific enthalpy and $\tau \equiv \rho^{-1}$, the specific volume. (The letter v, normally used to denote volume, is already associated with velocity.) Also

$$\frac{p^+ - p^-}{\rho^+ - \rho^-} = s_n^+ s_n^-. \qquad (42)$$

To do this, write $\mathbf{v} \equiv v_n\mathbf{n} + v_t\mathbf{T}$, where $\mathbf{T}$ is in the tangent plane to the shock surface. Take the $\mathbf{T}$ and $\mathbf{n}$ components of the momentum jump condition, remembering to use the constitutive equation for an inviscid fluid. Use the mass jump condition to show that $\mathbf{v}_t$ must be continuous; etc.

9. ‡(a) Show that for any shock s_n^+ and s_n^- must have the same sign.

(b) Show that the common sign of s_n^+ and s_n^- must be negative if fluid flows across the shock from $-$ to $+$. (See definitions and figure for Exercise 14.4.7.) The region $R-$ is then called the **front** of the shock.

(c) Extend Exercise 14.4.8 to show that for an inviscid fluid with no heat flux $[\![\rho s_n s]\!] \geq 0$, where s is specific entropy.

(d) Deduce $[\![s]\!] \geq 0$.

‡(e) The quantity $\tau^- - \tau^+$ is called the **shock strength**. It can be shown that this quantity is positive and that

$$s^+ - s^- = K(\tau^- - \tau^+)^3 + \cdots,$$

where K is a constant and $\cdots$ indicates omitted higher order terms. Show that the speed at which an infinitely weak shock moves relative to the fluid is the sound speed $(\partial p/\partial \rho)^{1/2}$. Start with the fact that there is an equation of state $p = p(\rho, s)$. Use (42).

10. Use Lagrange's method of characteristics (see Section 1.2) to derive (31).

15.4 Uniform Flow Past a Circular Cylinder

Determination of the force on an obstacle in a moving fluid has been a leading problem in the study of fluid motion for centuries. Although much progress has been made, major aspects of the problem remain unexplained. In this section we start a discussion of flow past obstacles by examining a two-dimensional inviscid flow past an infinite circular cylinder.

FORMULATION

We are trying to start with the simplest particular problem that might bear on the general phenomenon in question. Let us then consider a circular cylinder

$$x^2 + y^2 = a^2, \qquad -\infty < z < \infty,$$

immersed in a steady two-dimensional flow of a uniform-density fluid. Suppose that the fluid moves with uniform velocity $U\mathbf{i}$ far from the cylinder, so that the velocity vector $\mathbf{v}$ satisfies

$$\mathbf{v}(x, y) \to U\mathbf{i} \qquad \text{as } x^2 + y^2 \to \infty. \tag{1}$$

Making an assumption that will be justified toward the end of the section, we shall write the velocity as the gradient of a potential function ϕ. The assumption of uniform density requires that ϕ be harmonic if mass is to be conserved. For $\rho =$ constant and $D\rho/Dt + \rho\nabla \cdot \mathbf{v} = 0$ implies that $\nabla \cdot \mathbf{v} = \nabla^2\phi = 0$.

We shall employ the method of separation of variables to determine a harmonic function ϕ that satisfies appropriate boundary conditions. Once this velocity potential ϕ is determined, we can find the pressure p—for the inviscid version of momentum conservation links velocity and pressure. The integrated effect of pressure gives the total force on the body.

Since we are considering a circular cylinder, we should use a velocity potential Φ which depends on polar coordinates r and θ:

$$r = (x^2 + y^2)^{1/2}, \qquad \theta = \tan^{-1}\left(\frac{y}{x}\right),$$
$$\phi(r\cos\theta,\, r\sin\theta) \equiv \Phi(r, \theta). \tag{2}$$

We shall need an expression for the gradient in polar coordinates. From Equation 3 of Appendix 15.2 or Exercise 5, this is

$$\nabla\Phi(r, \theta) = \Phi_r \mathbf{e}^{(r)} + r^{-1}\Phi_\theta \mathbf{e}^{(\theta)}. \tag{3}$$

Here, the subscripts on Φ denote partial derivatives. Also $\mathbf{e}^{(r)} = \mathbf{e}^{(r)}(r, \theta)$ $[\mathbf{e}^{(\theta)} = \mathbf{e}^{(\theta)}(r, \theta)]$ is a unit vector at (r, θ) which points in the direction of increasing r (increasing θ). (See Figure 15.12.)

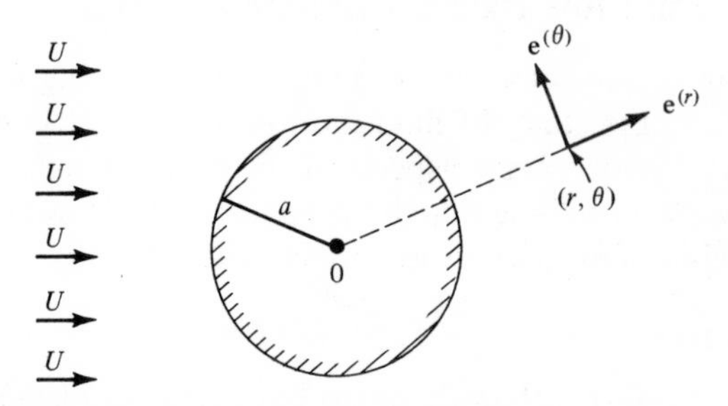

FIGURE 15.12. *Uniform flow at speed U past a right circular cylinder of radius a. Shown is a typical pair of unit vectors* $\mathbf{e}^{(r)}$ (r, θ) *and* $\mathbf{e}^{(\theta)}$ (r, θ) *that form the basis of polar coordinates.*

We are ready for the final formulation of the problem. According to Exercise 2 of Appendix 13.1 or other sources, Laplace's equation $\nabla^2\phi = 0$ in polar coordinates is

$$\boxed{\Phi_{rr} + r^{-1}\Phi_r + r^{-2}\Phi_{\theta\theta} = 0, \qquad r > a.} \tag{4}$$

From (3) and (1), we find that the requirement of uniform flow at infinity becomes

$$\boxed{\Phi_r \to U\cos\theta, \qquad r^{-1}\Phi_\theta \to -U\sin\theta \quad \text{as } r \to \infty.} \tag{5a, b}$$

Since no fluid can penetrate the cylinder, the radial component of $\nabla\Phi$ must vanish at $r = a$, so

$$\boxed{\Phi_r(a, \theta) = 0.} \tag{6}$$

One more condition must be imposed, because the point with polar coordinates (r_0, θ_0) also has the coordinates $(r_0, \theta_0 + 2n\pi)$, $n = \pm 1, \pm 2, \ldots$. We must certainly require that the velocity be the same whether we consider it at (r, θ) or at $(r, \theta + 2\pi)$. Thus $\nabla\Phi$ must have period 2π or θ, in

$$\boxed{\Phi_r(r, \theta + 2\pi) = \Phi_r(r, \theta), \qquad \Phi_\theta(r, \theta + 2\pi) = \Phi_\theta(r, \theta).} \tag{7a, b}$$

The boxed equations provide a definite mathematical problem. In solving it, we shall see that another condition must be prescribed if a unique solution is desired.

SOLUTION BY SEPARATION OF VARIABLES

Although a solution to (4)–(7) can almost be guessed, we shall solve these equations by the method of separation of variables. Those who have not had previous experience with this procedure should note that the six steps (A)–(F) that we apply to this particular separation of variables problem apply with little or no modification to any such problems. (Compare Section 4.1; see also Exercise 7.)

STEP A. *Assume a product solution:*

$$\Phi(r, \theta) = R(r)\Theta(\theta).$$

STEP B. *Substitute into the governing differential equation:* From (4),

$$R''\Theta + r^{-1}R'\Theta + r^{-2}R\Theta'' = 0 \qquad \left(R' \equiv \frac{dR}{dr},\ \Theta' \equiv \frac{d\Theta}{d\theta}\right). \tag{8}$$

STEP C. *Separate variables.* We write (8) in the form

$$\frac{R'' + r^{-1}R'}{r^{-2}R} = -\frac{\Theta''}{\Theta}. \tag{9}$$

The variables are now separated so that the left side is a function only of the independent variable r and the right side is a function only of the independent variable θ. Reasoning just as under (2.4), we deduce that both sides of (9) must equal a **separation constant** k so that we obtain the **separated equations**

$$R'' + r^{-1}R' = kr^{-2}R, \qquad \Theta'' = -k\Theta. \tag{10a, b}$$

STEP D. *Determine permissible values of the separation constant from an eigenvalue problem consisting of a separated equation and suitable homogeneous boundary conditions.* The equations of (10) are both second order, so we seek a pair of homogeneous boundary conditions. Trial and error indicates that of the various possibilities, only the conditions stemming from (7) are appropriate. From (7a)

$$R'(r)\Theta(\theta + 2\pi) = R'(r)\Theta(\theta),$$

so either

$$R' \equiv 0 \tag{11}$$

or

$$\Theta(\theta + 2\pi) = \Theta(\theta). \tag{12}$$

From (7b)

$$\Theta'(\theta + 2\pi) = \Theta'(\theta). \tag{13}$$

Since $R \equiv 0$ gives the trivial solution, there is no alternative to (13). As an eigenvalue problem to determine k, we therefore take (10b), (12), and (13). [We shall consider the possibility (11) separately.]

If $k > 0$, we write $k = \mu^2$, $\mu \neq 0$. The general solution to (10b) can then be written

$$\Theta = C_1 \cos \mu\theta + C_2 \sin \mu\theta. \tag{14}$$

From (12) and (13)

$$C_1[\cos \mu(\theta + 2\pi) - \cos \mu\theta] + C_2[\sin \mu(\theta + 2\pi) - \sin \mu\theta] = 0, \tag{15}$$

$$C_1[-\mu \sin \mu(\theta + 2\pi) + \mu \sin \mu\theta] + C_2[\mu \cos \mu(\theta + 2\pi) - \mu \cos \mu\theta] = 0. \tag{16}$$

Equations (15) and (16) have only the trivial solution $C_1 = C_2 = 0$ unless we require that the determinant of the coefficients vanishes. By the use of simple trigonometric identities, this requirement gives

$$\mu[\cos \mu(\theta + 2\pi) - \cos \mu\theta]^2 + \mu[\sin \mu(\theta + 2\pi) - \sin \mu\theta]^2 = 0,$$

or

$$2 - 2[\cos \mu(\theta + 2\pi) \cos \mu\theta + \sin \mu(\theta + 2\pi) \sin \mu\theta] = 0,$$

or

$$\cos 2\pi\mu = 1 \qquad \text{so } \mu = \pm 1, \pm 2, \pm 3, \ldots . \tag{17}$$

When μ is an integer, (15) and (16) are indeed satisfied for arbitrary constants C_1 and C_2.

If $k = 0$, the only nontrivial solution of (10b), (12), (13) is $\Theta = \theta_0$, where θ_0 is a nonzero constant. If $k < 0$, there are no nontrivial solutions of (10b), (12), and (13) [Exercise 1(a)].

STEP E. *Solve the remaining separated equation when the separation constant takes on the values determined in step* (D). When $k = n^2$ the equidimensional or Euler equation (10a) has as a general solution a linear combination of r^n and r^{-n}, $n > 0$. When $k = 0$ the general solution is a linear combination of 1 and $\ln r$ (Boyce and DiPrima, 1963, Sec. 4.4).

STEP F. *Superpose all possible product solutions and attempt to satisfy the remaining boundary conditions.* By "superpose" we mean "form a linear combination of." To include all possible product solutions in this combination, we must consider the possibility (11). Equations (11) and (8) imply that $\Theta'' = 0$ and lead to the conclusion that an arbitrary multiple of θ is a product solution of (8) and (7), which must be added to the list of such solutions already obtained. From this conclusion and the results of steps (D) and (E) we are led to assume for Φ the infinite series

$$\Phi(r, \theta) = A\theta + B_0 + B_1 \ln r$$
$$+ \sum_{n=1}^{\infty} [(C_n r^n + c_n r^{-n}) \cos (n\theta) + (D_n r^n + d_n r^{-n}) \sin (n\theta)]. \quad (18)$$

The condition at infinity (5) is satisfied if

$$C_1 = U; \quad D_1 = 0; \quad C_n = D_n = 0; \qquad n = 2, 3, 4, \ldots.$$

Boundary condition (6) then requires that

$$B_1 = 0; \quad c_1 = Ua^2; \quad d_1 = 0; \quad c_n = d_n = 0; \qquad n = 2, 3, 4, \ldots.$$

A solution to all equations and boundary conditions is thus

$$\Phi = A\theta + U(r + a^2 r^{-1}) \cos \theta, \tag{19}$$

as can easily be checked by direct substitution.

INTERPRETATION OF SOLUTION

The presence of the term $A\theta$ in (19) is noteworthy. This term is independent of U but depends on the arbitrary constant A, so there are an infinite number of uniform flows past the cylinder. The velocity vector corresponding to $A\theta$ is, using the polar coordinate gradient formula (3),

$$\nabla(A\theta) = Ar^{-1}\mathbf{e}^{(\theta)}. \tag{20}$$

The corresponding flow is purely circumferential. Its speed approaches zero as $r \to \infty$, so the uniform flow at infinity is unaffected by this circulatory motion.

As a measure of over-all rotation in a flow, Kelvin in 1869 introduced the concept of the **circulation** $K(C)$ **around a simple closed curve** C, which he defined in terms of a line integral as follows:

$$K(C) = \oint_C \mathbf{v} \cdot \mathbf{dr}. \tag{21}$$

When $\mathbf{v} = \nabla\Phi$ one can show without difficulty [as in Exercise 1(c)] that

circulation around C in a given direction	=	change of ϕ when C is traversed once in the given direction.

When Φ is given by (19), $K(C) = 2\pi A$ for any curve C circling the cylinder once in the counterclockwise direction [and $K(C) = 0$ for any curve C that does not enclose the cylinder].

Our separation of variables approach would thus have given a unique answer if we had required as $r \to \infty$ not only that the velocity approach a constant U but also that the circulation approach a constant $2\pi A$.

Let us return to the question that motivated our investigation: What is the force on the cylinder? To find the force we must compute the pressure from the momentum equation for steady motion. From (1.10) with $\mathbf{f} = \mathbf{0}$* we obtain

$$(\mathbf{v} \cdot \nabla)\mathbf{v} = -\rho^{-1}\nabla p \qquad \text{or} \qquad (\nabla\phi \cdot \nabla)\nabla\phi = -\nabla\left(\frac{p}{\rho}\right) \tag{22a, b}$$

The left side of (22) is known, so p can be computed by direct calculation. Even using polar coordinates, however, this is somewhat unpleasant. A simpler approach stems from the observation that (22) can be written

$$\nabla\left(\frac{1}{2}\nabla\phi \cdot \nabla\phi + \frac{p}{\rho}\right) = 0,$$

so

$$\frac{1}{2}|\nabla\phi|^2 + \frac{p}{\rho} = \text{constant}. \tag{23}$$

By considering (23) in the limit $r \to \infty$, we can determine the constant. Our final result is

$$\frac{1}{2}|\mathbf{v}|^2 + \frac{p}{\rho} = \frac{1}{2}U^2 + \frac{p_\infty}{\rho} \qquad \left(p_\infty \equiv \lim_{r\to\infty} p\right). \tag{24}$$

(This is a special case of Bernoulli's equation, proved in Exercise 1.7.) The force $\mathbf{F}$ on the cylinder of radius a is given in terms of the exterior normal $\mathbf{n}$ and arc length s by

$$\mathbf{F} = \int \mathbf{t}\, ds = -\int p\mathbf{n}\, ds = -\int_0^{2\pi} (\cos\theta\mathbf{i} + \sin\theta\mathbf{j})\, pa\, d\theta. \tag{25}$$

By using (24) and (19), a little calculation shows [Exercise 1(c)] that

$$\mathbf{F} = -\rho U K \mathbf{j} \qquad (K = 2\pi A = \text{counterclockwise circulation}). \tag{26}$$

With flow at infinity in the direction of increasing x, **lift** and **drag** are defined as the force components in the $\mathbf{j}$ and $-\mathbf{i}$ directions. We have found a lift proportional to the circulation, and no drag. To put it dramatically,

* If the effect of gravity were taken into account, there would merely be an additional lifting force due to buoyancy (Exercise 9).

the lack of drag can be interpreted as predicting that there would be no downward force exerted on your arm if you held it under Niagara Falls. Something is wrong somewhere.

Suspicion first falls on the assumption $\mathbf{v} = \nabla\phi$. But it is a central result of inviscid flow theory that $\nabla \wedge \mathbf{v} = 0$, and hence that this assumption is inevitable, *for flow starting from rest*—even if gravity is introduced and even if a varying density is considered, provided that the equation of state is of the form $p = p(\rho)$ (Exercise 1.6).

Perhaps there is another uniform flow past a circle that is not revealed by separation of variables. Perhaps the no-drag paradox holds only for the completely symmetric circular shape, and more sensible results emerge for more general shapes. But it can be shown that for two-dimensional uniform inviscid incompressible flow past *any* rigid bounded object,

$$\mathbf{v} = U\mathbf{i} + Ar^{-1}\mathbf{e}^{(\theta)} + O(r^{-2}) \qquad \text{as } r \to \infty, \tag{27}$$

and (26) continues to give the force (Kutta–Joukowsky theorem).

The three-dimensional counterpart of (27) is

$$\mathbf{v} = U\mathbf{i} + O(r^{-3})$$

and $\mathbf{F} = \mathbf{0}$, in some ways an even more striking result.

In Section 15.1 we stated that it appears reasonable to assume that shear stresses are always zero in water and air. This inviscid fluid assumption led to good results when used in our discussions of the stability of a stratified layer of an incompressible liquid, and of compression waves in gases. But we cannot accept the result that drag is absent in the flow past two- and three-dimensional obstacles. There are two valuable lessons here: (i) that a "reasonable" assumption may or may not turn out to be a good one; and (ii) that a given mathematical model may be perfectly adequate in some contexts and yet inadequate in others.

The predicted absence of drag is called **D'Alembert's paradox**.* One's first reaction on hearing such a counterintuitive result is that it was derived as a consequence of an error in reasoning. Only a mathematical proof can fully convince one that it is the basic assumptions which are at fault. A proof for the three-dimensional case is outlined in Appendix 15.1.

Resolution of D'Alembert's paradox requires the study of viscous fluids, wherein the possibility of shear stresses is considered. This is done in Chapter 3 of II. It is found that however small the viscosity, shear stresses are always important in thin layers located near solid boundaries and perhaps elsewhere. The thickness of these layers approaches zero as the viscosity tends to zero, but for "blunt" bodies there is an effect on drag that does not tend to zero.

* A proof for the two-dimensional case was given by Euler in 1745. D'Alembert "rediscovered or appropriated" it in 1752. Later (1768) "he reasserted it in sensational terms" (Truesdell and Toupin, 1960, p. 541.)

For streamlined bodies such as airfoils, on the other hand, the thin layers have little effect on the flow; the drag approaches zero as the viscosity goes to zero. For streamlined bodies, then, the D'Alembert "paradox" is virtually a correct result, for the drag on such bodies is very small. Furthermore, the two-dimensional force formula (26) is the basis for an explanation of the lift provided by an airfoil.

In summary, the inviscid model provides a good first approximation to the study of flow past streamlined bodies, but the flow past blunt bodies is grossly different from the simple predictions of classical inviscid flow theory. The most important elements in the theory of flow past streamlined bodies are discussed in Chapter 3 of II. The reader will search in vain, however, for a definitive treatment of flow past blunt bodies; the main questions here remain open.

EXERCISES

1. (a) Show that there are no nontrivial solutions of (10b), (12), and (13) if k is negative.

(b) Verify by direct substitution that (19) satisfies all required equations and boundary conditions.

(c) Verify (26).

(d) Use the relation between pressure and speed given by the Bernoulli equation (24) to show that (26) is plausible.

(e) Verify the relation between circulation and the change of ϕ that is stated below (21).

2. (a) We have found that the velocity potential, $\Phi(r, \theta)$, describing the uniform flow around a circular cylinder, is given by (19). Determine the corresponding **stagnation points**, i.e., the points at which $\mathbf{v} = 0$. Be sure to distinguish the cases $2aU > A$, $2aU = A$, and $2aU < A$. Indicate the positions of the stagnation points relative to the cylinder in each of these instances.

(b) Assume that $\mathbf{v}(\mathbf{x}, t)$ is differentiable and show that streamlines have a well-defined tangent except at stagnation points. Conclude that streamlines can cross only at such points.

3. Noting that the flow at infinity is uniform, that the circle must be part of a streamline, and that a streamline can branch at a stagnation point, one can make a surprisingly accurate sketch of the streamlines for this problem. Do so.

4. Suppose that a velocity potential ϕ exists. Show that $\phi(P_1) - \phi(P_0)$ equals the line integral of the velocity along any curve joining the points P_0 and P_1. A meaning is thus ascribed to the potential function.

5. (a) Suppose that a velocity potential $\phi(x, y)$ exists for the velocity field $\mathbf{v}(x, y)$. Consider the transformation into polar coordinates. Show that the relation between the radial velocity component $V_r(r, \theta)$

and the velocity potential in polar coordinates $\Phi(r, \theta)$ is $V_r = \partial\Phi/\partial r$. Do this by direct calculation in the following way. Since $\mathbf{e}^{(r)} = \cos\theta\mathbf{i} + \sin\theta\mathbf{j}$ is a unit vector making angle θ with positive x axis and $V_r = \mathbf{v}\cdot e^{(r)}$, then

$$V_r = \frac{\partial\phi}{\partial x}\cos\theta + \frac{\partial\phi}{\partial y}\sin\theta.$$

Change $\partial\phi/\partial x$ and $\partial\phi/\partial y$ into polar coordinates via the chain rule.

(b) Show that the azimuthal velocity component V_θ can be expressed as $V_\theta = r^{-1}(\partial\Phi/\partial\theta)$ by justifying and completing the following steps:

$$V_\theta = \nabla\phi\cdot\mathbf{e}^{(\theta)} = \frac{d\phi}{ds} = \frac{\partial\Phi}{\partial\theta}\frac{d\theta}{ds} = \cdots.$$

Here $d\phi/ds$ is the directional derivative in the direction of increasing θ.

‡6. Derive (26) by using the momentum transfer equation. (See Exercise 14.2.5.)

‡7. Consider potential flow ($\mathbf{v} = \nabla\Phi$), past a sphere of radius a, of an inviscid fluid of uniform density. Orient axes so that far from the sphere $\rho = a$ there is uniform flow of velocity $-U\mathbf{k}$, U a constant. Use spherical coordinates. (See Figure 15.13.) Look for axisymmetric flow

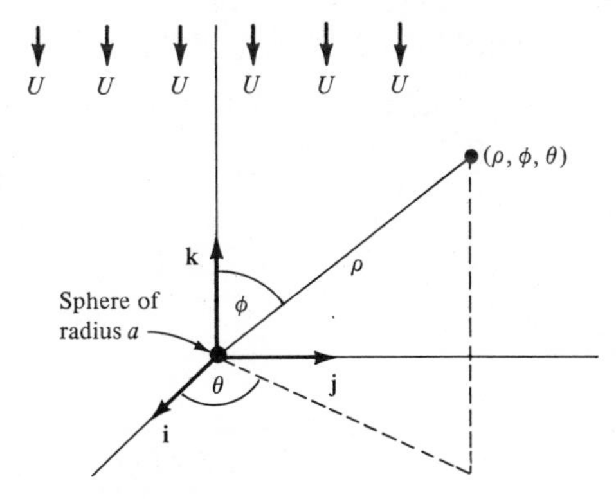

FIGURE 15.13. *Diagram of the situation discussed in Exercise 7.*

($\partial/\partial\theta = 0$), and show that the problem reduces to the following set of equations for $\Phi(\rho, \phi)$:

$$\sin\theta\frac{\partial}{\partial\rho}\left(\rho^2\frac{\partial\Phi}{\partial\rho}\right) = \frac{\partial}{\partial\phi}\left(\sin\phi\frac{\partial\Phi}{\partial\phi}\right) = 0,$$

$$\Phi \to -U\rho\cos\phi \quad \text{as } \rho\to\infty, \qquad \frac{\partial\Phi}{\partial\rho} = 0 \quad \text{at } \rho = a.$$

Find the most general solution that emerges from a separation of variables approach. You will need the following facts: the *Legendre equation* for $y(\phi)$ is

$$\frac{1}{\sin\phi}\frac{d}{d\phi}\left(\sin\phi\,\frac{dy}{d\phi}\right) + ky = 0.$$

Solutions to this equation remain finite for all ϕ only if $k = n(n+1)$, $n = 0, 1, 2, \ldots$. The corresponding solutions are multiples of $P_n(\cos\phi)$, where the P_n are (Legendre) polynomials of degree n, for example,

$$P_0(x) = 1, \qquad P_1(x) = x, \qquad P_2(x) = \tfrac{1}{2}(3x^2 - 1).$$

8. Find the net force exerted on the sphere by the flow of Exercise 7. Do this by integrating the local effects of pressure.

9. (a) Show how (23) would be altered if a body force term $\rho^{-1}\mathbf{f} = -g\mathbf{j}$ were added to the right side of (22a).

(b) Show that the effect considered in (a) would add a buoyancy term to the formula for $\mathbf{F}$ in (26).

Appendix 15.1 A Proof of D'Alembert's Paradox in the Three-dimensional Case

Consider three-dimensional irrotational inviscid fluid flow generated by the motion of a bounded object and suppose that the fluid is motionless at infinity. Let $\mathbf{v} = \nabla\phi$, where ϕ is the velocity potential. If the fluid is of uniform density, $\nabla\cdot\mathbf{v} = 0$, ϕ is harmonic, and (16.1.18) can be applied. This equation represents a harmonic function at a point P in terms of an integral over the boundary ∂R of a region R that surrounds P. If P is the origin, the equation reads

$$\phi(0) = -\frac{1}{4\pi}\iint_{\partial R}\left[\phi\frac{\partial}{\partial n}\left(\frac{1}{r}\right) - \frac{1}{r}\frac{\partial\phi}{\partial n}\right]d\sigma \quad (\partial/\partial n \equiv \text{exterior normal derivative}).$$

With a slight reinterpretation we can write

$$\phi(\mathbf{x}) = -\frac{1}{4\pi}\iint_{\partial R}\left[\phi\frac{\partial}{\partial n}\left(\frac{1}{r'}\right) - \frac{1}{r'}\frac{\partial\phi}{\partial n}\right]d\sigma, \tag{1}$$

where r' is the distance from $\mathbf{x}$ to the variable of integration that runs over ∂R.

Let R be a region that contains $\mathbf{x}$ and is bounded within by S, a surface containing the obstacle mentioned above; and is bounded without by Σ, a large sphere of radius ρ centered on $\mathbf{x}$ (Figure 15.14).

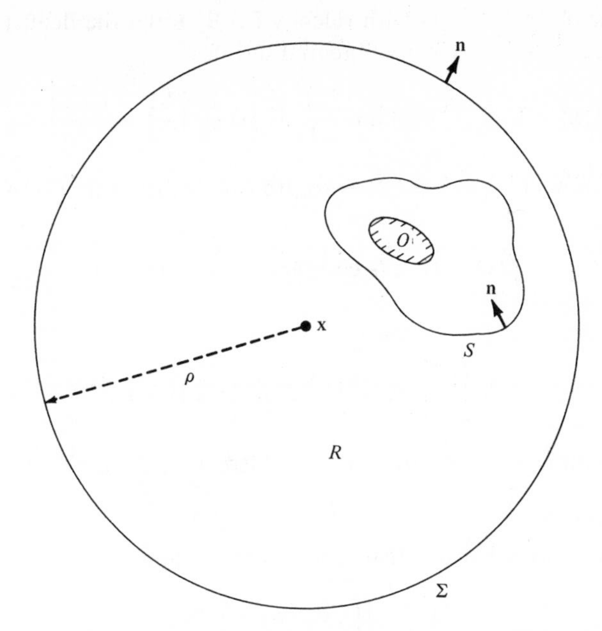

FIGURE 15.14. *Region R used in representing the velocity potential at* **x** *for flow past obstacle O. The vectors* **n** *are typical unit exterior normals to R.*

It is not difficult to show (Exercise 1) that (1) implies that

$$\phi(\mathbf{x}) = \frac{1}{4\pi\rho^2}\iint_{\Sigma} \phi \, d\sigma + \frac{1}{4\pi\rho}\iint_{\Sigma} [\nabla\phi \cdot \mathbf{n}] \, d\sigma + \frac{1}{4\pi}\iint_{S} \left[\phi \frac{\partial}{\partial n}\left(\frac{1}{r'}\right) - \frac{1}{r'}\frac{\partial \phi}{\partial n}\right] d\sigma. \tag{2}$$

In (2) and below, the time t is not explicitly designated, since it is regarded as fixed throughout our discussion.

The first integral in (2) is the mean value of ϕ over a sphere of radius ρ centered at $\mathbf{x}$. For the moment, let us take for granted the plausible assertion that this value is independent of $\mathbf{x}$ in the limit $\rho \to \infty$. We shall sketch a proof of this assertion below.

That the second integral in (2) vanishes in the limit can be deduced from the fact that $\nabla\phi \to \mathbf{0}$ at large distances from the origin, because of the assumption that the fluid is motionless at infinity.

Suppose now that the obstacle moves with a uniform velocity $-\mathbf{U}$. Take new coordinate axes that move with velocity $-\mathbf{U}$ relative to the old ones. (In the new coordinate system, of course, the obstacle is stationary and fluid far from the obstacle moves with velocity **U**.) By using the deductions made so far, we see that the velocity potential satisfies

$$\phi(\mathbf{x}) = \mathbf{U}\cdot\mathbf{x} + \text{constant} + \frac{1}{4\pi}\iint_S \left[\phi\frac{\partial}{\partial n}\left(\frac{1}{r'}\right) - \frac{1}{r'}\frac{\partial\phi}{\partial n}\right] d\sigma. \tag{3}$$

Exercise 2 shows that as $r \equiv |\mathbf{x}| \to \infty$, we can deduce from (3) that

$$\phi(\mathbf{x}) \to \mathbf{U}\cdot\mathbf{x} + \text{constant} - \frac{F}{4\pi r} + O(r^{-2}). \tag{4}$$

Here

$$F \equiv \iint_S \frac{\partial\phi}{\partial n}\,d\sigma = \iint_S \nabla\phi\cdot\mathbf{n}\,d\sigma = \iint_S \mathbf{v}\cdot\mathbf{n}\,d\sigma$$

is the net outflow of fluid through S. As there are no sources, we must have

$$F = 0. \tag{5}$$

To prove this, note that the divergence theorem shows that

$$F - \iint_{\text{obstacle}} \nabla\phi\cdot\mathbf{n}\,d\sigma = 0. \tag{6}$$

But the second integral in (6) is zero, since $\mathbf{v}\cdot\mathbf{n} = 0$ on the stationary obstacle.

From (4), (5), and Bernoulli's equation (4.24) we can conclude [Exercise 2(b)] that

$$\text{as } r \to \infty, \qquad \mathbf{v} = \mathbf{U} + O(r^{-3}), \quad p = p_\infty + O(r^{-3}). \tag{7}$$

Since we know that the speed and pressure approach their values at infinity in this fashion, it follows from Exercise 14.2.6 that there is no force on the obstacle.

NOTE. For a two-dimensional body, the estimate of behavior at infinity cannot be obtained by the argument used above: the velocity potential may be multiple-valued. One can utilize the Poisson integral representation of the solution,* or one can proceed from analytic function theory.

It remains to prove the lemma that the mean value of ϕ on a sphere $S_\mathbf{x}$ of radius ρ is independent of the position $\mathbf{x}$ of its center. Denote this mean value by $M_\rho(\mathbf{x})$, and let $\mathbf{u}$ denote a unit vector. Then

$$\begin{aligned} M_\rho(\mathbf{x} + a\mathbf{u}) - M_\rho(\mathbf{x}) &= \frac{1}{4\pi\rho^2}\iint_{S_{\mathbf{x}+a\mathbf{u}}} \phi\,d\sigma - \frac{1}{4\pi\rho^2}\iint_{S_\mathbf{x}} \phi\,d\sigma \\ &= \frac{1}{4\pi\rho^2}\iint_{S_\mathbf{x}} (\phi^* - \phi)\,d\sigma, \end{aligned} \tag{8}$$

* The Poisson integral representation is the two-dimensional version of (16.2.13), with a logarithmic singularity.

where ϕ^* is ϕ evaluated according to the change of variables necessary to shift the domain of integration in the first integral from $S_{\mathbf{x}+a\mathbf{u}}$ to $S_{\mathbf{x}}$. By the integral mean value theorem, we can deduce from (8) that

$$M_\rho(\mathbf{x} + a\mathbf{u}) - M_\rho(\mathbf{x}) = \phi^* - \phi = \nabla\phi \cdot (\text{vector of magnitude } a), \tag{9}$$

where $\phi^* - \phi$ and $\nabla\phi$ are evaluated at some point on S. Since $|\nabla\phi| \to 0$ as $\rho \to \infty$, we conclude that

$$M_\infty(\mathbf{x} + a\mathbf{u}) - M_\infty(\mathbf{x}) = 0$$

for arbitrary a and $\mathbf{u}$. Hence $M_\infty(\mathbf{x})$ is a constant.

EXERCISES

1. Verify equations (2), (3), and (6).
2. (a) Verify (4).
   (b) Verify (7).
3. Explicitly carry out the change of variables required in (8) and thereby obtain an expression for the "vector of magnitude a" mentioned in (9).

Appendix 15.2 Polar and Cylindrical Coordinates

Let $\mathbf{e}^{(r)}(r_0, \theta_0)$ be a unit tangent vector at the point (r_0, θ_0) to the "r-varying" coordinate curve $\theta = \theta_0$. Let $\mathbf{e}^{(\theta)}(r_0, \theta_0)$ be a unit tangent vector at the same point to the "θ-varying" coordinate curve $r = r_0$. $\mathbf{e}^{(r)}$ and $\mathbf{e}^{(\theta)}$ are to point in the directions of increasing r and θ, respectively (Figure 15.15). In polar coordinates the components of a vector $\mathbf{v}(r, \theta)$ are defined to be $v^{(r)}$ and $v^{(\theta)}$, where

$$\mathbf{v}(r, \theta) = v^{(r)}(r, \theta)\mathbf{e}^{(r)}(r, \theta) + v^{(\theta)}(r, \theta)\mathbf{e}^{(\theta)}(r, \theta). \tag{1}$$

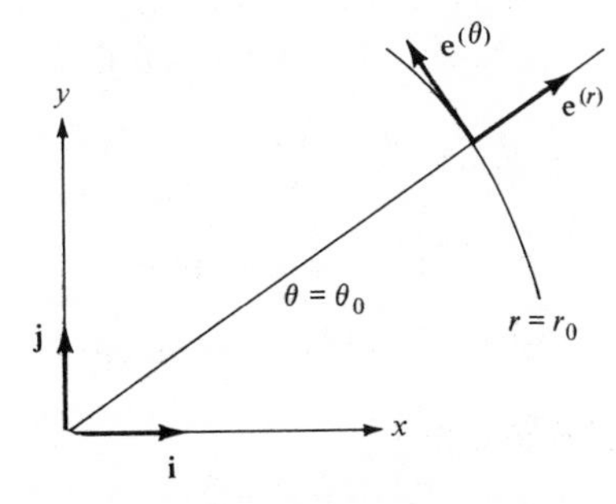

FIGURE 15.15. *Fixed cartesian unit vectors* $\mathbf{i}$ *and* $\mathbf{j}$, *and a typical pair of varying polar unit vectors* $\mathbf{e}^{(r)}$ *and* $\mathbf{e}^{(\theta)}$.

Since $\mathbf{e}^{(r)}$ and $\mathbf{e}^{(\theta)}$ are orthogonal unit vectors,

$$v^{(r)} = \mathbf{v} \cdot \mathbf{e}^{(r)}, \qquad v^{(\theta)} = \mathbf{v} \cdot \mathbf{e}^{(\theta)}. \tag{2}$$

In particular, when $\mathbf{v} = \nabla\Phi$ the relations

$$v^{(r)} = \frac{\partial \Phi}{\partial r}, \qquad v^{(\theta)} = r^{-1}\frac{\partial \Phi}{\partial \theta}, \tag{3}$$

follow almost immediately from the following formula for the directional derivative of Φ in the direction of the unit vector $\mathbf{e}$:

$$\frac{d\Phi}{ds} = \nabla\Phi \cdot \mathbf{e}.$$

The following formulas relate polar coordinates with cartesian (unit vectors $\mathbf{i}$ and $\mathbf{j}$, velocity components u and v). If

$$\mathbf{i} = C_1\,\mathbf{e}^{(r)} + C_2\,\mathbf{e}^{(\theta)},$$

then

$$\mathbf{e}^{(r)} \cdot \mathbf{i} = C_1, \qquad C_1 = \cos\theta; \qquad \mathbf{e}^{(\theta)} \cdot \mathbf{i} = C_2, \qquad C_2 = \cos\left(\theta + \frac{\pi}{2}\right).$$

Thus

$$\mathbf{i} = \cos\theta\,\mathbf{e}^{(r)} - \sin\theta\,\mathbf{e}^{(\theta)}. \tag{4}$$

Similarly,

$$\mathbf{j} = \sin\theta\,\mathbf{e}^{(r)} + \cos\theta\,\mathbf{e}^{(\theta)}, \tag{5}$$

and

$$\mathbf{e}^{(r)} = \cos\theta\,\mathbf{i} + \sin\theta\,\mathbf{j}, \tag{6}$$

$$\mathbf{e}^{(\theta)} = -\sin\theta\,\mathbf{i} + \cos\theta\,\mathbf{j}. \tag{7}$$

From (1), (6), and (7), since $\mathbf{v} = u\mathbf{i} + v\mathbf{j}$,

$$v^{(r)} = u\cos\theta + v\sin\theta, \tag{8}$$

$$v^{(\theta)} = -u\sin\theta + v\cos\theta. \tag{9}$$

In like manner, using (4) and (9),

$$u = v^{(r)}\cos\theta - v^{(\theta)}\sin\theta, \tag{10}$$

$$v = v^{(r)}\sin\theta + v^{(\theta)}\cos\theta. \tag{11}$$

In *cylindrical coordinates*, polar coordinates are used in the (x, y) plane, while the z-component is the same as in cartesian coordinates. For example, $\mathbf{e}^{(r)}(r_0, \theta_0, z_0)$ is a unit tangent vector at the point (r_0, θ_0, z_0) to the curve $r = r_0$, $z = z_0$. The components of the velocity vector $\mathbf{v}(r_0, \theta_0, z_0)$ are

$$[v^{(r)}(r_0, \theta_0, z_0), v^{(\theta)}(r_0, \theta_0, z_0), w(r_0, \theta_0, z_0)]$$

so

$$\mathbf{v} = v^{(r)}\mathbf{e}^{(r)} + v^{(\theta)}\mathbf{e}^{(\theta)} + w\mathbf{k}.$$

As usual, $\mathbf{k}$ denotes the unit vector along the z axis.

CHAPTER 16
Potential Theory

THE THEORY of gravitational potential, and its extension to the theory of electrostatics and to electromagnetic and sound waves, provides fine examples of the power of mathematical analysis. It also gives rise to a number of very subtle and challenging problems. In this chapter we shall discuss a few of the highlights of this subject to give a reader a feeling of the breadth of its implications and ramifications.

In Section 16.1, using the gravitational potential for motivation, we derive various results for the equations of Laplace and Poisson. Section 16.2 deals primarily with the construction and use of Green's functions; electrostatics provides physical background. The final section treats an inviscid flow problem that reduces to Helmholtz's equation: the problem of the diffraction of planar sound waves impinging on a hole in a screen.

16.1 Equations of Laplace and Poisson

The first major formula in this section, equation (7), gives the gravitational potential $V(\mathbf{x})$ due to mass that is distributed in various regions throughout space. We show that if $\mathbf{x}$ is outside the mass distributions, then $V(\mathbf{x})$ satisfies Laplace's equation (3), while if $\mathbf{x}$ is within a mass distribution, then $V(\mathbf{x})$ satisfies Poisson's equation (8). We then turn to a brief study of functions that satisfy these equations within bounded regions R, with the object of representing the functions by formulas which involve integrals over the boundary ∂R of R. For Laplace's equation, the representation leads to the mean value theorem and to the maximum principle for harmonic functions. We also conclude that (7) must be satisfied by a function which satisfies Poisson's equation *provided* that its behavior at infinity is specified in a certain way. This section closes with some remarks on uniqueness. A theorem-proof style seems appropriate.

GRAVITATIONAL POTENTIAL OF DISCRETE MASS DISTRIBUTIONS

As is discussed in introductory physics courses, the gravitational potential V at a field point $\mathbf{x}$, due to a point mass m located at the source point $\boldsymbol{\xi}$, is given by

$$V(\mathbf{x}) = -G\frac{m}{|\mathbf{x}-\boldsymbol{\xi}|}\,. \tag{1}$$

(Here G is the universal gravitational constant 6.67×10^{-8} dyne cm^2/g^2.) By definition, the potential is a scalar quantity whose negative gradient gives the gravitational force **g**. Thus from (1) we obtain the inverse square force due to a point mass m:

$$g_i(\mathbf{x}) = -\frac{\partial V}{\partial x_i} = -Gm\,\frac{x_i - \xi_i}{|\mathbf{x} - \boldsymbol{\xi}|^3}\,. \tag{2}$$

It is easy to verify [Exercise 1(b)] that, except when $\mathbf{x} = \boldsymbol{\xi}$, the gravitational potential satisfies the Laplace equation

$$\nabla^2 V = \sum_{i=1}^{3} \frac{\partial^2 V}{\partial x_i^2} = 0. \tag{3}$$

In complicated problems, e.g., in the presence of boundaries of all but the simplest shapes, it is not possible to guess an exact expression for the potential. It is usually best, in such cases, to work from the differential equation that the potential satisfies. Preparing ourselves for these cases, we observe that we can obtain (1) by seeking a solution of the partial differential equation (3) with a **singularity** at the point $\boldsymbol{\xi}$. Such a specification is really quite vague, since there are other solutions [e.g., the partial derivatives of (1)] that satisfy the same general requirement. One must be more precise in specifying the nature of the singularity. In fact, to recover (1), one should impose the restriction that

$$\lim_{r \to 0} rV = -Gm. \tag{4}$$

The gravitational potential due to a number of mass points $m_1, m_2, \ldots, m_s$ located at $\boldsymbol{\xi}_1, \boldsymbol{\xi}_2, \ldots, \boldsymbol{\xi}_s$ is

$$V(\mathbf{x}) = -G \sum_{k=1}^{s} \frac{m_k}{|\mathbf{x} - \boldsymbol{\xi}_k|}\,. \tag{5}$$

This function V also satisfies the differential equation (3) (when $\mathbf{x} \neq \boldsymbol{\xi}_k$). To recover (5) from (3), we have to stipulate

$$\lim_{\mathbf{x} \to \boldsymbol{\xi}_k} |\mathbf{x} - \boldsymbol{\xi}_k|\, V = -Gm_k, \qquad k = 1, 2, \ldots, s. \tag{6}$$

As we shall see later, there is also a condition at infinity to be specified.

GRAVITATIONAL POTENTIAL OF CONTINUOUS MASS DISTRIBUTIONS

If there is a mass distribution of density $\rho(\boldsymbol{\xi})$, the gravitational potential is given by

$$V(\mathbf{x}) = -G \iiint \frac{\rho(\boldsymbol{\xi})\, d\tau}{|\mathbf{x} - \boldsymbol{\xi}|}\,. \tag{7}$$

Here $d\tau$ is the element of volume; integration is with respect to $\boldsymbol{\xi}$. It is convenient to take infinite space as the region of integration, with the understanding that $\rho \equiv 0$ in regions devoid of mass. For $\mathbf{x}$ within such regions, the Laplace equation is again satisfied. What about a point that lies within a mass-filled region?

First, there is a question of the convergence of the integral, but this can be proved without difficulty. Second, if we wish to calculate the force vector by formally calculating the gradient of (7), the problem becomes more serious, but formal differentiation of (7) can still be justified. However, when we attempt to calculate the second derivatives of (7), the formal procedure is no longer justifiable. Indeed, we find that the gravitational potential (7) does not satisfy the Laplace equation (3), but satisfies the **Poisson equation**:

$$\nabla^2 V = 4\pi G\rho. \tag{8}$$

Although the formal derivation of this relationship is quite easy—as we shall see immediately—the proof of the existence of the second derivatives (for sufficiently smooth density distributions) is more complicated. [See Jeffreys and Jeffreys (1962), Sec. 6.04.] We begin the discussion with a famous formula that relates gravitational flux to mass.

Theorem 1 (Gauss). The *flux* F of the gravitational force field $\mathbf{g} = -\nabla V$ is defined by

$$F = \oint\!\!\oint_{\partial R} \mathbf{g} \cdot \mathbf{n}\, d\sigma. \tag{9a}$$

Here $\mathbf{n}$ is a unit exterior normal to ∂R, a surface (assumed to be smooth enough to permit various manipulations) that encloses a region R. The flux F is related to the total mass m enclosed within ∂R by the relationship

$$F = -4\pi GM. \tag{9b}$$

A mass point on the surface is considered half inside and half outside.

Proof. Consider first a single point mass. If the surface ∂R does not enclose it, then by the divergence theorem we have

$$\oint\!\!\oint_{\partial R} \mathbf{g} \cdot \mathbf{n}\, d\sigma = \iiint_R \nabla \cdot \mathbf{g}\, d\tau = -\iiint_R \nabla^2 V\, d\tau = 0. \tag{10}$$

If there is a mass point inside R, say at point P, the above theorem cannot be applied. But let us draw a spherical surface S of small radius ε, centered at the mass point P (Figure 16.1).*

* The use of a small sphere to exclude a singularity is a very useful device. We shall use it again subsequently.

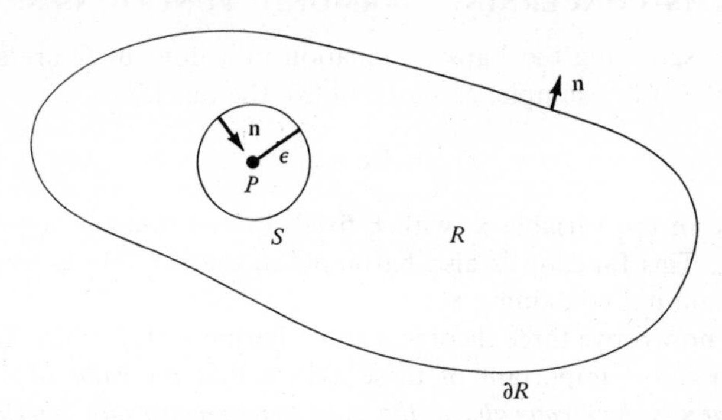

FIGURE 16.1. *Use of a sphere S of small radius ε to exclude a singularity point P. The vectors* **n** *are unit exterior normals to the region enclosed by S and ∂R.*

Since (10) holds if it is applied to the region between S and ∂R we have

$$\oiint_{\partial R} \mathbf{g} \cdot \mathbf{n}\, d\sigma + \oiint_{S} \mathbf{g} \cdot \mathbf{n}\, d\sigma = 0.$$

In this equation, the unit normals $\mathbf{n}$ to the surfaces ∂R and S are both directed *outward* from the region between these surfaces; in particular, the normal on S points toward the mass point. On the surface S, the law of inverse squares enables us to calculate the flux [Exercise 3(a)], and we immediately obtain the desired result (9b) for a single mass point. By the principle of superposition, it also holds in general. ▯

Theorem 2. If $\mathbf{x}$ is in a domain that contains continuously distributed mass of density ρ, then the potential $V(\mathbf{x})$ of (7) satisfies Poisson's equation (8).

Proof. We write (9) as

$$\oiint_{\partial R} \mathbf{g} \cdot \mathbf{n}\, d\sigma = -4\pi G \iiint_{R} \rho\, d\tau.$$

If the divergence theorem may be applied to the left-hand side of this equation, we have

$$\iiint_{R} (\nabla \cdot \mathbf{g} + 4\pi G\rho)\, d\tau = 0, \tag{11}$$

for an arbitrary region R. Thus, for any point with a neighborhood in which the integrand is continuous, we arrive at the desired result,

$$-\nabla \cdot \mathbf{g} = \nabla^2 V = 4\pi G\rho, \tag{12}$$

by using the Dubois–Reymond lemma (Section 4.1). ▯

THEOREMS CONCERNING HARMONIC FUNCTIONS

Functions satisfying the Laplace equation in a domain $\mathscr{D}$ are said to be **harmonic** in $\mathscr{D}$. For example, as stated in (3), the function

$$r^{-1} = |\mathbf{x} - \boldsymbol{\xi}|^{-1} \tag{13}$$

is harmonic in the variable $\mathbf{x}$, with $\boldsymbol{\xi}$ fixed, in any domain not containing the point $\boldsymbol{\xi}$. This function is also harmonic in the variable $\boldsymbol{\xi}$, with $\mathbf{x}$ fixed, in any domain not containing $\mathbf{x}$.

We shall now prove three theorems about harmonic functions. The second and perhaps most important of these asserts that *the value of a harmonic function ϕ at $\mathbf{x}$ is the mean value of the same function over any spherical surface centered around* $\mathbf{x}$ (and within the domain of the function). If ϕ is interpreted as steady state temperature (Section 4.1) this theorem is physically reasonable —for if the temperature at $\mathbf{x}$ were not equal to its average, one would expect heat to flow, in order to "smooth out" the heat distribution. If ϕ is given the probabilistic interpretation of Section 3.3, then the theorem states the highly plausible result that for a particle starting at $\mathbf{x}$ the probability of leaving a specified portion of the boundary is equal to the average of the corresponding probabilities for particles starting on a sphere around $\mathbf{x}$ (and within the domain of the function).

Theorem 3. A harmonic function $\phi(x)$ is determined within a region R by its values and those of its normal derivative on the boundary ∂R of R. The formula is given in (18). (We assume sufficient smoothness to carry out the steps of the proof.)

Proof. We start with the symmetric Green's formula

$$\iiint\limits_R (u\nabla^2 v - v\nabla^2 u)\, d\tau = \oiint \left(u\frac{\partial v}{\partial n} - v\frac{\partial u}{\partial n}\right) d\sigma, \tag{14}$$

where $\partial/\partial n$ denotes a derivative in the direction of the exterior normal. Formula (14) is an immediate consequence of the divergence theorem. (See Exercise 3.4.8.) In (14) we set

$$u = \phi, \qquad v = r^{-1}, \tag{15}$$

and regard $\boldsymbol{\xi}$ as the variable of integration. There is a singularity of $v = |\mathbf{x} - \boldsymbol{\xi}|^{-1}$ at the point P with coordinates $\mathbf{x}$. To exclude it, we again surround P with a sphere S of small radius ε (Figure 16.1). We then apply (14) to the region between R and S. Since the left-hand side of (14) vanishes identically,

$$\iint\limits_{\partial R} \left[\phi \frac{\partial}{\partial n}\left(\frac{1}{r}\right) - \frac{1}{r}\frac{\partial \phi}{\partial n}\right] d\sigma + \iint\limits_{S} \left[\phi \frac{\partial}{\partial n}\left(\frac{1}{r}\right) - \frac{1}{r}\frac{\partial \phi}{\partial n}\right] d\sigma = 0. \tag{16}$$

On S, the normal derivative satisfies $\partial/\partial n = -\partial/\partial r$, and hence the second integral becomes

$$\iint_S \left[\frac{\phi}{\varepsilon^2} + \frac{1}{\varepsilon}\frac{\partial \phi}{\partial r}\right] d\sigma = \int_0^{2\pi}\int_0^{\pi}\left[\frac{\phi}{\varepsilon^2} + \frac{1}{\varepsilon}\frac{\partial \phi}{\partial r}\right]\varepsilon^2 \sin \xi \, d\xi \, d\eta, \tag{17}$$

where ξ and η are the polar and azimuthal angles of spherical coordinates.

If we now let the radius ε of the sphere S approach zero, we obtain, from (16),

$$\iint_{\partial R}\left[\phi \frac{\partial}{\partial n}\left(\frac{1}{r}\right) - \frac{1}{r}\frac{\partial \phi}{\partial n}\right] d\sigma + 4\pi\phi(\mathbf{x}) = 0,$$

or

$$\phi(\mathbf{x}) = -\frac{1}{4\pi}\iint_{\partial R}\left[\phi \frac{\partial}{\partial n}\left(\frac{1}{r}\right) - \frac{1}{r}\frac{\partial \phi}{\partial n}\right] d\sigma. \; \square \tag{18}$$

The representation (18) is quite useful—indeed, we have already used it in the proof of D'Alembert's paradox (Appendix 15.1).

Theorem 4 (Mean value theorem for harmonic functions). Let $\mathbf{x}$ be the center of a sphere R of radius ρ. Suppose that ϕ is harmonic in a domain containing R. Then

$$\phi(\mathbf{x}) = \frac{1}{4\pi\rho^2}\oiint_{\partial R} \phi \, d\sigma.$$

Proof. Applying (18) and using the fact that $\partial/\partial n = \partial/\partial r$ on ∂R, we obtain

$$\phi(\mathbf{x}) = \frac{1}{4\pi\rho^2}\oiint_{\partial R} \phi \, d\sigma + \frac{1}{4\pi\rho}\oiint_{\partial R} \frac{\partial \phi}{\partial n} \, d\sigma.$$

Employing the divergence theorem, we see at once that the integral in the second term vanishes, for it equals

$$\iint_{\partial R} (\mathbf{n} \cdot \nabla\phi) \, d\sigma = \iiint_R \nabla^2\phi \, d\tau = 0. \; \square$$

The mean value theorem will now be used to show that a nontrivial harmonic function cannot have an interior maximum or minimum. (Compare the remarks in Section 4.1 concerning the maximum principle for the heat equation.)

Theorem 5 (Maximum principle for harmonic functions). Let ϕ be a function that is harmonic within a region R. Then either ϕ is identically constant, or the maximum and minimum values of ϕ occur on the boundary of R.

Proof. Suppose that ϕ has an interior local maximum M at a point $\boldsymbol{\xi}$, so that $|\phi(\mathbf{x})| \leq M$ for $|\mathbf{x} - \boldsymbol{\xi}| \leq \varepsilon$, ε sufficiently small. Apply Theorem 4, taking $\boldsymbol{\xi}$ as the center of a sphere of radius ε_1, $\varepsilon_1 \leq \varepsilon$. This gives

$$\phi(\boldsymbol{\xi}) = \frac{1}{4\pi\varepsilon_1^2} \iint \phi \, d\sigma \leq \frac{1}{4\pi\varepsilon_1^2} \iint M \, d\sigma = M.$$

But $\phi(\boldsymbol{\xi}) = M$, so the equality sign must be used in the above equation. Thus [Exercise 4(a)]

$$\phi(\mathbf{x}) \equiv M, \qquad |\mathbf{x} - \boldsymbol{\xi}| \leq \varepsilon. \tag{19}$$

Now take new circles, centered on $|\mathbf{x} - \boldsymbol{\xi}| = \varepsilon$, and repeat the argument. The desired result follows. The proof ruling out internal minima is essentially the same. □

INTEGRAL REPRESENTATION FOR THE SOLUTION TO POISSON'S EQUATION

Again let us employ Green's identity (14). As before, we set $v = r^{-1}$, but now we put $u = V$, a function satisfying Poisson's equation (8). We obtain the following integral representation for a function V that satisfies Poisson's equation within a bounded region R [Exercise 4(b)]:

$$V = -\frac{1}{4\pi} \oiint_{\partial R} \left[V \frac{\partial}{\partial n}\left(\frac{1}{r}\right) - \frac{1}{r}\frac{\partial V}{\partial n} \right] d\sigma - G \iiint_R \frac{\rho}{r} \, d\tau. \tag{20}$$

The last term in (20) is the volume integral (7), giving the gravitational potential due to the density distribution ρ. The other term represents contributions from the surface S, which can also be given physical interpretations. Such interpretations are quite real when we consider problems in electrostatics but are rather artificial for the gravitational problem. In the latter case the surface integral should really be regarded as giving contributions from mass distributions outside of R. We shall therefore consider the case where the surface $\partial R \to \infty$. In the limit we are dealing with mass distributed in an unbounded region, so it is expected that (7) would follow.

If (7) holds, we can make formal appraisals of the magnitudes of V and ∇V at infinity, and obtain

$$V \sim r^{-1}, \quad \nabla V \sim r^{-2}, \qquad \text{as} \quad r \to \infty. \tag{21}$$

The conditions (21) should be regarded as stipulations on the solutions we seek. Otherwise, there is no assurance of uniqueness; in fact, harmonic functions such as $r^n \cos n\theta$ could be added to V. With conditions (21) imposed, it is a matter of simple calculation to show that indeed the surface integral in (20) approaches zero as the surface ∂R recedes to infinity [Exercise 5(b)]. We thus obtain the following result.

Theorem 6 (A partial converse of Theorem 2). Let V satisfy Poisson's equation within a region R. If R is bounded and has a sufficiently smooth boundary, then the representation (20) applies. If R is unbounded and V has behavior (21) for large r, then (7) holds.

UNIQUENESS

Formula (18) suggests that a harmonic function is completely specified in a domain once the value of the function *and* its normal derivative are given on the enclosing surface. This is certainly true, but the function is actually uniquely determined by *either* the value of the function *or* its normal derivative. (In the latter case, there is still the freedom of an arbitrary additive constant.) If a harmonic function is sought with prescribed boundary values, it is customary to speak of the **Dirichlet problem** or the **first boundary value problem** for harmonic functions. If the normal derivative of the function is prescribed on the boundary, the literature refers to the **Neumann problem** or the **second boundary value problem.**

For both Dirichlet and Neumann problems (or a combination of them) uniqueness can be proved by following an argument similar to that used to prove uniqueness for heat conduction (Exercise 6). Alternatively, the maximum principle can be employed (Exercise 7). Extension of the proofs to the Poisson equation is very simple (Exercise 8).

EXERCISES

1. (a) Verify (2).
(b) Verify that the function V defined in (1) satisfies (3) except when $\mathbf{x} = \xi$.

2. Supply sufficient conditions to permit two differentiations to be taken under the integral sign in (7) and thereby prove that $V(\mathbf{x})$ satisfies Laplace's equation when $\mathbf{x}$ is in a domain that is devoid of mass.

3. (a) Complete formal verification of Theorem 1.
(b) Fill in the details that are required to complete the proof.

4. (a) Use the Dubois–Reymond lemma to verify (19).
(b) Verify (20).

5. (a) Provide formal calculations which indicate that (7) implies (21).
(b) Show that if the conditions of (21) are satisfied, then the surface integral in (20) approaches zero as the surface approaches infinity.

6. (a) Prove a uniqueness theorem for the Dirichlet problem by following the general approach that leads to (4.1.17). (The proof in the present case is actually easier.)
(b) For the Neumann problem, show that two solutions can differ at most by a constant. Give a physical argument which shows that this is a reasonable result.

(c) Prove a uniqueness theorem for the problem wherein a harmonic function is sought with prescribed values on part of the boundary and a prescribed normal derivative on the rest of the boundary.

7. (a) Use the maximum principle to prove that the Dirichlet problem has at most one solution.
   (b) Use the maximum principle to prove that a solution to the Dirichlet problem depends continuously on the boundary data.
8. Extend the various proofs so that they apply to the Poisson equation.

16.2 Green's Functions

In this section we construct a variety of Green's functions. Roughly speaking, these give the effect at a field point **x** of a unit source (of heat, electrostatic potential, gravitational potential, etc.) that is concentrated at $\boldsymbol{\xi}$. Explicit formulas for Green's functions are given for free space and for situations in which a planar or spherical boundary is present. We show how the corresponding general boundary value problems can be solved, once these functions are determined.

GREEN'S FUNCTION FOR THE DIRICHLET PROBLEM

To deepen our understanding of harmonic functions and to provide a useful method for solving specific problems, we should introduce the concept of Green's function. This is best done in the context of electrostatics.

Consider a cavity inside a perfect conductor, at potential zero ("grounded"). If there is a positive point charge e at an internal point $\boldsymbol{\xi}$ (the **source point**), there would be induced negative charges on the surface. These would be distributed in such a manner that the potential on the surface remains zero. The potential at any other internal point **x** (a **field point**) is then a well-defined quantity, and we shall denote it by $eG(\mathbf{x}, \boldsymbol{\xi})$. G is a **Green's function**. As is typical, *it can be regarded as the "effect"* (potential) *at* **x** *due to a unit "cause"* (charge) *located at* $\boldsymbol{\xi}$. Near the point $\boldsymbol{\xi}$ we would expect the potential directly due to the charge e to dominate effects of surface charges, and therefore we anticipate that

$$G(\mathbf{x}, \boldsymbol{\xi}) \sim r^{-1} \qquad \text{as} \quad r \equiv |\mathbf{x} - \boldsymbol{\xi}| \to 0. \tag{1}$$

Moreover, the potential (up to a factor e)

$$G_1(\mathbf{x}, \boldsymbol{\xi}) \equiv G(\mathbf{x}, \boldsymbol{\xi}) - r^{-1} \tag{2}$$

must be due to the induced surface charges only. Thus $G_1(\mathbf{x}, \boldsymbol{\xi})$ can have no contributions from interior singularities; this function must be harmonic everywhere inside the cavity.

The Green's function $G(\mathbf{x}, \boldsymbol{\xi})$ *for a domain D is thus mathematically defined as a solution of the Laplace equation for* $\mathbf{x} \neq \boldsymbol{\xi}$ *which* (i) *vanishes on the boundary of D and* (ii) *satisfies the condition* (1). This function is also known as the

fundamental solution (of the Laplace equation) for the domain D and represents a generalization of the function $1/r$, which is the fundamental solution for infinite space.

The existence of the Green's function is not easy to prove. Clearly, its existence proof is equivalent to that for the function $G_1(\mathbf{x}, \boldsymbol{\xi})$, which is harmonic in D and satisfies the boundary condition $G_1(\mathbf{x}, \boldsymbol{\xi}) = -1/r$. This special boundary condition does not make matters any better, so the existence proof for G is in general no easier than the existence proof for the Dirichlet problem itself.

REPRESENTATION OF A HARMONIC FUNCTION USING GREEN'S FUNCTION

We now show that the Green's function provides a representation for a harmonic function which, in contrast to (1.18), does not use values of both the function and its normal derivative on the boundary. We assert that the value of a harmonic function at any interior point of a region R may be expressed by

$$\phi(\mathbf{x}) = -\frac{1}{4\pi} \oiint_{\partial R} \phi \frac{\partial G}{\partial n} \, d\sigma. \tag{3}$$

To prove this relationship, we simply follow the same steps as in the proof of Theorem 1.3, with $1/r$ replaced by G. Since G vanishes on the boundary, the second term on the right-hand side of (1.18) is now replaced by zero.

Equation (3) thus defines the harmonic function in terms of its values on the boundary, in agreement with the uniqueness theorem.

SYMMETRY OF THE GREEN'S FUNCTION

In free space the potential at a point $\mathbf{x}$ due to a unit charge placed at the point $\boldsymbol{\xi}$ is the same as the potential at $\boldsymbol{\xi}$ due to a unit charge at $\mathbf{x}$, for the function $1/r$ is symmetric in $(\mathbf{x}, \boldsymbol{\xi})$. This is true also for potentials inside a cavity in a conductor, so we expect that

$$G(\mathbf{x}_1, \mathbf{x}_2) = G(\mathbf{x}_2, \mathbf{x}_1). \tag{4}$$

To prove this **reciprocity property**, we again follow the same steps as in the proof of Theorem 1.3. In Green's identity (1.14) we use the functions

$$u = G(\mathbf{x}_1, \boldsymbol{\xi}), \qquad v = G(\mathbf{x}_2, \boldsymbol{\xi}),$$

where $\boldsymbol{\xi}$ is the variable of integration. The detailed calculations are left to the reader (Exercise 1).

EXPLICIT FORMULAS FOR SIMPLE REGIONS

Green's function provides a key to the solution of the boundary value problem for the harmonic function. Unfortunately, it is not possible to write down the Green's function except for the simplest kind of regions. We shall now consider a pair of related cases in which such a function is known: (i) the semiinfinite space bounded by an infinite plane, and (ii) the region inside or outside a sphere. The geometry in these two cases is related by an inversion process.

Infinite plane. If there are two electrical charges of opposite sign situated in the position of mirror images relative to a plane $x_1 = 0$, the potential on the plane is zero. Thus the Green's function for $x_1 \geq 0$ is

$$G(\mathbf{x}, \boldsymbol{\xi}) = \frac{1}{|\mathbf{x} - \boldsymbol{\xi}|} - \frac{1}{|\mathbf{x} - \boldsymbol{\xi}'|}, \tag{5}$$

where $\boldsymbol{\xi}'$ is related to $\boldsymbol{\xi}$ by

$$(\xi_1', \xi_2', \xi_3') = (-\xi_1, \xi_2, \xi_3). \tag{6}$$

(The points $\boldsymbol{\xi}$ and $\boldsymbol{\xi}'$ are mirror images with respect to the plane $x_1 = 0$.) The solution of the Dirichlet problem for the right half-space, according to (3), is then found to be (Exercise 2)

$$\phi(x_1, x_2, x_3) = \frac{x_1}{2\pi} \iint \phi(0, \xi_2, \xi_3) |\mathbf{x} - \boldsymbol{\xi}|^{-3} \, d\sigma, \tag{7}$$

where the integration (with respect to $\boldsymbol{\xi}$) is taken over the infinite plane $\xi_1 = 0$.

Sphere. Consider a sphere with center O and radius a. In this case the "image" $\boldsymbol{\xi}'$ of a source point $\boldsymbol{\xi}$ turns out to be defined by the relationship

$$|\boldsymbol{\xi}| \cdot |\boldsymbol{\xi}'| = a^2, \tag{8}$$

plus the requirement that point and image lie on the same radial line. If the field point $\mathbf{x}$ is on the sphere, then the potential at $\mathbf{x}$ must be zero, by the definition of the Green's function. In this case, the geometric situation is depicted in Figure 16.2. The triangles $OQ'P$ and OPQ are similar so that

$$\frac{|\mathbf{x} - \boldsymbol{\xi}'|}{|\mathbf{x} - \boldsymbol{\xi}|} = \frac{a}{|\boldsymbol{\xi}|}. \tag{9}$$

To obtain the zero boundary condition, then, the charge placed at Q' must be opposite in sign, but smaller by a factor $a/|\boldsymbol{\xi}|$, compared with the charge placed at Q. Thus, for a point $\mathbf{x}$ outside the sphere, the Green's function is

$$G(\mathbf{x}, \boldsymbol{\xi}) = \frac{1}{|\mathbf{x} - \boldsymbol{\xi}|} - \frac{a/|\boldsymbol{\xi}|}{|\mathbf{x} - \boldsymbol{\xi}'|}, \tag{10}$$

while for a point $\mathbf{x}$ inside the sphere it is

$$G(\mathbf{x}, \boldsymbol{\xi}) = \frac{1}{|\mathbf{x} - \boldsymbol{\xi}'|} - \frac{|\boldsymbol{\xi}|/a}{|\mathbf{x} - \boldsymbol{\xi}|}. \tag{11}$$

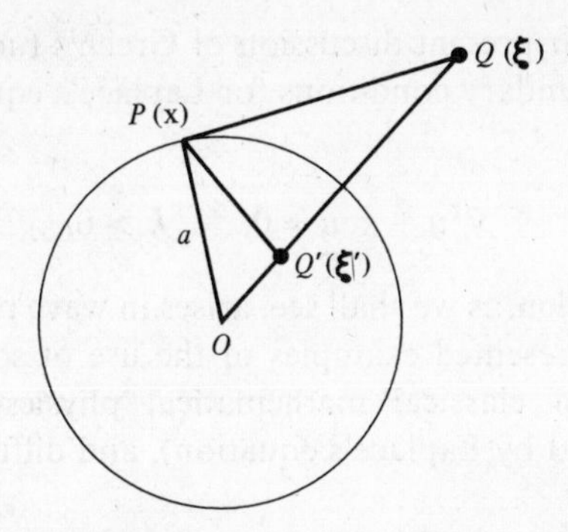

FIGURE 16.2. *Source point Q and image point Q′ for a conducting sphere of radius a. The field point* **x** *is on the sphere, so the potential there must be zero.*

From these functions it can be shown after some calculation (Exercise 4) that the solution for a point **x** outside the sphere is

$$\phi(\mathbf{x}) = \frac{|\mathbf{x}|^2 - a^2}{4\pi a} \oiint_{|\boldsymbol{\xi}|=a} \frac{\phi(\boldsymbol{\xi})\, d\sigma}{|\mathbf{x} - \boldsymbol{\xi}|^3}, \tag{12}$$

and (Exercise 5) for a point **x** inside the sphere is

$$\phi(\mathbf{x}) = \frac{(a^2 - |\mathbf{x}|^2)}{4\pi a} \oiint_{|\boldsymbol{\xi}'|=a} \frac{\phi(\boldsymbol{\xi}')\, d\sigma}{|\mathbf{x} - \boldsymbol{\xi}'|^3}. \tag{13}$$

Equations (12) and (13) are referred to as **Poisson integral formulas.**

WIDESPREAD UTILITY OF SOURCE, IMAGE, AND RECIPROCITY CONCEPTS

The general concepts that we have been discussing for the Dirichlet problem are widely applicable. We already have used them, in Chapter 3, for the diffusion equation. There, as here, we solved general problems with integrals involving source solutions. We also used the method of images. In particular, we mention that the reciprocity property (4) also holds for the fundamental solution $u_0(x - \xi, t)$ of the heat equation. [Recall from Section 3.4 that this solution is the temperature at point x, time t, which is present due to a unit amount of heat that is released at time zero from point ξ. As can be seen from the explicit formula (3.4.4), the value of $u_0(x - \xi, t)$ remains the same if x and ξ are interchanged.]

In Chapter 3 and the present chapter, then, we have seen reciprocity relationships that are inherent in heat flow, gravitational attraction, and electrostatics. But this is far from a complete picture. To mention one other example, the Nobel Prize in chemistry was awarded to L. Onsager in 1968 for his contributions to irreversible thermodynamics, especially for his discovery of certain reciprocity relationships in the transfer coefficients.

We now extend our present discussion of Green's functions by taking into account different boundary conditions for Laplace's equation and by dealing with a new equation,

$$\nabla^2 u + k^2 u = 0, \qquad k > 0. \tag{14}$$

This **Helmholtz equation**, as we shall see, arises in wave propagation problems. Thus we will have presented examples of the use of source solutions in the three great areas of classical mathematical physics—wave propagation, equilibrium (governed by Laplace's equation), and diffusion.

GREEN'S FUNCTION FOR THE NEUMANN PROBLEM

The boundary condition on the Green's function for the Dirichlet problem enabled us to obtain the solution (3), which depends only on the boundary values of ϕ, not on its normal derivative. There is an alternative to the choice of the boundary condition. We may choose to define a **Green's function of the second kind** $H(\mathbf{x}, \boldsymbol{\xi})$ satisfying the boundary condition

$$\frac{\partial H}{\partial n} = 0 \qquad \text{on the boundary.} \tag{15}$$

To be more precise, besides the condition (15), the function $H(\mathbf{x}, \boldsymbol{\xi})$ *with* $\boldsymbol{\xi}$ *fixed and* $\mathbf{x}$ *the running variable is a solution of the Laplace equation in an infinite domain D, except at the point* $\boldsymbol{\xi}$, *in whose neighborhood we have*

$$H(\mathbf{x}, \boldsymbol{\xi}) \sim r^{-1}, \qquad r = |\mathbf{x} - \boldsymbol{\xi}|. \tag{16}$$

If we follow through the reasoning of Section 16.1 (Exercise 6), we shall find, instead of (3), that

$$\phi(\mathbf{x}) = \frac{1}{4\pi} \iint_{\partial R} H \frac{\partial \phi}{\partial n} \, d\sigma. \tag{17}$$

A minor observation should be made here. If H is a Green's function of the second kind, then $H +$ constant is another. However, formula (17) is not influenced by this change, since the extra term is zero (as in the proof of Theorem 1.4). We thus have a formal solution of the boundary value problem of the second kind. We see that the harmonic function ϕ is completely specified by its normal derivative on the boundary. Bear in mind, however, that two functions which differ by a constant have the same normal derivative.

The physical interpretation of the Green's function of the second kind best emerges in the context of fluid dynamical examples. In Section 16.3 we shall discuss such an example in some detail.

The explicit construction of the Green's function of the second kind is also possible in the presence of simple boundaries like a plane or a sphere. In the case of a plane $x_1 = 0$, the method of images may again be applied (Exercise 7) to obtain the solution

$$H(\mathbf{x}, \xi) = \frac{1}{|\mathbf{x} - \xi|} + \frac{1}{|\mathbf{x} - \xi'|}, \tag{18}$$

where ξ and ξ' are image points related by (6). Here, unlike in the Dirichlet case (5), one has two terms of the *same* sign. The reader should work out the Green's function for the case of the sphere.

GREEN'S FUNCTION FOR THE HELMHOLTZ EQUATION

The concept of the Green's function can be extended to other partial differential equations. Consider the Helmholtz equation (14). In spherical coordinates (ρ, ϕ, θ) it is

$$\frac{1}{\rho^2}\left[\frac{\partial}{\partial \rho}\left(\rho^2 \frac{\partial}{\partial \rho}\right) + \mathscr{D}\right]u + k^2 u = 0, \tag{19}$$

where

$$\mathscr{D} = \frac{1}{\sin\phi}\frac{\partial}{\partial\phi}\left(\sin\phi\frac{\partial}{\partial\phi}\right) + \frac{1}{\sin^2\phi}\frac{\partial^2}{\partial\theta^2}. \tag{20}$$

If the solution has spherical symmetry, then $\mathscr{D} \equiv 0$ and we may rewrite (19) as

$$\frac{d^2}{d\rho^2}(\rho u) + k^2(\rho u) = 0. \tag{21}$$

This constant coefficient equation for ρu yields the solutions

$$u = \rho^{-1} e^{\pm ik\rho}. \tag{22}$$

Either of the solutions (22) corresponds to the fundamental solution r^{-1} in the case of potential theory $(k \to 0)$. We may therefore construct *two* Green's functions,

$$G^{(1)}(\mathbf{x}, \xi, k) \sim \frac{e^{ik|\mathbf{x} - \xi|}}{|\mathbf{x} - \xi|} \tag{23a}$$

and

$$G^{(2)}(\mathbf{x}, \xi, k) \sim \frac{e^{-ik|\mathbf{x} - \xi|}}{|\mathbf{x} - \xi|}. \tag{23b}$$

These are spherically symmetric solutions of the Helmholtz equation (14), except at the point ξ, and they satisfy the appropriate boundary condition $G \to 0$ as $r \to \infty$.

EXERCISES

1. (a) Prove (4).
2. (a) Verify that (5) has the required properties of a Green's function.
(b) Derive (7), by showing that an integral over an infinite hemisphere is negligible.
3. (a) Verify (9).
(b) Show that (10) and (11) are the appropriate Green's functions.
4. Derive (12).
5. Derive (13).
6. Derive (17).
7. (a) Verify that (18) gives the Green's function for the Neumann problem in a half-space.
(b) Generalize to the spherical case.

16.3 Diffraction of Acoustic Waves by a Hole

Light waves can propagate through a hole in a wall only as a narrow beam, whereas sound waves can propagate rather freely. The latter case is said to involve the *diffraction of waves* with wavelength much longer than the size of the hole. We shall now consider this phenomenon with the aid of a Green's function for a wave of arbitrary length. We shall see that the phenomenon of diffraction is most interesting when we are dealing with wavelengths comparable to the size of the hole. For waves of such lengths we derive the characteristic diffraction pattern, alternating rings of sound and silence.

FORMULATION

Consider acoustic waves governed by the following linearized equations of inviscid hydrodynamics, (15.3.9) and (15.3.10):

$$\frac{\partial \sigma}{\partial t} + \sum_{i=1}^{3} \frac{\partial v_i}{\partial x_i} = 0, \tag{1a}$$

$$\frac{\partial v_i}{\partial t} = -c_0^2 \frac{\partial \sigma}{\partial x_i}. \tag{1b}$$

Here c_0 is the speed of sound for the medium at rest, and $\mathbf{v}(\mathbf{x}, t)$ is the velocity of the medium at point $\mathbf{x}$, time t. The condensation $s(\mathbf{x}, t)$ (dimensionless density perturbation) is related to the density* $\sigma(\mathbf{x}, t)$ by the formula

$$\sigma = \sigma_0(1 + s), \tag{2}$$

* We are reserving the letter ρ, usually used to denote density, for a spherical coordinate.

where σ_0 is the density of the unperturbed medium. If we eliminate v_i from (1a) and (1b), we obtain the wave equation for the condensation s:

$$\frac{\partial^2 s}{\partial t^2} = c_0^2 \sum_{i=1}^{3} \frac{\partial^2 s}{\partial x_i^2}. \tag{3}$$

At a solid boundary the normal component of velocity vanishes. Hence, using (1b), we have

$$0 = \frac{\partial \mathbf{v}}{\partial t} \cdot \mathbf{n} = -c_0^2 \mathbf{n} \cdot \nabla \sigma = -c_0^2 \frac{\partial \sigma}{\partial n}.$$

That is,

$$\frac{\partial s}{\partial n} = 0. \tag{4}$$

Equations (3) and (4) may be used as the basis for the study of acoustic waves.

Alternatively, if we introduce the velocity potential ϕ such that

$$v_i = \frac{\partial \phi}{\partial x_i}, \tag{5}$$

we may derive, from (1b), the relation

$$\sigma = -c_0^{-2} \frac{\partial^2 \phi}{\partial t^2}. \tag{6}$$

Hence (1a) leads to the wave equation for ϕ:

$$\frac{\partial^2 \phi}{\partial t^2} = c_0^2 \sum_{i=1}^{3} \frac{\partial^2 \phi}{\partial x_i^2}. \tag{7}$$

The condition

$$\frac{\partial \phi}{\partial n} = 0 \qquad \text{at a solid boundary} \tag{8}$$

then follows from (4) and the definition of ϕ. We shall use (7) and (8) as the basic equations instead of (3) and (4).

To consider waves at a particular frequency ω, we study solutions of (7) that have the form

$$\phi = e^{-i\omega t}\psi(\mathbf{x}). \tag{9}$$

The spatial function ψ satisfies the Helmholtz equation

$$\nabla^2 \psi + k^2 \psi = 0, \qquad k^2 = \frac{\omega^2}{c_0^2}. \tag{10}$$

This is to be solved subject to the boundary condition

$$\frac{\partial \psi}{\partial n} = 0 \tag{11}$$

at a solid wall.

NOTE. From (10), k is the wave number and $2\pi/k$ is the spatial period of the waves under consideration. [Compare the discussion of (12.2.9.)]

Let us consider plane waves approaching a screen $x_3 = 0$ with a hole (Figure 16.3). We assume that the waves below the screen will be modified by a negligible amount. This idealization is equivalent to simulating the incoming plane wave at the hole by a vibrating piston. That is, we assume that the motion at the hole is given by

$$V_3 = ve^{-i\omega t}, \tag{12}$$

where v is a given function of x_1 and x_2. We are interested in obtaining a solution for the half-space above the screen. At the bounding screen, then,

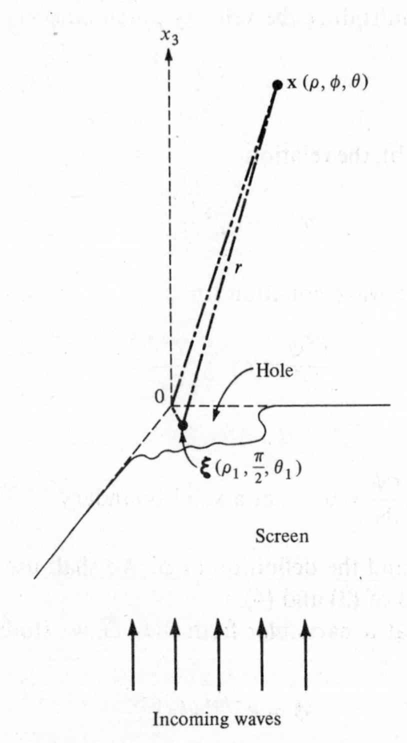

FIGURE 16.3. *Diffraction of waves by a hole in a screen. (A quarter of the screen is shown.) The effect at the field point* $\mathbf{x}$ *is caused by the superposition of influences from source points* $\boldsymbol{\xi}$ *that are distributed over the hole.*

the exterior normal points in the direction of the negative x_3 axis. Thus, over the hole we have the boundary condition

$$\frac{\partial \phi}{\partial n} = -\frac{\partial \phi}{\partial x_3} = -v_3 = -ve^{-i\omega t}, \tag{13}$$

and hence, by (9),

$$\frac{\partial \psi}{\partial n} = -v. \tag{14}$$

One other physical consideration must be introduced. We are interested in effects caused by the waves arriving at the screen from below and from no other cause. Such effects must be a superposition of *outward traveling waves only*. We thus impose the condition that the above italicized phrase holds. (Such a **radiation condition** has already been encountered in the discussion of longitudinal elastic waves in Section 12.2.)

Recapitulating, we present the final formulation of a mathematical problem that represents diffraction of plane acoustic waves by a hole in a screen.* We shall designate by Ω the set of (x_1, x_2) coordinates that correspond to the hole.

$$\nabla^2\psi + k^2\psi = 0, \quad x_3 > 0. \tag{10}$$

$$\frac{\partial \psi}{\partial n} = -v; \quad x_3 = 0, \quad (x_1, x_2) \in \Omega. \tag{14}$$

$$\frac{\partial \psi}{\partial n} = 0; \quad x_3 = 0, \quad (x_1, x_2) \notin \Omega. \tag{11}$$

Solution must be a synthesis of outward-moving waves.
k^2 and $v(x_1, x_2)$ are given.

SELECTION OF THE APPROPRIATE GREEN'S FUNCTION

In (2.22) we have recorded solutions to Helmholtz's equation (10) that possess spherical symmetry. Using (9), we are led to the following basic source solutions for the velocity potential ϕ:

$$\rho^{-1} \exp(-i\omega t + ik\rho) = \rho^{-1} \exp[ik(\rho - ct)], \tag{15a}$$

$$\rho^{-1} \exp(-i\omega t - ik\rho) = \rho^{-1} \exp[-ik(\rho + ct)]. \tag{15b}$$

Here ρ is the spherical coordinate, which measures distance from some origin, and $c = \omega/k$.

* Incorporation of the various subtleties connected with this problem remains a matter for continuing study. A recent paper, with references to earlier work, is A. V. Chinnaswamy and R. P. Kanwal's "Uniform Asymptotic Theory of Diffraction by a Plane Screen by the Method of Boundary Layers," *SIAM J. Appl. Math. 23*, 339–55 (1972).

The first of the above solutions corresponds to outgoing waves, the second to incoming. Both sets of waves move with speed c, and in both cases the effect can be regarded as due to a "cause" concentrated at the origin. We wish to consider "causes" that are located at any point $\boldsymbol{\xi}$ in the hole, and we have stipulated that only outgoing waves should be considered. Thus we make use of the Green's function having the singularity

$$\frac{e^{ik|\mathbf{x}-\boldsymbol{\xi}|}}{|\mathbf{x}-\boldsymbol{\xi}|}.$$

Since the problem involves given values of the normal derivatives, we choose the Green's function of the second kind,

$$H^{(1)}(\mathbf{x}, \boldsymbol{\xi}, k) = \frac{e^{ik|\mathbf{x}-\boldsymbol{\xi}|}}{|\mathbf{x}-\boldsymbol{\xi}|} + \frac{e^{ik|\mathbf{x}-\boldsymbol{\xi}'|}}{|\mathbf{x}-\boldsymbol{\xi}'|}. \tag{16}$$

Here $\boldsymbol{\xi}'$ is the reflection of $\boldsymbol{\xi}$ in the wall, so if $\mathbf{x}_w$ is a point on the wall, $\mathbf{x}_w - \boldsymbol{\xi} = -(\mathbf{x}_w - \boldsymbol{\xi}')$. By choosing the positive sign in (16), then, we have used the method of images to construct a Green's function whose normal derivative vanishes on the wall.

DERIVATION OF THE DIFFRACTION INTEGRAL

We shall now show that the solution is given by the "diffraction integral"

$$\psi(\mathbf{x}) = \frac{1}{4\pi} \iint_{\Omega} H^{(1)}(\mathbf{x}, \boldsymbol{\xi}, k) \frac{\partial \psi}{\partial n}\, d\sigma, \tag{17a}$$

where the variable of integration is $\boldsymbol{\xi}$ and integration is carried out over the hole in the screen. Here $\partial\psi/\partial n$ is assumed known. On the wall, $H^{(1)} = 2e^{ikr}/r$, where $r = |\mathbf{x}-\boldsymbol{\xi}|$. Thus the integral (17a) can also be written in the form

$$\psi(\mathbf{x}) = \frac{1}{2\pi} \iint_{\Omega} \frac{e^{ikr}}{r} \frac{\partial \psi}{\partial n}\, d\sigma. \tag{17b}$$

The physical interpretation of this formula is not difficult to discern. The amplitude of the wave at any point $\mathbf{x}$ is obtained by a superposition of waves generated from a "smear" of sources at the hole, of strength per unit area $(2\pi)^{-1}\, \partial\psi/\partial n$. Each source point $\boldsymbol{\xi}$ is associated with its own $r \equiv |\mathbf{x}-\boldsymbol{\xi}|$, so each effect acquires an appropriate phase delay during its propagation to the field point $\mathbf{x}$. The fact that complicated wave phenomena can be regarded as resulting from the superposition of waves from point sources, with appropriate phase differences and concomitant cancellation and reinforcement, is known as **Huygen's principle**.

The form of the solution (17) may be anticipated from (2.17), the solution of the boundary value problem of the second kind for the harmonic function. Derivation of (17) requires a more subtle discussion, however, because of conditions at infinity.

If we consider a domain bounded by the wall and a hemisphere S centered at some point on the hole in the screen [Exercise 3(b)], we may readily derive the following representation of the solution:

$$\psi(\mathbf{x}) = \frac{1}{4\pi} \iint_{\Omega} H^{(1)} \frac{\partial \psi}{\partial n}\, d\sigma + \frac{1}{4\pi} \iint_{S} \left(H^{(1)} \frac{\partial \psi}{\partial n} - \psi \frac{\partial H^{(1)}}{\partial n} \right) d\sigma. \tag{18}$$

When the radius of the hemisphere S is made infinitely large, it is expected—and it must be verified—that the integral over the hemisphere approaches zero. Equation (17a) then follows.

To verify that the contribution from S vanishes in the limit, let us note that the largest terms in $H^{(1)}$ and ψ have the following behavior:

$$H^{(1)} \sim \frac{e^{ikr}}{r}, \qquad \psi \sim \frac{e^{ikr'}}{r'}. \tag{19a, b}$$

Here r is the distance from a point ξ on the hemisphere S to a point $\mathbf{x}$ at a finite distance from the screen, and r' is the distance from $\mathbf{x}$ to the image point ξ' (see Figure 16.4). Thus, in the integrand $H^{(1)}(\partial\psi/\partial n) - \psi(\partial H^{(1)}/\partial n)$ that occurs in (18), the largest term arises from the differentiation of the exponential factor and is of the form

$$\frac{e^{ikr}}{r} \frac{e^{ikr'}}{r'} ik \left(\frac{\partial r}{\partial n} - \frac{\partial r'}{\partial n} \right), \tag{20}$$

other terms being $O(r^{-3})$ as $r \to \infty$.* Some calculation (Exercise 4) shows that

$$\frac{\partial r}{\partial n} - \frac{\partial r'}{\partial n} = O\left(\frac{1}{r}\right), \tag{21}$$

and hence the integrand over the spherical surface is $O(r^{-3})$. Since the area of the hemisphere is $O(r^2)$, the desired result holds.

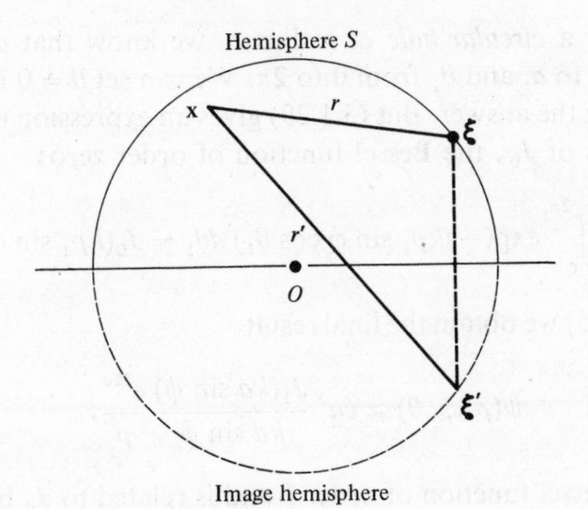

FIGURE 16.4. *Source point* ξ *and its image* ξ'.

* See Appendix 3.1 for the definition of the O symbol.

APPROXIMATE EVALUATION OF THE DIFFRACTION INTEGRAL

The diffraction integral (17b) can be simplified for easier evaluation. We shall employ spherical coordinates (ρ, ϕ, θ) for the field point $\mathbf{x}$, with the origin at a convenient point in the hole (Figure 16.3). In (17b) we are only concerned with source points $\boldsymbol{\xi}$ that lie in the plane of the screen; we take the spherical coordinates of such points to be $(\rho_1, \pi/2, \theta_1)$. We shall approximate the integral under the assumption that $\rho \gg \rho_1$; i.e., we consider only the field points which are many hole "diameters" away from the hole.

Using the law of cosines, we have

$$r = (\rho^2 + \rho_1^2 - 2\rho\rho_1 \cos\psi)^{1/2} = \rho\left[1 - \frac{\rho_1}{\rho}\cos\psi + O\left(\frac{\rho_1}{\rho}\right)^2\right], \tag{22}$$

where ψ is the angle between $\mathbf{x}$ and $\boldsymbol{\xi}$. Unit vectors in the direction of $\mathbf{x}$ and $\boldsymbol{\xi}$ are

$$\mathbf{i}\cos\theta\sin\phi + \mathbf{j}\sin\theta\sin\phi + \mathbf{k}\cos\phi \qquad \text{and} \qquad \mathbf{i}\cos\theta_1 + \mathbf{j}\sin\theta_1,$$

respectively. We obtain $\cos\psi$ by taking the scalar product of these vectors:

$$\cos\psi = \sin\phi(\cos\theta\cos\theta_1 + \sin\theta\sin\theta_1) = \sin\phi\cos(\theta - \theta_1). \tag{23}$$

Furthermore, in most cases, the hole is sufficiently small so that it is justifiable to regard $\partial\psi/\partial n$ as a constant v in the integrand. With this assumption we obtain, using (22) and (23),

$$\psi(\rho, \phi, \theta) \approx \frac{\exp(ik\rho)}{\rho}\frac{v}{2\pi}\iint_{\Omega} \exp[-ik\rho_1 \sin\phi\cos(\theta_1 - \theta)]\, d\sigma. \tag{24}$$

In the case of a *circular hole* of radius a, we know that $d\sigma = \rho_1\, d\rho_1\, d\theta_1$, ρ_1 varies from 0 to a, and θ_1 from 0 to 2π. We can set $\theta = 0$ in the integrand without affecting the answer. But (3.1.29) gives an expression for the resulting integral in terms of J_0, the Bessel function of order zero:

$$\frac{1}{2\pi}\int_0^{2\pi} \exp(-ik\rho_1 \sin\phi\cos\theta_1)\, d\theta_1 = J_0(k\rho_1\sin\phi). \tag{25}$$

Hence (Exercise 5) we obtain the final result

$$\psi(\rho, \phi, \theta) \approx va^2\, \frac{J_1(ka\sin\phi)}{ka\sin\phi}\frac{e^{ik\rho}}{\rho}. \tag{26}$$

Here J_1 is the Bessel function of order 1 and is related to J_0 by

$$\frac{d}{dx}[xJ_1(x)] = xJ_0(x). \tag{27}$$

We see from (26) that the wave amplitude vanishes on those directions ϕ for which $J_1(ka \sin \phi) = 0$. The direction $\phi = 0$ is an exception, since

$$\lim_{\phi \to 0} \frac{J_1(ka \sin \phi)}{ka \sin \phi} = \frac{1}{2}.$$

Consequently, "silent rings" are defined by the positive zeros of J_1:

$$ka \sin \phi = 3.832, 7.016, 10.173, 13.323, \text{etc.} \tag{28}$$

It is readily shown that the intensity is a maximum at $\phi = 0$, as one would expect from physical considerations.

Note that for there to be a real angle ϕ satisfying any of the equations (28), $ka \geq 3.832$. That is, the wavelength $2\pi/k$ must be smaller than $2\pi a/3.832 \approx 0.8(2a)$. *There are no "silent rings" unless the wavelength is somewhat less than the diameter of the hole.* On the other hand, if the wavelength is too small, the rings will be too close together for satisfactory resolution.

Many kinds of waves are governed by the wave equation (7) with the condition of vanishing normal derivative (8), and the above calculations apply with little or no modification. To demonstrate the wave nature of matter, G. P. Thompson sent electron beams through a small hole and obtained diffraction patterns of the kind just calculated.

EXERCISES

1. Show that introduction of a velocity potential [as in (5)] is justified in a problem governed by (1).

2. Write a paragraph or two on the advantages and limitations of the assumption (12). Can less restrictive assumptions be made without entailing too much additional effort?

3. (a) Verify that the normal derivative of $H^{(1)}$ vanishes on the wall.
(b) Derive (18).

4. Verify (19), (20), and (21).

5. (a) From the series definition of the Bessel functions in Exercise 3.1.21, prove that (27) holds.
(b) Obtain (26).

6. Discuss the location and magnitudes of the various maximum intensities given by (26).

7. Show that the image point ξ' is just one of the possible choices in (19b); any point behind the screen may be used to accomplish our purpose.

Bibliography

General Applied Mathematics

COURANT, R., and D. HILBERT. (1953) *Methods of Mathematical Physics*. Vol. 1. New York: Interscience Publishers, a division of John Wiley & Sons, Inc. 561 pp.

"Mathematical methods originating in problems of physics are developed and the attempt is made to shape results into unified mathematical theories." (Volume 2 does not have the classical style of Vol. 1 but is a compendium of important results in the theory of partial differential equations.)

FRIEDMAN, B. (1956) *Principles and Techniques of Applied Mathematics*. New York: John Wiley & Sons, Inc. 315 pp.

"This book was written in an attempt to show how the powerful methods developed by . . . abstract studies can be used to systematize the methods and techniques for solving problems in applied mathematics."

GENIN, J., and J. S. MAYBEE. (1970) *Introduction to Applied Mathematics*, Vol. 1. New York: Holt, Rinehart and Winston, Inc. 289 pp.

"We have tried to present the material in a unified way, motivated by more or less realistic problems drawn from a variety of fields. Moreover, we have tried to give a clear description of the nature and extent of the practical information that can be obtained with the techniques presented." Junior–senior level.

JEFFREYS, H., and B. JEFFREYS. (1962) *Methods of Mathematical Physics*. New York: Cambridge University Press. 716 pp.

"This book is intended to provide an account of those parts of pure mathematics that are most frequently needed in physics . . . . Abundant applications to special problems are given as illustrations."

KÁRMÁN, T. von, and M. A. BIOT. (1940) *Mathematical Methods in Engineering*. New York: McGraw-Hill Book Company. 505 pp.

"There are two ways of teaching the art of applying mathematics to engineering problems. One consists of a systematic course comprising selected branches of mathematics including a choice of appropriate examples for applications. The other chooses certain representative groups of engineering problems and demonstrates the mathematical approach to their solution . . . . This book might be considered an experiment in the second method."

POLLARD, H. (1972) *Applied Mathematics: An Introduction*. Reading, Mass.: Addison-Wesley Publishing Co., Inc. 99 pp.

"I wish to convince the advanced undergraduate, or beginning graduate student . . . that applied mathematics is interesting . . . by presenting a small selection of problems of major importance . . . and calling the attention of the student to failure as well as success."

SAGAN, H. (1961) *Boundary and Eigenvalue Problems in Mathematical Physics*. New York: John Wiley & Sons, Inc. 381 pp.

A senior–graduate text that provides a good alternative reference for much of the material in Part A.

Advanced Calculus

FRANKLIN, P. (1940) *A Treatise on Advanced Calculus.* New York: John Wiley & Sons, Inc., 1940. Reprinted by Dover Publications, Inc., New York, 1964. 595 pp.

HILDEBRAND, F. B. (1962) *Advanced Calculus for Applications.* Englewood Cliffs, N.J.: Prentice-Hall, Inc. 646 pp.

KAPLAN, W. (1952.) *Advanced Calculus.* Reading, Mass.: Addison-Wesley Publishing Co., Inc. 678 pp.

Differential Equations

BOYCE, W., and R. DIPRIMA. (1969) *Elementary Differential Equations and Boundary Value Problems,* 2nd ed. New York: John Wiley & Sons, Inc. 583 pp.

"This book is written from the ... viewpoint of the applied mathematician, whose interests in differential equations may at the same time be quite theoretical as well as intensely practical."

CODDINGTON, E., and N. LEVINSON. (1955) *Theory of Ordinary Differential Equations.* New York: McGraw-Hill Book Company. 429 pp.

Authoritative advanced work.

GARABEDIAN, P. (1964) *Partial Differential Equations.* New York: John Wiley & Sons, Inc. 672 pp.

A graduate text "written for engineers and physicists as well as mathematicians."

INCE, E. (1927) *Ordinary Differential Equations.* Essex, England: Longman Group Ltd. Reprinted by Dover Publications, Inc., New York, in 1956. 558 pp.

An excellent source for the classical theory.

KELLOGG, O. (1929) *Potential Theory.* Reprinted by Dover Publications, Inc., New York, 1953. 384 pp.

"It is inherent in the nature of the subject that physical intuition and illustration be appealed to freely, and this has been done. However, in order that the book may present sound ideals to the student, and also serve the mathematician, both for purposes of reference and as a basis for further developments, the proofs have been given by rigorous methods."

MOSER, J. (1973) *Stable and Random Motions in Dynamical Systems, with Special Emphasis on Celestial Mechanics.* Princeton, N.J.: Princeton University Press. 198 pp.

A modern and deep study.

TYCHONOV, A., and A. SAMARSKI. (1964) *Partial Differential Equations of Mathematical Physics,* Vol. I. San Francisco: Holden-Day, Inc. 390 pp. Vol. II, 1967, 250 pp.

Representative of a large class of books, this is relatively brief and has good illustrations of how the mathematics is used.

Dimensional Analysis

BIRKHOFF, G. (1960) *Hydrodynamics—A Study in Logic, Fact, and Similitude,* 2nd ed. Princeton, N.J.: Princeton University Press. Also reprinted by Dover Publications, Inc., New York, 1955. 186 pp.

Noteworthy for the use of group theory in unifying ideas of dimensional analysis and similarity theory.

BRIDGMAN, P. W. (1931) *Dimensional Analysis*, rev. ed. New Haven, Conn.: Yale University Press. (Paperback edition, 1963.) 113 pp.

Discusses fundamental aspects of the subject with clarity and profundity.

HUNTLEY, H. E. (1967) *Dimensional Analysis*. New York: Dover Publications, Inc. 158 pp.

A good, inexpensive general reference.

KLINE, S. J. (1965) *Similitude and Approximation Theory*. New York: McGraw-Hill Book Company. 229 pp.

A useful discussion of dimensional analysis and scaling.

LIGHTHILL, M. J. (1963) *Laminar Boundary Layers*, Chap. 1, L. Rosenhead (ed.). New York: Oxford University Press.

Part 3 of this chapter, in just 11 pages, contains an excellent discussion of dimensional reasoning and its application to the simplification of problems in fluid mechanics.

Perturbation Methods

COLE, J. (1968) *Perturbation Methods in Applied Mathematics*. Waltham, Mass.: Ginn/Blaisdell. 260 pp.

"This text is written very much from the point of view of the applied mathematician; much less attention is paid to mathematical rigor than to rooting out the underlying ideas, using all means at our disposal."

LAGERSTROM, P. A., and R. G. CASTEN. (1972) "Basic Concepts Underlying Singular Perturbation Techniques." *Society for Industrial and Applied Mathematics (SIAM) Review 14*, 63–120.

A readable and up-to-date survey.

NAYFEH, A. H. (1973) *Perturbation Methods*. John Wiley & Sons, Inc. 425 pp.

The latest and most comprehensive book on its subject.

VAN DYKE, M. (1964) *Perturbation Methods in Fluid Mechanics*. New York: Academic Press, Inc. 229 pp.

The first modern book on the subject of perturbation methods, it contains many insights.

Probability and Its Relation to Differential Equations

FELLER, W. (1968) *An Introduction to Probability Theory and Its Applications*, Vol. I. New York: John Wiley & Sons, Inc. 509 pp.

"It is the purpose of this book to treat probability theory as a self-contained mathematical subject rigorously, avoiding nonmathematical concepts. At the same time, the book tries to describe the empirical background and to develop a feeling for the great variety of practical applications.

WAX, N. (ed.) (1954) *Selected Papers on Noise and Stochastic Processes*. New York: Dover Publications, Inc. 337 pp.

A collection of six important papers. The text refers to S. Chandrasekhar's "Stochastic Problems in Physics and Astronomy" and M. Kac's "Random Walk and the Theory of Brownian Motion."

Foundations of Continuum Mechanics

ARIS, R. (1962) *Vectors, Tensors, and the Basic Equations of Fluid Mechanics.* Englewood Cliffs, N.J.: Prentice-Hall, Inc. 286 pp.

"Sets out to show that the calculus of tensors is the language most appropriate to the rational examination of physical field theories." The argument is illustrated by detailed consideration of fluid mechanical equations.

JAUNZEMIS, W. (1967) *Continuum Mechanics.* New York: Macmillan Publishing Co., Inc. 604 pp.

A graduate text that takes a "modern" rigorous approach to the subject. Emphasizes elasticity.

SOMMERFELD, A. (1950) "Mechanics of Deformable Bodies," in *Lectures on Theoretical Physics,* Vol. II. New York: Academic Press, Inc. 396 pp.

"My aim is to give the reader a vivid picture of the vast and varied material that comes within the scope of theory when a reasonably elevated vantage point is chosen." Unlike the other references, this requires no knowledge of tensors.

TRUESDELL, C., and W. NOLL. (1965) "The Non-linear Field Theories of Mechanics," in *Encyclopedia of Physics,* S. Flugge (ed.), Vol. III/3. New York: Springer-Verlag New York, Inc. 602 pp.

"The maximum mathematical generality consistent with concrete, definite physical interpretation is sought."

TRUESDELL, C., and R. TOUPIN. (1960) "The Classical Field Theories," in *Encyclopedia of Physics,* S. Flugge (ed.), Vol. III/1, pp. 226–790. New York: Springer-Verlag New York, Inc.

"We present the common foundation of the field viewpoint. We aim to provide the reader with a full panoply of tools of research, whereby he himself, put into possession not only of the latest discoveries but also of the profound but all too often forgotten achievements of previous generations, may set to work as a theorist. This treatise is intended for the specialist, not the beginner."

Elasticity

GREEN, A. E., and W. ZERNA. (1954) *Theoretical Elasticity.* New York: Oxford University Press, Inc. 442 pp.

"This book is mainly concerned with three aspects of elasticity theory which have attracted attention in recent years . . . finite elastic deformations, complex variable methods for two-dimensional problems . . . and shell theory."

LOVE, A. E. H. (1944) *A Treatise on the Mathematical Theory of Elasticity,* 4th ed. New York: Dover Publications, Inc. 643 pp.

"It is hoped . . . to present a fair picture of the subject in its various aspects, as a mathematical theory, having important relations to general physics, and valuable applications to engineering." Although not easy to read, this is the classical reference work.

PRESCOTT, J. (1924) *Applied Elasticity.* Essex, England: Longman Group Ltd. 666 pp. Reissued by Dover Publications, Inc., New York, 1946.

"In writing this book I have tried to see the subject from the point of view of the engineer rather [than] from that of the mathematician."

SOKOLNIKOFF, I. S. (1956) *Mathematical Theory of Elasticity*. New York: McGraw-Hill Book Company. 476 pp.

Perhaps the best single reference. "This book represents an attempt to present several aspects of the theory of elasticity from a unified point of view and to indicate, along with the familiar methods of solution of the field equations of elasticity, some newer general methods of solution of the two-dimensional problems."

TIMOSHENKO, S. P., and J. M. GOODIER. (1970) *Theory of Elasticity*, 3rd ed. New York: McGraw-Hill Book Company. 567 pp.

"The primary intention . . . [is] to provide for engineers, in as simple a form as the subject allows, the essential fundamental knowledge of the theory of elasticity together with a compilation of solutions of special problems that are important in engineering practice and design."

A leading journal is the *Journal of Applied Mechanics.*

Fluid Mechanics

BATCHELOR, G. K. (1967) *An Introduction to Fluid Dynamics*. New York: Cambridge University Press. 615 pp.

" A textbook which can be used by students of applied mathematics and which incorporates the physical understanding and information provided by past research."

LAMB, H. (1932) *Hydrodynamics*, 6th ed. New York: Cambridge University Press. 738 pp.

Also available from Dover Publications. An encyclopedic collection of classical results.

LANDAU, L. D., and E. H. LIFSCHITZ. (1959) *Fluid Mechanics*. Elmsford, N.Y.: Pergamon Press, Inc. 536 pp.

"The nature of the book is largely determined by the fact that it describes fluid mechanics as a branch of theoretical physics." Translated from the Russian by J. B. Sykes and W. H. Reid.

PRANDTL, L., and O. G. TIETJENS. (1934) *Applied Hydro- and Aeromechanics*. New York: McGraw-Hill Book Company. 270 pp. Also available from Dover Publications, Inc., New York, 1957, as *Fundamentals of Hydro- and Aero Mechanics.*

Brief and inexpensive, with considerable physical insight.

SEEGER, R. J., and G. TEMPLE (eds.). (1965) *Research Frontiers in Fluid Dynamics*. New York: Interscience Publishers, a division of John Wiley & Sons, Inc. 738 pp.

A collection of useful survey articles.

SERRIN, J. B. (1959) " Mathematical Principles of Classical Fluid Mechanics," in *Encyclopedia of Physics*, S. Flügge (ed.), Vol. VIII/1, pp. 125–263. New York: Springer-Verlag New York, Inc.

" Our intent . . . is to present in a mathematically correct way, in concise form, and with more than passing attention to the foundations, the principles of classical fluid mechanics."

The two principal journals are *Journal of Fluid Mechanics* and *Physics of Fluids.* Beginning in 1969, *Annual Reviews of Fluid Mechanics*, a collection of authoritative summaries of research progress, has been published by Annual Reviews, Inc., Palo Alto, Calif.

Numerical Analysis

RALSTON, A. (1965) *A First Course in Numerical Analysis.* New York: McGraw-Hill Book Company. 578 pp.
A standard introductory text.
WILKINSON, J. (1963) *Rounding Errors in Algebraic Processes.* London: Her Majesty's Stationery Office. 161 pp.
A thin book, with many ideas.

Handbooks and Tables

ABRAMOWITZ, M., and I. STEGUN. (1964) "Handbook of Mathematical Functions," *National Bureau of Standards Applied Mathematics Series 55*, 1046 pp. Reprinted by Dover Publications, Inc., New York, 1965.
An invaluable collection of formulas, graphs, and tables.
FREIBERGER, W. (ed.) (1960) *The International Dictionary of Applied Mathematics.* New York: Van Nostrand Reinhold Company. 1173 pp.
Defines the terms and describes the methods in the applications of mathematics to 31 fields of science and engineering. Includes French, German, Russian, and Spanish equivalents of the terms defined in the book.
MURPHY, G. (1960) *Ordinary Differential Equations and Their Solution.* New York: Van Nostrand Reinhold Company. 451 pp.
Contains a tabulation of 2315 equations and their solutions.

Hints and Answers

(for exercises marked with ‡)

SECTION 3.3

7(b). *Hint:* Separate off the terms $q[w(m+1, N-1) - 2w(m, N-1) + w(m-1, N-1)]$.

SECTION 3.4

9(a). $P(r) = (\ln R - \ln r)/(\ln R - \ln \varepsilon)$. **(b)** $P(\rho) = (R^{-1} - \rho^{-1})/(R^{-1} - \varepsilon^{-1})$.

SECTION 4.2

2(a). *Hint:* Reduce the problem to one in alternating series.
5. *Hint:* Use integration by parts.

SECTION 5.1

6. $\kappa = 0.004\ \text{cm}^2/\text{sec}$.
7. According to the given data, the cooling process would take about 3×10^6 years. It is now believed that heat given off by radioactive substances is the major source of the observed temperature gradient.

SECTION 6.2

3(c). $T/(L/g)^{1/2} = f(A/L)$, where A is the initial linear displacement of the bob. When A/L is small, $f(A/L)$ can be replaced by $f(0)$, so the result of this part reduces to the result of part **(b)**.
10(b). *Hint:* You must solve an equation of the form $[y, \partial f(y_1, \ldots, y_n)/\partial y_1]/f(y_1, \ldots, y_n) = \text{constant}$.

SECTION 6.3

2. $U = A \exp(-ab)$, $L = a^{-1}$.
9(b). From the zeroth order solution (7.1.2) with $a = 0$, the time scale should again be ω_0^{-1}; the scale for Θ^* should be $\Omega\omega_0^{-1}$. **(c)** The angle scale should be a (or $\Omega\omega_0^{-1}$) if a is considerably larger (or smaller) than $\Omega\omega_0^{-1}$. If both quantities are about the same size, then either scale or their average or geometric mean can be used.

SECTION 7.1

2(a). *Hint:* Use the definition of derivative.
4. *Hint:* Derive an equation expressing conservation of energy, and consider its consequences.

SECTION 7.2

1. *Hint:* If $\dot{x} = v$, $dv/dt = (dv/dx)(dx/dt) = v(dv/dx)$.
11(a). $\lambda_1^{(n)} = -(2n^2/\pi)\int_0^\pi x \sin^2 nx\, dx = -(2n^2/\pi)(\frac{1}{4}\pi^2 + \frac{1}{8}n^{-2})$.

Section 8.2

6(a). *Partial answer.* The steady state mass conservation requirement is $F_s(x) - F_s(x + \Delta x) = 0$ for arbitrary Δx so that F_s is constant.

Section 9.1

2(b). As the text indicates, begin by introducing $s = \varepsilon m$ into (1), where s can be safely assumed to have a Maclaurin expansion in powers of ε.
3. *Hint:* In expanding the square root, it is helpful to write $(4 - \varepsilon)^{1/2} = 2[1 - (\frac{1}{4})\varepsilon]^{1/2}$ and then to use the binomial theorem.
4(f). *Hint:* Introduce a new variable $y = x\varepsilon^{-1/2}$.
8(a). Divide (13) by εE^2. **(b)** Divide (13) by E and remember that $\bar{m}$ and hence E does not approach a nonzero limit as $\varepsilon \to 0$.

Section 9.2

2(c). $y_0(x) = \exp(-\frac{3}{2}x^{2/3})$, $y_1(\xi) = k\int_0^\xi \exp(-\frac{3}{4}s^{4/3})\, ds$; $k^{-1} = \int_0^\infty \exp(-\frac{3}{4}s^{4/3})\, ds$.
4(c). *Partial answer.* By balancing the first two terms in the differential equation, one sees that the time scale for t small is μk^{-1}. From the second initial condition it then follows that $I\mu^{-1}$ is the correct scale for the displacement.
5. *Hint:* There are boundary layers of $O(\varepsilon^{1/2})$ thickness near $x = 1$ and $x = -1$.
8. The limit is 1 if $\varepsilon x = o(1)$, e^{-k} if $\varepsilon x = k + o(1)$, and 0 if $1/x = o(\varepsilon)$.

Section 10.2

3(c). *Hint:* Use the method of undetermined coefficients.
5(a). *Hint:* Using (2), show that the equation for s_1 has a relatively simple integrating factor and can be written

$$\frac{d}{dt}\left(\frac{\kappa + s_0}{s_0} s_1\right) = \left[\frac{-\kappa}{s_0(s_0 + \kappa)} + \frac{\kappa\lambda}{s_0(s_0 + \kappa)^2}\right]\dot{s}_0 .$$

Completion of the solution thus requires computing

$$\int \left[\frac{-\kappa}{u(u + \kappa)} + \frac{\kappa\lambda}{u(u + \kappa)^2}\right] du,$$

which can be accomplished by employing partial fractions.
8(d). *Hint:* It is often helpful to multiply the numerator and denominator by the same parameter.

Section 11.2

3. *Hint:* A system quite similar to (20) appears. Again, the approach is to find an equation for the analogue of $A^2 + B^2$. When this is done, one of the functions involved will be found to be identically zero.
7(b). In establishing boundary conditions, ignore TST terms.

Section 12.1

6. *Hint:* First use Leibniz's rule to perform the differentiation. Then employ the integral mean value theorem to transform the integral. Finally, consider the limit $N \to M$.
12. *Hint:* Use the integral mean value theorem.

SECTION 12.2

14(b). *Hint:* First consider two special cases, where $f = 0$ and $g = 0$, respectively.
16(a). $Q = (1 - \kappa^2 + 2i\kappa)/(1 + \kappa^2)$, $B = 1$, $\delta = \tan^{-1}[2\kappa/(1 - \kappa^2)]$.

SECTION 13.2

1. $\mathbf{v} = \mathbf{i} + x_2\mathbf{j} + (x_1 - t)\mathbf{k}$.
2(b). *Hint:* The particle paths are given by $x = x(t)$, $y = y(t)$, where $dx/dt = y$, $dy/dt = x$. The solution of these equations can be obtained by eliminating either x or y.
3(a). $x = \frac{1}{2}t^2 + x_0$, $y = \frac{1}{6}t^3 + (x_0 + 1)t + y_0$. **(b)** $x^2 + x = (t/t_0)(yt_0 - y_0t_0 + x_0^2 + x_0)$.
4(a). $x = -\exp(-y)\cos t + \text{constant}$. **(b)** $x = \frac{1}{2}\exp(-t + t_0 - B)(\sin t - \cos t) + A - \frac{1}{2}\exp(-B)(\sin t_0 - \cos t_0)$, $y = t - t_0 + B$.

SECTION 13.3

1. *Partial answer.* $\delta(\mathbf{A}, t) = A_1 \exp(t) + A_2 \sin t \exp(-t)$.

SECTION 13.4

4(a). $J = 1$. **(b)** $J = \frac{1}{2}A_1^{-1/2}e^{2t}$.

APPENDIX 13.1

2. $\partial^2\tau/\partial r^2 + r^{-1}\partial\tau/\partial r + r^{-2}\,\partial^2\tau/\partial\theta^2$.

3(a). $\rho^{-2}\dfrac{\partial}{\partial\rho}\left(\rho^2\dfrac{\partial\tau}{\partial\rho}\right) + (\rho^2 \sin\phi)^{-1}\dfrac{\partial}{\partial\phi}\left(\sin\phi\dfrac{\partial\tau}{\partial\phi}\right) + (\rho^2\sin^2\phi)^{-1}\dfrac{\partial^2\tau}{\partial\theta^2}$.

APPENDIX 13.3

2. *Hints:* **(a)** Use the triangle inequality. **(b)** Use the fact that zero is an interior point of $R^{(n)}$.

SECTION 14.2

2. $(10)^{-1/2}(2T_{11} - 2T_{12} - T_{22})$.
3(c). Zero.
6. *Hint:* Integrate over a sphere of radius R. Use the fact that the absolute value of the integral is no greater than $4\pi R^2$ times the absolute value of the integrand.

SECTION 14.3

3(a). *Hint:* Use the integral mean value theorem.

SECTION 14.4

4. *Hint:* Write $\mathbf{v} = \nabla\phi + \mathbf{w}$. Show that $K^* = K +$ a nonnegative term.

SECTION 15.1

1. *Hint:* Transform $\oiint p\mathbf{n}\, d\sigma$ to a volume integral and use the hydrostatic expression for p.
4 and **5.** *Hint:* Use the identity curl (grad g) = $\mathbf{0}$.
5. $\rho^{-1}\nabla p = \nabla \int \rho^{-1}\, dp$.
8. *Hint:* Use Bernoulli's equation and a simple form of the continuity requirement.
10. *Hint:* Use a simple form of the continuity equation.

SECTION 15.2

5. The first steady solution is unstable, the second is stable.
9(a). *Hint:* If $Q \equiv \partial^2\hat{w}/\partial t^2$, show that $\partial^2 Q/\partial z^2 = k^2 Q$. **(b).** In case (i) the perturbation velocity maintains its initial value. There is neither growth nor decay, since there is no mechanism for altering the initial perturbation. (The perturbation would disappear in this situation if the slightest viscosity were present.) In case (ii) the slight variation in pressure, and the resulting pressure gradient, drive a flow that increases with time until linearization ceases to be valid.
10(f). Instability sets in at the minimum value of $\beta_c(m)$, namely, $(\sqrt{\gamma}+\sqrt{\kappa})^2$. This minimum occurs when $m = m_c = (\gamma\kappa)^{1/2}$, so the expected wavelength of the instability is $2\pi/(\gamma\kappa)^{1/4}$.
10(g). According to the possibility investigated here, instability would set in when β first exceeds the value $\gamma + \kappa$.

SECTION 15.3

4(c). The solution should be appropriate at a given point of an actual tube until the disturbance reaches the end of the tube and some reflected disturbance travels back to the point in question.
9(a). *Hint:* Use (41b) and introduce m, the common value of $\rho^+ s_n^+$ and $\rho^- s_n^-$ whose existence is guaranteed by (41a). **(e).** *Hint:* Employ Taylor's formula to approximate $p^+ - p^- = p(\rho^+, s^+) - p(\rho^-, s^-)$.

SECTION 15.4

6. *Hint:* Use as a control surface the cylinder $x^2 + y^2 = R^2$, $z = \pm 1$. Let $R \to \infty$.
7. $\Phi = -U\rho[1 - \frac{1}{2}(a/\rho)^3] \cos \phi$.

Authors Cited

Citations from the bibliography are not indexed.

Subject Index